Water Science and Technology Library

Volume 91

The aim of the *Water Science and Technology Library* is to provide a forum for dissemination of the state-of-the-art of topics of current interest in the area of water science and technology. This is accomplished through publication of reference books and monographs, authored or edited. Occasionally also proceedings volumes are accepted for publication in the series. *Water Science and Technology Library* encompasses a wide range of topics dealing with science as well as socio-economic aspects of water, environment, and ecology. Both the water quantity and quality issues are relevant and are embraced by *Water Science and Technology Library*. The emphasis may be on either the scientific content, or techniques of solution, or both. There is increasing emphasis these days on processes and *Water Science and Technology Library* is committed to promoting this emphasis by publishing books emphasizing scientific discussions of physical, chemical, and/or biological aspects of water resources. Likewise, current or emerging solution techniques receive high priority. Interdisciplinary coverage is encouraged. Case studies contributing to our knowledge of water science and technology are also embraced by the series. Innovative ideas and novel techniques are of particular interest.

Comments or suggestions for future volumes are welcomed.

Vijay P. Singh, Department of Biological and Agricultural Engineering & Zachry Department of Civil Engineering, Texas A and M University, USA
Email: vsingh@tamu.edu

More information about this series at http://www.springer.com/series/6689

Raj Mohan Singh · Prabhakar Shukla ·
Prachi Singh
Editors

Environmental Processes and Management

Tools and Practices

Springer

Editors
Raj Mohan Singh
Department of Civil Engineering
Motilal Nehru National Institute
of Technology, Allahabad
Prayagraj, Uttar Pradesh, India

Prabhakar Shukla
Department of Civil Engineering
Motilal Nehru National Institute
of Technology, Allahabad
Prayagraj, Uttar Pradesh, India

Prachi Singh
Department of Civil Engineering
Motilal Nehru National Institute
of Technology, Allahabad
Prayagraj, Uttar Pradesh, India

ISSN 0921-092X ISSN 1872-4663 (electronic)
Water Science and Technology Library
ISBN 978-3-030-38151-6 ISBN 978-3-030-38152-3 (eBook)
https://doi.org/10.1007/978-3-030-38152-3

This Springer imprint is published by the registered company Springer Nature Switzerland AG
The registered company address is: Gewerbestrasse 11, 6330 Cham, Switzerland

Preface

The environmental processes were usually managed and maintained by the nature itself. The emergence of modern industrial era has disturbed the natural ecological balance through heavy industrialization, technological revolution, faster growth by means of transportation, rapacious exploitation of resources, unplanned urbanization, etc. The Environmental Processes and Management book is an attempt to present a framework that helps to achieve Sustainable Development Goals (SDGs) through multifarious environment-related issues and processes, its consistent review, evaluation, and its contribution to environmental performance. The assumption is that this consistent review and evaluation will identify scope and opportunities for improving and implementing the environmental performance of society. This book gives an overview of various environmental processes and management in terms of their operation, complexities and natural and human interactions. This book offers research-based practices and strategies which integrate reading and writing and utilize technology.

This book is divided into three parts. Part I: Environmental Modeling consists of four chapters; Part II: Environmental Remediation contains eleven chapters; Part III: Groundwater Management includes seven chapters. Each chapter ends with a concluding remark, summarizing vital findings of the work. This book contains many illustrations providing exemplifying quantifications. The chapters demonstrate how the models, indicators, and ecological processes could be applied directly in the aforementioned environmental subdisciplines.

Overall, this book presents a range of case studies which demonstrate the complementary application of different modeling and other research techniques in combination with management-based thinking to the natural environment processes from the local level to international environmental regimes. With contributions from leading researchers in the field, this innovative volume provides a valuable teaching aid for students, as well as an insightful and practical reference tool for environmental practitioners. We are grateful to all the authors and reviewers for their contribution to this book. We also acknowledge the support of Springer and the staff of Springer Nature Editorial for their assistance during the preparation of this book.

Prayagraj, UP, India

Raj Mohan Singh
Prabhakar Shukla
Prachi Singh

Contents

Editors and Contributors

About the Editors

Dr. Raj Mohan Singh received Doctor of Philosophy (Ph.D.) degree from Indian Institute of Technology (IIT) Kanpur, Uttar Pradesh, India, in civil engineering: hydraulics and water resources in the year 2004. In the year 2005, he joined Motilal Nehru National Institute of Technology (MNNIT), Allahabad, India, as Lecturer in the Department of Civil Engineering. Presently, he is working as Professor in the Department of Civil Engineering, MNNIT Allahabad, Uttar Pradesh, India. His current research interests include water resources management, especially groundwater and climate change, conjunctive use of surface water and groundwater, optimization, and soft computing. He has published several technical papers in peer-reviewed national and international journals/seminars/conferences/symposia in the area of water resources management. He is also involved as an expert/PI in many esteemed research and consultancy projects of Government of India. He is Lifetime Member of many professional organizations/societies like American Society of Civil Engineers (ASCE), New York City, USA; The Indian Society for Hydraulics (ISH), Pune, India; Indian Water Resources Society (IWRS), Roorkee; Indian Association of Hydrologists (IAH), Roorkee; and International Association of Hydrological Sciences (IAHS), UK. Also, he is Corporate Member of The Institution of Engineers (India).

Dr. Prabhakar Shukla was born in Prayagraj, Uttar Pradesh, India. He obtained Bachelor of Technology (B.Tech.) degree in civil engineering from Sardar Vallabhbhai National Institute of Technology (SV NIT), Surat, Gujarat, India, in the year 2012. He attained the Master of Technology (M.Tech.) degree from Indian Institute of Technology (IIT) Roorkee, Uttarakhand, India, in the year 2014. He was awarded Doctor of Philosophy (Ph.D.) in civil engineering: water resources management at Motilal Nehru National Institute of Technology (MNNIT), Allahabad, in the year 2019. His research interests are hydrological modeling, climate change, vulnerability, and risk assessments in the water sector, and GIS applications in water resources management. He has published numerous research papers, technical notes, chapters, and books related to his interests. He is associated as Member of editorial board and Reviewer in Science Citation Index (SCI), SCI Expanded (SCIE), Scopus, Web of Science-indexed journals. Also, he is Active Member of reputed international professional bodies such as American Society of Civil Engineers (ASCE), New York City, USA; International Society for Development and Sustainability (ISDS), Japan; American Academy of Environmental Engineers and Scientists (AAEES), USA; International Association of Hydrological Sciences (IAHS), UK; National Ground Water Association (NGWA), USA; and Engineers Australia (EA), Australia. He is associated as Hub Scientist (Technical–Water Resources/GIS) with United Kingdom Research and Innovation (UKRI) Global Challenges Research Fund (GCRF) Water Security and Sustainable Development Hub at the Department of Civil Engineering, Indian Institute of Technology (IIT) Delhi, India.

Dr. Prachi Singh received her Bachelor of Engineering (B.E.) degree in the year 2013 under the discipline of civil engineering from University Institute of Technology—Rajiv Gandhi Proudyogiki Vishwavidyalaya (UIT—RGPV), Bhopal, Madhya Pradesh, India. She obtained Master's degree in environmental engineering from Maulana Azad National Institute of Technology [MANIT], Bhopal, India, in the year 2015. She acquired her Doctor of Philosophy (Ph.D.) degree in civil engineering: water resources and environmental engineering from Motilal Nehru National Institute of Technology [MNNIT], Allahabad, Uttar Pradesh, India, in the year 2019. Her current research interests include water resources management, especially groundwater and climate change, conjunctive use of surface water and groundwater, optimization, and soft computing. She has published several technical papers in peer-reviewed national and international journals/seminars/conferences/symposia in the area of water resources management. Also, she has contributed abundant books and chapters related to environmental engineering under the edges of internationally recognized and reputed publishers. She is a fellow of editorial board member and reviewer in numerous standard peer-reviewed Science Citation Index (SCI), SCI Expanded (SCIE), Scopus, and Web of Science-indexed journals. She is Lifetime Member of many professional organizations/societies like American Society of Civil Engineers [ASCE], New York City, USA; International Society for Development and Sustainability [ISDS], Japan; American Academy of Environmental Engineers and Scientists [AAEES], USA; International Association of Hydrological Sciences [IAHS], UK; and Engineers Australia [EA], Australia. She is an active regional- and national-level chess player and champion since 2002.

Contributors

Madhu Agarwal Department of Chemical Engineering, MNIT, Jaipur, Rajasthan, India

Ishtiyaq Ahmad Department of Civil Engineering, National Institute of Technology Raipur, Raipur, India

Sameer Arora Department of Civil Engineering, Indian Institute of Technology Delhi (IIT Delhi), New Delhi, India

Saima Aslam Department of Biotechnology, Baba Ghulam Shah Badshah University, Rajouri, J&K, India

Samir Bajpai Department of Civil Engineering, National Institute of Technology, Raipur, Chhattisgarh, India

Kakoli Banerjee Department of Biodiversity and Conservation of Natural Resources, Central University of Orissa, Landiguda, Koraput, India

Poorna B. Bhagawati Department of Electronics and Instrumentation Engineering, Basaveshwar Engineering College, Bagalkot, Karnataka, India

Prashant Basavaraj Bhagawati Department of Civil Engineering, Annasaheb Dange College of Engineering and Technology, Ashta, Maharashtra, India

Volker Birke Mechanical Engineering/Process Engineering and Environmental Engineering, Hochschule Wismar—University of Applied Sciences, Technology, Business and Design, Wismar, Germany

Mamta Bisht Groundwater Hydrology Division, National Institute of Hydrology, Roorkee, Uttarakhand, India

Isha Burman Department of Environmental Science and Engineering, Indian Institute of Technology (ISM), Dhanbad, Jharkhand, India

Sumedha Chakma Department of Civil Engineering, Indian Institute of Technology (IIT) Delhi, New Delhi, India

P. K. Chaudhari Department of Chemical Engineering, National Institute of Technology, Raipur, Chhattisgarh, India

Mridusmita Debnath Civil Engineering Department, Indian Institute of Technology Guwahati, North Guwahati, Assam, India;
Land and Water Management Division, ICAR-Research Complex for Eastern Region Patna, Patna, India

Subhajit Dey Department of Civil and Environmental Engineering, Indian Institute of Technology Patna, Bihta, Patna, India

Swati Dubey Department of Chemical Engineering, Banasthali Vidyapeeth, Tonk, Rajasthan, India

Khurshid Ahmad Ganai Department of Botany, Islamia College of Science & Commerce, Srinagar, J&K, India

Harsh Ganapathi New Delhi, India

Sahajpreet Kaur Garewal National Institute of Technology, Raipur, India

Aniruddha D. Ghare Visvesvaraya National Institute of Technology, Nagpur, India

N. C. Ghosh Groundwater Hydrology Division, National Institute of Hydrology, Roorkee, Uttarakhand, India;
Bengal Institute of Technology, Kolkata, West Bengal, India

Nadia Gul Department of Biotechnology, Baba Ghulam Shah Badshah University, Rajouri, J&K, India

A. B. Gupta Department of Civil Engineering, MNIT, Jaipur, Rajasthan, India

Kruti Jethwa Department of Civil Engineering, National Institute of Technology, Raipur, Chhattisgarh, India

Namrata D. Jhariwala SVNIT, Surat, India

Ashok K. Keshari Department of Civil Engineering, Indian Institute of Technology Delhi (IIT Delhi), New Delhi, India

Anjali K. Khambete SVNIT, Surat, India

Gopal Raj Khemundu Department of Biodiversity and Conservation of Natural Resources, Central University of Orissa, Landiguda, Koraput, India

Gopal Krishan Groundwater Hydrology Division, National Institute of Hydrology, Roorkee, Uttarakhand, India

Avanish Kumar Department of Chemical Engineering, Jaipur National University, Jaipur, India

Chandan Mahanta Civil Engineering Department, Indian Institute of Technology Guwahati, North Guwahati, Assam, India

Abhijit Maiti Department of Polymer and Process Engineering, Indian Institute of Technology Roorkee, Saharanpur, Uttar Pradesh, India

Darshan Mehta CED, Dr. S & S. S. G. G. E. C, Surat, India

Saurabh Mishra Department of Polymer and Process Engineering, Indian Institute of Technology Roorkee, Saharanpur, Uttar Pradesh, India

Yogendra D. Mishra L&T-Sargent & Lundy Limited, Vadodara, Gujarat, India

Satish G. Muttagi Scientific and Industrial Research Centre, Bangalore, Karnataka, India

Seema A. Nihalani SVNIT, Surat, India;
Civil Engineering Department, PIET, Parul University, Vadodara, Gujarat, India

Epari Ritesh Patro Department of Civil and Environmental Engineering, Politecnico di Milano, Milan, Italy

Mayuri Phukan New Delhi, India

Keyur Prajapati CED, S. S. A. S. I. T, Surat, India

Om Prakash Department of Civil and Environmental Engineering, Indian Institute of Technology Patna, Bihta, Patna, India

S. K. Pramada Department of Civil Engineering, NIT Calicut, Kozhikode, India

Gokul Prasad Groundwater Hydrology Division, National Institute of Hydrology, Roorkee, Uttarakhand, India

Nihar Ranjan Sahoo Department of Biodiversity and Conservation of Natural Resources, Central University of Orissa, Landiguda, Koraput, India

Arup Kumar Sarma Civil Engineering Department, Indian Institute of Technology Guwahati, North Guwahati, Assam, India

Sandip S. Sathe Department of Civil Engineering, Annasaheb Dange College of Engineering and Technology, Ashta, Maharashtra, India

Anupma Sharma Groundwater Hydrology Division, National Institute of Hydrology, Roorkee, Uttarakhand, India

Rahul Singh Department of Civil Engineering, Indian Institute of Technology (IIT) Delhi, New Delhi, India

K. V. Sruthi Groundwater Hydrology Division, National Institute of Hydrology, Roorkee, Uttarakhand, India;
KSCSTE—Centre for Water Resources Development and Management, Kozhikode, Kerala, India

Kim Hyun Su The Earth and Environmental Science System Research Center, Chonbuk National University, Jeonju, Republic of Korea

Shahid Ul Islam Department of Civil Engineering, Baba Ghulam Shah Badshah University, Rajouri, J&K, India

Avinash D. Vasudeo Visvesvaraya National Institute of Technology, Nagpur, India

Sowmya Venugopal Department of Civil Engineering, NIT Calicut, Kozhikode, India

Mukesh Kumar Verma Department of Civil Engineering, National Institute of Technology Raipur, Raipur, India

Sahita Waikhom CED, Dr. S & S. S. G. G. E. C, Surat, India

S. M. Yadav CED, SVNIT, Surat, India

Abhijit M. Zende Department of Civil Engineering, Daulatrao Aher college of Engineering, Karad, Maharashtra, India

Abbreviations

%	Percentage
[C_2mim][Oac]	1-Ethyl-3-Methylimidazolium Acetate
[C43MPy][BF4]	1-Butyl-3-Methylpyridinium Tetrafluoroborate
[C4Py][BF4]	N-Butylpyridinium Tetrafluoroborate
[C4Py][SCN]	N-Butylpyridiniumthiocyanate
[C4Py][Tf2N]	N-Butylpyridinium Bis(trifluoromethylsulfonyl)imide
[C63MPy][BF4]	1-Hexyl-3-Methylpyridinium Tetrafluoroborate
[C_6Py][BF_4]	N-Hexylpyridinium Tetrafluoroborate
[C83MPy][BF4]	1-Octyl-3-Methylpyridinium Tetrafluoroborate
[C8Py][BF4]	N-Octylpyridinium Tetrafluoroborate
[Et2NEmim] [PF6]	1-(2-Diethyl-Aminoethyl)-3-Methylimidazolium Hexafluorophosphate
[Et2NEmpyr][PF6]	1-(2-Diethyl-Aminoethyl)-1-Methylpyrrolidinium Hexafluorophosphate
[HMIM][Tf2N]	Hexyl-3-Methylimidazolium Bis(trifluoromethylsulfonyl) imide
[NTf2]−	Bis(trifluoromethanesulfonyl)imide
Δt	Time of Simulation
°C	Degree Celsius
η	Effective Porosity
λ	Coefficient of Initial Abstraction
λmax	Principal Eigenvalue
A	Aquifer Media
ABY	Atal Bhujal Yojana
ACC	Air-Cooled Condenser
ADE	Advection-Dispersion Equation
ADER	Advection-Dispersion Equation With First-Order Reaction
ADERS	Advection-Dispersion Equation With First-Order Reaction and Sink/Source
ADMI	American Dye Manufacturers' Institute

AEI	Agro-Ecological Indicator
AESA	Agro-Ecological System Attributes
AHP	Analytic Hierarchy Process
AMC	Antecedent Moisture Condition
ANFIS	Adaptive Neuro-Fuzzy Inference System
ANN	Artificial Neural Network
ANOVA	Analysis of Variance
APCA	Absolute Principal Component Analysis
APTI	Air Pollution Tolerance Index
ASP	Activated Sludge Process
AV	Ammonium Volatilization
AWC	Activated Wood Charcoal
b	Bottom Width
B	Top Width
B(a)P	Benzo[A]Pyrene
B1 and B2	Blank CWs
BbF	Benzo[B]Fluoranthene
BDL	Below Detectable Limit
BghiP	Benzo [GHI] Perylene
BkF	Benzo[K]Fluoranthene
BMIM	1-Butyl-3-Methylimidazolium
BMIM[$MeSO_4$]	1-Butyl-3-Methylimidazolium Methyl Sulfate
BNR	Biological Nutrient Removal
BOD	Biochemical Oxygen Demand
Br+	Bromine
Bt	Bacillus thuringiensis
C	Concentration Well
Ca	Calcium
CCOST	Chhattisgarh Council of Science & Technology
CFBBR	Circulating Integrated Fluidized Bed Bioreactor
CFD	Computational Fluid Dynamics
CGWB	Central Groundwater Board
CH4	Methane
CI	Consistency Index
Cl^-	Chlorine
cm	Centimeter
CMB	Chemical Mass Balance
CO2	Carbon Dioxide
COC	Cycle of Concentration
COD	Chemical Oxygen Demand
COPREM	Constrained Physical Receptor Model
CPCB	Central Pollution Control Board
CPL	Caprolactam
CPU	Central Processing Unit
CR	Consistency Ratio

Cr(VI)	Chromate
CRN	Controlled Release N Fertilizer
CT	Cooling Tower
CVOCs	Chlorinated Volatile Organic Carbons
CW	Cooling Water
CWC	Central Water Commission
CWRD	Chhattisgarh Water Resources Department
CWRDM	Centre for Water Resources Development and Management, Kozhikode
CWs	Constructed Wetlands
D	Depth
D_{16}	16% Pass Particle Size
D_{50}	50% Pass Particle Size
D_{60}	60% Pass Particle Size
D_{84}	84% Pass Particle Size
DB151	Direct Blue 151
DEM	Digital Elevation Model
DM water	Demineralized Water
DMEAH	Dimethyl Ethanolammoniumdimalate
DNA	Deoxyribonucleic Acid
DO	Dissolved Oxygen
DR31	Direct Red 31
DS	Domestic Sewage
E	Easting
EAS	Evolutionary Annealing Simplex
EC	Electrical Conductivity
ECS	Glutamylcysteine
EF	Enrichment Factor
EI	Environmental Indicator
EILs	Eutectic Ionic Liquids
EM	Electromagnet
EMA	Environmental Management for Agriculture
EMIm	1-Ethyl-3-Methyl Imidazolium
EMP	Environment Management Planning
EP	Ecopoint
EPA	Environmental Protection Act
ERF	Ethylene-Responsive Factor
ERT	Electrical Resistivity Tomography
ESA	Electrostatic Spraying Absorber
ESP	Electrostatic Precipitator
FA	Factor Analysis
FAO	Food and Agricultural Organization
FCM	Fuzzy C-Mean
FD	Finite Difference
FDM	Finite Difference Method

FEMWATER	Finite Element Model of Water Flow Through Saturated-Unsaturated Media
FGD	Flue Gas Desulfurization
FICCI	Federation of Indian Chambers of Commerce and Industry
FIS	Fuzzy Inference System
FSI	Farmer Sustainability Index
GA	Genetic Algorithm
GBILs	Guanidine-Based Ionic Liquids
GC-MS	Gas Chromatography–Mass Spectrometry
GDEM	Global Digital Elevation Model
GEAC	Genetic Engineering Approved Committee
GIS	Geographical Information System
GP	Genetic Programming
GPS	Global Positioning System
GQI	Groundwater Quality Index
ha	Hectare
HCO_3	Hydrogen Carbonate
HCSD	High Concentrated Slurry Disposal System
HEC-HMS	Hydrologic Engineering Centre-Hydrologic Modeling System
HEMM	Heavy Earth Moving Machinery
HF	Hollow Fiber
HMIm	1-Hexyl-3-Methyl Imidazolium
HRT	Hydraulic Retention Time
HSG	Hydrological Soil Group
HVAC	Heating, Ventilation, and Air Conditioning
I	Inflow
Ia	Initial Abstraction
ICP	Insecticidal Crystal Protein
ICW	Integrated Constructed Wetland
IDC	Initial Dye Concentration
IFA	International Fertilizer Industry Association
IFS	Indicators of Farm Sustainability
ILs	Ionic Liquids
IM	Imidazole
IMD	India Meteorological Department
IMSD	Integrated Mission for Sustainable Development
IVI	Intrinsic Vulnerability Index
JSA	Jal Shakti Abhiyan
K	Hydraulic Conductivity
K_2	Re-aeration Coefficient
kg	Kilogramme
L	Basin Lag Time
LASAC	Lightweight Alum Sludge Aggregate Concrete
LCAA	Life Cycle Analysis for Agriculture

LCAE	LCA for Environmental Farm Management
LFL	Landfill Leachate
Ls	Length of the Channel in the Sub-basin
Lu	Land Use
m	Meter
M1A and M2A	Althernanthera Sessilis Polygonum
M1C	Canna indica
M1Co	Colocasia Esculenta
M1O	Ocimum Americanum L
M1P	Pistia Stratiotes
M1T	Typha Latifolia
m^2	Square Meter
M2C	Canna indica
M2Co	Colocasia Esculenta
M2O	Ocimum Americanum L
M2P	Pistia Stratiotes
M2T	Typha Latifolia
MATLAB	Matrix Laboratory
MBBR	Moving Bed Biofilm Reactor
MBC	Microbial Biomass Carbon
MBR	Membrane Bioreactor
MC	Monte Carlo
MCM	Million Cubic Meter
MEAL	monoethanolaminium lactate
MF	Microfiltration
MFC	Microbial Fuel Cell
MFs	Membership Functions
Mg	Magnesium
MG	Malachite Green
MLE	Multilinear Engine
MLP	Multilayer Perceptron
MLSS	Mixed Liquor Suspended Solids
mm	Millimeter
MNN	Modular Neural Network
MODFLOW	Modular Three-Dimensional Finite-Difference Groundwater Flow Model
MoEF	Ministry of Environment and Forest
MOP	Multi-objective Parameter
MRL	Maximum Residue Limit
MSL	Mean Sea Level
MSW	Municipal Solid Waste
MT3DMS	Modular Three-Dimensional Multispecies Transport Model
n	Manning's Constant
N	Nitrogen
Na^+	Sodium

NAAQS	National Ambient Air Quality Standards
NaCl	Sodium Chloride
NADH-DCIP	Nicotinamide adenine dinucleotide-dichlorophenolindophenol
NBSS	National Bureau of Soil Survey
NBSS & LUP	National Bureau of Soil Survey and Land Utilisation Planning
ND	Not Defined
NFI	Non-flooded Irrigation
NGB	Naphthol Green B
NO_3	Nitrate
Nox	Oxides of Nitrogen
NPS	Non-point Source
NRCS	Natural Resources Conservation Service
NRSC	National Remote Sensing Centre
NSGA-II	Non-shorting Genetic Algorithm-II
NUFER	Nutrient Flow in Food Chains, Environment and Resource Use
NWR	Northwest Region
O	Outflow
OS	Operationalizing Sustainability
OT	Operational Time
P	Phosphorus
P1 and P2	Typha latifolia, Colocasia esculenta, Alternanthera sessilis, Polygonum, Canna indicia, Ocimum americanum L, Pistia stratiotes
PACl	Poly Aluminum Chloride
PAHs	Polycyclic Aromatic Hydrocarbons
PC	Principal Component
PCA	Principal Component Analysis
PCs	Phytochelatins
pH	Potential of Hydrogen
PHE	Phenol
PHED	Public Health Engineering Department
PLRV	Potato Leafroll Virus
PM	Particulate Matter
PMF	Positive Mass Factorization
PR	Pathogenesis-Related
PRB	Permeable Reactive Barrier
psu	Practical Salinity Unit
PUC	Pollution Under Control
PUF	Polyurethane Foam
PVY	Potato Virus Y
Q	Discharge
QNDDP	Quasi-Newton Approximations

R	Recharge
R^2	Coefficient of Determination
RBC	Rotating Biological Contractor
RBF	Radial Basis Function
RCI2A	Rare Cold-Inducible 2a
RG19	Reactive Green 19
RI	Randomness Index
RM	Receptor Model
RMSE	Root-Mean-Square Error
RO	Reverse Osmosis
RO16	Reactive Orange 16
RS	Remote Sensing
RSPM	Respirable Suspended Particulate Matter
S1	Segment-1
S2	Segment-2
S3	Segment-3
S4	Segment-4
SA	Simulated Annealing
SAR	Sodium Adsorption Ratio
S_{avg}	Average Bed Slope
S_b	Bed Slope
SBR	Sequencing Batch Reactor
SC	Subtractive Clustering
SCS-CN	Soil Conservation Service Curve Number
SD	Solagro Diagnosis
SEAWAT	A Computer Program for Simulation of Multispecies Solute and Heat Transport
SEC	Sustainability of Energy Crop
SEM	Scanning Electron Microscopy
SI	Susceptibility Index
SILPs	Supported Ionic Liquid Phase absorbents
SKK	The Siam Cement (kaeng Khoi) Co., Ltd.
SO2	Sulfur Dioxide
SOD	Sediment Oxygen Demand
SOI	Survey of India
SRP	Soluble Reactive Phosphorus
SRT	Solid Residence Time
SS	Specific Storage
SSNM	Site-Specific Nutrient Management
SUTRA	Saturated-Unsaturated Transport
T	Topography
TBAB	Tetrabutyl ammonium bromides
Tc	Time of Concentration
TCE	Trichloroethylene
TCLP	Toxicity Characteristic Leaching Procedure

TDS	Total Dissolved Solids
TE	Trifluoroethanol
Temp.	Incubation Temperature
TERT	Time-Lapse Electrical Resistivity Tomography
TF	Trickling Filter
TFs	Transcription Factors
TKN	Total Kjeldahl Nitrogen
TMG	Tetramethylguanidinium
TMGL	Tetramethylguanidinium Lactate
TN	Total Nitrogen
TNT	Trinitrotoluene
TOC	Total Organic Carbon
TP	Total Phosphorus
TPP	Thermal Power Plant
TS	Takagi-Sugeno
TSS	Total Suspended Solids
UASB	Up-Flow Anaerobic Sludge Bed Reactor
UF	Ultrafiltration
UK	United Kingdom
USBR	Up-Flow Sludge Blanket Reactor
USDA	United States Department of Agriculture
VES	Vertical Electrical Sounding
VFA	Volatile Fatty Acid
Vs	Average Velocity of Flow in the Channel
WCC	Water-Cooled Condenser
WDT	Wavelet De-noising Technique
WHO	World Health Organization
WoRMS	World Register of Marine Species
WRD	Chhattisgarh Water Resources Department
WT	Wild Type
WTP	Water Treatment Plant
WW	Wastewater
y	Average Watershed Slope
z	Side Slope
ZLD	Zero Liquid Discharge
ZSI	Zoological Survey of India

Part I
Environmental Modeling

Chapter 1
Monte Carlo Simulation and Fuzzy Modelling of River Water Quality for Multiple Reaches Using QUAL2kw

Sameer Arora and Ashok K. Keshari

Abstract Water quality of the river is associated with various processes, defining its complexity in nature. Monitoring of surface water can produce uncertain results due to spatial and temporal variations, fluxes in river cross section, and depth of flow. Computation and modelling of water quality from a large number of parameters also characterize uncertainties with the random selection of input parameters. Parameters also vary significantly due to the different scale of measurement and different influence of each parameter on to the output. A combined study of monitoring, modelling and sensitivity of parameters has been conducted to identify the fluxes in river water using QUAL2Kw model. Adaptive neuro-fuzzy inference system (ANFIS) is applied with the different composition of input parameters to simulate the DO, BOD and COD of Yamuna River, Delhi. ANFIS algorithm comprises of grid partitioning and subtractive clustering to identify the most dominant parameters. To evaluate the uncertainty in the simulation results, sensitivity analysis is carried out using Monte Carlo simulation. Results obtained from the MC simulation applied to QUAL2Kw, to evaluate the fluxes, as the result of simulation act as the decision-maker for the design of water resource management policies and the application of various pollutant reduction strategies.

Keywords ANFIS · Water quality · Uncertainty analysis · QUAL2Kw

S. Arora (✉) · A. K. Keshari
Department of Civil Engineering, Indian Institute of Technology Delhi (IIT Delhi), Hauz Khas, New Delhi 110016, India
e-mail: sameer_arora01@yahoo.co.in

A. K. Keshari
e-mail: kesari@civil.iitd.ac.in

R. M. Singh et al. (eds.), *Environmental Processes and Management*, Water Science and Technology Library 91,
https://doi.org/10.1007/978-3-030-38152-3_1

1.1 Introduction

River water quality monitoring is primarily required for the detection of status and trend of water quality parameters and identification of factors generating variations among parameters. In last few decades, the commonly observed environmental problems associated with rivers are low freshwater flow, the flux of organic matter, oxygen imbalance, the flow of excess nutrients with a high concentration of heavy metals and eutrophication (Mulholland et al. 2005). Stretch of rivers passing through the cities carries pollutants at a worrying level. Discharge from domestic and industrial activities, along with the flow of pesticides from agricultural fields, leads to the reduction in dissolved oxygen concentration of river below the value required for the sustainability of aquatic life. Rivers have become the large size drains of an urban area, which disperse the wastewater to the downstream area (Djaouida et al. 2019). Assimilate capacity of the river also get reduced with the low freshwater flow, as most of the water is being diverted to the nearby urban agglomerations for fulfilling to meet the demand of water supply for both the domestic and industrial activities (Parmar and Keshari 2012). The impact of wastewater discharged in the downstream area is commonly determined by the dispersion and sorption process (Chiffre et al. 2016). However, the velocity of flow also plays a significant role in estimating the concentration of various pollutants (Keller et al. 2014) due to the dilution of wastewater along with the flow and velocity. The low velocity of flow, near to the banks of the river or in the stagnant water, raises another problem associated with water quality. Eutrophication is a commonly observed environmental concern of inland and coastal waters, which causes major damage to the aquatic biological diversity along with the enriching water body with high contaminant concentration (Zhao et al. 2012). The major effects of eutrophication are high production of plankton algae, excessive growth of weeds and macroalgae, leading to oxygen deficiency, which ultimately results in the death of fishes. Hence, serious efforts are needed to reduce the pollutant discharge and improvement of river water quality.

Besides the identification of the status and trends through the monitoring activities, water quality, hydraulic and meteorological data at regular interval are also required for the prediction of water quality using different stochastic and deterministic models (Arheimer and Olsson 2003; Cox 2003). Water quality models are preferred to identify the source of pollution, principal parameter causing maximum degradation of water quality and identification of trend of water quality parameters. Predictive models are selected over other models to analyse the changes in quality incurred by either single or combination of parameters for the application of various management policies. This enables efficient environmental control and the development of management practices through the prediction of variation in the parameters.

Pollution and acidification are the most important factors behind the development of water quality models. In context with pollution, the model includes the parameters influencing aquatic health as the flow of excessive nutrients, organic matter, heavy metals and oxygen balance of the riverine system. Dissolved oxygen (DO), biochemical oxygen demand (BOD) and chemical oxygen demand (COD) are the commonly used parameters for defining the health of the aquatic ecosystem. Concentration of DO, BOD and COD are significantly influenced by the presence of nutrients, organic content, algal biomass and suspended solids within the water body (Zhang et al. 2012; Wen et al. 2017). Depending upon the available flow in the river during the monsoon and non-monsoon period, and respective velocity of flow, tremendous alterations are observed in water quality spatially and temporally (Arora and Keshari 2018). Integration of soft computing techniques such as artificial neural network, genetic algorithm, support vector machine and fuzzy logic along with the water quality modelling has improvised the results considerably and was effectively utilized by several researchers. Akkoyunlu et al. (2011) had applied ANN in the Lake Iznik, Turkey, to identify the concentration of DO for one day using physical and chemical parameter as input to the model (Akkoyunlu et al. 2011). Guru and Jha (2013) utilized multilayer perceptron (MLP) neural network to analyse the concentration of DO and BOD in Mahanadi River, Odisha (Guru and Jha 2013). Chang and Chang 2006 used ANFIS to study the water level in the Shihmen Reservoir, Taiwan (Chang and Chang 2006). Ahmed et al. (2013) used several modelling techniques such as MLP-ANN, radial basis function neural network (RBF-ANN), ANFIS and WDT-ANFIS to predict the water quality and concluded that ANFIS and WDT-ANFIS performed exceptionally well over ANN model (Ahmed et al. 2013). Tiwari et al. (2018) designed water quality index using two ANFIS techniques: SC-ANFIS and FCM-ANFIS for the determination of accuracy in prediction (Tiwari et al. 2018).

Along with the prediction of water quality parameters, uncertainties generated in the river are also playing an important role. Uncertainty analysis guides with the upper and lower flux in water quality data that is required for the development or selection of treatment technologies. Uncertainty could arise either from the parameter selection for modelling or from structure model or from result data uncertainty (Melching and Bauwens 2001; Lindenschmidt et al. 2007). Several studies have been performed on the uncertainty analysis, whereas quantitative assessment of parameters based on their impact and prioritization in affecting the water quality is weakly studied. Researchers have applied the principle of data uncertainties commonly on data collection (Harmel et al. 2009), quantification of techniques comparing similar data from various sources (Ramsey 1998) or assessment of physical, chemical and hydraulic parameter of water body (Horowitz 1997; Rode and Suhr 2007). Few studies have found on the site-specific characterization of river water quality data using uncertainties techniques. Harmel et al. (2009) conducted a study to assess the uncertainty in flow of stream by dividing the stream into four categories (Harmel et al. 2009), but the findings were restricted to nutrients and small streams. Various other methods are also available to quantify the uncertainty in the data such as genetic algorithm, Monte Carlo simulation, first-order analysis, etc., but sensitivity analysis is more advantageous over other as it removes the uncertainty from the model parameters and produces the simulated results using modified parameter (Zou et al. 2002; Newham et al. 2003; Wu et al. 2006).

The objective of this study is to develop a generalized model to identify uncertainties in river water quality data. In this study, ANFIS model has been developed for DO, BOD and COD and used to predict the water quality through QUAL2Kw model. The uncertainty of the model parameters is determined through Monte Carlo simulation. Parameters have also been prioritized also according to their impact and prone to the spatial and temporal variation.

1.2 Study Area

Yamuna River is the largest tributary of the Ganges in India and travels 1376 km before its confluence in the Ganges. Yamuna originates from Yamunotri in Uttarakhand, India, at 31°01′0.12″ N and 78°27′0″ E having basin area of 3,66,223 km^2. After travelling 375 km river enters Delhi at Palla and has a total length of 48 km falls in Delhi. First 24 km of the stretch of river in Delhi till Wazirabad barrage contains the sufficient quantity as well as adequate water quality. Water is being abstracted at the Wazirabad barrage and diverted for the supply to meet the water supply-demand of the city. During the non-monsoon period hardly, any water flows in the downstream of the Wazirabad barrage and only wastewater flow in the river channel. Yamuna River travels 22 km before the Okhla barrage, where again abstracted to meet the irrigation demand of Haryana and Uttar Pradesh. In between the 22 km of stretch, Yamuna receives the wastewater from 16 major drains and four minor drains (CPCB 2006).

Stretch of Yamuna River from the downstream of Wazirabad till Okhla barrage is considered for the study. Figure 1.1 shows the study area indicating the location of Wazirabad and Okhla barrage and points of the confluence of drains mainly containing untreated/partially treated wastewater. Najafgarh drain joins the river at the 500 m downstream of Wazirabad barrage and carries the maximum discharge even five times to the remaining flow in the river. Najafgarh drain causes a tremendous rise

Fig. 1.1 Yamuna River in the stretch of Delhi

in the BOD load of the river as 120 tons/day. The second-largest drain is Shahdara drain but joins the river in the downstream of Okhla Barrage. Other drains contribute around 24% of the wastewater discharge, whereas only Najafgarh drain is responsible for the 76% of the wastewater flow into the river. Three another drains carrying the wastewater load of Delhi join the Agra Canal. Before the Wazirabad barrage, water quality of river is classified in class C, whereas after joining of Najafgarh drain and other subsequent drains, water quality falls in class E. This implies that the Yamuna in the stretch of Delhi is not fit for any purpose as the DO concentration falls to zero in most of its part (CPCB 2006).

1.3 Material and Methodology

1.3.1 River Segmentation

The study area is divided into different segments based on point source (drains) joining the river, quantity and quality. Average monthly data was collected from each segment for 60 months. Segment-1 (S1) is from the downstream of Wazirabad barrage to the confluence of Najafgarh drain. The total length of S1 is 0.5 km, and Najafgarh drain joins the river 300 m downstream of the barrage. As the Najafgarh drain carries the maximum discharge about five times to the discharge in the river after abstraction at barrage (CPCB 2006), a separate segment is designed to study the impact of Najafgarh drain only. Segment-2 (S2) is proposed from the Najafgarh drain to ITO Bridge. In S2, around eight small drains join the river as shown in Fig. 1.2. Segment-3 (S3) is proposed from the ITO Bridge to Nizammudin and receives the

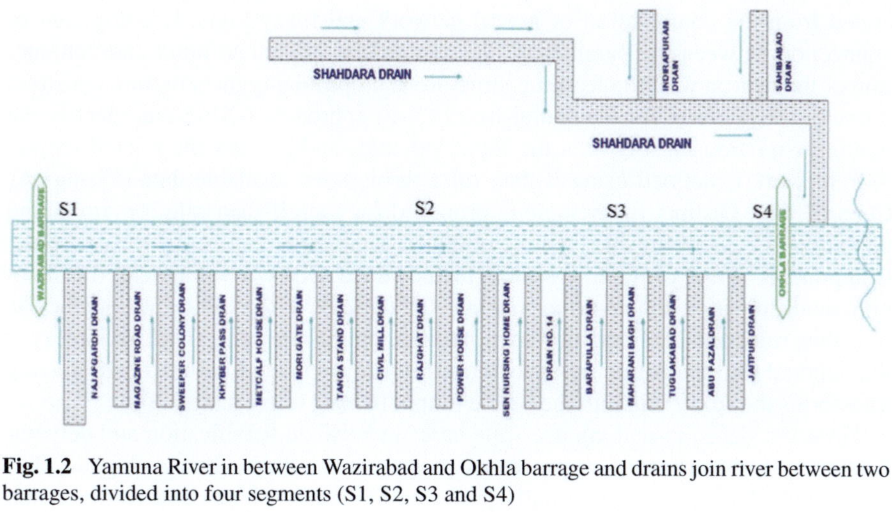

Fig. 1.2 Yamuna River in between Wazirabad and Okhla barrage and drains join river between two barrages, divided into four segments (S1, S2, S3 and S4)

wastewater from small drains. On the right bank of the river, effluent from the power plant also joins the river. Segment-4 (S4) is in between Nizammudin and Okhla Barrage. This segment receives the wastewater from small drains and Hindon cut canal. Length of S2 is approximately 12 km and S3 is 5 km, whereas last segment S4 having a length of 6 km. River water in the last segment receives the significant discharge from Hindon cut canal and hence get diluted. Again at the Okhla barrage, water abstracted at Agra canal for the irrigation requirement in Haryana and Uttar Pradesh.

Physio-chemical parameters have been tested out in each segment as per the standard procedure of central pollution control board (CPCB 2006). Temperature (°C) and pH are measured on-site, whereas other parameters such as DO (mg/L), BOD(mg/L), COD(mg/L), ammonia (μS/cm), total Kjeldahl nitrogen (μS/cm), conductivity (μmol/cm), faecal coliform (MPN/100 ml) and total coliform (MPN/100 ml) are measured in laboratory. Samples are preserved with the preferred reagent to conduct the off-site testing. All the parameters were analysed using standard and recommended method (APHA 2005).

Variation in DO, BOD and COD along the study period is shown in Fig. 1.3. Delhi receives the monsoon in July–September, and during this period, river carries maximum discharge. BOD of river falls to less than 30 mg/L as the standard limit for the inland water discharge (CPCB 2006). However, in the pre-monsoon period and post-monsoon period, BOD and COD load of water body increase rapidly as the flow subsides.

1.3.2 Adaptive Neuro-Fuzzy Inference System (ANFIS)

Adaptive neuro-fuzzy inference system (ANFIS) is the modelling technique generated from the combination of neural network and fuzzy logic. Development of connection between the parameters like neurons in the human brain and conjunction of the neuron with the learning structure and optimizing the structure construct provides the platform for the neural-fuzzy (N-F) approach. ANFIS can identify the nonlinear relationship between the input variables and process the related output. A framework is defined using if-then rules through the available data (Chang and Chang 2006). Distinct fuzzy logic is proposed for each if-then rule for simulating the output. Greater efficiency can be obtained using large number of the rule base. Components of fuzzy inference system include the database development, fuzzification, defuzzification and decision-making. Database development is the conjunction of if-then rules and membership functions forming the rule base. Input variables are then altered into the brittle input to form the fuzzy sets, and defuzzification again transforms the fuzzy sets into the brittle output (Chang and Chang 2006).

However, decision-making uses rule bases in between fuzzification and defuzzification to perform inference processes (Nayak et al. 2005). Takagi-Sugeno (TS) model and Mamdani are the usually preferred FIS model (Zounemat-Kermani and Teshnehlab 2008; Galavi et al. 2013). TS generates the output using the first-order

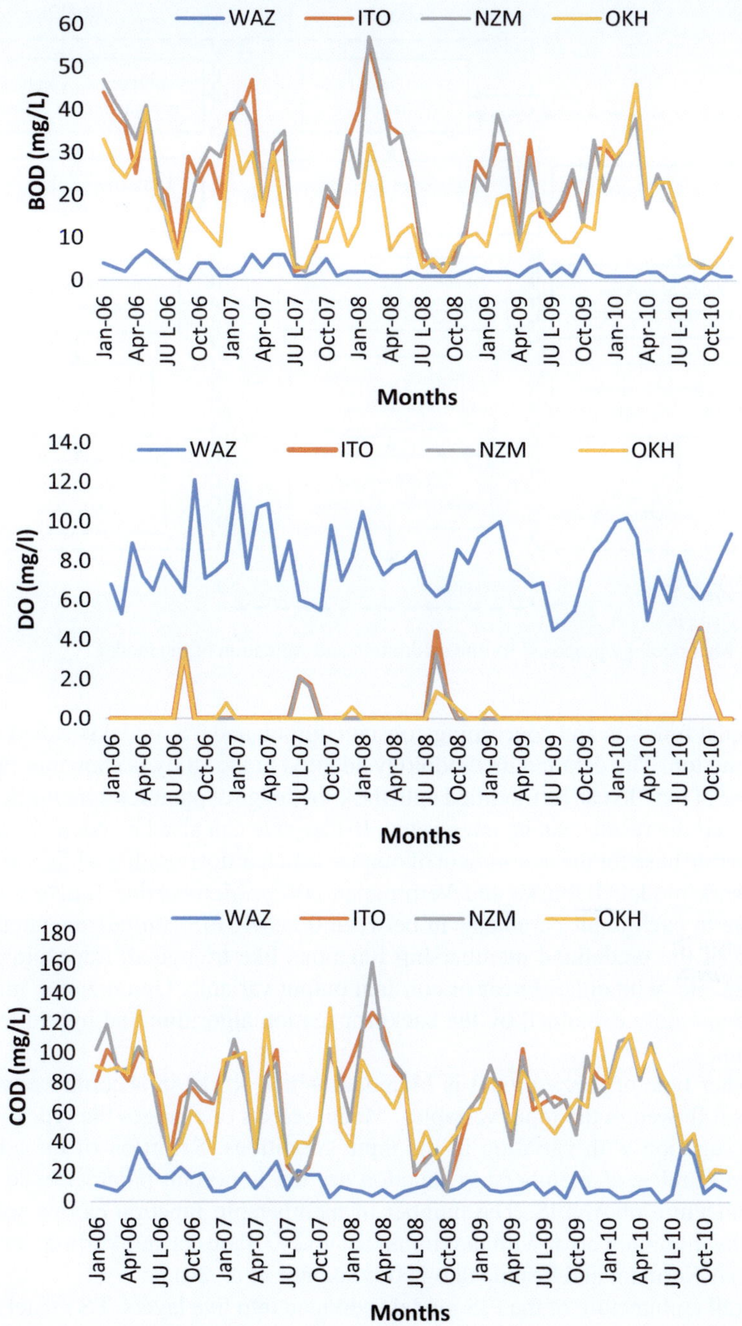

Fig. 1.3 Variation in BDO, DO and COD in all the four sampling sites for five years in Yamuna River

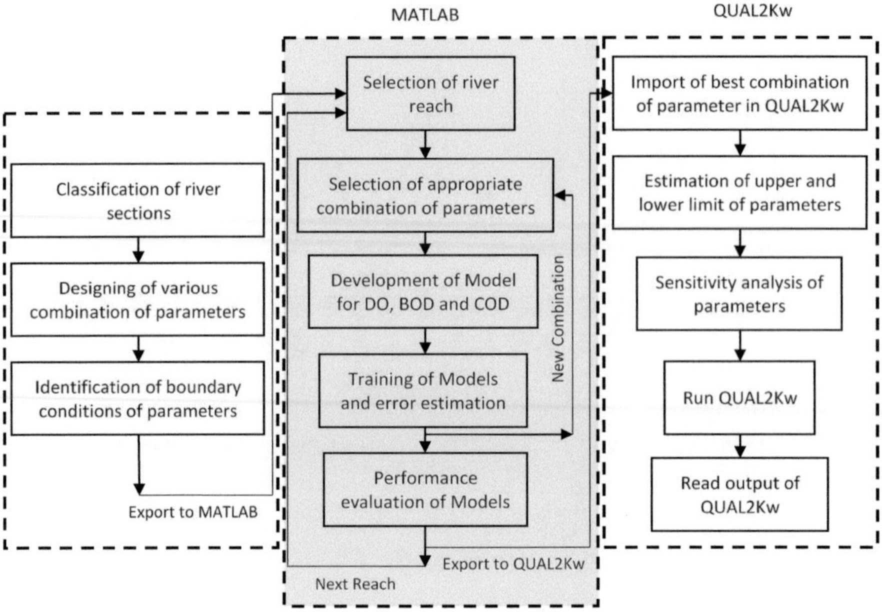

Fig. 1.4 Methodology proposed for the calibration and validation of the model

polynomial function and concerning constant output, and TS model is called a zero-order function. The detailed methodology adopted in the study is shown in Fig. 1.4. Rule base of the FIS can be modified with the knowledge of parameters using data processing and the results can be interpreted. If-then rule can also be added or removed from the rule base for the assessment of output, which is not possible while using neural network model (Babuska and Verbruggen 2003). Membership function defines the value to each input parameter in between 0 and 1. FIS can be generated using any one of the predefined membership functions like triangular, trapezoidal, bell, Gaussian, etc. with either linear or constant output variable. Optimization model of FIS is commonly explained by the backpropagation algorithm and hybrid learning algorithm.

Another type of fuzzy model is Mamdani which works in accordance with the changes followed in the input variables. Model selects or changes the type of membership function with variation in the input conditions. Selection of the adequate type and number of membership function are the important parameters to define the architecture of ANFIS. The number of membership function can be varied to obtain the minimum error with maximum accuracy (Babuska and Verbruggen 2003; Zounemat-Kermani and Teshnehlab 2008; Sonmez et al. 2017).

Overall architecture of the FIS model is divided into five layers. TS model is considered for the FIS structure mapping with the neural network. As shown in Fig. 1.5, two input variables (x_1 and x_2) are considered for representing the architecture along with output (f). Two membership functions are proposed for each input variable, i.e.

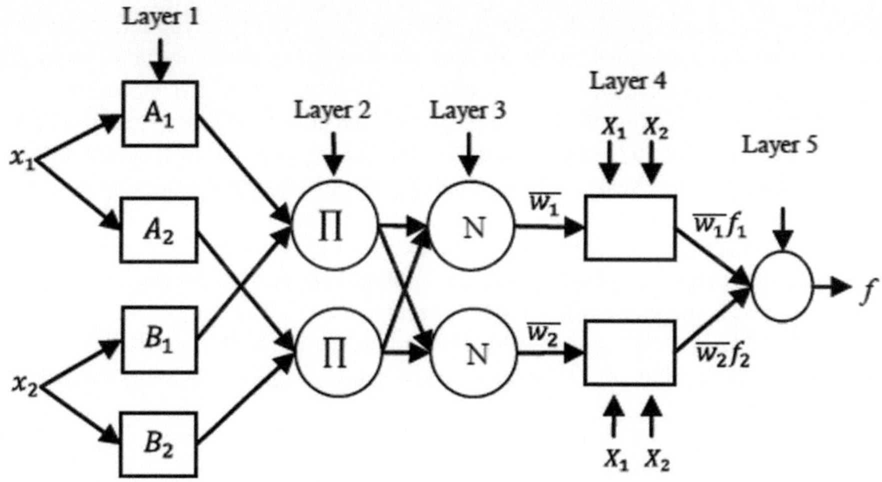

Fig. 1.5 ANFIS architecture (Aqil et al. 2007)

$A1$ and $A2$ for input x_1; $B1$ and $B2$ for input x_2. Rules formed through if-then rule base are as illustrated in Eqs. (1.1) and (1.2):

$$f_1 = p_1x_1 + q_1x_2 + r_1 \tag{1.1}$$

$$f_1 = p_1x_1 + q_1x_2 + r_1 \tag{1.2}$$

where p, q, r are the coefficient of linear output of the rule base. Output from this layer is the weighted average of rules. In the second layer, AND fuzzy operator is applied to each input, and the output from this layer represents the strength of each layer. In the third layer, strength is standardized from each input neuron, and in the next layer, the involvement of each input for the respective rule is determined in developing the output in the model. In the last layer, signals from all the layers are summarized to form the overall output.

1.3.3 Water Quality Model

QUAL2Kw is selected for the assessment of water quality and sensitivity analysis of water quality parameters. Model is initiated with the water quality parameter data obtained through the experimental analysis. Water quality parameters for all the segments are allocated to the model along with the hydraulic parameters (DJB 2005). Boundary conditions of the head location of the study area are incorporated in the Headwork sheet along with the concentration of various pollutants. Mode of simulation, solution method, variables for simulation, step of calculation, etc. are selected in the main QUAL2Kw sheet, defining the state of the model. Point and non-point load are submitted to the desired sheet of the model. Parameter required for the estimation is selected based on water quality processes which are assessed for simulation. Amount of organic load in the river is the dominant factor that affects the concentration of DO and BOD (Chapra 2008). DO content of the river is replenished with the re-aeration, which occurs through the natural processes such as transformation at air–water interface, photosynthesis and oxidation–reduction process, etc, whereas BOD is the major consumer of the DO during the process of organic matter decomposition in the water body. However, another important parameter is COD that represents the chemically oxidized rate of organic matter. The concentration of DO, BOD and COD is greatly influenced by re-aeration coefficient (K_2), sediment oxygen demand (SOD), benthic oxygen demand, photosynthesis, nitrogen compound, algae concentration, etc.

1.3.4 Model Calibration

Several parameters were studied for the water quality assessment; however, parameters which are significant to the DO, BOD and COD are considered for the simulation. Boundary conditions of the parameters were defined in the model for the calibration. QUAL2Kw (Brown and Barnwell 1987; Pelletier et al. 2006) is used to identify the upper and lower range of significant parameters, and results obtained from the ANFIS models are also used for the calibration of the model. All the segments of the river are calibrated with one segment at a time until the whole study area is calibrated.

1.3.5 Model Validation

The major part of data is used in the calibration, and one-year data is utilized for the validation of the model. Coefficient of determination (R^2) and root mean square error (RMSE) are evaluated to compare the predicted and observed parameters obtained from the validation of the model. As the value of RMSE approached towards zero, reflects the optimum state of the model and when R^2 approaches towards one indicate the high correlation between the observed and predicted parameters. Equations used in the performance assessment of models are shown in Table 1.1.

1.3.6 Uncertainty Analysis

Uncertainty analysis used to determine the flux that can occur in the significant parameters, i.e. DO, BOD and COD. The significant variation is observed in the major parameter both spatially and temporally, in the study area as shown in the previous

Table 1.1 Performance parameters used for model accuracy analysis

Model performance parameter	Equations	Ideal value
Root mean square error (RMSE)	$\text{RMSE} = \sqrt{\frac{\sum_{i=1}^{n}(p_i - o_i)^2}{n}}$	0
Coefficient of determination (R^2)	$R^2 = 1 - \frac{\sum_{i=1}^{n}(p_i - o_i)^2}{\sum_{i=1}^{n}(p_i - \bar{o}_i)^2}$	1

p_i is the predicted value, o_i is the observed value and $\bar{o}_i$ is average observe value

section and same have been reflected in the literature also (Parmar and Keshari 2012; Chaudhary et al. 2017). Monte Carlo simulation is the probability analytics method and operates by selecting a random parameter in the multiple iterations. The number of parameter selected approaches for the equal number of iterations as parameter. MC simulation generates the large number of samples f of parameters X up to k, i.e. $k = 1, 2, 3, \ldots, n$. Y is the scalar quantity determined by the number of parameters. Approximate result of interest 'A' forms equation:

$$A = \frac{1}{n}\sum_{k}^{n} YX_k \tag{1.3}$$

Monte Carlo function also generates the random and statistical error in estimation, and to reduce the error, variance reduction is applied. Variance reduction also ensures the reduction in computational time of the model simulation. YASAI function based on the MC simulation is used in QUAL2Kw for the uncertainty analysis of models.

1.4 Result and Discussion

1.4.1 Performance Analysis of ANFIS

An appropriate combination of parameters is selected for the development of ANFIS models. DO, BOD and COD model are developed using a different combination of parameters. Accuracy of the FIS model majorly depends upon the number of input parameters, number of membership functions (MFs) and type of MFs. Takagi-Sugeno (TS) algorithm is used for the development of the model. Respective elevation in rule base and MFs, FIS generates a huge memory and function with large operational time. For DO and BOD models, three MFs are selected for each input parameters with Gaussian function type. Constant MF is selected for the output generation. To generate the FIS, grid portioning is selected for the DO and BOD models, whereas COD model is developed using subtractive clustering as shown in Table 1.2, as it has the tendency to select the rule base and MFs automatically based on flux in

Table 1.2 ANFIS model architecture used for designing the model

Model	Type of MF	Optimization model	Fuzzy rules
DO model	Grid partitioning	Hybrid learning	2187
BOD model	Grid partitioning	Hybrid learning	2187
COD model	Subtractive clustering	Hybrid learning	37

Table 1.3 Performance analysis of ANFIS models

Model	Training		Testing	
	RMSE	R^2	RMSE	R^2
DO model	0.3178	0.90	0.3349	0.89
BOD model	0.2151	0.96	0.2218	0.95
COD model	0.3065	0.90	0.3349	0.88

parameters and has lower operational time also. Subtractive clustering is preferred when there is no clear idea about the number of clusters that can be obtained in the model. In subtractive clustering, rule base formed is equivalent to the membership function formed. Training data is used to train the ANFIS model and testing data for the validation of the ANFIS model. 80% of the data is used for the training, and the rest 20% is used for the testing of the model. Performance of the models is identified using RMSE and R^2 listed in Table 1.3.

Hybrid learning technique is used for the optimization of model. Several epochs are performed on each model until the error observed gets constant or reduced to the minimum. ANFIS calculate the sum of the square difference between the observed and predicted parameters (Chang and Chang 2006). It was clear from the results that predicted and observed parameters are close to each other and same has been evident in the Fig. 1.6. However, the output in Fig. 1.6 represents the DO, BOD and COD in 1.6a, b, c, respectively, and index indicates the number of data points. Minimum RMSE was found in BOD model and maximum R^2 of 0.96 and 0.95 in training and testing of the model, respectively.

Surface plot of the model has also been designed with the principal parameter among other parameters used for designing the model. Temperature, BOD and conductivity are found as the principal parameters causing the significant variation in the concentration of DO. The optimum condition for saturation of DO is observed at a temperature in the range of 20–22 °C with zero BOD, whereas, along with the conductivity, optimum DO was observed at the temperature of 18–20 °C with approximately zero conductivity as shown in Fig. 1.7. In the surface plot designed for the BOD model, DO, COD, ammonia and conductivity are the significant parameters. At the zero DO content, BOD can rise to the 300 mg/L against the 30 mg/L of allowable disposal standard as per CPCB (2006).

Whereas, ammonia as high as ten μmol/L with zero DO, BOD can rise up to 150 mg/L as shown in Fig. 1.8. However, DO, BOD, temperature and ammonia are observed as the significant parameters for the COD prediction. Zero DO content

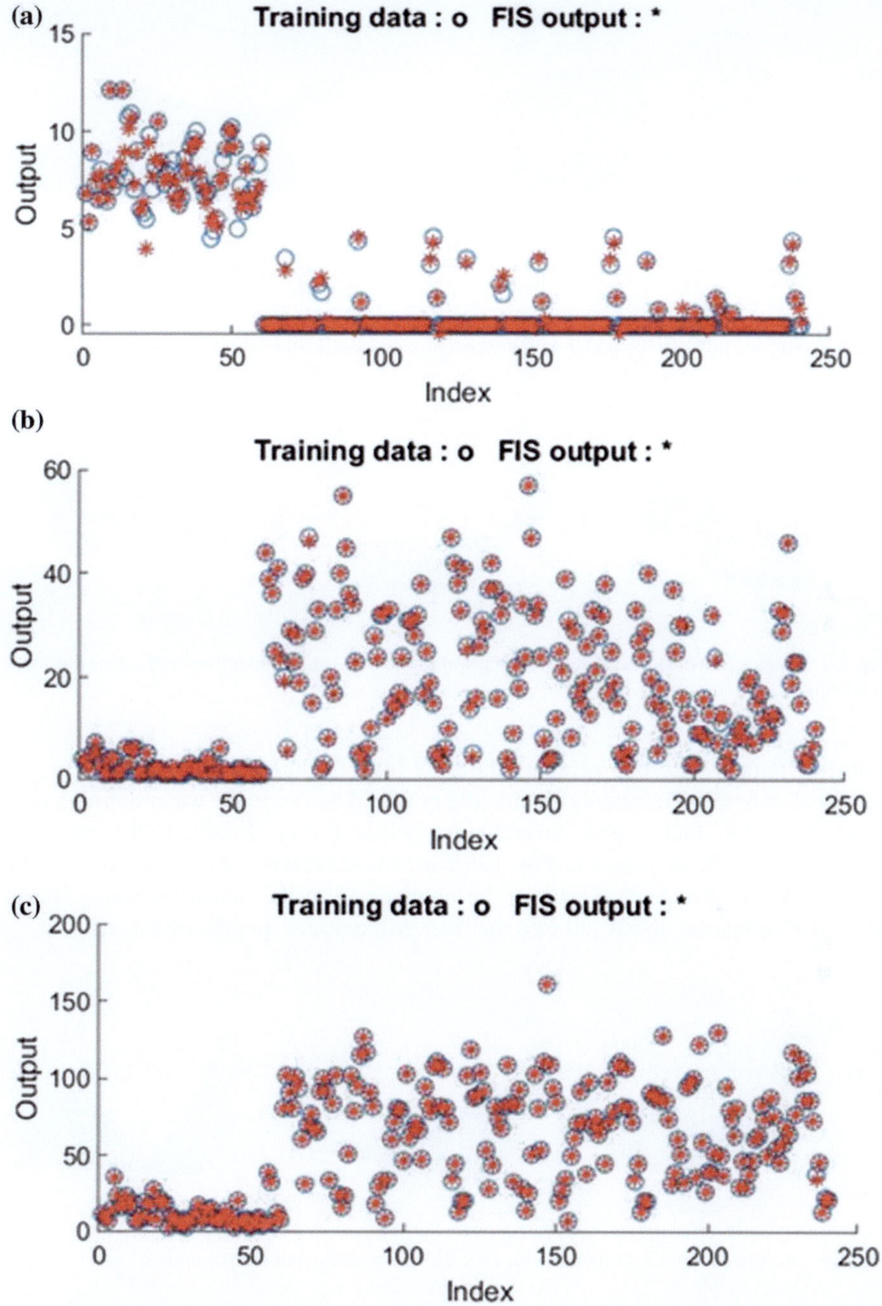

Fig. 1.6 Correlation between observed and predicted parameters using ANFIS model **a** DO model, **b** BOD model and **c** COD model

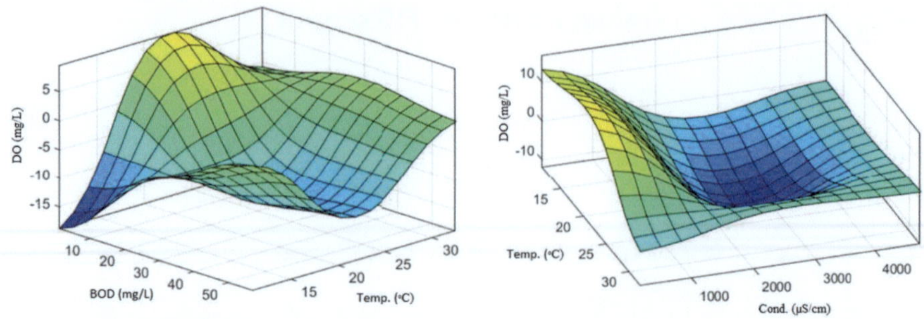

Fig. 1.7 Surface plot of DO model with temperature and BOD, indicating the variation in DO

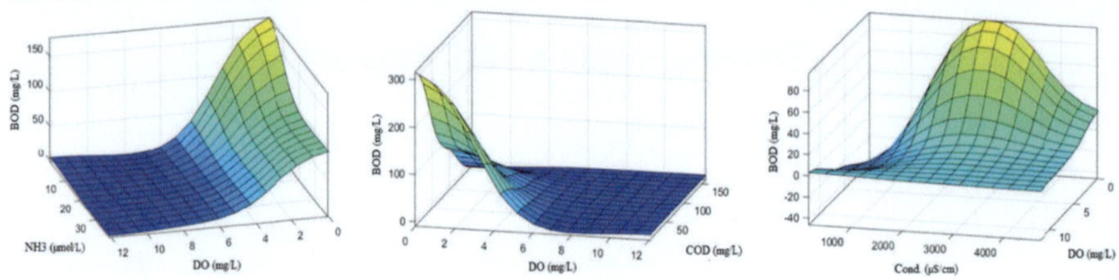

Fig. 1.8 Surface plot of BOD model with DO, ammonia, COD and conductivity indicating variation in BOD

with low temperature can generate the even high of 500 mg/L of COD. Whereas condition, where we have only chemically oxidized matter in the water defining zero BOD and at zero DO content, there can be the possibility of rising COD content as high as 600 mg/L as shown in Fig. 1.9, which will be the worse condition for any water body. Such type of waste can be obtained from the industrial drains and in streams if industrial drains fall into the river without any specific treatment.

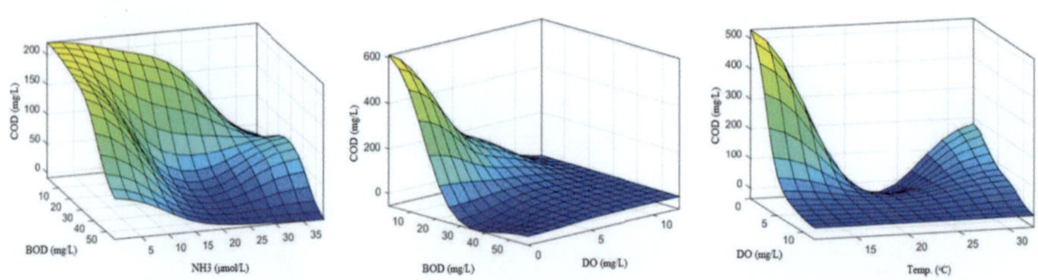

Fig. 1.9 Surface plot of COD model with DO, BOD, ammonia and temperature

1.4.2 Calibration and Validation of the Model

DO, BOD and COD predicted in the ANFIS are used for the calibration of the model. Upper and lower bound of model parameters are entered in the model. Performance of the model parameters and hydraulic parameters used for the calibration are shown in Table 1.4. K_2 (day^{-1}) of Yamuna River is found varying from 0.04 to 0.78. At the downstream of Wazirabad barrage, the low re-aeration rate is observed due to the abstraction of almost all the water for the city's drinking water demand and no freshwater remains in the channel at the downstream of the barrage (Arora and Keshari 2018). Variation in the model parameters along all the reaches has been used for the calibration of QUAL2Kw. Sufficient DO is observed at $S1$ as there is no wastewater drain that joins the river up to significant distance in the upstream direction. Whereas, at the other three segments, zero minimum DO is found and even maximum concentration of DO approaches well below the saturation limit of the river. Concentration of BOD is found low at $S1$ and considerably higher in $S2$, $S3$ and $S4$. Likewise, a pattern is observed in COD also due to the high organic content discharged by several drains in the river. Combination of parameters selected in the ANFIS is used for the calibration of the model. Three models have been designed using DO, BOD and COD. Simulated value of all three models obtained with the combination of best parameters is shown in Fig. 1.10, indicating the high correlation between the simulated and observed parameters.

Performance of the models is validated using one year of data, and simulation results are shown in Fig. 1.11 signifying good correlation of predicted values with

Table 1.4 Lower and upper limit of the model parameter for calibration

Parameter	Segment	Lower limit	Upper limit
Re-aeration (day^{-1})	S1	0.04	0.66
	S2	0.10	0.78
	S3	0.10	0.73
	S4	0.10	0.75
DO (mg/L)	S1	4.5	12.1
	S2	0.0	4.6
	S3	0.0	4.6
	S4	0.0	4.4
BOD (mg/L)	S1	1.0	7.0
	S2	2.0	55.0
	S3	2.0	57.0
	S4	2.0	46.0
COD (mg/L)	S1	3.0	39.0
	S2	9.0	127.0
	S3	7.0	161.0
	S4	13.0	130.0

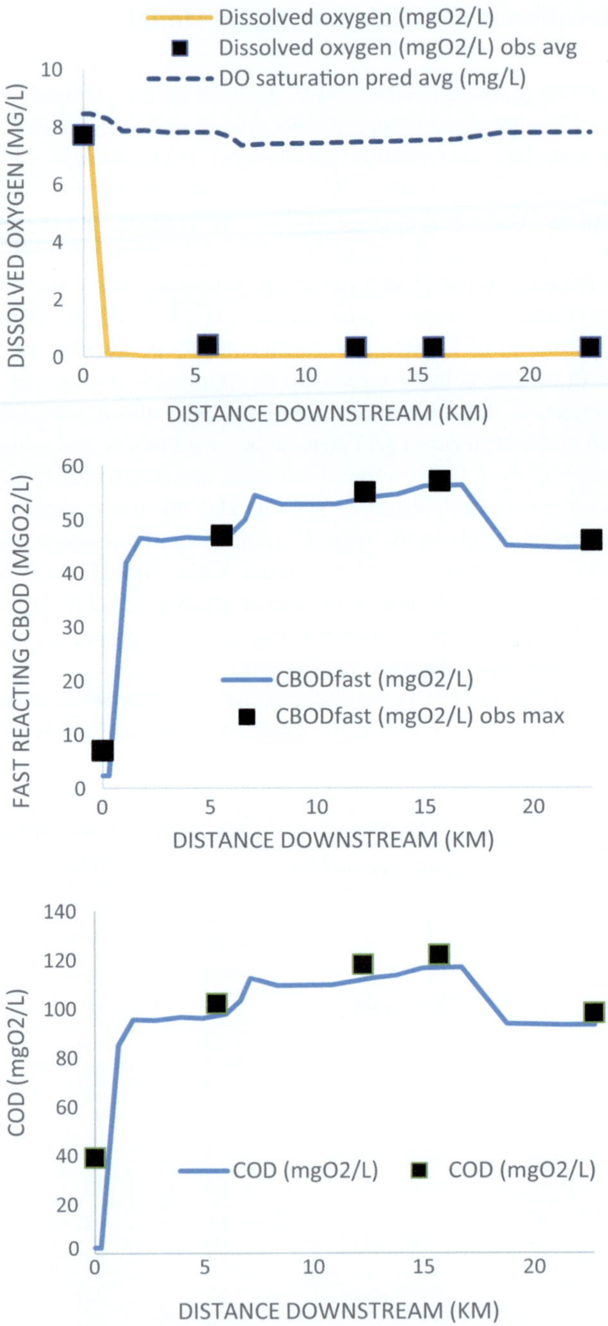

Fig. 1.10 Calibration results of DO, BOD and COD models developed using QUAL2Kw

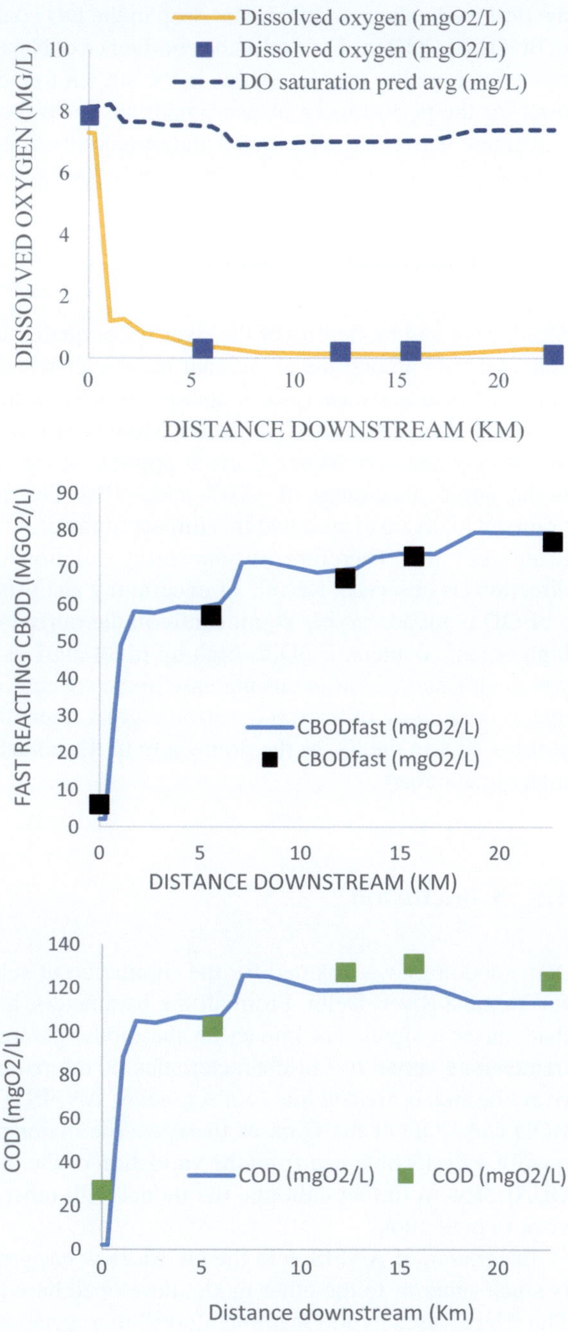

Fig. 1.11 Validation results of DO, BOD and COD models developed using QUAL2Kw

the observed values. A significant drop in the DO concentration and subsequent rise in BOD and COD is observed at the confluence of Najafgarh drain, which changes the river water quality characteristics in the stretch tremendously. Both the approaches used for the performance measurement are performed well and resulted in a close agreement between the predicted and observed variables.

1.4.3 Uncertainty Analysis

Monitoring and modelling of the river water quality clearly indicate that DO, BOD and COD are among the significant parameter which rules the overall river water quality. Uncertainty analysis is also performed on these parameters to identify the flux in the concentration of parameter. Monte Carlo simulation method is used for the uncertainty analysis. Monte Carlo is applied on the validated model of QUAL2Kw in the significance range of +10% and −10%. Due to the confluence of Najafgarh drain, DO falls up to zero and this impact of the drain last longs throughout the river in the study area. Therefore, no noteworthy variation in both the positive and negative directions is observed. Results of uncertainty analysis are shown in Fig. 1.12.

BOD is found varying significantly in the range of +10% and −10% due to the high organic content. BOD as high up to 88 mg/L is observed at +10% in the S4, which indicates a continuous increase in the concentration of BOD along the length due to the joining of several drains at regular intervals. COD content is found high at the +10% in the S2, as the drains join in S2, containing maximum discharge and high organic load.

1.5 Conclusion

A methodology was applied for the simulation of selected water quality parameter for Yamuna River, Delhi. From all the parameters, a set of parameters are selected that causes a significant impact on the model parameters. As the study area have tremendous variability in characteristics of the parameter along the length of the river, the area is divided into four segments. ANFIS is used for the modelling of DO, BOD and COD of the river, as these are the major parameters affecting the water quality. Results obtained from the validation of the model in ANFIS are used in the QUAL2Kw to further calibrate the model with other parameters and minimize the error in prediction.

Integration of ANFIS with the QUAL2Kw has proved that the developed model is much superior to the other model developed based on only water quality model. The ANFIS uses hybrid learning algorithm to generate the optimized model for the identification of most influential input parameters. Results from the ANFIS exported to QUAL2Kw, and the model is calibrated sequentially for each segment. These two

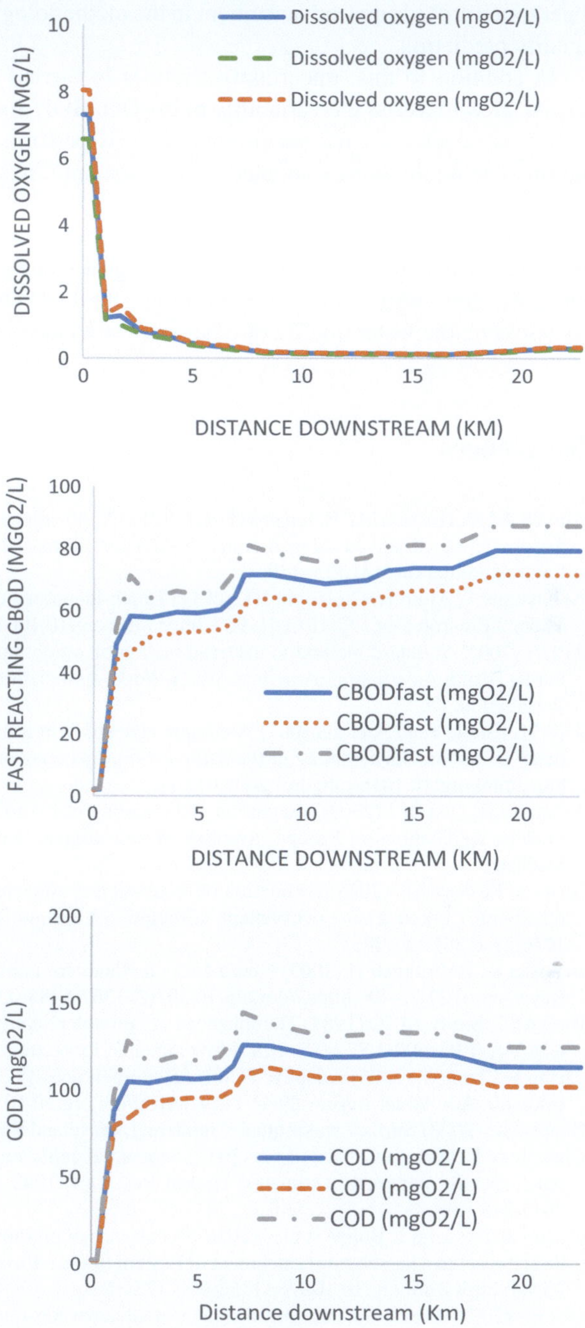

Fig. 1.12 Uncertainty analysis of DO, BOD and COD models performed on QUAL2Kw

operations provide new advancement in the methodology for the assessment of water quality prediction.

In addition to this, uncertainty analysis is carried to the validation models of QUAL2Kw to assess the variability of the DO, BOD and COD. Range of $+10\%$ and -10% is selected for the uncertainty analysis performed using Monte Carlo simulation. The methodology adopted derives dynamic results, improves the efficiency of the model and proved that can be used for the several reaches/segments of the river and any number of parameters. Such models can be used for the development of water management policies and planning of strategies related to water resource categorization, utilization and monitoring activities. The model can also be utilized in assessing the water quality of other river stretches.

References

Ahmed AAM, Hossain MI, Rahman MT et al (2013) Application of artificial neural network models for predicting dissolved oxygen concentration for Surma River Bangladesh. J Appl Technol Environ Sanitation 3(3):135–140

Akkoyunlu A, Altun H, Cigizoglu HK (2011) Depth-integrated estimation of dissolved oxygen in a lake. J Environ Eng 137(10):961–967. https://doi.org/10.1061/(ASCE)EE.1943-7870.0000376

APHA (2005) Standard method to the examination of water and wastewater, 21st edn. American Public Health Association, American Water-Works Association, Water Environment Federation, Washington, DC

Aqil M, Kita I, Yano A et al (2007) A comparative study of artificial neural networks and neuro-fuzzy in continuous modeling of the daily and hourly behaviour of runoff. J Hydrol 337:22–34. https://doi.org/10.1016/j.jhydrol.2007.01.013

Arheimer B, Olsson J (2003) Integration and coupling of hydrological models with water quality models: applications in Europe. Swedish Meteorological and Hydrological Institute (SMHI), Sweden

Arora S, Keshari AK (2018) Estimation of re-aeration coefficient using MLR for modelling water quality of rivers in urban environment. Groundwater Sustain Dev 7:430–435. https://doi.org/10.1016/j.gsd.2017.11.006

Babuska R, Verbruggen H (2003) Neuro-fuzzy methods for nonlinear system identification. Annu Rev Control 27:73–85. https://doi.org/10.1016/S1367-5788(03)00009-9

Brown LC, Barnwell TO (1987) The enhanced stream water quality models QUAL2E and QUAL2E-UNCAS (EPA/600/3-87-007). U.S. Environmental Protection Agency, Athens, GA, 189

Chang F, Chang Y (2006) Adaptive Neuro-fuzzy inference system for prediction of water level in reservoir. Adv Water Resour 29(1):1–10. https://doi.org/10.1016/j.advwatres.2005.04.015

Chapra SC (2008) Surface water quality modeling. Waveland Press, Long Grove, IL

Chaudhary S, Dhanya CT, Kumar A (2017) Sequential calibration of a water quality model using reach-specific parameter estimates. Hydrol Res 49(4):1042–1055. https://doi.org/10.2166/nh.2017.246

Chiffre A, Degiorgi F, Buleté A et al (2016) Occurrence of pharmaceuticals in WWTP effluents and their impact in a karstic rural catchment of Eastern France. Environ Sci Pollut Res 23(24):25427–25441. https://doi.org/10.1007/s11356-016-7751-5

Cox BA (2003) A review of currently available in-stream water-quality models and their applicability for simulating dissolved oxygen in lowland rivers. Sci Total Environ 314–316(1):335–377. https://doi.org/10.1016/S0048-9697(03)00063-9

CPCB (2006) Water Quality Status of Yamuna River (1999–2005). Central Pollution Control Board, Ministry of Environment & Forests, Assessment and Development of River Basin Series: ADSORBS/41/2006-07

Delhi Jal Board (2005) Delhi water supply and sewerage project. Final report—Project preparation study, Part C: Sewerage

Djaouida B, Somia H, Arab S et al (2019) Impact of pollution on the quality of water of Oued El Harrach (Algeria). Advances in sustainable and environmental hydrology, hydrogeology, hydrochemistry, and water resources. Adv Sci Technol Innov 141–144. https://doi.org/10.1007/978-3-030-01572-5_35

Galavi H, Mirzaei M, Shui LT et al (2013) Klang River-level forecasting using ARIMA and ANFIS models. J Am Water Works Assoc 105(9):E496–E506. https://doi.org/10.5942/jawwa.2013.105.0106

Guru N, Jha R (2013) Simulation of BOD-DO modeling in mahanadi river system lying in Odisha using ANN, India. IOSR J Environ Sci Toxicol Food Technol 2:52–57. https://doi.org/10.9790/2402-0245257

Harmel RD, Smith DR, King KW et al (2009) Estimating storm discharge and water quality data uncertainty: a software tool for monitoring and modeling applications. Environ Model Softw 24(7):832–842. https://doi.org/10.1016/j.envsoft.2008.12.006

Horowitz AJ (1997) Some thoughts on problems associated with various sampling media used for environmental monitoring. Anal 122:1193–1200. https://doi.org/10.1039/A704604I

Keller VDJ, Williams RJ, Lofthouse C et al (2014) Worldwide estimation of river concentrations of any chemical originating from sewage-treatment plants using dilution factors. Environ Toxicol Chem 33(2):447–452. https://doi.org/10.1002/etc.2441

Lindenschmidt K, Drastig K, Baborowski M (2007) Structural uncertainty in a river water quality modelling system. Ecol Modell 204(3–4):289–300. https://doi.org/10.1016/j.ecolmodel.2007.01.004

Melching CS, Bauwens W (2001) Uncertainty in coupled nonpoint source and stream water-quality models. J Water Resour Plan Man-ASCE 127(6):403–413. https://doi.org/10.1061/(ASCE)0733-9496(2001)1276(403)

Mulholland PJ, Houser JN, Maloney KO (2005) Stream diurnal dissolved oxygen profiles as indicators of in-stream metabolism and disturbance effects: Fort Benning as a case study. Ecol Indic 5:243–252. https://doi.org/10.1016/j.ecolind.2005.03.004

Nayak PC, Sudheer KP, Rangan DM et al (2005) Short-term flood forecasting with a neuro-fuzzy model. Water Resour Res 41(4). https://doi.org/10.1029/2004WR003562

Newham LTH, Norton JP, Prosser IP et al (2003) Sensitivity analysis for assessing the behavior of a landscape-based source and transport model. Environ Model Softw 13:741–751. https://doi.org/10.1016/S1364-8152(03)00076-8

Parmar DL, Keshari AK (2012) Sensitivity analysis of water quality for Delhi stretch of the River Yamuna India. Environ Monit Assess 184(1487–1508):80. https://doi.org/10.1007/s10661-011-2055-1

Pelletier GJ, Chapra SC, Tao H (2006) QUAL2Kw—a framework for modeling water quality in streams and rivers using a genetic algorithm for calibration. Environ Model Softw 21:419–425. https://doi.org/10.1016/j.envsoft.2005.07.002

Ramsey MH (1998) Sampling as a source of measurement uncertainty: techniques for quantification and comparison with analytical sources. J Anal At Spectrom 13:97–104. https://doi.org/10.1039/A706815H

Rode M, Suhr U (2007) Uncertainties in selected river water quality data. Hydrology and earth system sciences discussions. Eur Geosci Union 11(2):863–874. hal-00305056

Sonmez AY, Kale S, Ozdemir RC et al (2017) An adaptive Neuro-Fuzzy Inference System (ANFIS) to Predict Cadmium(Cd) Concentrations in the Filyos River, Turkey. Turk J Fish Aquat Sci 126:119–126. https://doi.org/10.4194/1303-2712-v18

Tiwari S, Babbar R, Kaur G (2018) Performance Evaluation of two ANFIS models for predicting water quality index of River Satluj (India). Adv Civil Eng 2018:10. https://doi.org/10.1155/2018/8971079

Wen Y, Schoups G, van de Giesen N (2017) Organic pollution of rivers: combined threats of urbanization, livestock farming and global climate change. Sci Rep. 7:43289. https://doi.org/10.1038/srep43289

Wu J, Zou R, Yu SL (2006) Uncertainty analysis for coupled watershed and water quality modeling system. J Water Resour Plan Man 132(5):351–361. https://doi.org/10.1061/(ASCE)0733-9496(2006)132:5(351)

Zhang R, Qian X, Yuan X et al (2012) Simulation of water environmental capacity and pollution load reduction using QUAL2K for water environmental management. Int J Environ Res Public Health 9(12):4504–4521. https://doi.org/10.3390/ijerph9124504

Zhao F, Xi S, Yang X et al (2012) Purifying eutrophic river waters with integrated floating island systems. Ecol Eng 40:53–60. https://doi.org/10.1016/j.ecoleng.2011.12.012

Zou R, Lung W, Guo H (2002) Neural network embedded Monte Carlo approach for water quality modeling under input information uncertainty. J Comput Civil Eng 16(2):135–142. https://doi.org/10.1061/(ASCE)0887-3801(2002)16:2(135)

Zounemat-Kermani M, Teshnehlab M (2008) Using adaptive neuro-fuzzy inference system for hydrological time series prediction. Appl Soft Comput 8(2):928–936. https://doi.org/10.1016/j.asoc.2007.07.011

Chapter 2
Stable Channel Design of Tapi River Using HEC-RAS for Surat Region

Darshan Mehta, S. M. Yadav, Sahita Waikhom and Keyur Prajapati

Abstract Stable channel design plays an important role for the research of stream restoration. Since sediment transport rates are highly variable in observed stream, methods for channel design must study a wide range of transport relationships to define stability conditions. The objective of study is to design stable channel for the reach of the Tapi River located between Sardar Bridge and Magdalla Bridge using hydrodynamic software. Stable channel design has carried out using Copeland method which is a part of Army Corp of Engineer (HEC-RAS) software. In the present analysis, existing sections are compared with the design sections for 2006 flood having magnitude of 9.09 cusecs (25,760 cumecs). For study purpose, Tapi River reach is selected having 31 cross sections. The bed material samples from the study reach were collected and analyzed to determine the actual grain size. This paper presents a detailed design step of cross section using Copeland method including approximating the upstream channel as trapezoidal. Results indicate that the model adequately predicts the bank shape and significant dimensions of stable channels.

Keywords Flood events · HEC-RAS · Stable channel design · Tapi River

2.1 Introduction

Stable channel design is a major topic of stream restoration research. The Copeland (1994) is commonly used for stable channel design. This method uses flow continuity and resistance equations with a sediment transport capacity equation to compute

D. Mehta (✉) · S. Waikhom
CED, Dr. S & S. S. G. G. E. C, Surat 395001, India
e-mail: darshanmehta2490@gmail.com

S. M. Yadav
CED, SVNIT, Surat 395007, India
e-mail: shivnam27@gmail.com

K. Prajapati
CED, S. S. A. S. I. T, Surat 395006, India
e-mail: keyurprajapati34@gmail.com

R. M. Singh et al. (eds.), *Environmental Processes and Management*, Water Science and Technology Library 91,
https://doi.org/10.1007/978-3-030-38152-3_2

cross-section geometry: bed slope, channel width, and flow depth. The Copeland design method is implemented in the SAM Lane (1995) and HEC-RAS Mehta et al. (2013) software packages.

A further limitation of the Copeland method is that channel stability is determined only from the balance between upstream sediment supply and local transport capacity (Neary and Korte 2001; Richard et al. 1984). However, this balance does not ensure stability because a channel can still adjust its geometry by the interaction between bed and banks. True stability rarely exists in natural systems. Channel slope along the axis of flow reflects the combined constraints of cross-section shape and plan form. Shelly and Parr (2009) describe a stable channel downstream profile design approach based on cross-section geometry and meander plan form.

Estimation of channel roughness parameter is of key importance in the study of open-channel flow, particularly in hydraulic modeling (Subramanya 2006). Channel roughness is a highly variable parameter which depends upon numbers of factor like surface roughness, vegetation, channel irregularities, channel alignment, etc (Vijay et al. 2007). Several researchers including John Shelley and A. David Parr, Krovak, Gary W. Brunner, P. E. and Stanford Gibson, Vincent S. Neary and Nic Korte, Agnihotri P. G, and Patel J. N has calibrated channel roughness for different rivers for the development of the hydraulic model. Datta et al. estimated single-channel roughness value for open-channel flow using optimization method, taking the boundary condition as constraints. The channel roughness is not a constant parameter and it varies along the river depending upon variation in channel characteristic along with the flow (Firenzi et al. 2000). This clearly demonstrates the variation in channel roughness along the natural river (Garde and Raju Ranga 2000). The Surat city, being a coastal city, had been seen susceptible to major floods and undergone huge damage in the past.

2.2 Objective of Study

To analyze the stability of a segment of lower reach approximately 7.75 km length of Tapi river between two bridges (i.e., Sardar bridge and Magdalla Bridge) by evaluating its capacity and stability in response to discharge and sediments using HEC-RAS software.

2.3 Scope of the Study

Following are the broad scope of work:

- To collect bed material samples.
- To carry out sieve analysis for various samples collected.

- To carry out analysis of stable channel design using HEC-RAS software and comparing the existing section with a computed section for a flood event of the year 2006.
- To determine the adequacy of existing sections.

2.4 Study Area

Tapi River, being one of the largest west-flowing rivers of India, is 724 km long and has its origin in Madhya Pradesh. It then flows through Maharashtra, Gujarat and ultimately meets the Arabian Sea which is approximately 20 km west of Surat. In the last reach, at downstream of Ukai, the river enters the flat and fertile land of Gujarat. The study reach, located between Sardar Bridge and Magdalla Bridge, approximately 7.75 km long with 31 cross sections, is shown in Fig. 2.1 Surat, being a coastal city, had been susceptible to major floods and undergone huge damages in the past. The river reach selected for the present study is extremely important as 80% of the total population of Surat is settled on either side of the bank. Major business centers for diamond industries, textile industries, and industrial area of Hazira are within 1 km radius of the study reach (Agnihotri and Patel 2011).

The Tapi River receives several tributaries on both the banks and there are 14 major tributaries having length more than 50 km. On the right bank, four tributaries namely Vaki, Gomai, Arunavati, and Aner join the Tapi River. On the left bank, ten important tributaries namely the Nesu, Amaravati, Buray, Panjhra, Bori, Girna, Waghur, Purna, Mona, and Sipna drain into the main river channel. The drainage system on the left bank of the Tapi river is, therefore, more extensive as compared to the right bank area. The Purna and Girna, the two important left-bank tributaries, together account for nearly 45% of the total catchment area of the Tapi River.

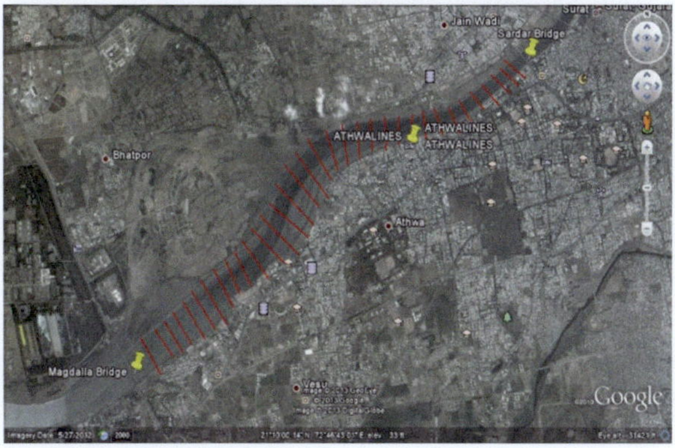

Fig. 2.1 Cross sections in the study reach, Tapi River, Surat

2.5 HEC-RAS Modeling Concepts

Hydraulic Engineering Center—River Analysis System (HEC-RAS) is a freely available computer program that models the hydraulics of water flow through natural rivers and other channels. The model has been developed by the US Department of Defense; Army Corps of Engineers is made available for public use in 1995. The program is one dimensional, meaning that there is no direct modeling of the hydraulic effect of cross-section shape changes, bends, and other two- and three-dimensional aspects of flow. It is used in modeling of water flowing through systems of open channels and computing water surface profiles. This model has a wide application in flood management studies.

2.5.1 HEC-RAS Input Parameters for Steady Flow Analysis

According to steady flow analysis, the model is intended for calculating water surface profiles for steady gradually varied flow. The steady flow component us capable of modeling subcritical, supercritical, and mixed flow regime water surface profiles. The basic computational procedure is based on the solution of the 1-Dimensional energy equation. Also, it is capable of assessing the change in water surface profiles due to levees. In this mathematical model, the state variables for the numerical scheme are flow and stage, which are computed are stored at each cross section. Accordingly, various parameters are used to do the analysis for the same. At particular flood discharges, steady flow simulation has been carried out by using various parameters on the developed model in HEC-RAS. The parameters are conducted for analysis listed below:

- Manning's $n = 0.022$.
- Contraction coefficient = 0.1, expansion coefficient = 0.3.
- Downstream reach length cross section to cross section (left bank, channel, right bank).
- Avg. bed slope $S_{avg} = 0.00842$.

2.5.2 Geometric Data

A reach herein is described as a river, lake, stream, channel, or a portion of these drawn in the geometry data window. A reach is comprised of at least two user-entered cross sections. Cross sections can be depicted by a maximum of 500 station and elevation coordinates with the first station being zero. All stations should be entered from left to right looking upstream. Data such as Manning's *n* values, bank stations, reach lengths, and expansion/contraction coefficients are required for each cross section. Manning's n values are used primarily for calibration purposes. The study reach consists of 53 cross sections. The detailed topographic features of study reach were collected from Surat Municipal Corporation (SMC) and Surat Irrigation Circle (SIC), Govt. of Gujarat, India.

2.5.3 *Cross-Sectional Data*

Boundary geometry for the analysis of flow in natural streams is specified in terms of ground surface profiles (cross sections) and the measured distances between them (reach lengths). Cross sections should be perpendicular to the anticipated flow lines and extend across the entire flood plain. Cross sections are required at locations where changes occur in discharge, slope, shape or roughness; at locations where levees begin or end and at bridges or control structures such as weirs. Each cross section is identified by a reach and river station label. The cross section is described by entering the station and elevations (x–y data) from left to right, with respect to looking in the downstream direction.

2.6 Flood Conveyance Performance

For evaluation of flood performance, past flood data are collected from Flood Cell, Surat was used. The flood frequency analysis results were based on data that coincides with the upstream limit of the project reach. Major flood events took place in the year 1883, 1884, 1942, 1944, 1945, 1949, 1959, 1968, 1994, 1998, 2006, 2007, 2012, and 2013. The summary of the flood is given in Table 2.1.

2.7 Methodology

The following step has been carried out for the design of stable channel using HEC-RAS software:

Step 1: **Field reconnaissance**.

- Perform a cross-sectional survey at a stable upstream riffle or inflection point.
- Determine the slope of the upstream section from rifle to riffle or inflection point to an inflection point.
- Slope $S_b = 0.00842$.
- Estimate the bank-full stage by geomorphologic observation.
- Take a bed-sediment sample to determine the sediment gradation of the upstream cross section, consisting of the D_{16}, D_{50}, D_{60}, and D_{84} (sieve size for which, 16, 50, and 84% of the material is finer by weight).

Table 2.1 Flood history of Surat city

S. No	Year	Discharge (cumecs)	Discharge (lakhs cusecs)
1	1882	2095.5	7.40
2	1883	2845.8	10.05
3	1884	23,956	8.46
4	1944	33,527	11.84
5	1945	28,996	10.24
6	1949	23,843	8.42
7	1959	36,642	12.94
8	1968	43924.2	15.50
9	1998	29817.27	10.53
10	2006	25,788	9.10
11	2012	9508.04	3.35
12	2013	13,178	4.65

Step 2: **Calculate the design bank-full flow using HEC-RAS** (Fig. 2.2).
Step 3: **Choose a value for Manning's 'n' for the channel sidewalls and bed.**
Step 4: **Create a prismatic hydraulic model for the upstream reference cross section.**
Step 5: **Calculate a Copeland stable channel design.**

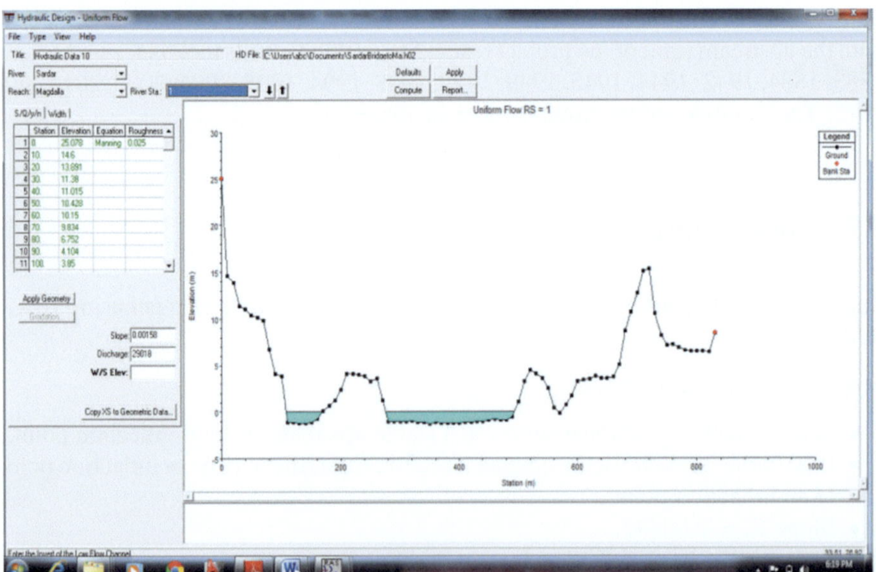

Fig. 2.2 Uniform flow window in HEC-RAS software

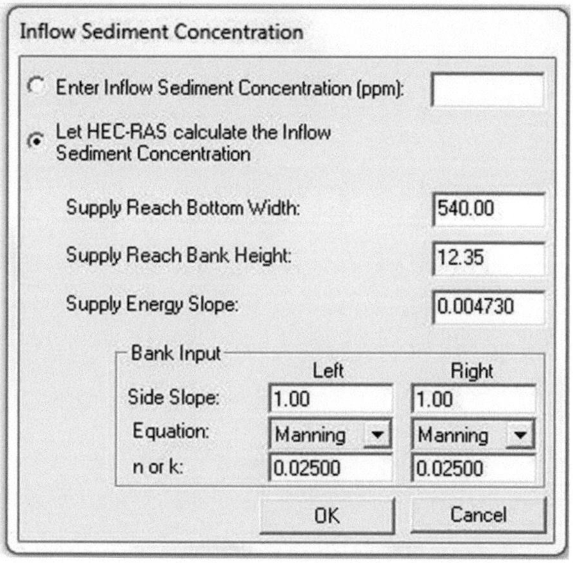

Fig. 2.3 Inflow sediment concentration window in HEC-RAS

Step 6: Enter the bottom width (*b*), bank height (*y*), energy slope (*m*-channel slope), and side slopes (*z*), calculated earlier for the upstream representative channel. If the upstream section is a meandering channel and the channel design is a meandering channel, the same *n* value for the side slopes should be used in both the upstream and the design section. When the upstream section is stable, but non-meandering and a meandering channel will be designed downstream or when a straight channel is designed downstream from a meandering channel, then the *n* values should be different as shown in Fig. 2.3.
Step 7: After completing all the steps, with the help of all parameters and by computing, HEC-RAS will give trapezoidal shape using Copeland method as shown in Fig. 2.4.

After doing this, comparing the above section with the existing section and check:

- Whether the section is sufficient or not, if it is not sufficient then at that particular section cutting and filling can be done.
- Whether the section is overtopping or not.

2.8 Result and Discussion

Part A: **Bed material analysis**

As discussed in methodology bed material samples were collected from the study reach. Following graph were obtained after doing a sieve analysis on bed material sample (Figs. 2.5, 2.6, 2.7, and 2.8).

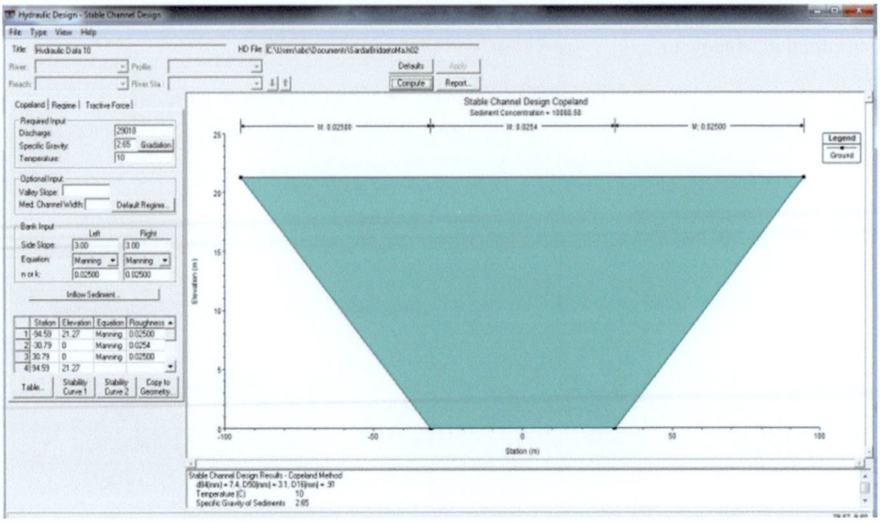

Fig. 2.4 Stable channel design (hydraulic design window)

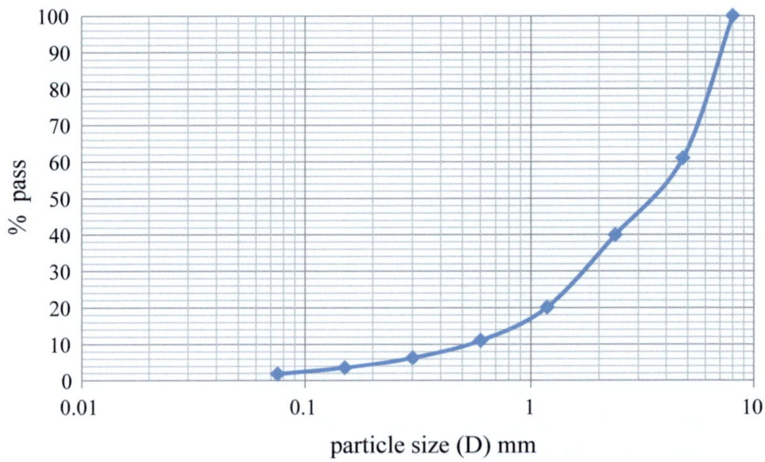

Fig. 2.5 Grain size distribution curve

After plotting semi-logarithmic curve between grain size and percentage finer, the following values are obtained: $D_{16} = 0.91$ mm, $D_{50} = 3.1$ mm, $D_{60} = 4.54$ mm, $D_{84} = 7.4$ mm.

Part B: **Geomorphic channel design and analysis**

In 1-D, steady state, gradually varied flow analysis, the following assumptions are made:

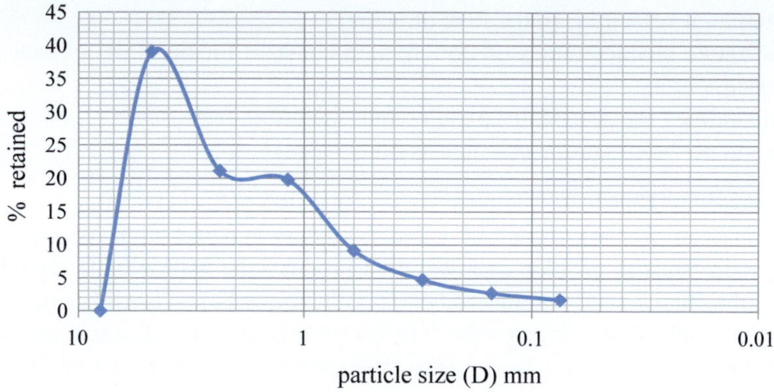

Fig. 2.6 Frequency curve

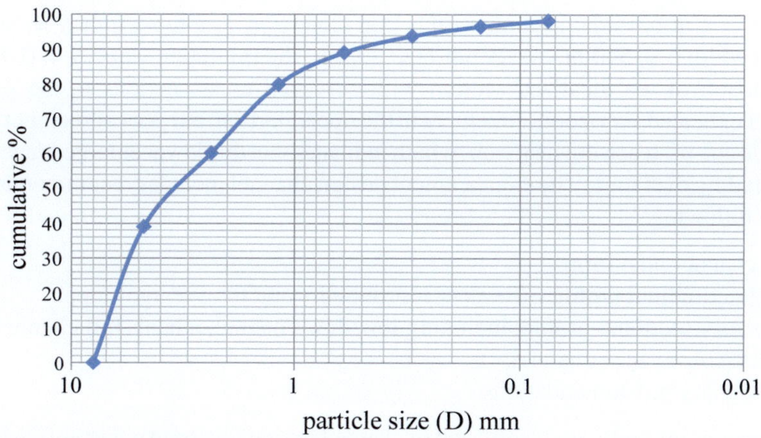

Fig. 2.7 Cumulative frequency curve

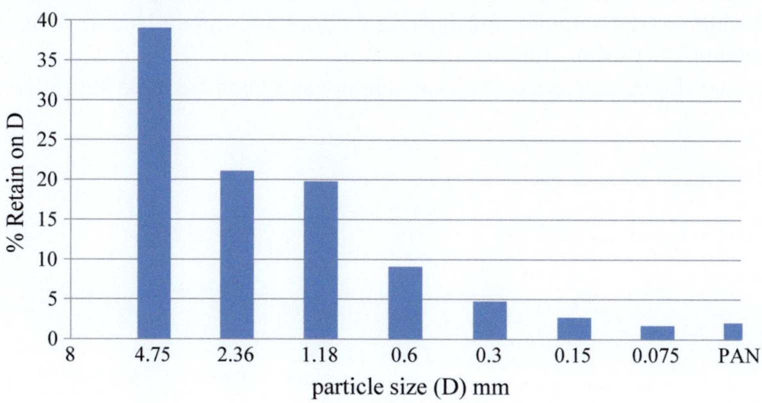

Fig. 2.8 Histogram showing percentage of particle size retained

- Dominant velocity is in the flow direction.
- Hydraulic characteristics of flow remain constant for the time interval under consideration.
- Streamlines are practically parallel and, therefore, hydrostatic pressure distribution prevails over the channel section.

The Copeland (2001) was employed to estimate the preferred hydraulic geometry of the bank-full channel. Hydraulic geometry theory is based on the concept that a river system tends to develop in a way that produces an approximate equilibrium between the channel and the inflowing water and the sediment load. Chang method which is rational was developed, especially for active gravel-bed channels, as opposed to threshold channel that has no significant transport at bank-full discharge. It predicts the top width *B*, depth *D,* and slope *S* of the bank-full channel for a given water discharge *Q*, particle size D_{50}, and side slope z. The water discharge $Q = 25{,}788\ \text{m}^3/\text{s}$, the particle size $D_{50} = 3.1$ mm, and the side slope $z = 1$. The resulting geometry channel is a trapezoidal section with a side slope $z = 1$, a top width $B = 800$ m, bottom width $b = 540$ m, and depth $D = 2.9$ m. The minimum slope $S = 0.00158$. Comparison of the idealized section with surveyed cross sections in the project channel indicated very good agreement in the upper portion, but poor agreement in the enlarged cut-off reach downstream, which indicates that this lower enlarged reach will be aggrading over time. The enhancement design channel includes recommendations for the following:

- The channel plan form.
- The longitudinal profile riffle-pool morphology and the average slope.
- The cross-sectional profile, including low flow channel, a bank-full channel, and a flood conveyance channel.
- Preliminary recommendations.

In the present study, analysis of stable channel design is carried out using Copeland method as explained in the methodology and the existing section is compared with the computed section using 2006 flood data. By following the procedure described in methodology the computed and existing sections are summarized in Table 2.2 for cross section 1–31 using 2006 flood data.

From the above table, sections which are not sufficient are to be modified.

Table 2.2 Summary of designed section and comparison with sections of flood event of the year 2006

S. No.	Cross section	Computed section					Area of existing cross section (m^2)	Comparison
		b	B	D	Side slope 1 in	Area (m^2)		
1	CS-1	470	650.33	8.21	200	3520.84	3800.56	Sufficient
2	CS-2	550	636.74	9.41	200	3428.12	3620.12	Sufficient
3	CS-3	650	751.57	9.9	200	3687.12	3425.12	Insufficient
4	CS-4	630	846.43	15.4	200	5825.91	5210.12	Insufficient
5	CS-5	730	923.34	14.33	200	6462.83	6100	Insufficient
6	CS-6	750	910.22	13.66	200	6643.92	6200.32	Insufficient
7	CS-7	650	972.36	14.33	200	7471.03	7007.03	Insufficient
8	CS-8	700	908.54	13.06	200	7291.72	7000.02	Sufficient
9	CS-9	820	1065.77	11.96	200	5579.99	5479.56	Sufficient
10	CS-10	850	1051.61	11.53	200	6837.32	6788.32	Sufficient
11	CS-11	230	589.82	12.09	200	4911.67	4735.36	Sufficient
12	CS-12	230	539.72	12.08	200	4409.97	4136	Sufficient
13	CS-13	230	456.13	10.64	200	3574.45	3600.12	Insufficient
14	CS-14	200	444.71	10.45	200	3333.24	3215.12	Insufficient
15	CS-15	190	488.52	11.8	200	4151.69	3215.12	Insufficient
16	CS-16	146.1899	453.32	11.58	200	3365.81	3265.1	Sufficient
17	CS-17	190	364.3	10.47	200	2904.33	2900	Sufficient
18	CS-18	220	480.3	12.41	200	3771.7	3500.71	Insufficient
19	CS-19	200	514.93	13.3	200	4211.34	4000.25	Insufficient
20	CS-20	100	555.77	13.76	200	5039.03	4756.23	Insufficient
21	CS-21	130	568.77	14.52	200	5531.6	5012.17	Insufficient
22	CS-22	290	577.24	14.12	200	5978.79	5725.85	Sufficient
23	CS-23	250	514.53	13.96	200	5628.73	5625.7	Sufficient
24	CS-24	200	630.3	13.94	200	6404.25	6305	Sufficient
25	CS-25	250	640.31	14.24	200	6602.62	6102	Insufficient
26	CS-26	300	637.32	14.7	200	6790.23	6690	Sufficient
27	CS-27	380	682.36	15.88	200	7457.44	7412.25	Sufficient
28	CS-28	400	639.84	14.78	200	7395.46	7393.25	Sufficient
29	CS-29	420	750.88	15.06	200	8509.26	8505.25	Sufficient
30	CS-30	450	756.42	14.89	200	8339.22	8336.21	Sufficient
31	CS-31	520	729.18	14.87	200	8370.04	8367.03	Sufficient

2.9 Conclusion

1. Based on available data, it is reasonable to believe that this channel would be in dynamic equilibrium with its sediment load, would restore main channel—flood plain dynamics, can incorporate physical variability for low flow conditions, including variation in depth, velocity, and roughness, and can achieve an acceptable level of the 2006 year flood performance; the degree of performance-based partly on the landscape plan and resultant channel roughness considered acceptable.
2. It is strongly recommended that the sections found critical in the present study require restoration work/design and needs to be raised as per requirement.
3. It is strongly recommended that no new construction is allowed in the flood plain area.
4. It is strongly recommended that the width of the river in no case encroach as already sections are sensitive to high floods, encroachment will result in flooding of the study region.

References

Agnihotri PG, Patel JN (2011) Modification of channel of Surat city over Tapi river using HEC-RAS software. Int J Adv Eng Technol 2:231–238

Firenzi AL, Watson CC, Bledsoe BP (2000) Stable channel design for mobile gravel bed rivers. J Water Res Prot 10:1–9

Garde RJ, Raju Ranga KG (2000) Mechanics of Sediment Transportation and alluvial stream's problems. New Age International publishers (P) Ltd., New Delhi, India

Lane EW (1995) Design of stable channels. Trans. Am Soc Civil Eng 120:191–197

Mehta D, Yadav SM, Waikhom S (2013) Geomorphic channel design and analysis using HEC-RAS hydraulic design functions. Paripex Int J Glob Res Anal 2(4):90–93

Neary VS, Korte N (2001) Preliminary channel design of Blue River reach enhancement in Kansas city. Am Soc Civil Eng 1:31–42

Richard D, Hey M, Thome Colin R, Aff M (1984) Stable channels with mobile gravel bed. J Am Soc Civil Eng 112(8):671–689

Shelly J, Parr DA (2009) Hydraulic design functions for Geomorphic channel design and analysis using HEC-RAS. J World Environ Water Res Congr 2:41–50

Subramanya K (2006) Flow in open channels. Tata McGraw-Hill Publishing Company Limited, New Delhi, India

Vijay R, Sargoankar A, Gupta A (2007) Hydrodynamic simulation of river yamuna for riverbed assessment: a case study of delhi region. Environ Monit Assess 130(1–3):381–387

Chapter 3
Nutrient Fluxes from Agriculture: Reducing Environmental Impact Through Optimum Application

Mridusmita Debnath, Chandan Mahanta and Arup Kumar Sarma

Abstract Unmindful of the increasing environmental risk exerted by agriculture, farmers are often liberal in application of chemical fertilizers, herbicides, pesticides in order to enhance farm productivity. Use of pesticide, herbicides like glyphosate has risen over the past few decades in most countries some of which are identified carcinogenic which affect human as well as many wildlife forms besides polluting freshwater resources. Agriculture mainly livestock sector also contributes to anthropogenic causes of NH_3 emissions. Nutrient emissions from agriculture due to more and more fertilizer application have been the cause of major concern nowadays. Eutrophication is caused by accounting to excess nitrogen (N) and phosphorus (P) in the water bodies. Furthermore, paddy fields due to its water-loving character are much more vulnerable to nutrient outflows which results in eutrophication in water bodies. Nutrient balance method can be used as a land-based indicator for measuring the sustainability of agro-ecosystem. Various models like NUTMON have been developed for nutrient balance calculation, depending on the site-specific fertilization management practices of agriculture soils. Results across various countries including Vietnam, China and Thailand showed that nutrient loss through various means like infiltration, leaching and runoff from paddy fields is in significant amount, especially for N and P contributing more than 50% of the total input. However, irrigating with the nutrient-loaded water minimized the use of fertilizer, especially N fertilizer up to 20%. Furthermore, it was also found that use of sewage water for irrigation not only increased N uptake by 29, 23, 18 and 37% in food grain, agroforestry, fodder and vegetable production systems, respectively, while the corresponding values were 28, 21, 29 and 35% for P uptake but also increased soil microbial biomass carbon (MBC) and activities of dehydrogenase, urease and phosphatase. The results revealed that

M. Debnath (✉) · C. Mahanta · A. K. Sarma
Civil Engineering Department, Indian Institute of Technology Guwahati, North Guwahati, Assam 781039, India
e-mail: mridu.ars590@gmail.com

M. Debnath
Land and Water Management Division, ICAR-Research Complex for Eastern Region Patna, Patna 800014, India

R. M. Singh et al. (eds.), *Environmental Processes and Management*,
Water Science and Technology Library 91,
https://doi.org/10.1007/978-3-030-38152-3_3

emphasizing on growing organic foods devoid of chemical input should be of priority to keep the soil health intact along with sound local environment. Further, it is recommended that two or more indicators should be accompanied for robust testing of sustainability of the agro-environment.

Keywords Agro-environment · Fertilizer · Emissions · Eutrophication · Indicator · Nitrogen · Phosphorus

3.1 Introduction

Input of nutrients to crops in the form of chemical fertilizers, pesticides, herbicides, etc., is mandatory for raising a healthy crop and increase farm output. However, the overloading of these inputs by the farmers, in long term, affects the health of local environment through nutrient emissions to the local soil, water and air environment. Liberal use of pesticide including compounds like herbicides, insecticides, fungicides, nematicides, rodenticides and various such other products has risen over the past few decades in most countries including India. Some of which are identified as carcinogenic which affect human as well as many wildlife forms besides polluting freshwater resources. Agriculture including livestock sector also contributes to anthropogenic causes of greenhouse gas emissions. About 81% of NH_3 emission comes from agriculture. Out of which 13.3% comes from livestock and 43.3% from agricultural soils and crops (Sutton et al. 2013). It was further projected by FAO that by the year 2030 there would be 60% more ammonia and methane emissions from livestock sector. However, animal feed composition and feed management have a strong influence on animal performance as well as the chemical composition of dung and urine, and ultimately on the NH_3 emission. On the other hand, various mitigation strategies are adopted for crop field like deep placement of nitrogen (N) fertilizer and irrigation in cropland.

Nutrient emissions from agriculture due to fertilizer application have been the cause of major concern nowadays. Increasing population along with cropland expansion globally has led to more input of fertilizers in crop field. Data from International Fertilizer Industry Association (IFA) and Food and Agricultural Organization (FAO) indicated that globally N and phosphorus (P) fertilizer use rates increased by approximately 8 times and 3 times, respectively, since the year 1961–2013. In addition to this, N fertilizer use was replaced by Asia in the early twentieth-first century from the USA and Western Europe in the 1960s, whereas P fertilizer input exhibited the same trend with additional hotspot in Brazil (Lu and Tian 2016). However, less nutrient uptake efficiency of the crop plants and more fertilizer application has accounted to nutrient emission from agriculture to soil, water and air. Moreover, eutrophication is caused by accounting to excess N and P that are transported to the water bodies. Paddy fields due to its water-loving character are much more vulnerable to nutrient outflows which results in eutrophication in water bodies. Various lakes in and around paddy fields of paddy growing countries of Asia suffer from severe eutrophication. High N dose of 300–350 kg/ha was applied than the recommended dose 185–270 kg/ha in China (Wang et al. 2004; Huang et al. 2007; Xia and Yan 2011) and ranging from 26 to 190 kg N ha^{-1} as in India (Alivelu et al. 2006). In addition to this N losses also occur through means like ammonia volatilization (AV), denitrification, surface runoff and leaching among which AV contributes to 10–60% of N applied to paddy soils (Liang et al. 2007). Further, the amount of AV losses depends on the type of N fertilizer and N application methods in flooding condition. Environmental indicator is term defined as a variable in environment which is used to supply knowledge about other variables in environment that is difficult to quantify and as a result these indicators can be used for policy and decision-making process.

The impact of agriculture on environment is assessed through two types of variable: means-based and effect-based (Fig. 3.1). Means-based variables or indicators are primarily determined from the production methods of farmers, whereas effect-based indicators are from the effects caused from the agricultural system. Though agricultural impact on environment is more through production practices of farmers, but effect-based indicators are more relevant as well as linked to the objective that is used to quantify the environmental impact. Means-based indicators are not directly linked to the emissions to the environment as various other climatic factors such as rainfall and temperature also play a role.

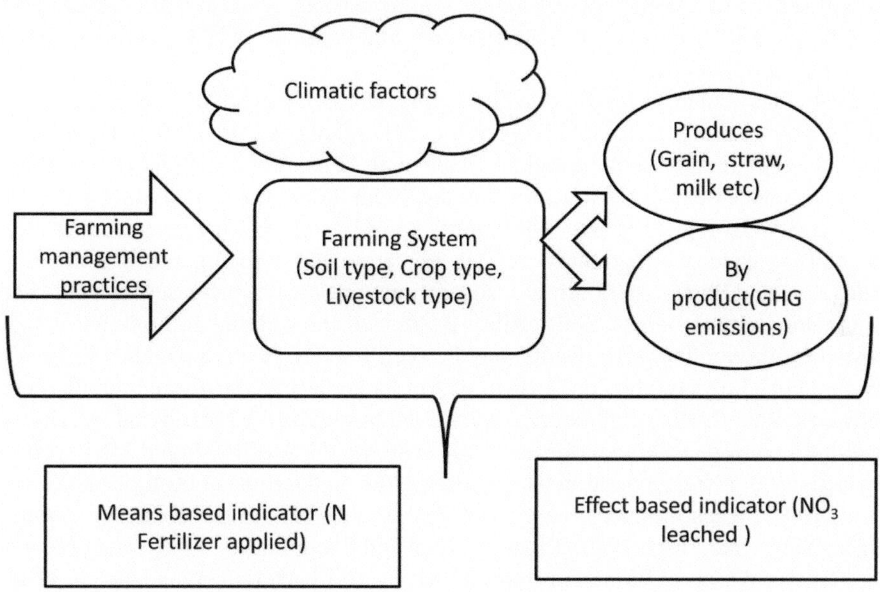

Fig. 3.1 Representation of the natural as well as human-made factors affecting the farming practice and the linked production of emission along with farm produces as well as the indicators for environmental impact

3.2 Description of Pollution Causing Drivers in Agriculture

3.2.1 Role of Pesticides

Abuse of pesticides with the thought that "if little application is little good, more application will be better" has brought disaster in the balance of agricultural ecosystem. Recently, India is the second-largest producer of pesticides in Asia and twelfth largest in the world (Mathur and Tannan 1999). Moreover, pesticide poisoning is more in developing countries and some groups in some countries (WHO 1990). It numbers to 1 million every year (Environews Forum 1999). Certain environmental chemicals, including pesticides termed as endocrine disruptors, are known to bring out hazardous effects in human, and their long term, low-dose exposure is increasingly related to various human ailments like immune suppression, hormone disruption, diminished intelligence, reproductive abnormalities as well as cancer (Brouwer et al 1999; Crisp et al 1998; Hurley 1998). The results support the impression that dioxin, an environmental pollutant is carcinogenic to humans and validate the hypotheses that it is related to cardiovascular and endocrine-related effects (Pier et al. 2001). For the agricultural food commodities investigated, it was found that around 34% had pesticides residue at or below maximum residue limit (MRL). Furthermore, analysis of results of varieties of vegetables confirmed that beans (1.9%), followed by mandarins (1.8%), pears (1.3%), and bananas and potatoes (0.5%) showed pesticide residues more than MRL.

Pesticides contaminate soil, water by killing beneficial insects or weeds also harming host of other organisms including non-target organisms. Almost all streams globally are contaminated with pesticides (Kole et al. 2001). For example, twenty-three pesticides were detected in waterways in the Puget Sound Basin, including 17 herbicides (USA Geological Survey 1999). Survey in Bhopal, India has revealed that 58% of drinking water from groundwater was contaminated with Organochlorine pesticide (Kole and Bagchi 1995). In soil, adsorption of pesticides increases with increase in organic matter content. Soil fertility is affected through the non-optimal use of pesticides though lucrative results may be seen for few years (Savonen 1997). For example, triclopyr inhibits the growth of soil bacteria that transform ammonia into nitrite (Pell et al. 1998); glyphosate was found to reduce the growth and activity of free-living nitrogen-fixing bacteria in soil (Santos and Flores 1995) and 2,4-D retards the activity of nitrogen fixation bacteria that grow on the roots of bean plants (Fabra et al. 1997), it also reduces the growth and activity of nitrogen-fixing blue-green algae (Singh and Singh 1989; Tözüm-Çalgan and Sivaci-Güner 1993), and prevents the transformation of ammonia into nitrates by soil bacteria (Frankenberger et al. 1991; Martens and Bremner 1993). It was found that oryzalin and trifluralin both retarded the growth of some species of mycorrhizal fungi (Kelley and South 1978).

3.2.2 NPS from Agriculture and Livestock Sector

Nonpoint source (NPS) pollution from agricultural and livestock sector to water bodies has been the major concern nowadays (Rong et al. 2017; Volk et al. 2016; Duncan 2014). Use of various models like ECA and SWAT identified that 82% of TN was mainly from livestock production, while the TP was from both livestock production and crop cultivation contributing 52% and 42%, respectively (Hua et al. 2019). In addition, using NUtrient Flow in food chains, Environment and Resources use (NUFER) model in Haihe Basin, China it was revealed that total nitrogen (TN) was 54% from cropland about 1079 Gg N in 2012, total phosphorus (TP) emission to water bodies was 208 Gg P, of which livestock accounting 78% (Zhao et al. 2019). Furthermore, spatial variation in terms of pollutant load could be seen accounting for 80% of TN and TP from plain areas relative to mountainous areas. The transportation and transformation of particles of environmental pollutants within watersheds consisting of agricultural land and many other pollutions causing infrastructures can be examined from the modelling outputs. Also, direct discharge of livestock manure is the main pathways of agricultural N and P emission in surface and groundwater system. Hence, required management measures in crop field and prevention of direct discharge of livestock manure, which was the primary source of agricultural P emission, plays a vital role in mitigating eutrophication. Scenario analysis recommended that TN and TP emissions can be reduced up to 45% and 77%, respectively, in 2030, via improved agricultural and environmental policies, technologies and managements (Zhao et al 2019).

3.2.3 NH_3 Emissions from Agriculture and Livestock Sector

Various mitigation measures adopted such as deep placement, incorporation, injection and irrigation was seen to reduce the emission of NH_3 by 98.7, 53.2, 67.5 and 73.6%, respectively, compared with the control. Overall, reduction in land application strategies was 60.7% (Ti et al. 2019). Irrigation water applied just after N fertilization reduced NH_3 emission up to 70%. It is due to the fact that irrigation water washes the N into the soil converting to NH_4 + to be adsorbed on cation exchange sites (Oenema and Velthof 1993). Alternate wet–dry cycles in the non-flooded irrigation (NFI) paddy took more N into the soil, leading to lower ammonium volatilization (AV) losses. Furthermore, NH_3 emission from cropping systems could be reduced by the application of N fertilizers such as ammonium nitrate and ammonium sulphate as compared to the application of urea fertilizers.

Day by day, there is increase in livestock production due to change in human diet and this necessitates adoption of simultaneous management measures to reduce emissions. It has been found that the effect of land application strategies on reducing NH_3 emission was dependent on the animal type: the highest was for poultry (74.3%) and the lowest was for cattle (58.9%). On the other hand, mitigation measures through management of fodder, was higher in pig (42.2%) than in poultry (39.7%), while for cattle was not significant. Mitigation through housing strategy reduced NH_3 emission by 32.2, 48.9 and 53.5% for cattle, poultry and pig, respectively. Furthermore, manure cover and application of acidifiers and additives, reduced NH_3 emission by 92.5 and 29.3% (Ti et al. 2019).

3.2.4 Leaching of N Fertilization

Leaching of excess N fertilization in cropland contributes to groundwater pollution as well as unavailability of N to crops when beyond the root zone resulting in underutilization of applied N. Leaching is dependent on factors like type of irrigation, fertilization as well as soil type. It was found that drip fertigation reduced the amount of leaching as compared to surface flooding fertigation. Also, leaching was less in clay ranging from 5.7 to 9.6% as compared to 16.2 and 30.4% in sandy loam soil. In addition, mineral N loss was more in nitrate fertilizer than in urea or ammonium fertilizer (Zhou et al. 2006). Further, it was found that soluble forms of nitrogen (N) fertilizers caused more leaching of nitrate N (NO_3–N) as compared to controlled-release N fertilizer (CRN). Results revealed that leachate accounted for 28, 12, 6 and 5% of the total N applied as urea, Poly-S, Meister and Osmocote, respectively (Paramasivam and Alva 1997).

3.2.5 Recycling of Irrigation Water

It was also found that of sewage water for irrigation not only increased N uptake by 29, 23, 18 and 37% in food grain, agroforestry, fodder and vegetable production systems, respectively, while the corresponding values were 28, 21, 29 and 35% for P uptake. Thus, recycling of nutrient-loaded irrigation water also increased nutrient uptake (Lal et al. 2015).

3.2.6 Environmental Indicators (EIs)

The production practices like tillage operations, sowing and fertilization are carried out in order to optimally combine various inputs based on natural capital like soil, solar energy, rain, fossil energy as well as human-made capital inputs like fertilizers, seeds, pesticides that yield desirable outputs (farm produce) along with undesirable pollution to the environment. Till 1970, raising the productivity was the major concern. And these situations led to the development of various methods for evaluating the impacts of agriculture on environment. These methods take into account many environmental problems like soil erosion, emission, volatilization, surface and groundwater quality, etc. Hence, these tools are necessary for the implementation of sustainable agriculture. Nutrient balance method can be used as a land-based indicator for measuring the sustainability of agro-ecosystem. Gross balance in a agriculture system is mainly the difference between the nutrient entering into the system (fertilizer + manures + nutrient stocks) and nutrient output leaving the system (harvested part). N and P budget in a hydro system consisting of reservoir catchment revealed that 66 and 79% of total N ad P input were deposited or eliminated in the reservoir.

Both values and scores can be used as indicator output. However, values are more preferable for balancing against other values. Scores differ in their range are of dimensionless units (Table 3.1).

Table 3.1 Description of evaluation methods based on indicators

Method	Description	Authors	Indicators	Output format
The farmer sustainability index (FSI)	It reflects ecological sustainability. Method was developed in Malaysia	Taylor et al. (1993)	Means	Scores
Sustainability of energy crops (SEC)	Assessment of ecological and economic sustainability of growing and conversion of crops to energy in Europe	Biewinga and van der Bijl (1996)	Effects	Values and scores
Ecopoints (EP)	Evaluates the farmer production practices and landscape maintenance in lower Austria	Mayrhofer et al. (1996)	Means	Scores
LCA for agriculture (LCAA)	Identification of methodological difficulties for the application of LCA to agricultural production in Europe	Audsley et al. (1997)	Effects	Values
Agro-ecological indicators (AEI)	Evaluation of the effects of farmer production practices on various components of the agro-ecosystem in France	Girardin et al. (2000)	Effects	Scores

(continued)

Table 3.1 (continued)

Method	Description	Authors	Indicators	Output format
Agro-ecological system attributes (*AESA*)	ECOPAT, mass balance software is used in Philippine in smallholder rice (*Oriza* sp.) farms	Dalsgaard and Oficial (1997)	Effects	Values
Operationalizing sustainability (*OS*)	Design of environmentally friendly flower bulb production methods in Netherlands	Rossing et al. (1997)	Effects	Values
Multi-objective parameters (*MOP*)	Design of integrated and ecological arable farming systems. It takes into account a set of ecological, economic and social objectives in Europe	Vereijken (1997)	Effects	Values
Environmental management for agriculture (*EMA*)	Comparison of actual farmer production practices and site-specific details with what is perceived to be best practice for that site in UK	Lewis and Bardon (1998)	Means	Scores
Solagro diagnosis (*SD*)	Evaluates the number of production systems (animal and/or crop) within the farm, diversity of crops grown, management of inputs and management of space. The method can be applied to all agricultural production systems in France	Pointereau et al. (1999)	Means and effects	Values
LCA for environmental farm management (*LCAE*)	The method is used for identification of the main pollution sources and the evaluation of possible modifications of the farms or farming methods in Switzerland	Rossier (1999)	Effects	Values

(continued)

Table 3.1 (continued)

Method	Description	Authors	Indicators	Output format
Indicators of farm sustainability (IFS)	The method is used for the evaluation of agro-ecological, socio-territorial and economic sustainability of different farm types in France	Vilain (1999)	Means and effects	Scores

3.3 Measures Adopted for Sustainable Agro-Environment

3.3.1 *Pesticides*

Many pesticides are prohibited for use by the government of India, due to their harmful effects on human and living organisms. According to the Directorate of Plant Protection of India, Quarantine and Storage, 148 of the 414 MT of weedicides consumed in 2014–2015 was glyphosate which has been banned in many states during the year 2018 by the Department of Agriculture Cooperation and Farmers' welfare, along with other hazardous pesticides. As reported in various recent studies, many pesticides in soil can also be replaced by application of strong acids.

3.3.2 *NPS*

According to Zhao et al. 2019, N export was equally from crop and livestock production, whereas the source for P export was primarily from livestock production. Furthermore, due to low flow in rivers of the plain regions in the Haihe Basin (HB) (Xia and Zhang 2008), concentration of discharged manure N and P was high in the rivers of the HB. This leads to pollution of groundwater due to the seepage of polluted water. Mitigation can occur through prohibiting the direct discharge of animal manure and replacing chemical fertilizer application via manure application in crop fields. High N and P emission are more often in the plains part of river than in the mountainous part. This is mainly because in the plain areas, anthropogenic factors contribute to N and P pollution, whereas in the mountainous areas, both anthropogenic and natural factors contribute to the N and P emissions.

3.3.3 *Efficient Use of Fertilizer*

Site-specific nutrient management (SSNM) is an important technique for efficient management of fertilizer, thereby increasing its efficiency without various possible losses including leaching, ammonia volatilization and surface runoff. Various models like NUTMON, NUBalM have been developed for nutrient balance calculation, depending on the site-specific fertilization management practices of agriculture soils. In addition to this, accounting to complex nature of soil the corresponding changes from fertilizer to plant-available nutrient is variable (Debnath et al. 2016). While, efficient irrigation management, such as NFI (Peng 1992; Peng et al. 2011), saturated soil culture (Borell et al. 1997), alternate wet–dry irrigation (Tabbal et al. 2002; Belder et al. 2004; Peng and Bouman 2007), the system of rice intensification (Stoop et al. 2002; Uphoff et al. 2011), ground cover systems (Tao et al. 2006) and aerobic cultivation (Bouman and Tuong 2001) have been found to be useful for increasing water use efficiency, increased nutrient use efficiency is only possible with efficient water management. However, nutrient cycling though reuse of the drained water from paddy fields helps in the reuse of nutrient through increased soil microbial biomass carbon (MBC) and activities of dehydrogenase, urease and phosphatase (Lal et al. 2015). Further, excess soil nutrient levels from the previous crop could be beneficially utilized in the subsequent crop cultivation. Volatilization losses can be reduced to a greater extent by increasing the water depth as well as the duration of flooding in the first wet–dry cycle after fertilization or deep placement of N into a shallow rhizosphere in case of paddy (Xu et al. 2012).

3.3.4 *Environmental Objectives*

Objectives are the environmental issues required to address the impact of agriculture on environment (Table 3.2). The number of objectives should be large to decrease the probability of error and creation of any new issues. However, objectives should be as few as possible for feasibility of the solution.

3.4 Conclusion

Excessive use of fertilizers, pesticides by the farmers has been increasing day by day globally which may in long term affect the health of local environment. Hence, many of which are dangerous for human and animal health, are prohibited for use. Emission of NH_3 is dependent more on the type of fertilizer than land application strategy while in livestock sector land application strategies are more prominent for emission reduction than feeding and housing strategies. However, there should be trade-off between NH_3 emission due to manure application and N emission due to reduced

Table 3.2 Consideration of environmental objectives in various evaluation methods

Objectives	Methods											
	FSI	SEC	EP	LCAA	AEI	AESA	OS	MOP	EMA	SD	LCAE	IFS
Farm input (human and natural capital) related												
Non-renewable energy		•		•	•				•	•	•	•
Soil erosion		•	•							•		
Land		•		•		•					•	
Water		•							•	•		•
Nitrogen fertilizer[a]	•		•							•		
Phosphorus[a]	•		•							•		
Pesticide[a]	•		•				•			•		
Emission (farm by product) related												
GHG		•		•							•	
Pesticides		•									•	
Waste production		•		•							•	
Waste utilization		•		•							•	
State of system (local environment) related												
Landscape quality		•	•		•			•		•		•
Air quality					•			•				
Soil quality					•				•	•		•
Water quality					•			•				
Animal welfare									•			

[a]Optimal use

chemical fertilization in crop field. Combination of NFI and CRN treatments in some places decreased AV losses from a rice paddy. In addition to this, cyclic irrigation in paddy or other agricultural fields during the monsoon can help in reduced application of fertilizer, in case of highly soluble N. Moreover, the use of untreated wastewater can reduce fertilizer costs as well as contamination by adjusting NP doses in crop fields. This was mainly because of lower ammonium contents in surface water and topsoil solutions. Furthermore, NPS pollutant is primarily concentrated in the downstream of river, thus, it is utmost importance to improve water quality through strengthening the control in agricultural source of pollution in the middle and downstream part of any river in a watershed. Nevertheless, nutrient balance models developed based on nutrient balance equation has been found helpful in optimum nutrient management practices. In addition to this, building of high bund around the agricultural field can prevent nutrient outflows from the field. Furthermore, reduction of nutrient losses can be minimized if water-saving irrigation is implemented and overdose of fertilizer is prohibited.

The varied trait of the nutrient fluxes owing to the complex and heterogeneous nature of soil-plant system makes the quantification of environmental indicators difficult and complex. Also, no direct relationship could be established among the EIs, however, two or more indicators should be accompanied for robust testing of sustainability of the agro-environment. The economic feasibility for implementing the management practices should be studied through the cost-benefit analysis for sustainable agro-environment. Nevertheless, consideration of both anthropogenic and natural factors is required when disseminating and applying the new knowledge, while emphasizing on growing organic foods devoid of chemical input should be of priority to keep the soil health intact along with sound local environment.

References

Alivelu K, Rao AS, Sanjay S, Singh KN, Raju NS, Madhuri P (2006) Prediction of optimal nitrogen application rate of rice based on soil test values. Eur J Agron 25(1):71–73

Audsley E, Alber S, Clif R et al (1997) Harmonisation of environmental life cycle assessment for agriculture. In: Final report concerted action AIR3-CT94-2028. Silsoe Research Institute, Silsoe, UK

Belder P, Bouman BAM, Cabangon R et al (2004) Effect of water-saving irrigation on rice yield and water use in typical lowland conditions in Asia. Agric Water Manage 65:193–210

Biewinga EE, van der Bijl G (1996) Sustainability of energy crops. In: A methodology developed and applied, report no. 234. Centre for Agriculture and Environment (CLM), Utrecht, The Netherlands

Borell A, Garside A, Shu FK (1997) Improving efficiency of water for irrigated rice in a semi-arid tropical environment. Field Crops Res 52:231–248

Bouman BAM, Tuong TP (2001) Field water management to save water and increase its productivity in irrigated lowland rice. Agric Water Manage 49:11–30

Brouwer A, Longnecker MP, Birnbaum LS et al (1999) Characterization of potential endocrine-related health effects at low-dose levels of exposure to PCBs. Environ Health Perspect 107(4):639–649

Crisp TM, Clegg ED, Cooper RL et al (1998) Environmental endocrine disruption: an effects assessment and analysis. Environ Health Perspect 106(1):11–56

Dalsgaard JPT, Oficial RT (1997) A quantitative approach for assessing the productive performance and ecological contributions of smallholder farms. Agric Sys 55:503–533
Debnath M, Mahanta C, Sarma AK (2016) Fertilizer inputs and corresponding changes in plant nutrient availability in a rice cultivated soil of Assam, India. In: Precision farming and resource management. 2nd international conference on emerging technologies in agricultural and food engineering. IIT Kharagpur, Excel India Publishers, New Delhi, p 351, Dec 2016
Duncan R (2014) Regulating agricultural land use to manage water quality: the challenges for science and policy in enforcing limits on non-point source pollution in New Zealand. Land Use Policy 41:378–387
Environews Forum (1999) Killer environment. Environ Health Perspect 107:A62
Fabra A, Duffard R, Evangelista DDA (1997) Toxicity of 2,4-dichlorophenoxyacetic acid in pure culture. Bull Environ Contam Toxicol 59:645–652
Frankenberger WT, Tabatabai MA, Tabatabai MA (1991) Factors aff ecting L-asparaginase activity in soils. Biol Fert Soils 11:1–5
Girardin P, Bockstaller C, van der Werf HMG (2000) Assessment of potential impacts of agricultural practices on the environment: the AGRO ∗ ECO method. Environ Impact Assess Rev 20:227–239
Hua L, Li W, Zhai L, Yen H, Lei Q, Liu H, Ren T, Xia T, Zhang Y, Fan X (2019) An innovative approach to identifying agricultural pollution sources and loads by using nutrient export coefficients in watershed modeling. J Hydrol 571:322–331
Huang JB, Fan XH, Zhang SL, Ge GF, Sun YH, Feng X (2007) Investigation on the economically-ecologically appropriate amount of nitrogen fertilizer applied in rice production in Fe-leaching-Stagnic Anthrosols of the Taihu Lake region. Acta Ecol Sin 27(2):588–595
Hurley PM (1998) Mode of carcinogenic action of pesticides inducing thyroid follicular cell tumors in rodents. Environ Health Perspect 106(8):437–445
Kelley WD, South DB (1978) In vitro effects of selected herbicides on growth and mycorrhizal fungi. Weed Sci Soc America Meeting. Auburn University, Auburn, Alabama, pp 38
Kole RK, Bagchi MM (1995) Pesticide residues in the aquatic environment and their possible ecological hazards. J Inland Fish Soc India 27(2):79–89
Kole RK, Banerjee H, Bhattacharyya A (2001) Monitoring of market fish samples for endosulfan and hexachlorocyclohexane residues in and around Calcutta. Bull Environ Contam Toxicol 67(4):554–559
Lal K, Minhas PS, Yadav RK (2015) Long-term impact of wastewater irrigation and nutrient rates II. Nutrient balance, nitrate leaching and soil properties under peri-urban cropping systems. Agric Water Manage 156:110–117
Lewis KA, Bardon KS (1998) A computer-based informal environmental management system for agriculture. Environ Model Softw 13:123–137
Liang XQ, Chen YX, Hua LI, Tian GM, Zhang ZJ, Ni WZ, He MM (2007) Nitrogen interception in floodwater of rice field in Taihu region of China. J of Environ Sci 19(12):1474–1481
Lu C, Tian H (2016) Global nitrogen and phosphorus fertilizer use for agriculture production in the past half century: shifted hot spots and nutrient imbalance. Earth Syst Sci Data 9(1):181–192
Martens DA, Bremner JM (1993) Infl uence of herbicides on transformations of urea nitrogen in soil. J Environ Sci Health B 28:377–395
Mathur SC, Tannan SK (1999) Future of Indian pesticides industry in next millennium. Pesticide inf 24(4):9–23
Mayrhofer P, Steiner C, Gärber E et al (1996) Regionalprogramm Ökopunkte Niederösterreich. Informationsheft, NÖ Landschaftsfonds, Wien, Austria
Oenema O, Velthof GL (1993) Ammonia volatilization from compound nitrogen-sulfur fertilizers. In: Optimization of plant nutrition, pp 341–349
Paramasivam S, Alva AK (1997) Leaching of nitrogen forms from controlled-release nitrogen fertilizers. Commun Soil Sci Plant Anal, pp 17–18
Pell M, Stenberg B, Torstensson L (1998) Potential denitrification and nitrification tests for evaluation of pesticide effects in soil. Ambio 24–28

Peng SZ (1992) The new characters of rice water requirement in water-saving irrigation. China Rural Water Hydropower 10(3):401–412

Peng S, Bouman BAM (2007) Prospects for genetic improvement to increase lowland rice yields with less water and nitrogen. In: Spiertz JHJ, Struik PC, van Laar HH (eds) Scale and complexity in plant systems research: gene–plant–crop relations. Springer, Netherlands, pp 251–266

Peng SZ, Yang SH, Xu JZ et al (2011) Nitrogen and phosphorus leaching losses from paddy fields with different water and nitrogen managements. Paddy Water Environ 9(3):333–342

Pier AB, Consonni D, Bachetti S et al (2001) Health effects of dioxin exposure: a 20-year mortality study. Am J Epidemiol 153(11):1031–1044

Pointereau P, Bochu JL, Doublet S et al (1999) Le diagnostic agri-environnemental pour une agriculturerespectueuse de l'environnement. Trois méthodes passées àla loupe. Travaux et Innovations. Société Agricole et Ruraled'Edition et de Communication, Paris, France

Rong Q, Cai Y, Chen B, Yue W, Yin XA, Tan Q (2017) An enhanced export coefficient based optimization model for supporting agricultural nonpoint source pollution mitigation under uncertainty. Sci Total Environ 580:1351–1362

Rossier D (1999) L'écobilan, outil de gestion écologique de l'exploitation agricole? Revue suisse Agric 31(4):179–185

Rossing WAH, Jansma JE, de Ruijter FJ et al (1997) Operationalising sustainability: exploring options for environmentally friendly flower bulb production systems. Eur J Plant Pathol 103:217–234

Santos A, Flores M (1995) Effects of glyphosate on nitrogen fixation of free-living heterotrophic bacteria. Lett Appl Microbiol 20(6):349–352

Savonen C (1997) Soil microorganisms object of new OSU service. In: Good fruit grower. http://www.goodfruit.com/archive/1995/6other.html. Accessed 23 Apr 2019

Singh JB, Singh S (1989) Eff ect of 2,4-dichlorophenoxyacetic acid and maleic hydrazide on growth of bluegreen algae (cyanobacteria) Anabaena doliolum and Anacystis nidulans. Sci Cult 55:459–460

Stoop W, Uphoff N, Kassam A (2002) A review of agricultural research issues raised by the system of rice intensification (SRI) from Madagascar: opportunities for improving farming systems for resource-poor farmers. Agric Sys 71:249–274

Sutton MA, Reis S, Riddick SN et al (2013) Towards a climate-dependent paradigm of ammonia emission and deposition. Philos Trans R Soc B-Biol Sci 368

Tabbal DF, Bouman BAM, Bhuiyan SI et al (2002) Onfarm strategies for reducing water input in irrigated rice; case studies in the Philippines. Agric Water Manage 56:93–112

Tao HB, Brueck H, Dittert K et al (2006) Growth and yield formation of rice (*Oryza sativa* L.) in the water-saving ground cover rice production system. Field Crops Res 95:1–12

Taylor DC, Mohamed ZA, Shamsudin MN, Mohayidin MG et al (1993) Creating a farmer sustainability index: a Malaysian case study. Am J Alt Agric 8:175–184

Ti C, Xia L, Chang SX, Yan X (2019) Potential for mitigating global agricultural ammonia emission: a meta-analysis. Environ Pollution 245:141–148

Tözüm-Çalgan SRD, Sivaci-Güner S (1993) Eff ects of 2,4-D and methylparathion on growth and nitrogen fixation in cyanobacterium, Gloeocapsa. Int J Environ Stud 23:307–311

Uphoff N, Kassam A, Harwood R (2011) SRI as a methodology for raising crop and water productivity: productive adaptations in rice agronomy and irrigation water management. Paddy Water Environ 9:3–11

US Geological Survey (1999) The quality of our nation's waters—nutrients and pesticides. Circular 1225. USGS, Reston VA. http://water.usgs.gov/pubs/circ/circ1225. Accessed 10 May 2019

Vereijken P (1997) A methodical way of prototyping integrated and ecological arable farming systems (I/EAFS) in interaction with pilot farms. Eur J Agron 7:235–250

Vilain L (1999) De l'exploitation agricole à l'agriculture durable. Aide méthodologique à la mise en place de systèmes agricoles durables, Educagri éditions, Dijon, France

Volk M, Bosch D, Nangia V et al (2016) SWAT: Agricultural water and nonpoint source pollution management at a watershed scale. AgricWater Manage 175:1–3

Wang DJ, Liu Q, Lin JH, Sun RJ (2004) Optimum nitrogen use and reduced nitrogen loss for production of rice and wheat in the Yangtse Delta region. Environ Geochem Health 26(2):221–227
WHO (1990) Public health impact of pesticides used in agriculture. World Health Organization, Geneva, pp 88
Xia J, Zhang Y (2008) Water security in north China and countermeasure to climate change and human activity. Phys Chem Earth Parts A/B/C 33(5):359–363
Xia Y, Yan X (2011) Economic optimal nitrogen application rates for rice cropping in the Taihu Lake region of China: taking account of negative externalities. Biogeosci Discuss 8(4)
Xu J, Peng S, Yang S, Wang W (2012) Ammonia volatilization losses from a rice paddy with different irrigation and nitrogen managements. Agric Water Manage 104:184–192
Zhao Z, Qin W, Bai Z, Ma L (2019) Agricultural nitrogen and phosphorus emissions to water and their mitigation options in the Haihe Basin, China. Agric Water Manage 212:262–272
Zhou JB, Xi JG, Chen ZJ et al (2006) Leaching and transformation of nitrogen fertilizers in soil after application of N with irrigation: a soil column method. Pedosphere 16(2):245–252

Chapter 4
An Experimental Study on Benzo[a] Pyrene Concentration in Particulate Matter at Industrial Area of Bangalore

Prashant Basavaraj Bhagawati, Satish G. Muttagi, Poorna B. Bhagawati, Sandip S. Sathe and Abhijit M. Zende

Abstract Assessment of benzo[a]pyrene concentrations, with particularly preferred metals in polluted air, is significant issue for estimating counter health effects. Polycyclic aromatic hydrocarbons (PAHs) recognized one of the major crucial toxic air pollutants in industrials cum urban regions. With partial combustion of organic materials such as motor oils, gasoline, tobacco, cooking oils, butter, and other food leads to the formation of PAHs. With considering the importance of tracing the concentration of PAHs the experimental analysis work is carried out, to recognize the adverse effect on atmosphere. The air samples were collected (as 24 h sample once in a month) from eight specific spots within Peenya industrial monitored network Bangalore, India. The qualitative and quantitative analysis tests were carried out with modern scientific instrumental tool GC-MS. The concentration of B[a]P in eight measuring locations ranged from below detectable limit (BDL) to 0.0490 ng/m^3. From the results obtained, there was noticeable variation in B[a]P concentration, with respect to measuring locations.

Keywords Polycyclic aromatic hydrocarbons · benzo[a]pyrene · Particulate matter

---

P. B. Bhagawati (✉) · S. S. Sathe
Department of Civil Engineering, Annasaheb Dange College of Engineering and Technology, Ashta, Maharashtra, India
e-mail: prashantbhagawati@gmail.com

S. G. Muttagi (✉)
Scientific and Industrial Research Centre, Bangalore, Karnataka, India
e-mail: researchin.env@gmail.com

P. B. Bhagawati
Department of Electronics and Instrumentation Engineering, Basaveshwar Engineering College, Bagalkot, Karnataka, India

A. M. Zende
Department of Civil Engineering, Daulatrao Aher college of Engineering, Karad, Maharashtra, India

R. M. Singh et al. (eds.), *Environmental Processes and Management*,
Water Science and Technology Library 91,
https://doi.org/10.1007/978-3-030-38152-3_4

4.1 Introduction

Particulate matter (PM) plays a central role in aerosol physics and chemistry. It is main uncertain factor in climate variational calculations. In day today's life, climatic variation adversely affects on human health by making carcinogenic penetration on human body results in long term vision-related problems. The whole world facing a giant environmental problem where industrial air pollution is the main component of the air pollution. In countrywide, manufacturing and processing industries release vary dangerous toxic gases to the atmosphere. In every country, there exist the problem of toxic gases that are released into the air everyday, during the manufacturing of products. The process industries like petroleum, chemical, sugar, and cement manufacturing plants have significant adverse effects on environment. Thus, we can say that industry is one of the major culprits in polluting air. Quality of the air is an important concern for both government and citizens. In last few decades, particularly in urban cities, atmospheric quality has been deteriorated to a larger extent. Worldwide, in order to assess the quality of air continuously, many metropolitan urban cities have been selected. Modern analytical instruments are used to measure and record the concentration of air pollutants at various highly polluted zones. However, with all existed instrumental approaches there are a lack of information about particulate matter (PM10) composition. The process of understanding the mechanism linking particulate air pollution with health hazards still remains a major challenge. Presently, with an increase in population, heavy traffic, and growth of industrialization made all most all urban cities to suffer from air pollution problems (Jain and Khare 2008).

The increased levels of PM10 and toxic metals such as As, Cd, Cu, Co, Cr, Fe, Mn, Ni, Pb, and Zn nearby the sponge iron industries were reported (Barman et al. 2010). In an urban environment, polycyclic aromatic hydrocarbons (PAHs) are recognized as crucial deadly air pollutants. With partial combustion or pyrolysis of organic materials, these PAHs are formed. During the survey, it was reported that vehicular pollution contributes to 70% of the total pollution in Delhi, 52% in Mumbai, and 30% in Kolkata. India has 23 major cities of over 1 million people and ambient air pollution levels exceed in many of them (Gupta et al. 2002). Even electric power generating is also a main cause for polluting the air. In the world, coal-based thermal power plants are the main contributor for the production of electric power. The gas and oil-based plants also contribute for the production of electricity but in smaller extent. In thermal power plants, after the combustion of coal fly ash is the remained portion that leaves through the smokestack and becomes part of the effluent plume (Zhang et al. 2003). Pulverized coal boilers are the dominant combustion technology used in power plants. Reddy and Venkataraman (2002) prepared an Indian aerosol inventory. They found that the emission factors of PM for the boiler capacities of 30, 60, 110, 210, and 500 MW (representative of Indian utilities coal boilers) were 0.46–0.71 g/kg. They reported that high PM2.5 emission fluxes (1000–2000 kg/km^2) were located over north India, West Bengal, Maharashtra, Gujarat, and Tamil Nadu. In India, due to anthropogenic sources like domestic cooking, industrialization, electric power generation, uncontrolled increase of vehicles, garbage burning, and rapid urbanization leads to progressive deterioration of ambient air quality (Dash and Dash 2017). Shi et al. (2015) stated that the degradation of air quality became an important environmental risk factor for lung cancer and cardiopulmonary disturbances. Dash and Dash (2018) examined the air pollution tolerance index (APTI) of 20 plant species near an industrial complex and out of those plant air pollution tolerance index of Dalbergia sisoo (25.75) and Artocarpus (11.40) were found highest and lowest, respectively.

Across the globe, according to the World Health Organization (WHO), urban air pollution is responsible for approximately 800,000 deaths annually. Plants are also affected by various air pollutants. Excessive SO_2 makes the plant cell inactive and finally they are killed (Sahoo et al. 2017). When water combines with SO_2 leads to a chemical action results in the production of sulfuric acid, which is extremely corrosive. Various metals such as copper, iron, and aluminum are corroded upon exposing them with contaminated air.

In the present scenario, the existence of PM in the atmosphere and its destructive effects on (living being) human health and the environment are of great concern across the globe. From the lower troposphere to the upper stratosphere even at relatively low concentration PM, it has a significant impact on physical and chemical process of atmosphere. In 1962, Lodge scientifically defined PM as it was larger than single molecule, capable to disperse in the atmosphere and can stay for longer period of time. A mixture of solid particles and liquid vapors found in the air is known as ambient particulate matter (Oliver and Fairbridge 1987). Examples for solid particles which are responsible for the formation of ambient particulate matter (PM) are dirt, soot, and smoke. The sources for the rapid rise in the concentration of PM in ambient air are fast construction development, remarkable increase in the vehicle number, and many other ambient factors.

4.2 Concentrations of B[a]P in the Indian Environment

Industrial process carried out at glass production, construction, blast furnaces, sinter plants, cement production, lime production, and quarrying are main contributors for the formation of particulate matter (PM) in the air. In Asian regions particularly in India, B[a]P concentrations were reported in similar to China and higher than Taiwan, Japan, Korea, and Hong Kong. Since from last two decades, literature survey says that most of the research work was mainly focused on PAHs. According to a literature survey, in India at various different segmental levels, such as water, sediments, air, and soil, B[a]P concentration was reported. In seven various agricultural sites of Delhi during January, 2006, the concentrations of PAHs were reported in soil (Agarwal et al. 2009). The concentration of PAHs was in between 18 and 71 μg/kg with a mean of 49 μg/kg. With respect to their discussion concentration of PAHs was 2–8 times greater in urban areas than in rural agricultural sites.

4.2.1 Ambient Air

Chemical characterization of PM10 concentration was carried out and reported that there was a seasonal variation in PM10 concentration at residential and industrial sites of Kolkata between November 2003 and November 2004 (Karar and Gupta 2006). In their discussion, it found that at residential site B[a]P concentration was 12.9 ng/m^3 and 14.3 ng/m^3 at industrial sites, respectively. The concentration of B[a]P was higher in winter and followed by summer and monsoon. (Herlekar et al. 2012) reported the concentrations of PAHs associated with PM10 in different sites of Mumbai during 2007–2008. In this study, they found that the contribution of B[a]P in the total PAHs concentration was in the range between 8.1 and 10.2%. The carcinogenic PAHs contributed 23.3–29.2% to the total PAHs concentration. This suggests that B[a]P is major carcinogen found in this study at the specified sites in Mumbai.

4.2.2 Soil

The concentration of B[a]P was examined in various sites of Agra representing residential, industrial, agricultural, and roadside areas (Masih and Taneja 2006). They revealed that B[a]P was measured only at roadside site and below the detection limit at the remaining sites. At roadside site, the concentration of B[a]P was detected to be 0.39 μg/g and seasonal variation studies show a larger amount of B[a]P in winter than in the other seasons.

Ray et al. (2012) reported the concentration of B[a]P in the rural, background, and urban soil of Delhi. The mean concentration of B[a]P measured in the urban, rural, and background areas was 99.6, 17.0, and 2.06 μg/kg, respectively. This observation suggests that the B[a]P levels were almost 50 times higher in urban areas than the background sites; which is a frightening issue in the public health due to its carcinogenic nature.

4.2.3 Water and Sediments

Identification of B[a]P concentration in the samples of wastewater at three different industrial plants; steel industries located in Jharkhand state, petroleum refinery in West Bengal, and petrochemical located in Gujarat (Krishnamurthi et al. 2008). In the industrial effluent and wastewater samples, there work forecasted the presence of B[a]P and other pollutants. In India, the research work on measurement of B[a]P concentration at different water media was not thoroughly carried out. Due to its carcinogenic nature and bioaccumulation into other aquatic animals, such as fish, it is an important environmental issue for researchers (Unnikrishnan 2006). According to Ma et al. (2011) the concentration of PAHs in the environment can be controlled by introducing environmental friendly policies, and also by controlling the PAHs sources in the environment.

4.3 Health Risks Associated with PM

The vital health issues associated with PM such as cardiovascular disease, changes in lung function, aggravation of respiratory, premature death, and increased respiratory symptoms. Because of carcinogenic nature, the study on PAHs has attracted many environmental researchers. According to Melikian et al. (1999), Abbas et al. (2018) total structure of PAHs more than 80% consisted of 5-ring and 6-ring PAHs such as benzo[a]pyrene, benzo[b]fluoranthene (BbF), benzo[k]fluoranthene (BkF), and benzo[g,h,i]perylene (BgP), particularly benzo[a]pyrene, a PAH composition is extremely cancer-causing agent. Park et al. (2002) revealed about the health effects caused by the direct inhalation of pollutant air, drinking of polluted water, and contaminated processed food. Jamhari et al. (2014) carried out the work using high volume sampler (HVS) with collecting PM10 samples on glass fiber filter paper for the duration of 24 h. Between September 2010 and April 2011 at two sites within the Klang Valley, Kuala Lumpur (urban, KL) and Petaling Jaya (industrial, PJ), and one site outside the Klang Valley, Bangi (semi-urban, BG). Unique sampling method was developed by combining a high volume air sampler (HV) with a PM2.5 impactor for collecting large quantities of PM2.5 (Sugita et al. 2019), and the study focused on measurement of micropollutants present in the air. Nagarajappa et al. (2012) done the analysis of benzo[a]pyrene in suspended particles by collecting air samples from highly polluted location (as 24 h sample once in a month) within the Peenya industrial monitoring network. From their experimental results, it observed that B[a]P levels were high during winter in almost all monitoring locations. Rajput and Lakhani (2010) collected residential cum industrial site samples of Agra, India over the period of December 2005 to December 2006, for measuring 16 PAHs. The average total PAHs concentration was 115 ± 17 ng/m^3, which found to be lower than those reported from other cities in India like Delhi, Chennai, Kanpur, Kolkata, and Mumbai. Yang et al. (2012) used filtered samples and selected three locations in Tanggu District at Tianjin Binhai New Area, China for the monitoring of PAHs concentrations in PM 2.5 and PM10 during the period of November 2008.

Karnataka is a hub of many industries such as iron and steel, pulp and paper, cement, chemicals, fertilizers, silk, textiles, mining, machine tools, and pharmaceuticals. The emissions and industrial effluent characteristics are mainly depend upon the nature of raw material used and manufacturing process. Immuno modulatory effects of chemicals adsorbed to particles with aerodynamic diameter below 0.49 m (PM 0.5) collected in winter 2001 at three sampling points (industrial area (LPIn), traffic-influenced urban area (LPCi), and control area (LPCo) of La Plata, Argentina were investigated (Wichmann et al. 2009).

Sparse research in open literature is available on determination concentration level of air pollutant like benzo[a]pyrene. The present research work objectives are as follows: i) to carry out comprehensive literature survey of benzo[a]pyrene, ii) to determine the concentration level of air pollutants like benzo[a]pyrene, at Peenya Industrial area in Bangalore City.

4.4 Description of Sampling Site

The average temperature of the city 23 °C increases from February to April indicating summer season. The wind speed was observed to be high in February and the peak value observed 15 km/h. Relative humidity was observed to be high in April. Peenya industrial area spread over an area of about 40 km^2 comprising about 4000 small-scale industries and 50 medium- and large-scale industries. The Peenya industrial area is primarily divided into four stages and surrounded by the number of industries. The first stage of Peenya industrial area is surrounded by electroplating, chemical, turbine industry, metal extracting industries, some commercial shopping centers, and restaurants. First stage of Peenya industrial hub is the significant producer of benzo[a]pyrene. The second stage is crowded by lead, brewery, textile, chemical industries, electroplating, and pharmaceuticals industry. It is also surrounded by many numbers of residential houses and commercial shops. In the third stage of the Peenya industrial area, oil and grease industry, smelting and processing of metal industries and lead industries are located. At the fourth stage of Peenya industrial area, the chemical, turbine, metal, textile, and steel industries are situated and also surrounded by residential houses, with a number of commercial shops. Around four lakh workers are working in Peenya industrial area.

4.5 Methodology

According to the Central Pollution Control Board (CPCB), the methods prescribed for the benzo[a]pyerene pollutant and the particulate pollutants are very sensitive ones, yet the percentage of errors is very less. Determination of Benzo[a]pyrene concentration in ambient air was analyzed by solvent extraction and GC method. The respirable dust sample was collected in February 2011–April 2011 for 24 h. Polyurethane foam (PUF) was used as a sorbent for gas-phase enrichment. Other Adsorbent materials (porous polymer), such as TENAX (XAD-2, XAD-4 and Carbopack C) were employed occasionally during the analysis. All filters were maintained in a condition of 50% RH and 25 °C for over 48 h and weighted before sampling. High flow rate and high volume samplers were operated at 34–1250 L/min. The respirable dust sampler introduced by Envirotech technology was used for PM10 sample collection.

4.6 Results and Discussion

One of the reasons that PAHs are lower during the summer is because semi-volatile PAHs are present in the gaseous phase, due to high temperature. While in the winter season as ambient temperature is much lower, hence PAHs are higher. B[a]P was found to be a relatively unstable compound and it is a good indicator of total PAHs or carcinogenic PAHs. During the study period of 2011, in the months of February–April the air quality of Peenya was monitored for the parameter benzo[a]pyrene. The results are summarized in Table 4.1 and Fig. 4.1 represents graphical variation of the selected parameters for study period February 2011–April 2011. The highest concentration of benzo[a]pyrene found in Kongovi electroplating industry. This is evident because of electroplating; textile and metal industries situated in that area and of high traffic flow.

Table 4.1 Average level of B[a]P monitored over the period of three months (February–April 2011)

Location	Avg B[a]P (ng/m^3)		
	February	March	April
Surya Hard Chrome	0.018518519	0.009259259	0.008454106
Microtex Energy	0.029671717	0.010732323	0.009469697
Kongovi Industry	0.049013688	0.016414141	0.01010101
Ace Designers	0.025719148	0.009259259	0
Swan Silk	0.030604963	0.011904762	0.011363636
Arun Industry	0.031483311	0.013888889	0.011994949
Naptha Resins	0.032932848	0.009469697	0.007575758
Triveni Turbines	0.039168863	0.0120772295	0.011243386

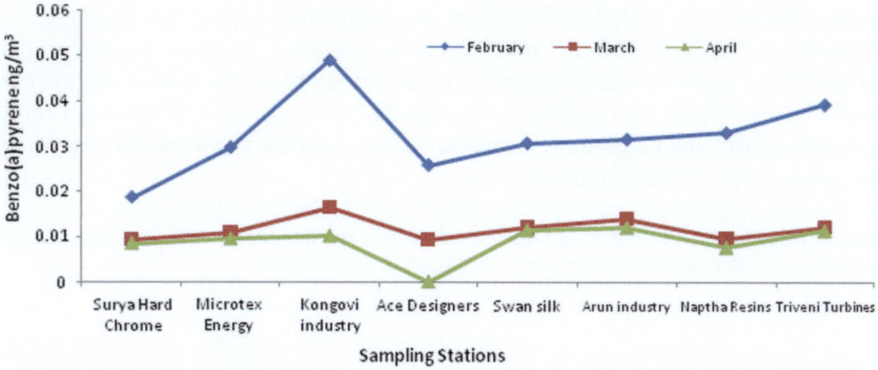

Fig. 4.1 Variant levels of B[a]P in ambient air monitored at sampling stations over the period of three months (February–April 2011)

The concentration of B[a]P reduced from February to April in all locations. B[a]P concentration shown a significant seasonal variation. B[a]P concentration was greater in the month of February due to metrological factors, with conditions of low temperature and low irradiation. From March to April, B[a]P concentration decreased due to metrological factors and increased wind speed. This leads to increase in dispersion and dilution. Variation of B[a]P in ambient air at sampling station during study period February 2011–April 2011 is shown in Fig. 4.1. The variation in concentration of benzo[a]pyrene was taken at eight locations at different date over a period of three months at Peenya industrial area. During study period, it was observed that benzo[a]pyrene ranged from BDL to 0.0490 ng/m^3 in the month of February. Kongovi electroplating location has higher concentration compare to the other locations.

4.7 Conclusions

The industrial and transportation zones are the main culprits for the high pollution load in the ambient air of the study area. The existence of high particulate pollutants level has a considerable harmful effect on the ambient air of the study area. This study focuses on the acerbity of the level of B[a]P concentration nearby Peenya industrial hub, Bangalore. The average temperature of the city 23 °C increased from February to April indicating summer season. Relative humidity was observed to be high in April. The wind speed was observed to be high in February and the peak value observed was 15 km/h. In this study, the concentration of B[a]P was analyzed to investigate the seasonal variation and recorded results were correlated with National Ambient Air Quality Standards (NAAQ) of Central Pollution Control board (CPCB). It was found that benzo[a]pyrene level varies in between BDL and 0.0490 $\mu g/m^3$ which was less than allowable limit of 1 $\mu g/m^3$. We observed that benzo(a)pyrene concentration was within the tolerable limit. This work carried for a period of three months and earlier data was not available. Therefore, it is difficult to predict the air quality of Peenya industrial area. So, it requires better and accurate studies to be carried out by collecting more number of samples from various locations around the Peenya industrial area. From the results, it is obvious that modernization of industrial zone, transport sector, and community consciousnesses can play a key role in improving the ambient air of Peenya industrial area.

Acknowledgements The authors acknowledge Scientific and Industrial Research Centre, Bangalore, India, to carry out the analysis and evaluation of obtained results.

References

Abbas I, Badran G, Verdin A, Ledoux F, Roumié M, Courcot D, Garçon G (2018) Polycyclic aromatic hydrocarbon derivatives in airborne particulate matter: sources, analysis and toxicity. Environ Chem Lett 16(2):439–475

Agarwal T, Khillare PS, Shridhar V, Ray S (2009) Pattern, sources and toxic potential of PAHs in the agricultural soils of Delhi, India. J Hazard Mater 163:1033–1039

Barman SC, Kumar N, Singh R, Kisku GC, Khan AH, Kidwai MM, Murthy RC, Negi MPS, Pandey P, Verma AK, Jain G, Bhargava SK (2010) Assessment of urban air pollution and it's probable health impact. J Environ Biol 31(6):913–920

Dash AK, Dash SK (2017) Atmospheric pollution load assessment through air quality index: a case study. Ind J Env Prot 37(9):736–741

Dash SK, Dash AK (2018) Air pollution tolerance index to assess the pollution tolerance level of plant species in industrial areas. Asian J Chem 30(1):219–222

Gupta HK, Gupta VB, Rao CVC, Gajghate DG, Hasan MZ (2002) Urban air quality and its management strategy for a metropolitan city of India. Bull Environ Contam Toxicol 68:347–354

Herlekar M, Joseph AE, Kumar R, Gupta I (2012) Chemical speciation and source assignment of particulate (PM10) phase molecular markers in Mumbai. Aerosol Air Qual Res 12:1247–60

Jain S, Khare M (2008) Urban air quality in mega cities: a case study of Delhi city using vulnerability analysis. Environ Monit Assess 136(1–3):257–265

Jamhari AA, Sahani M, Latif MT, Chan KM, Tan HS, Khan MF, MohdTahir N (2014) Concentration and source identification of polycyclic aromatic hydrocarbons (PAHs) in PM 10 of urban, industrial and semi-urban areas in Malaysia. Atmos Environ 86:16–27

Karar K, Gupta AK (2006) Seasonal variations and chemical characterization of Ambient PM10 at residential and industrial sites of an urban region of Kolkata (Calcutta), India. Atmos Res 81:36–53

Krishnamurthi K, Devi SS, Hengstler JG, Hermes M, Kumar K, Dutta D, Vannan SM, Subin TS, Yadav RR, Chakrabarti T (2008) Genotoxicity of sludges, wastewater and effluents from three different industries. Arch Toxicol 82(12):965–971

Ma WL, Sun DZ, Shen WG, Yang M, Qi H, Liu LY, Shen JM, Li YF (2011) Atmospheric concentrations, sources and gas-particle partitioning of PAHs in Beijing after the 29th olympic games. Environ Pollut 159:1794–1801

Masih A, Taneja A (2006) Polycyclic aromatic hydrocarbons (PAHs) concentrations and related carcinogenic potencies in soil at a semi-arid region of India. Chemosphere 65:449–456

Melikian AA, Sun P, Prokopczyk B, El-Bayoumy K, Hoffmann D, Wang X, Waggoner S (1999) Identification of benzo[a]pyrene metabolites in cervical mucus and DNA adducts in cervical tissues in humans by gas chromatography-mass spectrometry. Cancer Lett 146(2):127–134

Nagarajappa DP, Muttagi SG, Lokeshappa B, Dikshit AK (2012) Study on evaluation of Benzo(a)Pyrene soluble fraction in respirable suspended particulate matter in Peenya industrial area by GCMS. Am J Environ Eng 2(6):182–187

Oliver JE, Fairbridge RW (1987) Climatic variation, historical record. In: Oliver JE (ed) The encyclopedia of climatology. Van Nostr and Reinhold, New York, pp 293–305

Park SS, Kim YJ, Kang CH (2002) Atmospheric polycyclic aromatic hydrocarbons in Seoul, Korea. Atmos Environ 36(17):2917–2924

Rajput N, Lakhani A (2010) Measurements of polycyclic aromatic hydrocarbons in an urban atmosphere of Agra, India. Atmosfera 23(2):165–183

Ray S, Khillare PS, Kim KH, Brown RJC (2012) Distribution, sources, and association of polycyclic aromatic hydrocarbons, black carbon, and total organic carbon in size-segregated soil samples along a background-urban-rural transect. Environ Eng Sci 29(11):1008

Reddy MS, Venkataraman C (2002) Inventory of aerosol and sulphur dioxide emissions from India: I-Fossil fuel combustion. Atmos Environ 36(4):677–697

Sahoo D, Dash AK, Sahu SK (2017) International journal of engineering sciences & research technology ambient air quality monitoring and health impact study of air pollution near Joda of keonjhar, Odisha, india. Int J Eng Sci Res Technol 6(1):429–434
Shi X, Liu H, Song Y (2015) Pollutional haze as a potential cause of lung cancer. J Thoracic Disease 7(10):E412–E417
Sugita K, Kin Y, Yagishita M, Ikemori F, Kumagai K, Ohara T, Nakajima D (2019) Evaluation of the genotoxicity of PM2.5 collected by a high-volume air sampler with impactor. Genes Environ 41(1)
Unnikrishnan L (2006) Polycyclic aromatic hydrocarbons and toxic trace metals in seawater and sediments from the west coast of India and their bioaccumulation in some selected species of fishes. In: Ph.D. Thesis submitted to Cochin University of Science and Technology, Cochin, India
Wichmann G, Franck U, Herbarth O, Rehwagen M, Dietz A, Massolo L, Ronco A, Müller A (2009) Different immune modulatory effects associated with sub-micrometer particles in ambient air from rural, urban and industrial areas. Toxicology 257(3):127–136
Yang D, Qi S, Devi NL, Tian F, Huo Z, Zhu Q, Wang J (2012) Characterization of polycyclic aromatic hydrocarbons in PM2.5 and PM10 in Tanggu district, Tianjin Binhai New Area, China. Frontiers Earth Sci 6(3):324–330
Zhang J, Sun R, Wu S, Chen B, Li Z, Qin Y (2003) An experimental and numerical study on swirling combustion process in a 200 MW pulverized coal fired boiler. Proc CSEE

Part II
Environmental Remediation

Chapter 5
Environmental Impact of Landfill Leachate and Its Remediation Using Advanced Biological Methods

Isha Burman

Abstract This paper presents recent developments in advanced biological treatment technologies for landfill leachate (LFL) treatment that are attracting increasing attention and have a high treatment potential in the near future. Landfills are designed to dispose high quantities of waste at economical costs with potentially less environmental effects; however, improper landfill management may pose serious environmental threats through discharge of high-strength polluted wastewater also known as leachate. The generated leachate must be appropriately treated before being discharged into the environment. LFL, is very difficult to treat using conventional biological processes. This paper focused on achievements on landfill leachate treatment by different advanced biological treatment technology. Study also explores the fundamental principles as well as the applicability of the advanced biological treatment technologies which are finding increasing application worldwide for their compactness, low footprint, and high efficiency.

Keywords Landfill leachate · Conventional treatment · Biological treatment · Membrane bioreactors · Wastewater

5.1 Introduction

Landfilling of municipal waste is still a very important issue of the waste management system in all over the world. Some alternative methods such as recycling, composting, and incineration are nowadays very much encouraged but even incinerations create residue of approximately 10–20% that must be ultimately landfilled (Wiszniowski et al. 2006). As the developing countries are beginning to adopt modern solid waste management practices with the up-gradation of the existing dumpsites and unsanitary landfills, leachate management would require a highly specialized approach in

I. Burman (✉)
Department of Environmental Science and Engineering, Indian Institute of Technology (ISM), Dhanbad, Jharkhand 826004, India
e-mail: ishaburmanjsr@gmail.com

R. M. Singh et al. (eds.), *Environmental Processes and Management*,
Water Science and Technology Library 91,
https://doi.org/10.1007/978-3-030-38152-3_5

dealing with the constituent pollutants in it. Landfill leachate treatment is an integral part of municipal solid waste (MSW) management that in turn has a skin-blood relationship with urban infrastructure. To ensure the protection of our ecosystem, environmental health, and foster sustainable development, the waste generated by the increasing urban population requires treatment and disposal in an environmentally sound manner. MSW from the urban habitat is disposed off in dumpsites (a crude form of disposal) or in sanitary or engineered landfills. The constituents of the MSW undergo biological and chemical degradation after disposal resulting in emissions of landfill gas and discharge of leachate, which is a highly polluted form of wastewater when discharged into the environment, and would cause potential damages to environmental health and the ecosystem (Visvanathan et al. 2004).

Landfill leachate is the dark colored liquid with strong smell produced by natural humidity and water present in the residue of organic matter. It released as a result of the biological degradation of organic matter present and by water infiltration in the covering and inner layers of landfill cells, supplementing dissolved or suspended material originating from the residue mass (Yao 2017). Leachate is considered a very complex wastewater containing mixture of different heavy metals, organics, refractory organics, toxic, color, and odor. One of its characteristic features is an aqueous solution in which four groups of pollutant are present: dissolved organic matter (volatile fatty acid and more refractory organic matter such as humic substances), macro inorganic compounds (Ca^{2+}, Mg^{2+}, Na^{+}, K^{+}, NH^{4+}, Fe^{2+}, Mn^{2+}, HCO^{-3}), heavy metals (Cd^{2+}, Cr^{3+}, Cu^{2+}, Pb^{2+}, Ni^{2+}, Zn^{2+}), and xenobiotic organic compounds originating from chemical and domestic residue present at low concentrations (aromatic hydrocarbons, phenols, pesticides, etc.) (Christensen and Kjeldsen 1989), and microorganisms that indicate, predominantly total and thermotolerant coliform (Moravia et al. 2013).

As time proceeds, LFL goes through the successive aerobic, acetogenic, methanogenic, and stabilization stages of organic waste degradation, in which its properties such as the chemical oxygen demand (COD), biological oxygen demand (BOD), BOD/COD ratio, ammonium nitrogen (NH_3–N), and pH vary widely (Kjeldsen et al. 2002). Table 5.1 summarizes the classification of landfill leachate according to the composition changes with age. As shown in Table 5.1, the parameters have their typical ranges in association with the age of LFL, which is commonly classified into three stages: young (<5 years), medium (5–10 years), and stabilized (>10 years). Among the different LFL properties, the ratio BOD/COD is commonly recognized to be the most representative of LFL age because it is directly related to the biodegradability of LFL. Young LFL is characterized by high concentrations of biodegradable organic matters such as volatile fatty acids (VFAs) and as a result, it has a high BOD/COD ratio. Most of the BOD, which is the biodegradable portion of the COD, would have been decomposed in the stabilization process. Therefore, the BOD/COD ratio decreases with time because the non-biodegradable portion of COD will largely stay unchanged in this process. The BOD/COD ratios of young, medium, and old LFL are in the ranges of 0.5–1.0, 0.1–0.5, and less than 0.1, respectively. It is, however, worth mentioning that there is not a clear cut off between the medium and old LFLs and BOD/COD ratios less than 0.2 may also be treated as old LFL

Table 5.1 Landfill leachate classification according to age

	Young	Medium	Stabilized
Age (years)	<5	5–10	>10
pH	6.5	6.5–7.5	>7.5
COD (mg/L)	>10,000	4000–10,000	<4000
BOD5/COD	>0.3	0.1–0.3	<0.1
TOC/COD	<0.3	0.3–0.5	>0.5
Organic compounds	80% volatile fatty acids (VFA)	5–30% VFA + humic and fulvic acids	Humic and fulvic acids
Ammonia nitrogen (mg/L)	<400	–	>400
Heavy metals	Low–medium	Low	Low
Biodegradability	Important	Medium	Low

Source Renou et al. (2008), Yao (2017)

by some researchers (Bohdziewicz and Kwarciak 2008). Stabilized leachate is also characterized by high concentrations of NH_3–N and recalcitrant matter (e.g. humic acids), which has profound implications to the effectiveness of different biological treatment technologies.

It is, therefore, important to apply reliable and effective treatment technology for leachate treatment (Chaudhari and Murthy 2010). There are several methods for leachate treatment such as precipitation, electrocoagulation, membrane processes, adsorption, and biosorption or combination of the above (Gotvajn et al. 2009; Laitinen et al. 2006; Mohan and Gandhimathi 2009). Increasing pressure to meet more stringent discharge standards or not being allowed to discharge treated effluent has led to implementation of a variety of advanced biological treatment processes in recent years (Mittal 2011).

The principal biological process (activated sludge and biological filter) has been known quite well and is successfully applied for domestic wastewater. However, for leachate, the conventional approach for treatment requires some modifications. Depending on the wastewater and the standards which they have to meet, different process design and/or operational control parameters must be considered. At first, the laboratory-scale approach is needed.

Today, the strictness of landfill regulations, controls and managements hamper efficient conventional treatments which appears under-dimensioned or does not allow to reach the specifications required by the legislator. So that, advanced biological processes offers the best solution, and have been proved to be the more efficient, adaptable such as up-flow sludge blanket reactor (USBR), rotating biological contractor (RBC), membrane bioreactors (MBR) (aerobic/anaerobic), sequencing batch reactors (SBR), moving bed biofilm reactor (MBBR), and other emerging biological processes.

5.2 Evolution of Landfill Leachate Treatments

Conventional landfill leachate treatments can be classified into five major groups: (a) natural treatment system (b) leachate transfer: recycling and combined treatment with domestic sewage, (c) biological treatment: biodegradation by aerobic and anaerobic processes, and (d) chemical and (e) physical methods: chemical oxidation, adsorption, chemical precipitation, coagulation/flocculation, sedimentation/flotation and air stripping (Renou et al. 2008; Yu 2007) (Table 5.2).

Table 5.2 Overview of leachate treatment technologies

Technology	Operation	Application
Natural treatment systems	Assimilation/Infiltration	Irrigation
		Overland flow
		Constructed wetlands
		Aquatic systems
Leachate transfer	Recycling	
	Combined treatment with domestic sewage	
Physical treatment	Adsorption	Activated carbon
	Ion exchange	Peat filter
	Membrane filtration	MF and UF
		Reverse osmosis
	Evaporation	
	Stripping	Ammonia stripping
Chemical treatment	Precipitation/Flocculation/Sedimentation chemical oxidation/Reduction	
Biological treatment	Aerobic process	Activated sludge
		SBR
		Aerated lagoon
		RBC
		Suspended biofilm
		Trickling filter
	Anaerobic process	Anaerobic filter
		Up-flow Anaerobic Sludge Bed Reactor (UASB)
		Recirculation
	Biological nitrogen reduction	Nitrification
		Denitrification

Source Renou et al. (2008), Yao (2017)

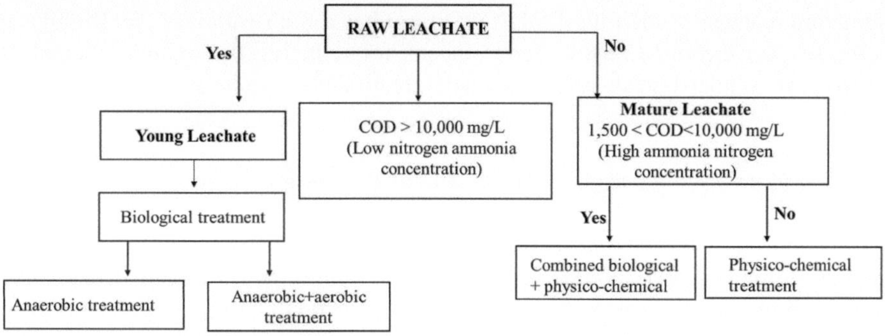

Fig. 5.1 Selection of the appropriate treatment techniques for leachate. Adapted from Renou et al. (2008), Costa et al. (2019)

Few years ago, a common solution was to treat the leachate together with municipal sewage in the municipal sewage treatment plant. It was preferred for its easy maintenance and low operating costs (Ahn et al. 2002). However, this option has been increasingly questioned due to the presence of organic inhibitory compounds in the leachate with low biodegradability and heavy metals that may reduce treatment efficiency and increase the effluent concentrations (Cecen and Aktas 2004). Recycling leachate back through the tip has been largely used in the past decade because it was one of the least expensive options available (Lema et al. 1988).

The schematic in Fig. 5.1 shows which parameters should be evaluated in the choice of a treatment for a landfill leachate. According to Fig. 5.1, for a leachate containing a high concentration of organic material (>10,000 mg L^{-1}), the most appropriate approach is biological treatment. However, for leachates with a high concentration of ammoniacal nitrogen and a low biodegradability, the most suitable approach is a physical–chemical process, possibly in combination with biological treatment.

5.2.1 *Biological Treatment Systems*

Amongst the treatment classifications, biological treatment is worldwide and the most common practice for leachate treatment because of its reliability, simplicity and high cost-effectiveness. Biological treatment is commonly used for the removal of the bulk of leachate containing high concentrations of BOD. Biological systems can be divided in anaerobic and aerobic treatment processes. Both can be realized by using different plant concepts (Kumar et al. 2013). Biodegradation is carried out by microorganisms, which can degrade organics compounds to carbon dioxide and sludge under aerobic conditions and to biogas (a mixture comprising chiefly CO_2 and CH_4) under anaerobic conditions (Lema et al. 1988). Biological processes have been shown to be very effective in removing organic and nitrogenous matter from immature leachates when the BOD/COD ratio has a high value (>0.5). Biological processes are the most cost-effective means for reducing the organic content of leachate, particularly when complete onsite treatment is required.

5.2.1.1 Conventional Biological Treatment Processes

A Conventional aerobic treatment should allow a partial abatement of biodegradable organic pollutants and should also achieve the ammonium nitrogen nitrification. Aerobic biological processes based on suspended-growth biomass, such as aerated lagoons, conventional activated sludge processes (ASP) and SBR, have been widely studied and adopted (Abbas et al. 2009). However, this method has been shown in the more recent decades to be inadequate for handling landfill leachate treatment (Lin et al. 2000). Attached-growth systems have recently attracted major interest: MBBR and biofilters. The combination of membrane separation technology and aerobic–anaerobic bioreactors, most commonly called membrane bioreactor, has also led to a new focus on leachate treatment. Even if processes were proved to be effective for the removal of organic carbon, nutrients and ammonia content, too much disadvantages tend to focus on others technologies: inadequate sludge settleability and the need for longer aeration times, high energy demand and excess sludge production, microbial inhibition due to high ammonium nitrogen strength.

Anaerobic treatment is the biological treatment without use of air or elemental oxygen. Many applications are directed toward the removal of organic pollution in wastewater, leachate (containing high concentrations of organic acids), slurries, and sludge. An anaerobic digestion treatment of leachates allows to end the process initiated in the tip, being thus particularly suitable for dealing with high-strength organic effluents, such as leachate streams from young tips. Contrary to aerobic processes, anaerobic digestion conserves energy and produces very few solids, but suffers from low reaction rates. Moreover, it is possible to use the CH_4 produced to warm the digester that usually works at 35 °C and, under favorable conditions, for external purposes. The main advantage of the anaerobic treatment processes is the low energy requirement, because no oxygen has to be supplied. Technical anaerobic processes need adequate temperatures of 35–55 °C (Kumar et al. 2013).

5.2.1.2 ASP

The activated sludge process is a suspended-growth and biological treatment process that uses aerobic microorganisms to biodegrade organic contaminants in leachate. With conventional activated sludge treatment, the leachate is aerated in an open tank basin with diffusers or mechanical aerators.

After the aeration phase, the mixed liquor (the mixture of microorganisms and the treated water) is pumped to a gravity clarifier to settle out the microorganisms. A high percentage of the settled biomass is recycled to the aeration tank to maintain the design mixed liquor suspended solids level, and the excess sludge is wasted. Variations in the conventional activated sludge process have been developed to provide greater tolerance for shock loadings, to improve sludge-settling characteristics, and to achieve higher BOD removals. Process modifications include complete mixing, step aeration, modified aeration, extended aeration, contact stabilization, and the use of pure oxygen (EPA 1982).

5.2.1.3 Trickling Filters

This method has been investigated for the biological nitrogen lowering from municipal landfill leachate. The trickling filter is an attached-growth; aerobic biological treatment process in which leachate is continuously distributed over a bed of rocks or plastic medium that supports the growth of microorganisms. The wastewater trickles through the filter bed, contacts the slime layer formed on the medium, and is collected by an under-drain system. The microorganisms assimilate and oxidize substances in the leachate; as the microorganisms grow, the slime layer increases. Periodic sloughing of the slime layer into the under-rain system results from organic and hydraulic loadings on the filter, and a new slime layer begins to grow. Sloughed solids are separated from the treated effluent by settling.

Trickling filters operate under short hydraulic retention times that do not allow for complete biodegradation of organics; as a result, effluent recirculation is required to increase the net contact time of the leachate with the biomass and achieve high organic removal efficiency. Recirculation also provides a constant hydraulic loading and dilutes high-strength leachates (EPA 1982). Effluent recirculation is essential for trickling filters constructed with plastic medium, which has a high percentage of void space, to ensure that the medium is thoroughly wetted and will sustain microbial growth and promote effective sloughing.

5.2.2 Advanced Biological Treatment Techniques

Due to main problems of sludge bulking or inadequate separability in conventional aerobic systems, a number of innovative treatment processes, called attached-growth biomass systems, using biofilm, have been recently developed. These systems present the advantage of not suffer from loss of active biomass (Renou et al. 2008).

5.2.2.1 Rotating Biological Contractor (RBC)

The RBC is an attached-growth, aerobic biological treatment process which consists of a series of closely spaced plastic disks on a horizontal shaft. The assemblage is mounted in a contoured-bottom tank containing the water to be treated so that the disks are partially (about 40%) immersed. The disks, which eventually develop a slime layer 2–4 mm thick over the entire wetted surface, rotate slowly through the water and alternately contact the biomass with the organic matter in the wastewater and then with the atmosphere for adsorption of oxygen. Excess biomass on the media is stripped off by rotational shear forces, and the stripped solids are held in suspension with the wastewater by the mixing action of the disks. The sloughed solids are carried with the effluent to a clarifier, where they are settled and separated from the treated waste (Abbas et al. 2009). A COD removal of about 52% is obtained at an HRT of 24 h and a rotational speed of 6 rpm. Even it is cheaper in operating it, so it is suitable for a developing country like India. However, it requires infinitesimal amount of energy for running the operation (Kumar et al. 2013).

5.2.2.2 SBR

The SBR is an activated sludge, biological nutrient removal (nitrification/denitrification) process, based on a cycle of operation. Unlike conventional, continuous-flow, activated sludge systems, which have separate tanks for equalization, aeration, and clarification, the SBR performs all operations in a single tank. According to E.P.A (1992), SBR process has widespread application where mechanical treatment of small wastewater flows is desired. Because it provides batch treatment, it is ideally suited for wide variations in flow rates operation in the "fill and draw" mode prevents the "washout" of biological solids that often occurs with extended aeration system. Another advantage of SBR systems is that they require less operator attention yet produce a very high quality effluent. SBR is ideally suited to nitrification–denitrification processes since it provides an operation regime compatible with concurrent organic carbon oxidation and nitrification. The greater process flexibility of SBR is particularly important when considering landfill leachate treatment, which has a high degree of variability in quality and quantity (Kennedy and Lentz 2000).

5.2.2.3 Moving Bed Biofilm Reactor (MBBR)

MBBR process is based on the use of suspended porous polymeric carriers, kept in continuous movement in the aeration tank, while the active biomass grows as a biofilm on the surfaces of them. During the operation process, characteristics of attached-growth media play a key role in MBBR performance. In recent years, different kinds of media have been employed in MBBRs for wastewater treatment, including plastic media, polyurethane foam, activated carbon (granular and powdered), natural occurring materials (e.g. sand, zeolite, earth, light expended clay aggregate, etc.), non-woven carriers, ceramic carriers, modified carriers (e.g. BIO-CONS carrier), bioplastic-based moving bed biofilm carriers, and wood chips (Deng et al. 2016). MBBR provides a long biomass retention time and accommodates high loading rates without any problems of clogging. In a MBBR, the bacteria are fixed in a biofilm on a carrier. The carrier is suspended and moves freely in the reactor. The

MBBR has been applied for organic matter removal, for nitrification, and for nutrient (N and P) removal (Tawfik et al. 2010). Mains advantages of this method compared to conventional suspended-growth processes seems to be: higher biomass concentrations, no long sludge-settling periods, lower sensitivity to toxic compounds, and both organic and high ammonia removals in a single process (Loukidou and Zouboulis 2001).

5.2.2.4 Up-Flow Anaerobic Sludge Blanket (UASB) Reactor

The UASB reactor technology is considered a breakthrough in the development and application of anaerobic high-rate technology for wastewater treatment UASB process is a modern anaerobic treatment that can have high treatment efficiency and a short hydraulic retention time. UASB reactors, when they are submitted to high volumetric organic loading rate values have exhibited higher performances compared to other kinds of anaerobic reactors. In addition, the UASB lends itself to a design where liquid, gas, and solid phases can be separated within the one vessel. The process temperatures reported have generally been 20–35 °C for anaerobic treatment with UASB reactors. These promising results show that high-rate treatment at low temperature may minimize the need for heating the leachate prior to treatment, which may thus provide an interesting cost-effective option. The main disadvantages of such a treatment stay sensitivity to toxic substances (Renou et al. 2008).

5.2.2.5 Membrane Bioreactors

Recently, membrane bioreactor (MBR) technology, an advanced biological treatment process that replaces the traditional secondary clarifier of an activated sludge process with a membrane separation unit, has emerged as a promising alternative. The combination of membranes to biological processes for treatment has led to the emergence of membrane bioreactors for the separation and retention of solids, for bubble-less aeration within the bioreactor, and for extraction of priority organic pollutants from industrial contaminated water (Stephenson et al. 2000). MBRs have been demonstrated to be particularly advantageous in treating old leachate. The advantages of MBRs over conventional biological processes are well known and include better effluent quality, process stability, smaller footprint, increased biomass or mixed liquor suspended solids (MLSS) retention, and low sludge production (Ahmed and Lan 2012). Several early research efforts investigated the potential of using MBRs for leachate treatment (Costa et al. 2019; Deng et al. 2016; Eldyasti et al. 2011; EPA Manual 1992). Recently, more research studies are being conducted on the feasibility and performance of MBRs for LFL treatment.

As shown in Table 5.3, high COD removals between 60 and 99% were achieved, regardless of the leachate age or the operating conditions used. Such high removal of the biodegradable organic matter is in accordance with most biological treatments of LFL.

Table 5.3 Performance of MBR in COD removal from landfill leachate in different scale

Scale	Landfill location	Reactor type	Leachate characteristics COD (mg L^{-1})	Membrane configuration	COD (%)	References
Pilot	Thailand	Two-stage MBR(anoxic tank + aerobic MBR)	2605–7318	Sub (HF)	60–78	Chiemchaisri et al. (2011)
Laboratory	Tychy (Poland)	Mixture of 10% LFL + synthetic WW/SBR	3000–3500	Sub (MF/Cap)	>98–89	Puszczało et al. (2010)
Bench	Diyarbakir (Turkey)	Mixture of LFL + domestic WW	8500–14,200	Sub (HF)	72–99	Hasar et al. (2009)
Full	Dorset (United Kingdom)	Three aerobic biological	5000	Ext (UF/Tub)	>96	Robinson (2007)

LFL landfill leachate, *WW* wastewater, *SBR* sequencing batch reactors, *Ext* external, *Sub* submerged, *UF* ultrafiltration, *MF* microfiltration, *HF* hollow fiber, *Tub* tubular, *Cap* capillary.
Source Ahmed and Lan (2012)

5.2.2.6 Circulating Integrated Fluidized Bed Bioreactor (CFBBR)

Compared to conventional physical, chemical, and biological treatment processes for industrial wastewater, the CFBBR system has numerous advantages including small footprint with elimination of clarifiers, high biomass retention resulting in long solid residence times (SRTs) and relatively short hydraulic retention times (HRTs), enhanced mass transfer, and lower sludge production rate (Eldyasti et al. 2011). An extensive pilot-scale investigation of the patented CFBBR for biological nutrient removal (BNR) from municipal wastewater and landfill leachate has been reported by Nakhla et al. (2005). The CFBBR employs attached microbial films resulting from biodegradation of both organics and nutrients within an integrated system comprising an anoxic column in a fast fluidization regime and an aerobic column in a conventional fluidization regime. This new promising patented technology combines the compactness and efficiency of a fixed-film process with excellent organics, nitrogen, and phosphorus removal efficiencies of 85%, 80%, and 70%, respectively.

5.3 Conclusions

Optimal leachate treatment, in order to fully reduce the negative impact on the environment, is today's challenge. But, the complexity of the leachate composition makes it very difficult to formulate general recommendations. Variations in leachates, in particular their variation both over time and from site to site, means that the most appropriate treatment should be simple, universal and adaptable. The various methods presented in the current study, offer their treatment technique with respect to certain facets of the problem. There has been a steady progress of new and advanced sustainable landfill leachate treatment which proven to be a promising alternative. Utilization of advanced biological treatment technologies may be suitable to mitigate the hazard created by landfill leachate. Though there are still uncertainties whether these techniques could enhance environmental sustainability and safety of human being, more efforts should be carried out to ensure a livelihood of human being and earth coexistence. Therefore, a holistic approach is essential for finding a suitable leachate treatment opportunity in order to safeguard environmental and human being livelihood, as a whole.

References

Abbas AA, Jingsong G, Ping LZ, Ya PY, Al-Rekabi WS (2009) Review on land fill leachate treatments. Am J Appl Sci 6(4):672–684

Ahmed FN, Lan CQ (2012) Treatment of landfill leachate using membrane bioreactors: a review. Desalination 287:41–54. https://doi.org/10.1016/j.desal.2011.12.012

Ahn WY, Kang MS, Yim SK, Choi KH (2002) Advanced landfill leachate treatment using an integrated membrane process. Desalination 149(1–3):109–114. https://doi.org/10.1016/S0011-9164(02)00740-3

Bohdziewicz J, Kwarciak A (2008) The application of hybrid system UASB reactor-RO in landfill leachate treatment. Desalination 222:128–134. https://doi.org/10.1016/j.desal.2007.01.137

Cecen F, Aktas O (2004) Aerobic co-treatment of landfill leachate with domestic wastewater. Environ Eng Sci 21:303–312

Chaudhari LB, Murthy ZVP (2010) Treatment of landfill leachates by nanofiltration. J Environ Manage 91(5):1209–1217. https://doi.org/10.1016/j.jenvman.2010.01.007

Chiemchaisri C, Chiemchaisri W, Nindee P, Chang CY, Yamamoto K (2011) Treatment performance and microbial characteristics in two-stage membrane bioreactor applied to partially stabilized leachate. Water Sci Technol 64:1064–1072

Christensen TH, Kjeldsen P (1989) Basic biochemical processes in landfills. In: Sanitary landfilling: process, technology, and environmental impact. Academic Press, New York, pp 29–49

Costa AM, Greice R, Marotta DS, Campos JC (2019) Land fill leachate treatment in Brazil—An overview. 232(October 2018):110–116

Deng L, Guo W, Ngo HH, Zhang X, Wang XC, Zhang Q, Chen R (2016) New functional biocarriers for enhancing the performance of a hybrid moving bed biofilm reactor-membrane bioreactor system. Biores Technol 208:87–93. https://doi.org/10.1016/j.biortech.2016.02.057

Eldyasti A, Andalib M, Hafez H, Nakhla G, Zhu J (2011) Comparative modeling of biological nutrient removal from landfill leachate using a circulating fluidized bed bioreactor (CFBBR). J Hazard Mater 187:140–149. https://doi.org/10.1016/j.jhazmat.2010.12.115

EPA (1982) Treatability Data. Volume 1 EPA-600/2-82-OOla

EPA Manual (September 1992) Wastewater treatment/disposal for small communities. Office of Research & Development, Office of water

Gotvajn AŽ, Tišler T, Zagorc-Končan J (2009) Comparison of different treatment strategies for industrial landfill leachate. J Hazard Mater 162(2–3):1446–1456. https://doi.org/10.1016/j.jhazmat.2008.06.037

Hasar H, Ipek U, Kinaci C (2009) Joint treatment of landfill leachate with municipal wastewater by submerged membrane bioreactor. Water Sci Technol 60:3121–3127
Kennedy KJ, Lentz EM (2000) Treatment of landfill leachate using sequencing batch and continuous flow upflow anaerobic sludge blanket (UASB) reactors. Water Res 34:3640–3656
Kjeldsen P, Barlaz MA, Rooker AP, Baun A, Ledin A, Christensen TH (2002) Present and long-term composition of MSW landfill leachate: a review. Crit Rev Environ Sci Technol 32(4):297–336. https://doi.org/10.1080/10643380290813462
Kumar S, Katoria D, Singh G (2013) Leachate treatment technologies 1, 4(5):439–444
Laitinen N, Luonsi A, Vilen J (2006) Landfill leachate treatment with sequencing batch reactor and membrane bioreactor. Desalination 191(1–3):86–91. https://doi.org/10.1016/j.desal.2005.08.012
Lema JM, Mendez R, Blazquez R (1988) Characteristics of landfill leachates and alternatives for their treatment: a review. Water Air Soil Pollution 40:223–250
Lin CY, Chang FY, Chang CH (2000) Co-digestion of leachate with septage using a UASB reactor. Biores Technol 73:175–178
Loukidou MX, Zouboulis AI (2001) Comparaison of two biological treatment process using attached-growth biomass for sanitary landfill leachate treatment. Environ Pollut 111:273–281
Mittal BA (2011) Biological wastewater treatment, 2–9
Mohan S, Gandhimathi R (2009) Removal of heavy metal ions from municipal solid waste leachate using coal fly ash as an adsorbent. J Hazard Mater 169(1–3):351–359. https://doi.org/10.1016/j.jhazmat.2009.03.104
Moravia WG, Amaral MCS, Lange LC (2013) Evaluation of landfill leachate treatment by advanced oxidative process by Fenton's reagent combined with membrane separation system. Waste Manag 33(1):89–101. https://doi.org/10.1016/j.wasman.2012.08.009
Nakhla G, Zhu J, Cui Y (2005) Liquid–solid circulating fluidized bed wastewater treatment system for simultaneous removal of carbon, nitrogen, and phosphorus, US Patent No. 7261811 (2004) Int'l PCT patent awarded
Puszczało E, Bohdziewicz J, Świerczyńska A (2010) The influence of percentage share of municipal landfill leachates in a mixture with synthetic wastewater on the effectiveness of a treatment process with use of membrane bioreactor. Desalination Water Treat 14:16–20
Renou S, Givaudan JG, Poulain S, Dirassouyan F, Moulin P (2008) Landfill leachate treatment: review and opportunity. J Hazard Mater 150(3):468–493. https://doi.org/10.1016/j.jhazmat.2007.09.077
Robinson T (2007) Membrane bioreactors: nanotechnology improves landfill leachate quality. Filtr Sep 44:38–39
Stephenson T, Judd S, Jefferson B, Brindle K (2000) Membrane bioreactors for wastewater treatment. IWA Publishing, ISBN 1-90-022207-8
Tawfik A, El-Gohary F, Temmink H (2010) Treatment of domestic wastewater in an up-flow anaerobic sludge blanket reactor followed by moving bed biofilm reactor. Bioprocess Biosyst Eng 33(2):267–276
Visvanathan C, Tränkler J, Zhou G (2004) State of the art review landfill leachate treatment, 93
Wiszniowski J, Robert D, Surmacz-Gorska J, Miksch K, Weber JV (2006) Landfill leachate treatment methods: A review. Environ Chem Lett 4(1):51–61. https://doi.org/10.1007/s10311-005-0016-z
Yao P (2017) Perspectives on technology for landfill leachate treatment. Arab J Chem 10:S2567–S2574
Yu D (2007) Landfill leachate treatment case study, 82

Chapter 6
Biological Methodologies for Treatment of Textile Wastewater

Saurabh Mishra and Abhijit Maiti

Abstract Globally, the enormous amount of coloured wastewater generated from textile industries is considered as highly toxic composition that causes severe biotic risk to ecosystem, when discharged untreated. Textile wastewaters are mostly saline in nature and contain both organic and inorganic contaminants, which could easily pass through conventional treatment process and remain unaffected. In this regard, the bio-based treatment processes have been found to be very efficient for decolourization/detoxification of coloured textile wastewater. Although, the performance of treatment system varies with type of organism, growth, and activity of organism, complexity of contaminants and operational parameters such as pH, temperatures, dissolved oxygen, and nutrient composition. In the chapter, a critical review of available literature focused on the use of bioremediation agents such as bacteria, fungi, algae, plants, and their derived enzymes to decolourize wastewater containing either individual or mixed dyes and simultaneously remove heavy metals has been elaborately discussed. The effect of operational parameters on the activity of bioremediation agents, their limitations, and suitability of soluble or immobilized state for effective functioning of the system has been explained in detail. Alternatively, an overview of application of bio-treatment system for generation of bio-energy has been elucidated to extract future scope of research.

Keywords Bioremediation · Dye · Bio-decolourization · Bio-energy · Wastewater

6.1 Introduction

Nowadays, the decline in water quality and scarcity of freshwater has become a serious concern of research for the protection of life on earth. The rapid increase in human population, industrialization, and anthropogenic activities around the freshwater bodies is the major responsible factors for current scenario of water problem

S. Mishra (✉) · A. Maiti
Department of Polymer and Process Engineering, Indian Institute of Technology Roorkee, Saharanpur Campus, Saharanpur, Uttar Pradesh 247001, India
e-mail: saurabhmishra20057@gmail.com

R. M. Singh et al. (eds.), *Environmental Processes and Management*, Water Science and Technology Library 91,
https://doi.org/10.1007/978-3-030-38152-3_6

(Singh et al. 2017; Dalakoti et al. 2018). The water utilizing industries extract a huge amount of water from water bodies and generate wastewaters containing hazardous contaminants, which causes severe biotic risk/stress and ecological disturbances when discharged untreated/partially treated (Singh et al. 2015a, b; Maiti et al. 2019). Considerably, dyeing industries (textile, food, tannery, cosmetics, paper, and pharmaceutical) are one of the greatest generators of wastewater, due to large quantity of water utilization in the dyeing process (Mishra and Maiti 2018a). According to the Federation of Indian Chambers of Commerce and Industry (FICCI) report, China is the largest producers of dyestuffs and supplies about 40% of the textile products in the world, followed by India as the second largest producer that accounts ~9% of total world dyestuffs production. A brief idea about the demographical distribution of textile producers and consumers in the world is shown in Fig. 6.1.

The coloured wastewater generated from textile industries contains most complex toxic chemical composition that includes dyes and heavy metals, which are highly mutagenic and carcinogenic promoters and cause severe diseases to human health (Mishra and Maiti 2018b). Annually, about 280,000 tonnes of synthetic textile dyestuffs are discharged in such industrial effluents worldwide (Saratale et al. 2011). In the recent past, numerous chemical, physical, and advanced chemical oxidation wastewater treatment technologies have been investigated for removal and detoxification of recalcitrant contaminants from textile wastewater (You et al. 2010), as shown in Table 6.1.

However, these treatment techniques have their own limitations in practical implementation like ineffective removal of contaminants, generation of huge amount of secondary sludge, and high operational cost (Mishra and Maiti 2018c). The bio-based methodologies that involve bacteria, fungi, algae, higher plants, and enzymes are being investigated for decolourization of coloured wastewater and removal of toxic inorganic contaminants (Telke et al. 2015; Naseer et al. 2016; Roy et al. 2018). The bio-based techniques have several beneficial features such as convenience in handling, selectivity, specificity, higher stability and reusability, improved activity,

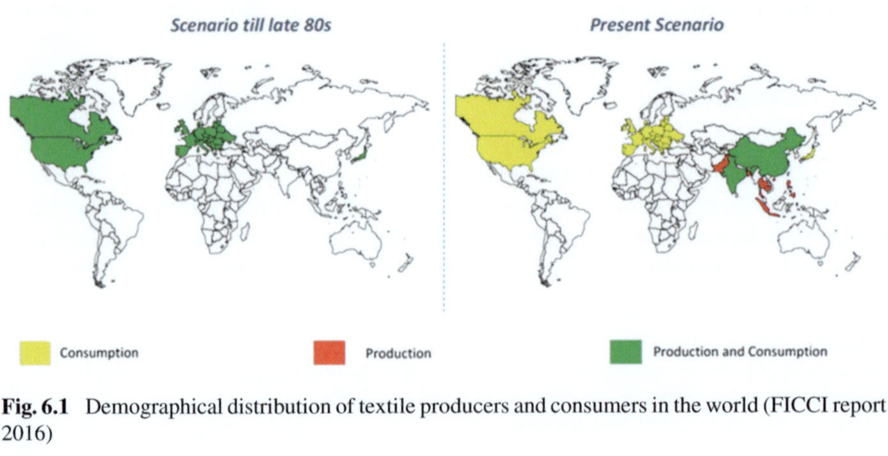

Fig. 6.1 Demographical distribution of textile producers and consumers in the world (FICCI report 2016)

Table 6.1 Advantages and disadvantages of physical and chemical techniques for wastewater treatment

Methods	Disadvantages	Advantages	References
Ozonation	Short half-life (20 min), ineffective removal of dye	No change in volume due to gaseous state	Wu et al. (2008)
Photochemical	Generation of secondary toxic waste products	Effective decolourization	Rezaee et al. (2008)
Electrochemical oxidation	High operational cost	Non-toxic end products	Jovic et al. (2013)
Activated carbon	Costlier process	Good sorption of various dyes	Nabil et al. (2014)
Fentons reagent	High sludge generation	Effectively remove both soluble and insoluble dyes	Xu et al. (2012)
Electrokinetic coagulation	High sludge production	Cost effective	Can et al. (2003)
Membrane filtration	Highly concentrated sludge production	Effective decolourization	Wu et al. (1998)
Ion exchange	Not applicable for all dyes	Good reusability for the removal of dyes	Vijayaraghavan et al. (2013)

and ease separation of enzymes from a reaction mixture (Bilal et al. 2019). Moreover, the selection of effective biotic component is an essential foremost step for effective bio-treatment of textile wastewater (Ji et al. 2012). The bio-decolourization of dye compounds occurs either through biosorption, bioaccumulation, biotransformation, or biodegradation in cellular metabolic/respiratory pathways under optimal operational condition (Sun et al. 2015). Although, biotic components are sensitive to operational condition, thus the effect of operational parameters (like pH, temperature, dissolved oxygen concentration, and nutrient dose) on growth and activity must be investigated for effective performance of the system (Ayed et al. 2017). In this regard, numerous studies have been reported in the literature related to optimization of process condition for decolourization of dyes either in individual, or mixed state, using in synthetic dye aqueous solution (Narayanan et al. 2015). Chang et al. (2001) investigated the decolourization of reactive red 22 dye by *Pseudomonas luteola* and reported that the maximum 90% decolourization of initial dye concentration of 300 mg/L within 20 h is under static-incubation condition. In another study, simultaneous decolourization of mixed reactive dyes (i.e. reactive yellow 2, reactive black 5, reactive red 120, and reactive orange 16) and removal of Cr(VI) ion using *Pseudomonas aeruginosa* ZM130 has been investigated by Maqbool et al. (2016). The results revealed that the bacteria exhibited 76.6–98.7 and 51.9–91.1%, decolourization of dyes and removal of Cr(VI) ion, respectively, within 180 h of incubation under static conditions. Also, they found that the degradation of dyes was mediated by nicotinamide adenine dinucleotide-dichlorophenolindophenol (NADH-DCIP) reductase and laccase enzymes produced by in the bacteria. Arabaci and Usluoglu (2014a, b) investigated the enzymatic decolourization of individual textile dyes (direct blue 86, reactive yellow 39, and acid black 194) by immobilized polyphenol oxidase enzyme, which was isolated from leaves of quince (*Cydonia oblonga*). They reported the 20–80% decolourization of initial respective dyes concentration of 100 mg/L by enzymes system. Saratale et al. (2011) reviewed the articles related to bacterial-mediated biodegradation of azo dyes and suggested that anaerobic–aerobic culture of bacteria might be appropriate for the effective degradation of azo dye laden wastewaters, required, the various physiochemical parameters that influence the performance are to be optimized. Mishra and Maiti (2019a) reviewed the articles related to the applicability of enzymes for biodegradation of textile dyes and concluded that the enzyme belonging to oxidoreductase class is chief biological agents involved in bio-treatment of textile dyes. The importance of bioremediation agents for textile wastewater treatment has also been explored through other studies like Mahmood et al. (2015), Singh et al. (2015a, b), Mishra and Maiti (2018d), Sreedharan and Rao (2019), and others. Those review articles are mainly focused on microbial (bacteria and fungi) or enzymatic degradation of especially azo dyes or other groups of textile synthetic dyes.

Despite of these available literatures, a comprehensive study to represent a comparative assessment of dye decolourization by different biotic agents like bacteria, fungi, algae, plants, and enzymes is not explored. Considering this gap in available literature, a review study has been explored in this chapter, focusing the use of bioremediation agents such as bacteria, fungi, algae, plants, and their derived enzymes to decolourize textile wastewater (either synthetic or real) containing either individual dye, mixed dyes, or dyes in the presence of heavy metals. In the latter section, research advancement in the field of bio-decolourization of dyes contaminated wastewater coupled with bio-energy generation has been explored to identify the future scope of research. This study will provide a good support to the researchers for investigation of potential bio-technique for bio-treatment of synthetic textile dyes laden wastewater.

6.2 Dyes and Complexity of Real Textile Wastewater

Generally, various types of dyes are used as or colourants during dyeing process in textile industries. It has been estimated that about 10,000 commercial dyes that account 7×10^5 tonnes of dyestuffs are produced annually for dyeing process (Mishra and Maiti 2018a). Production of dyestuff and pigments in India is close to 80,000 tonnes, nearly 9% of the world's production (Abraham 2013). India is now the world's second largest exporter of dyestuffs and its intermediates among developing countries, after China. The commercial dyes are classified in terms of colour, structure, and application method in the Colour Index by the Society of Dyers and Colourists and the American Association of Textile Chemists and Colourists. On the basis of chromophoric functional groups found in the chemical structure, dyes have been classified as anthraquinone, azo, triarylmethane, phthalocyanine, oxazine, formazan, and others (Sudha et al. 2014). The Colour Index discerns 14 different application classes as follows: acid dyes, reactive dye, metal complex dyes, direct dyes, basic dyes, mordant dyes, disperse dyes, pigment dyes, vat dyes, anionic dyes, ingrain dyes, sulphur dyes, solvent dyes, and fluorescent brighteners. Among the commercial dyes, azo and anthraquinone dyes are extensively used due to the fastness in binding with fabrics, better chemical stability, and bright colour appearance ((Mishra and Maiti 2019b). Considerably, salts (NaCl) are used to improve the binding affinity of dye molecules to the fibres. In addition, dyes (especially anthraquinone dyes) require metallic salts to impart colour to the fibres. Chromium salt (as chromate (Cr(VI)) is commonly used as mordant in dyeing process (Liu et al. 2017). In textile processing, the sequential major steps are as desizing, bleaching, mercerizing, dyeing, printing, and finishing. The chemicals used in these textile processing steps are also found in produced wastewater, enlisted in Table 6.2.

However, approximately >15% of dyestuffs remain unfixed during the dyeing process and are lost in the wastewater (Wang et al. 2017). The types of chemical dyestuffs applied in textile processing stages determine the variable characteristics of wastewater in terms of dissolved solids, pH, organic, and inorganic compounds. Based on practical estimation, about 45% of material is lost during preparatory step, and 22% is reprocessed (Vigneswaran et al. 2014). Thus, the textile wastewaters generated from different processing steps are mostly saline in nature and contain both organic compound (as dyestuffs) and inorganic metal ions (as Cr(VI)), which are hazardous chemical composition and well beyond the regulatory standard for industrial wastewater. The characteristics of real textile wastewater produced after chemical processing are enlisted in Table 6.3. Generally, the concentration of individual dye from aqueous dye solution or wastewater can be easily analysed using UV–visible absorption spectroscopy, at their respective maximum absorption wavelength. However, when coloured solution or wastewater contains more than one type of dye compound, then it is difficult to analyse the individual concentration of dyes. In this regard, the UV–visible transmittance data of mixed dye solution is used to evaluate American Dye Manufacturers Institute (ADMI) value that represents total dye content in the solution.

Table 6.2 Some of the important chemicals used in this textile processing

S. No.	Process	Chemicals
1.	Desizing	Starch, enzymes, and wax
2.	Bleaching	Organic stabilizer, sodium silicate, surfactants, and H_2O_2
3.	Mercerizing	Wax and NaOH
4.	Dyeing	Salts, dye, surfactant, soda ash, urea, and Cr(VI)
5.	Printing	Dye, pigment, soda ash, binder, thickener, and urea
6.	Finishing	Wax, resins, formaldehyde, and others

Table 6.3 Characteristics of textile industrial wastewater

Parameters	Permissible limit	Cotton industry wastewater	Wool industry wastewater	Synthetic fibre industry wastewater
pH	5.5–9	7.5–12	3–10	7–9
BOD (mg/L)	30	150–800	4500–8000	150–300
COD (mg/L)	250	250–2500	9000–20,000	300–700
TDS (mg/L)	–	2000–8000	10,000–14,000	1000–1200
TOC (mg/L)	–	200–400	5000–8000	150–350
Cr(VI) (mg/L)	0.1	10–20	10–25	10–15
Total alkalinity (mg/L)	–	20–30	50–100	15–20
Total solid (mg/L)	–	3000–8500	15,000–22,000	2000–6000

6.3 Bacterial-Mediated Decolourization of Dyes

The decolourization of textile dyes through bacterial (whole cell)-mediated process has been proved to be an effective bio-based technique. It is well known that the significant growth and activity of bacteria could be achieved, only when the process condition is optimized. Considerably, the use of single bacterial species in bio-treatment system provides an ease to assess the effect of process parameters (like temperature, pH, dissolved oxygen, electron donor, and some others) on the performance of the system. The dyes belonging to azo, triphenylmethane, and anthraquinone class with the complex aromatic structure are known to be recalcitrant to microbial degradation. However, the numerous researchers have identified and investigated various bacterial species that exhibit high potential to decolourize textile dyes, effectively. Meerbergen et al. (2018) worked out the degradation of azo dyes reactive orange 16 (RO16) and reactive green 19 (RG19) by *Klebsiella* strain ST16.16/034 and *Acinetobacter* strain ST16.16/164, isolated from activated sludge from coloured wastewater treatment plant. They reported that the bacteria *Klebsiella sp.* exhibited maximum 97.9% decolourization of RO16 dye, while the RG19 dye was effectively decolourized by *Acinetobacter sp.* by 93.4% for initial dye concentration 100 mg/L. Both of these salt-tolerant bacteria could retain good decolourization capability at a wider temperature range of 10–40 °C and at alkaline pH medium, under static condition. Parmar and Shukla (2017) investigated the biodegradation of anthraquinone dye reactive blue 4 by non-pathogenic *Staphylococcus hominis* subsp. hominis DSM 20328, which was isolated from sludge and collected from the dye-affected area. They reported that the bacteria could decolourize 97% of dye (initial concentration 50 mg/L) within 24 h at temperature 37 °C and pH 7 in glucose-supplemented medium, under anaerobic/static conditions. During the analysis of end metabolites from treated water, they found the conversion of dye molecules into non-toxic compounds like catechol, which promotes the germination of *Triticum aestivum* and

Vigna radiata seeds. Generally, the decolourization of anthraquinone dyes occurs through bioaccumulation mechanism by bacterial cell cultured under aerobic condition. While, biodegradation of theses dyes occurs by bacteria under anaerobic condition. Cheng et al. (2019) studied the biodegradation of metal complex dye—naphthol green B (NGB) dye by marine bacteria *Pseudoalteromonas sp.* CF10-13. They reported that the bacteria could exhibit the decolourization potential of 94% for initial dye concentration of 100 mg/L in saline medium supplemented with acetate at pH 7.5, under anaerobic condition. The authors suggested that NGB degradation could have occurred in the presence of redox mediators synthesized by the bacteria, which promoted the breakdown of NGB chromophore groups $N = O$ in anaerobic condition. The authors have indicated the formation of iron–sulphur nanoparticles as an end metabolite produced during the biodegradation of NGB, which reduces the dangers of H_2S and ferric iron and helps in dye resources recovery. Neetha et al. (2019) investigated the decolourization of direct blue 14 dye by an alkaliphilic endophyte *Bacillus fermus* isolated from *Centella asiatica*. They reported that the bacteria exhibited maximum of 92.76% biodegradation of initial dye concentration ~69 mg/L in sucrose-supplemented medium within 72 h, under optimized process condition. They have suggested that the dye degraded metabolites were less toxic to the growth of *Allium cepa* and induced negligible chromosomal aberrations than raw dye solution. The proposed pathway for degradation of direct blue 14 dye is shown in Fig. 6.2.

In many textile industries, the salts of heavy metals (like chromium, cobalt, and copper) are extensively used to stabilize dye onto the fibres. The wastewater generated from such industries contains toxic metal ions along with dyes, a hazardous composition. In this regard, to detoxify and remove contaminants from wastewater, the studies are being conducted to isolate and identify potential bacterial species that could simultaneously decolourize dyes and remove metal ions. Chaudhari et al. (2013) investigated the simultaneous decolourization of reactive orange-M2R and removal of Cr(VI) by *Lysinibacillus sp.* KMK-A under static anoxic condition. The authors have reported that the bacteria could effectively decolourize dye of initial concentration 2000 mg/L in the presence of 250 mg/L Cr(VI), under optimal condition. The bacteria could completely reduce Cr(VI) into non-toxic form Cr(III). The authors suggested that the dye decolourization process is mediated by enzymes azoreductase and chromate reductase in optimum environmental condition. In order to improvise the efficiency of dye decolourization process, the studies are being conducted to develop and investigate efficient bacterial consortium. Bacterial consortium represents the culture of two or more bacterial species that can actively propagate in a single culture medium. The complexity of bacterial consortia could provide a better system for effective removal of diverse contaminants from wastewater. Lalnunhlimi and Krishnaswamy (2016) studied the decolourization of direct blue 151 (DB151) and direct red 31 (DR31) by bacterial consortium of *Bacillus flexus* NBN2, *Bacillus cereus* AGP-03, *Bacillus cytotoxicus* NVH 391-98, and *Bacillus* sp. L10, which were isolated from saline soil. The authors reported that the consortium could decolorize initial dye concentration of 200 mg/L for individual DB151 and DR31 dyes by 97.57% and 95.25%, respectively, under optimal condition. They suggested

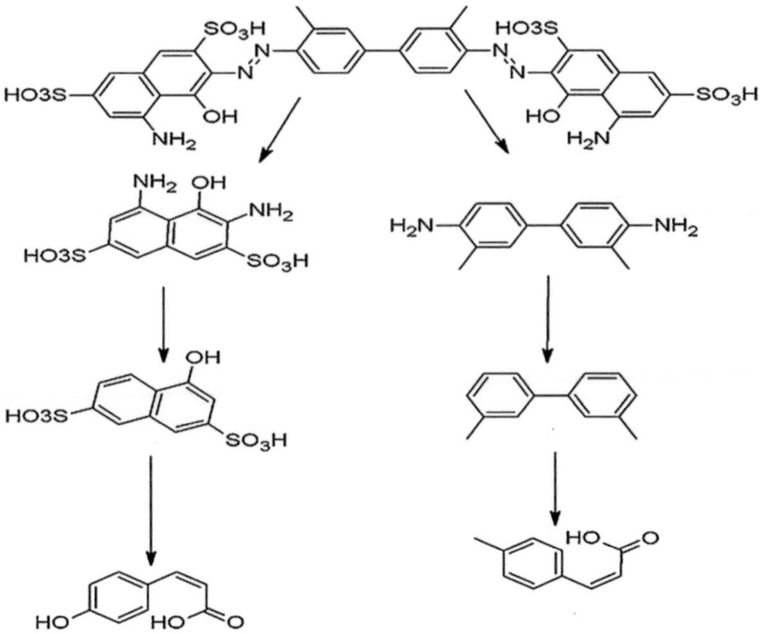

Fig. 6.2 Proposed metabolic pathway for degradation of direct blue 14 dye (modified from Neetha et al. 2019)

that the decolourization process could be enhanced by supplementation of additional nitrogen and carbon source in the form of yeast extract and sucrose. Karim et al. (2018) investigated the comparative dye decolourization potential of mono culture and bacterial consortium of *Neisseria* sp., *Vibrio* sp., *Bacillus* sp., *Bacillus* sp., and *Aeromonas* sp. against commercially available textile reactive dyes novacron brilliant blue FN-R, novacron orange FN-R, bezema yellow S8-G, novacron super black G, and bezema red S2-B. They authors reported that the monoculture of bacteria could exhibit highest 90% decolourization of dyes (except bezema red S2-B); whereas, the decolourization potential of bacterial consortium varied considerably in the range of 65–90% for individual dye and dyes mixture of initial concentration 100 mg/L within 6 d incubation period. The decolourization process was enhanced by the presence of co-substrate like glucose and yeast extract in the culture medium. The authors suggested that bacterial consortium was proved as efficient dye decolourizer than monocultures. In most of the studies, it has been a common practice that the potential bacteria for dye decolourization were isolated from dye-contaminated areas. In the recent trend, the numerous non-pathogenic endophyte bacteria are being investigated for satisfactory decolourization of dyes. Shang et al. (2019) investigated the decolourization of malachite green (MG) dye by *Klebsiella aerogenes* S27, which was isolated from the leaves of plant *Suaeda salsa*. They have reported that the bacteria synthesize enzyme triphenylmethane reductase, which mediated the complete degradation of MG dye (initial concentration 100 mg/L) within the incubation period of 8 h at temperature 30 °C, under neutral pH medium. The end metabolites produced after degradation MG dye were found as non-toxic and could promote the germination of Chinese cabbage seeds. The available literatures related to bacterial efficacy for decolourization of textile dyes are illustrated in Table 6.4.

The available literature reveals that bacterial species have high potential to decolourize wide variety of textiles dyes. However, the dye decolourization potential of different bacterial species varies, and the environmental process parameters play a significant role in the performance of the system. Hence, the selection of potent bacteria and optimization of process condition are the foremost steps in bacterial-mediated dye decolourization system. The complex dyes could be biodegraded either in microaerobic static or two stages anaerobic to aerobic conditions. This indicates the significant effect of dissolved oxygen in the dye degradation process. The bacteria gratuitously mineralize the dyes in the anaerobic condition that produces intermediate amines, which are utilized as a nutrient by bacteria in aerobic conditions.

Table 6.4 Studies related to decolourization of dyes by bacterial species

S. No.	Individual bacteria species/consortium	Dyes and heavy metals	Optimized process condition (IDC, pH, Temp., OT)	Removal efficiency	References
1.	*Pseudomonas aeruginosa* NGKCTS	Reactive red-BS	Static condition (300 mg/L, 7, 30–40 °C, 5.5 h)	91%	Sheth and Dave (2009)
2.	*Bacillus cohnii* RAPT1	Reactive red 120	Immobilized state (200 mg/L, 8, 35 °C, 4 h)	~100%	Padmanaban et al. (2016)
3.	*Enterobacter* sp. F NCIM 5545	Reactive blue 19	Anaerobic conditions (50 mg/L, 10, 37 °C, 24 h, 592)	99.7%	Holkar et al. (2014)
4.	*Pseudomonas desmolyticum* NCIM 2112	Vat red 10	Aerobic conditions (100 mg/L, 9, 25 °C, 23 d)	55.5%	Gurav et al. (2011)
5.	*Pseudomonas putida* MTCC 4910	Acid blue 93, and basic violet 3	Anaerobic condition (50 mg/L, 6–7, 37–45 °C, 48 h)	~100%	Khan et al. (2015)

(continued)

Table 6.4 (continued)

S. No.	Individual bacteria species/consortium	Dyes and heavy metals	Optimized process condition (IDC, pH, Temp., OT)	Removal efficiency	References
6.	*Shewanella sp.* IFN4	Direct red 81, reactive black 5, and acid red 88	Static condition (200 mg/L, 8.5, 35 °C, 4 h)	>95%	Imran et al. (2016)
7.	*Lactobacillus paracase* CL1107	Acid black, and Cr (VI)	Aerobic medium (100 mg/L, 6, 25 °C, 7 d)	92.3% dye and 95.8% Cr(VI)	Huang et al. (2015)
8.	*Acinetobacter baumannii* MN3 and *Pseudomonas stutzeri* MN1	Congo red (CR) and gentian violet (GV)	Static condition (100 mg/L, 8, 37 °C, 5 d)	97% (for CR) and 95% (for GV)	Kuppusamy et al. (2017)
9.	*Bacillus* sp1, *Bacillus* sp2, *Citrobacter* sp., *Acinetobacter* sp., and *klebsiella* sp.	Acid blue 25	Static condition (300 mg/L, 8, 37 °C, 48 h)	98%	Aruna et al. (2015)
10.	*Aeromonas hydrophila*	Basic violet 3, basic violet 14, and acid blue 90	Aerobic condition (50 mg/L, 7–8, 35 °C, 24 h)	72–96%	Ogugbue and Sawidis (2011)

Note IDC is the initial concentration of dye; *Temp.* is the incubation temperature; *OT* is the operational time

6.4 Myco-Decolourization of Dyes

Fungi are known to be the dominant microorganism that could be exploited for bio-treatment of textile wastewater. The fungi (whole cell)-mediated dye decolourization is performed in aerobic process condition, which is an advantageous step and reduces the cost of operation. The decolourization of dye is catalysed by oxidoreductase enzymes synthesized by the fungal cells cultured in dye-containing media. Thus, the decolourization occurs due to biosorption followed by degradation of dye molecules by fungal cell. In this regard, various potent fungi have been investigated for satisfactory decolourization of wide variety of dyes. Bankole et al. (2018a, b) investigated the bio-decolourization of acid red 88 dye by a filamentous fungus, *Achaetomium strumarium*. The authors reported that the fungi could decolourize 99% of initial dye concentration 10 mg/L in slightly acidic medium (pH 4), at temperature 40 °C, within time period of 96 h. The fungal decolourization mechanism involved adsorption followed by degradation of dye, which was mediated by NADH-DCIP reductase and laccase enzymes that asymmetrically cleave, dehydroxylate, and desulphonate dye molecules. The myco-transformation of adsorbed dye molecules resulted into the production of metabolites like naphthalen-2-ol, sodium naphthalene-1-sulphonate, and 1,4-dihydronaphthalene, which are non-phyto-toxic end products. He et al. (2018) worked out the decolourization acid red 3R dye by fungi *Trichoderma tomentosum*, isolated from degrade wood wafers. The authors reported the maximum decolourization of 99.2% for initial dye concentration of 85.5% within 72 h under optimal condition. The decolourization of dye was mediated by manganese peroxidase and lignin peroxidase enzymes that catalyse the dye molecules into non-phyto-toxic aromatic amines. Chen and Ting (2015) investigated the decolourization recalcitrant triphenylmethane dyes by a white rot fungus *Coriolopsis* sp. isolated from compost of oil palm. The authors reported that the fungi could decolourize initial concentration of dyes crystal violet (100 mg/l), methyl violet (100 mg/l), cotton blue (50 mg/l), and malachite green (50 mg/L) by 94, 97, 91, and 52%, respectively, within 9 d of incubation period. The dye removal was presumably via biosorption on fungal biomass followed by biodegradation of dye molecules mediated by laccase, lignin peroxidase, and NADH-DCIP reductase enzymes synthesized by fungi. In another study, Chen et al. (2019) investigated the bio-decolourization of same triphenylmethane dyes by a non-white rot fungus *Penicillium simplicissimum*, isolated from metal-rich wastewater. The authors reported that the highest decolourization 98.7, 97.5, 97.1, and 96.1% were obtained for crystal violet, methyl violet, malachite green, and cotton blue dye initial concentration 50 mg/L, in slightly acid medium pH 5 and at temperature 25 °C. The decolourization mechanism was mediated by enzymes manganese peroxidase, tyrosinase, and triphenylmethane reductase that catalysed and degraded the dye molecules. However, the treated end products were also found to be phytotoxic and inhibited the germination of *Vigna radiata* seed, thus require additional processes for detoxification. Tan et al. (2016) investigated the decolourization of an azo dye acid scarlet 3R by a salt-tolerant yeast *Scheffersomyces spartinae*, isolated from marine mud. The authors reported that the fungi could decolourize >90% of initial 80 mg/L dye, within 16 h under optimal process condition as: sucrose 2 g/L, yeast extract 0.08 g/L, $(NH4)_2SO_4$ 0.6 g/L, NaCl 6–30 g/L, 160 r/min, temperature 30 °C, and pH 5.0–6.0. After the analysis of treated water, they proposed the degradation of dye molecules through azo reduction, deamination, and desulfonation into non-toxic end product. Similar study has been reported by Song et al. (2017), who have investigated the decolourization of acid red B dye by salt-tolerant yeast *Pichia occidentalis*. The authors reported that more than >90% decolourization of initial 50 mg/L dye occurred within 16 h under similar optimal condition, which was mediated by NADH-DCIP reductase enzyme that promoted the degradation of dye molecules. Adnan et al. (2017) investigated the decolourization anthraquinone dyes alizarin Red S (polar) and Quinizarine Green SS (non-polar) by fungi *Trichoderma lixii* F21, isolated from decayed bark of trees. The authors reported that the fungi could decolourize 77.78 and 98.31% of polar and non-polar dyes, via adsorption and enzymatic degradation with 7 d of the treatment period. The degradation of dye was to enhance in presence of glucose and yeast extract in the culture medium that promotes the production of laccase and catechol 1,2-dioxygenase enzymes for dye degradation. Ning et al. (2018) investigated the bio-decolourization of azo dyes (direct blue 71 and direct blue 86) and anthraquinone dye (reactive blue 19) by fungi *Aspergillus flavus* A5p1. The authors reported that the fungi could decolourize >90% of all dyes for initial high concentration up to 500 mg/L, under optimal condition. Interestingly, the effective decolourization of azo dyes was primarily due to biosorption mechanism; whereas, decolourization of RB19 dye was via biodegradation mechanism. The available literatures related to fungal decolourization of textile dyes are illustrated in Table 6.5.

Based on the available literature, it is revealed that fungi cells exhibit the good potential to decolourize a wider class of dyes. However, the decolourization of most of the dyes by fungi occurs through biosorption mechanism that might increase the amount of sludge produced at the end of process and require additional cost for sludge management. The biosorption of dyes reduces the reusability of the fungal cell. Compared to bacteria, the fungal cells require more time to achieve desired growth and activity, which limits the performance of the system. Sen et al. (2016) reviewed the articles related to decolourization of azo dyes by fungi and suggested that the fungal decolouration occurs either via adsorption, enzymatic biodegradation mechanism, or combination of both processes. Unlike bacteria, fungal cells possess high potential to degrade complex organic compounds, which is catalysed by producing extracellular ligninolytic enzymes such as laccase, lignin peroxidase, and manganese peroxidase. Moreover, the potential of fungi to decolourize dyes in the presence of heavy metals from real textile wastewater needs to be investigated.

6.5 Phyco-Decolourization of Dyes

Algal species are highly versatile in nature and exhibit good potential for the decolourization of various synthetic textile dyes under certain process conditions. Algal species potentially produce oxidoreductase enzymes (such as peroxidase and azoreductase) that could catalyse the degradation of dye molecules in the treatment system. In this regard, various potent algal species have been investigated for satisfactory decolourization of wide variety of dyes. El-Sheekh et al. (2018) investigated the biodegradation of disperse orange 2RL dye by green alga *Chlorella vulgaris*, isolated from dye-contaminated soil. The authors have reported that the alga could degrade 55.22% of initial dye concentration 20 mg/L within 7 d of incubation period. The degradation of dye molecule by alga was mediated by enzyme azoreductase, which cleaves the azo bond and converts dye into aromatic amine. Arteaga et al. (2018) studied the decolourization aniline blue dye using microalgae *Chlorella vulgaris*, from synthetic aqueous solution. The authors reported that the algae could remove 53% of initial dye concentration 25 mg/L through biosorption mechanism after 11 d contact period. They suggested that the algae could be used for bleaching aniline

Table 6.5 Studies related to decolourization of dyes by fungal species

S. No.	Individual fungal species/consortium	Dyes	Optimized aerobic process condition (IDC, pH, Temp., OT)	Removal efficiency (%)	References
1.	*Aspergillus* sp.	Direct violet	100 µg/mL, 7, 28 °C, 5 d	71.1–93.3	El-Rahim et al. (2017)
2.	*Candida* sp. VITJASS	Reactive green	100 mg/L, 7, 30 °C, 96 h	84	Sinha et al. (2018)
3.	Fungal strain VITAF-1	Reactive green	250 mg/L, 7, 30 °C, 96 h	97.9	Sinha and Osborne (2016)
4.	*Chaetomium globosum* IMA1	Indigo dyes containing real textile wastewater	ND, 7, 30 °C, 5 d	~100	Manai et al. (2016)
5.	*Peyronellaea prosopidis*	Scarlet RR	10 mg/L, 6, 35 °C, 5 d	90	Bankole et al. (2018a, b)
6.	*Phanerochaete chrysosporium*	Amido black 10B	200 mg/L, 7, 37 °C, 3 d	98	Senthilkumar et al. (2014)
7.	*Ganoderma sp.* En3	Reactive orange 16	100 mg/L, 7, 28 °C, 96 h	95.1	Ma et al. (2014)
8.	*Cordyceps militaris* MTCC3936	Reactive yellow 18	300 mg/L, 5.5, 28 °C, 48 h	73.1	Kaur et al. (2015)
9.	*Trametes ljubarskyi*	Reactive violet 5	80 mg/L, 7, 25 °C, 7d	97.92	Goh et al. (2017)
10.	Fungi consortium of 21 fungal strains	Reactive black 5	30 mg/L, 10, 30 °C, 12 h	80	Yang et al. (2009)

Note IDC is the initial concentration of dye; *Temp.* is the incubation temperature; *OT* is the operational time; *ND* is not defined)

blue dye in tannery industries. Lebron et al. (2018) investigated the biosorption of methylene blue from aqueous synthetic solution by biomass of *Chlorella pyrenoidosa* and *Spirulina maxima*, having high content of protein and polysaccharide. They reported that algae *C. pyrenoidosa* exhibited maximum biosorption capacity of 101.75 mg/g and dye removal of 98.2%, which was comparably better than that of *S. maxima* (145.34 mg/g biosorption capacity and 94.19% dye removal). Moghazy (2019) investigated the removal of methylene blue dye from aqueous solution using activated biomass of *Chlamydomonas variabilis* as an efficient biosorbent. The authors reported that the maximum removal of 98% for initial dye concentration 82.4 mg/L was obtained by using 1 g/L activated biomass with a maximum adsorption capacity of 115 mg/g. They suggested that the biosorption mechanism is a chemisorption reaction, and the activated biomass is a promising biosorbent for the effective removal of dye. Rosa et al. (2018) investigated the biosorption of rhodamine B dye using microalgae *Chlorella pyrenoidosa*, from dyeing stones effluent. The authors reported that the microalgae biomass exhibited the highest biosorption capacity of 89% for initial dye concentration of 100 mg/L, biomass dose 0.1 g at pH 8.0 and temperature of 25 °C. They have suggested that the dye biosorption process was inhibited by the increase in temperature of the medium. The result revealed that the microalgae have good biosorption potential for rhodamine B removal. Aragaw and Asmare (2018) have studied the bio-decolourization of textile wastewater by microalgae consortium of *Scenedesmus* sp., *Chlorella* sp., *Synedra* sp., and *Achnanthidium* sp. using photo bioreactor system. The authors reported that the algal system could reduce COD, BOD, TDS and colour of the wastewater by 91.50%, 91.90%, 89.10%, and 82.6%, respectively, within 20 d at temperature 30 °C, in slightly alkaline medium. Authors suggested that the activity of microalgae is suppressed by increase in temperature >30 °C and pH of the medium either below 4 or above 8. Mahajan et al. (2019) investigated the phyco-decolourization potential of *Chara vulgaris* for treatment of real textile wastewater. The authors reported that the alga exhibited maximum reductions in COD (78%), BOD (82%), TDS (68%), and EC (86%) for 10% concentrated wastewater. While, no remarkable change was observed in water quality parameters for 75 and 100% textile effluent that showed toxicity stress on algal growth and activity.

Based on the literature available, it is proved that the algal species exhibit good potential for decolourization of textile dyes. However, the decolourization of dye is mainly due to biosorption mechanism by algal biomass, which could increase the amount of secondary sludge generated after treatment. Compared to bacteria and fungi, the efficacy of algal-mediated bio-decolourization system is inferior in terms of performance, treatment duration, and sustainability in real textile wastewater treatment.

6.6 Phyto-Decolourization of Dyes

The bio-treatment of textile wastewater using higher plant species has been successfully investigated and reported in numerous studies. The wetland construction is a well-known efficient technique, commonly used for wastewater treatment. Numerous higher plant species have been found to exhibit high potential to treat textile wastewater, suitably in wetlands. Watharkar and Jadhav (2014) investigated the decolourization of textile dye mixture using plantlets of *Petunia grandiflora* and *Gaillardia grandiflora*. The authors reported that the individual plant species could exhibit reduction in ADMI by 76% (*P. grandiflora*) and 62% (*G. grandiflora)* for simulated dyes mixture wastewater. While the consortium of these plants exhibited ADMI reduction of 94% for simulated wastewater, under same process condition. In this case, they suggested that the phyto-decolourization of dyes was mediated by enzymes like veratryl alcohol oxidase, laccase, tyrosinase, riboflavin reductase, and lignin peroxidase, which were synthesized in the root cell of the plants. Due to the synergistic effect of enzymes in consortium, ADMI reduction was effective by consortium as compared to the individual plant species treatment of wastewater. Rane et al. (2015) investigated the decolourization of sulphonated remazol red dye from wastewater using *Alternanthera philoxeroides*, a macrophyte. The authors reported that the macrophyte could completely remove initial dye concentration of 70 mg/L within 72 h, under hydroponic cultivation system. The researchers found the induction of oxido-reductive enzymes in stem and root tissues of the plant that were involved in degradation of dye molecules. However, there was a marginal decrement in chlorophyll and carotenoid contents in leaves of the plant after treatment. Hussain et al. (2018) investigated the integrative treatment of textile wastewater using endophyte-assisted horizontal flow constructed wetlands at pilot-scale. The authors reported that the system could reduce BOD and COD from 190 to 42 mg/L and 493 to 70 mg/L, respectively, after 48 h. The treated water was found to be non-toxic and supported the growth of plants in bacterial augmented system. The schematic diagram of constructed wetland system is shown in Fig. 6.3.

Chandanshive et al. (2018) investigated the phytoremediation of textile dyes from wastewater using independently cultivated plants like *Tagetes patula, Aster amellus, Portulaca grandiflora,* and *Gaillardia grandiflora*, grown in ridges of constructed wetland. The respective plants could reduce ADMI value by 59%, 50%, 46%, and 73%, respectively, within 30 d of the cultivation period, mediated by oxido-reductive enzymes such as lignin peroxidase, veratryl alcohol oxidase, laccase, azo reductase, and tyrosinase induced in the root tissues of the plants. The dyes were found to be accumulated in the root tissues of plants for subsequent degradation that confirms the removal of dyes from wastewater. Kooh et al. (2018) investigated the phyto-extraction potential of *Azolla pinnata* for removal of toxic methyl violet 2B dye from wastewater. The authors reported that the water fern exhibited highest dye removal efficiency of 93% for initial concentration of 10 mg/L, within 7 d of incubation. The dye exhibited toxicity effect on plant, which was revealed from lower chlorophyll content in plant leaves than the control. Sinha et al. (2019) studied the phyto-decolourization

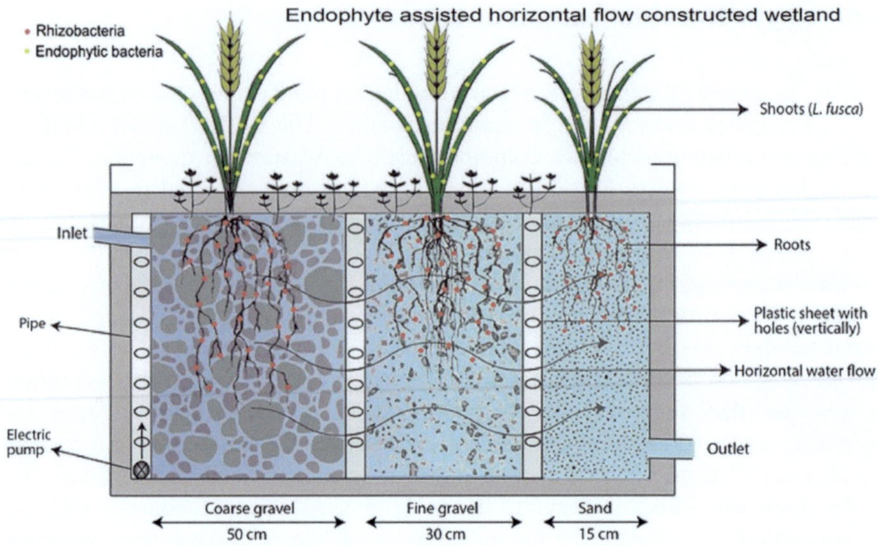

Fig. 6.3 Schematic diagram of endophyte-assisted horizontal flow constructed wetland system (modified from Hussain et al. 2018)

of reactive green dye using *Alternanthera philoxeroides* plantlets coupled with an augmentation of *Klebsiella* sp. VITAJ23 to the rhizosphere. The authors reported that the contaminated soil with initial 1000–3000 mg/kg dye contamination was decolourized by 79% through phyto and rhizoremediation treatments, mediated by enhanced oxidoreductase enzyme activity. The augmented bacteria supported the growth of plant with a significant increase in the chlorophyll content, root, and shoot length. Khandare and Govindwar (2015) reviewed the research articles related to phytoremediation of textile dyes and suggested that some of the macrophytes as *Rheum rhabarbarum*, ornamental plants, (like *Aster amellus, Portulaca grandiflora,* and others) many ferns, and aquatic plants have advocated the potential for dye degradation. The phyto-degradation of dyes is mediated by oxidoreductases enzymes such as lignin peroxidase, laccase, tyrosinase, and others.

Based on the literature available, it is revealed that the phyto-treatment of textile wastewater containing a wide variety of dyes is suitably easy and cost effective. However, the duration of phyto-treatment is comparatively more than action period of bacteria and fungi. In addition, phyto-treatment system requires a large landscape for the propagation of plants and a good skill for management of plant species to get effective functioning of the system.

6.7 Enzyme-Mediated Decolourization of Dyes

In the recent past, numerous oxidoreductase enzymes have been isolated from both intracellular and extracellular culture extracts of bacteria, fungi, algae, and higher plants, which have been exploited for degradation of textile dyes by many researchers. Such catalytic proteins/enzymes exhibit beneficial characteristics like higher stability and reusability, prevention of interactions with interfaces, rigidification via multipoint covalent attachment, prevention of subunit dissociation, resistance to inhibitions, and even an improved purity. Moreover, the selection and extraction of potential enzyme from biota are a foremost step of biological aspect to achieve desired and effective decolourization/degradation of textile wastewater. Wanyonyi et al. (2019) investigated the decolourization of reactive black 5 dye using crude enzyme protease extracted from bacteria *Bacillus Cereus* Strain KM201428. The authors reported results that displayed over 97% decolourization efficiency for initial dye concentration of 1×10^{-4}M. Considerably, the decolourization efficiency was highly dependent of contact time, pH, and dye concentration, which were optimized as 120 h and pH 9 at temperature 25 °C. The dye molecules were degraded by enzymes with Michaelis–Menten constant of 548.06 mg/L. Gioia et al. (2018) investigated the decolourization of azo dyes using laccase enzyme (either in insoluble or immobilized state), extracted from fungi *Trametes villosa*. They found that the insoluble enzyme removed 97% of the colour of acid red 88 and 92% of acid black 172 in 24 h at temperature 22 °C in the presence of redox mediators, vanillin (0.1 mM), and violuric acid (1.0 mM), respectively. In the immobilized state of enzyme, the decolourization efficiency of 78% was maintained after three cycles of use. The end metabolites produced after treatment were found to be non-phytotoxic in nature. Britos et al. (2018) investigated the decolourization of textile dyes using extracellular bacterial laccase, isolated from *Proteus vulgaris* ATCC 6896. The authors reported that the immobilized biocatalyst could decolorize bromothymol blue (59%), methyl violet 10B (52%), coomassie brilliant blue R (72%), trypan blue (85%), and remazol brilliant blue R (51%), within short reaction time period. Considerably, this result was obtained without the addition of mediators and reused for 160 cycles retaining the enzyme activity. The enzyme activity was enhanced up to five times with the addition of metal ions like Fe^{2+}, Cu^{2+}, Zn^{2+}, and others. Verma et al. (2019) investigated the decolourization of azo dyes methyl red, amaranth, and methyl orange using azoreductase enzyme, extracted from *Chromobacterium violaceum*. The authors reported that the three azo dyes were used as substrates by the enzyme with NADH as a co-substrate in the dye decolourization medium. Thus, the complete degradation of dyes was achieved at optimal temperature 30–37 °C and at pH 4 to 10. The degraded end metabolites had reduced toxicity on fibroblast cell lines as compared to raw dye. Qi et al. (2016) investigated the decolourization of methyl red dye using flavin-containing oxygen-insensitive azoreductase enzyme, extracted from *Rhodococcus opacus* 1CP. The authors reported that the degradation was initially catalysed by enzyme in the presence of NADH as an electron donor, at optimal temperature of 53 °C. The enzyme remained stable at temperature range from 10 to 40 °C, while the maximum unfolding temperature was verified at 55 °C, limits the performance of system. Chiong et al. (2016) investigated the potential of *Luffa acutangula* (luffa) peroxidase and soybean peroxidase to degrade azo dye methyl orange from wastewater. The enzymatic activities of luffa and soybean peroxidase were 0.355 and 0.373 U/mL, respectively. The authors reported that the soybean peroxidase (0.5 mL) exhibited maximum decolourization efficiency of 81.4% for 30 mg/L initial methyl orange within 1 h, at temperature 30 °C and pH 5.0, while luffa peroxidase (1.5 mL) showed maximum decolourization potential of 75.3% for initial 10 mg/L dye within 40 min at temperature 40 °C and pH 3.0. In recent decades, the application of crude plant peroxidases for dye degradation has received significant attention of research as a cost-effective approach for textile wastewater treatment. The available literatures related to enzymatic decolourization of textile dyes are illustrated in Table 6.6.

Based on the reviewed articles, it has been observed that the dye decolourization efficiency of enzymes varies with respect to their class and production origin in biotic species. Enzymes exhibit higher activity at optimal temperature (in range from 15 to 60 °C) and pH (in range from 3 to 10) of culture medium. The immobilized enzymes have shown better response than soluble enzyme for effective decolourization of dyes. However, the feasibility of enzyme for the treatment of real textile wastewater is still limited and requires further investigation. Furthermore, the major problems observed during bio-treatment of textile wastewater include the unavailability of efficient engineered tools and techniques in recombinant DNA technology and genetic engineering to produce valuable enzymes with high yield, robustness, and decolourization efficacy under industrial process conditions. Moreover, few studies have been reported (Beg et al. 2001; Uday et al. 2016) to improvise these features of enzymes by introducing genes and developing genetically engineered organism, but responses are not so absolute.

Table 6.6 Summary of the literature survey related to isolated enzymes-mediated degradation of dye

Name of enzyme Producing bacterial species	Dyes	Process condition (IDC, pH, Temp., OT)	Enzyme activity	Removal efficiency (%)	References
Azoreductase *Lysinibacillus* sp. KMK-A	Orange M2R	200 mg/L, 7, 37 °C, 48 h	1.65 nmoles/min-mg	98	Chaudhari et al. (2013)
Cot-laccase *Streptomyces ipomoeae* CECT 3341	Indigo carmine	50 μM, 8, 35 °C, 24 h	0.4 U/mL	98.4	Blánquez et al. (2018)
Laccase *Peroneutypa scoparia*	Acid red 97	Soluble state (100 mg/L, 8, 40 °C, 6 h)	60 U/mL	75	Pandi et al. (2019)

(continued)

Table 6.6 (continued)

Name of enzyme Producing bacterial species	Dyes	Process condition (IDC, pH, Temp., OT)	Enzyme activity	Removal efficiency (%)	References
Manganese peroxidase *Irpex lacteus* F17	Malachite green	200 mg/L, 3.5, 40 °C, 1 h	923.1 U/L	96	Yang et al. (2016)
Peroxidase *Coelastrella sp.*	Rhodamine B	Batch culture (100 mg/L, 8, 30 °C, 20 d)	2.1 µmol/min-mg	80	Baldev et al. (2013)
Laccase *Trametes versicolour*	Sulphur blue 15	Immobilized on chitosan beads (200 mg/L, 6.5, 30 °C, 12 h)	2.24 U/L	81.6	Nguyen et al. (2016)
Catalase *Dermatocarpon vellereceum*	Navy blue HE22	Continuous flow reactor (50 mg/L, 8, 40 °C, 28 h)	0.5×10^{-8} U/mL	98	Kulkarni et al. (2018)
Peroxidase *Brassica rapa*	Crystal ponceau 6R	Soluble state (30 mg/L, 4, 30 °C, 1 min)	0.14 U/mL	97	Almaguer et al. (2018)
Lignin peroxidase, Catalase, Veratryl alcohol oxidase *Salvinia molesta*	Rubine GFL	Soluble state (100 mg/L, 7, 30 °C, 72 h)	716%, 194% and 193%, respectively	97	Chandanshive et al. (2016)
Polyphenol oxidase *Cydonia oblonga*	Telon yellow ARB	Immobilized on sodium alginate (100 mg/L, 4, 30–35 °C, 1 h)	1.5 E U/mL	72.68	Arabaci and Usluoglu (2014a, b)

Note IDC is the initial concentration of dye; *Temp.* is the incubation temperature; *OT* is the operational time)

6.8 Factors Affecting the Decolourization of Dyes

The biological process involves the use of biotic organism for treatment of textile wastewater. The bio-decolourization processes are significantly influenced by the parameters like temperature, pH, availability of growth supporting nutrients in culture medium, and others. In this regard, the numerous studies have been carried out to investigate the effect of each factor to improve the decolourization efficiency of the bio-treatment process. The quantity and type of each nutrient should be optimized to achieve desired result. Mishra and Maiti (2018d) reviewed the articles related to factors affecting the bacterial-mediated bio-decolourization process and suggested the major influential factors (like dissolved oxygen, pH, electron donor co-substrate, incubation temperature, dye structure, and concentration) for decolourization of dyes. The dissolved oxygen has a strong redox potential than dye molecules. When the electron generated from oxidation of co-substrate preferentially binds with oxygen molecule than dye, the dye molecules are oxidized in the presence of oxygen that results in darker appearance of colour. Although oxygen is required for propagation of aerobic organism, its concentration should be carefully controlled in the medium. The redox mediators like glucose, lactate, yeast extract, and others apparently enhance the mineralization of dye molecules and promote the dye degradation potential of microorganism. Kagalkar et al. (2015) investigated the role of co-substrates in plant laccase enzyme-mediated degradation of brilliant blue R dye. The enzyme was isolated from *Blumea malcolmii* Hook. The authors reported that the supplementation of 2,20-azinobis-3-ethylbenzothiazoline-6-sulphonic acid in reaction medium enhanced the enzyme catalytic performance and resulted into complete decolourization of dyes. Qi et al. (2017) investigated the change in activity of azoreductase in the presence of electron donor. Considerably, the azoreductase is an oxygen-insensitive enzyme and incompetent biocatalyst at acidic pH medium. They found that the decolourization activity of azoreductase was enhanced in the presence of electron donor like 1-benzyl-1,4-dihydronicotinamide. Moreover, the simple structure and lower molecular weight dyes (like acid red 88, direct red 80, and others) could be degraded easily. While the degradation of high molecular weight dyes is a complex task, it contains electron withdrawing substitution groups such as $-SO_2NH_2$, and $-SO_3H$, when present at para position of aromatic ring (especially in azo and anthroquinone dyes) (Pandey et al. 2007). The higher concentration of such dyes (>50 mg/L) could be toxic to the cell in the culture medium that ultimately decreases the decolourization rate. In addition, the optimum temperature for satisfactory viability of biotic organism needs to be defined. Generally, most of biotic organisms could function in optimum temperature of 25–40 °C and at pH 5–9. The enzymes could sustain at incubation temperature of ~60 °C, but loses their activity in short duration. Moreover, the physiology and activity of biotic organism vary with respect to dye structure, concentration, and environment process of culture media. Therefore, the selection of potent species is an important step in bio-treatment of textile wastewater to achieve effective decolourization.

6.9 Toxicity Assessment of the Bio-Treated Textile Wastewater

The toxicity assessment of the raw wastewater and treated water must be performed to demonstrate their environmental impact, before final discharge into the surrounding environment. The toxic effect of textile wastewater on surrounding environment is represented in Fig. 6.4.

In this regard, numerous biotic organisms as pollution indicator have been investigated to classify the toxicity of wastewater. Vigneshpriya et al. (2019) investigated the bio-treatment of brilliant green dye using seaweed *Sargassum wightii* and confirmed the non-toxicity of treated water through bacterial and fungal activity tests. The authors reported that the treated dye did not exhibit microbial toxicity as no inhibition zones were observed on solid medium culture, which might be due to the complete absence or transformation of the dye molecules. The cytotoxicity study of *Allium cepa* bulbs grown in treated water exhibited comparatively effective root growth, increased mitotic index, and decrease in chromosomal damage than untreated water. Bayramoglu et al. (2019) carried out the toxicity assessment of cibacron blue 3GA dye and its biodegraded products, using test organisms *Daphnia magna* and microalgae *Chlorella vulgaris*. The authors reported that the initial dye molecules were more toxic than their degraded end products produced after treatment through laccase enzyme, which was verified by the mobility test of *Daphnia magna* and

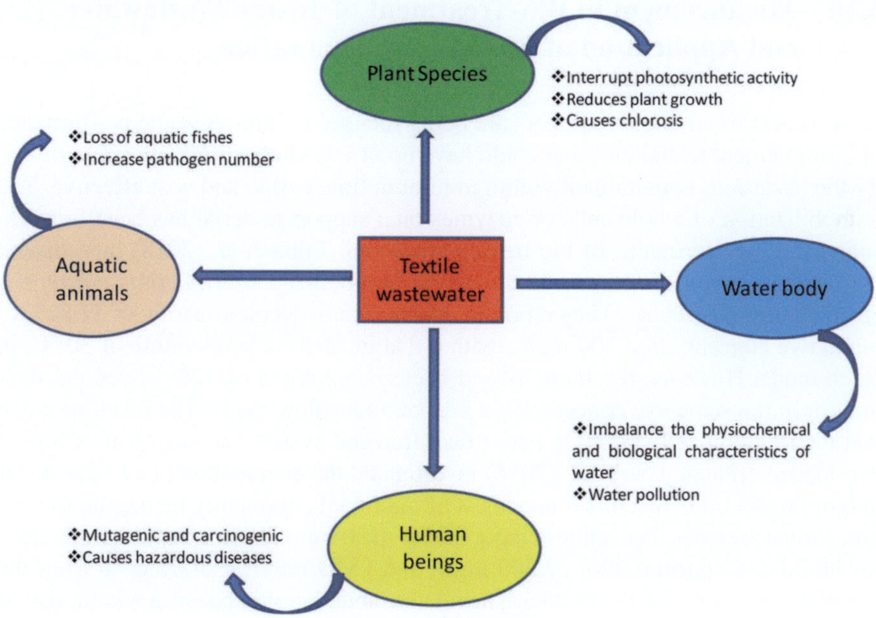

Fig. 6.4 Toxic effect of textile wastewater on the environment

chlorophyll content examination in microalgae. Khan and Malik (2018) assessed the toxicity of textile effluents using mung bean seed germination test. The authors reported that the exposure of textile effluent to mung bean seeds indicated a significant reduction in seed germination rate and radical–plumule growth. Haq et al. (2018) performed the phyto-toxicity, genotoxicity, and cytotoxicity assessment of azure-B dye and biodegraded by-products using *Vigna radiata, Allium cepa,* and kidney cell line (NRK-52E). The phyto-toxicity assay using *Vigna radiata* revealed that bacterial (*Serratia liquefaciens*)-treated dye end products were non-toxic and promoted the better germination of seeds than raw dye. The genotoxicity assay with *Allium cepa* for raw dye solution exhibited reduced mitotic index and various chromosomal abnormalities in the root tip cells, while bacterial-treated water induced relatively lesser genotoxic effects. Considerable cell survivability was also notified in kidney cell line (NRK-52E) when exposed to the same treated water. Ventura-Camargo et al. (2016) investigated the cytotoxic, genotoxic, and mutagenic effects black dye commercial product in *Allium cepa* cells and reported that biodegraded dye products were more genotoxic than the raw dye itself. Gita et al. (2019) reported the toxicity effect of dyes like optilan yellow, drimarene blue, and lanasyn brown on photosynthetic pigments, elemental composition, protein content, and growth of *Chlorella vulgaris*. They reported the negative effect of dye compounds on microalgae growth and biomass accumulation.

6.10 Advancement in Bio-Treatment of Textile Wastewater and Application of Bio-Energy Generation

In the recent trend, the researchers are being focused to improvise the performance of bio-treatment techniques that could have no or less sludge production and detoxify the hazardous contaminant within minimum time period and cost effective. The immobilization of whole cells or enzymes on a support material has been found to enhance the performance of bio-treatment system. Talha et al. (2018) investigated the biodegradation of Congo red dye using immobilized *Brevibacillus parabrevis* on coconut shell biochar. They reported the maximum decolourization of 77.82% of initial dye concentration 100 mg/L, with 6 d at incubation temperature of 30 °C, in batch mode. However, the immobilized bacteria exhibited 88.92% of decolourization for initial same dye concentration in continuous flow mode. The result revealed that the immobilized system is better than free cell system for treatment of higher dye-loaded effluent. Cao et al. (2018) investigated the degradation of azo dye X-3B in biofilm electrode reactors that consist of microbial community belonging to phylum *Proteobacteria*. The authors reported the maximum decolourization of 66.26% for initial dye concentration of 200 mg/L and COD removal of 75.64% from the saline wastewater (NaCl 0.33 g/L). The relative abundance of bacteria was increased from 34.93 to 35.34% after treatment of dye molecules, which was affected by process parameters like the presence of electrons, temperature, and an anaerobic environment. However, a detailed study is still required to achieve better dye removal efficiency of this system. Thanavel et al. (2018) investigated the decolourization of textile dyes using combination of biological and advanced oxidation process that involves *Aeromonas hydrophila* SK16 and AOPs-H_2O_2. The authors reported that the combo treatment process exhibited 100% decolourization reactive red 180, reactive black 5, and remazol red dyes, which was comparatively better than individual treatment system. The combo system was found to be reliable, economically feasible, solar energy-dependent, efficient, and requires less maintenance for treatment of textile wastewater. Lade et al. (2012) investigated the biodegradation of Rubine GFL and treatment of real textile wastewater using bacterial–fungal consortia of *Pseudomonas* sp. SUK1 (bacteria) and *Aspergillus ochraceus* NCIM-1146 (fungi). The authors reported that enhancing the decolourization of synthetic dye by 95% in 30 h and ADMI removal of real wastewater by 98% in 35 h, without production of toxic aromatic amines (as secondary product) under microaerobic conditions, was comparatively higher than individual performance of same microorganism in a separate medium. Induction of veratryl alcohol, laccase, peroxidase, NADH-DCIP reductase, and azo reductase in the consortia media indicates synergetic reactions of bacterial and fungal cultures for higher decolourization response. Kurade et al. (2015) investigated the biodegradation of reactive red 198 dye using consortium of *Brevibacillus laterosporus* and *Galactomyces geotrichum*, a bacterial–yeast consortium. The authors reported that the consortia could exhibit dye decolourization of 92% for dye concentration of 50 mg/L within 18 h (at pH-7, 40 °C, in static condition) as compared to 42 and 58% decolourization using *Galactomyces geotrichum* and *Brevibacillus laterosporus* alone, respectively, under same experimental conditions. In another study, Waghmode et al. (2019) investigated the sequential photocatalytic and biological treatment using same microbial consortium for degradation of methyl red dye. The authors reported that the combine treatment system completely decolourized initial dye concentration of 500 mg/L with 4 h, which was better that individual photocatalytic treatment of 70% dye decolourization. The authors suggested that the single microorganism promotes partial mineralization of dye molecules, while the consortia advantageously induced complete degradation of dye molecules.

Alternatively, in order to make the process cost effective, the integrative technologies are being investigated by the researchers. The treatment processes are integrated to produce bio-energy, which could be a good initiative and sustainable method to convert waste into money. The schematic diagram of commonly used microbial fuel cell set-up is shown in Fig. 6.5. Behl et al. (2019) investigated the dye decolourization potential of microalgae *Chlorella pyrenoidosa* in biochar-supplemented medium. The authors reported that the algae could decolourize 80.12% of initial direct red 31 dye concentration 10 mg/L after 180 min of contact time, at temperature 30 °C. They suggested that the biochar promoted the algal growth and enhanced the decolourization efficiency of algae as well as biofuel production of 5.680 g/m^{3}d. To an estimate, the total percentages of hexadecanoic acid to octadecanoic acid were found to be more than 50%, and saturated fatty acid content was higher than unsaturated fatty acid in

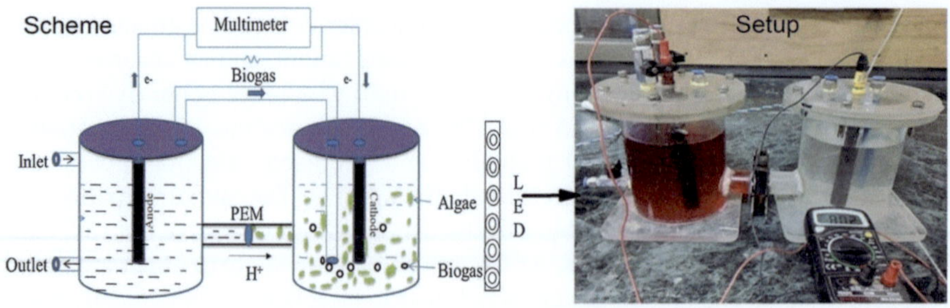

Fig. 6.5 Schematic diagram and set-up of microbial fuel cell (MFC) system for dye decolourization and generation of bioelectricity simultaneously

produced biofuel. Das and Mishra (2019) explored the simultaneous degradation of textile and generation of bioelectricity from integrated microbial fuel cell (MFC) using bacterial consortium of *Bacillus pumilus* HKG212, *Zobellella taiwanensis* AT 1-3, and *Enterococcus durans* GM13. They reported that the consortium could exhibit complete degradation of remazol navy blue dye with 12 h of MFC operation. Compared to traditional static decolourization of dyes, a rapid decolourization of dye was found in anodic chamber of MFC with a maximum power density of 52.5 mW/m^2 for initial dye concentration of 100 mg/L. Considerably, the external resistance in MFC was found as a key parameter that affects the decolourization dyes. The optimal external resistance was found as 1000 Ω, giving maximum decolourization efficiency of MFC system.

Based on the literature available, the bacterial-mediated bio-decolourization of textile wastewater is comparatively more effective technique than other bioremediation agents. The bacteria belonging to *Pseudomonas* sp. have high potential to degrade different classes of dye. However, the studies are required to investigate suitable culture medium for simultaneous degradation of mixed dyes and heavy metals from real textile wastewater. The recent trend of research is targeted to utilize the waste material for betterment of human life with less ecological risk. In this regard, the investigation of efficient and cost-effective integrative wastewater treatment technologies that could produce bio-energy, which will help to meet the growing energy demand and to increase the sustainable industrial growth.

6.11 Concluding Remarks and Future Trends

The bio-based techniques that involve bacteria, fungi, algae, higher plants, and their isolated enzymes are commonly used for bio-treatment of textile wastewater. Literatures based on dye decolourization reveal that the decolourization mediated by enzymes or whole cell facultative bacteria and fungi have high potential to decolourize wide variety of textile dyes. However, the consortia of bacteria–fungi could enhance the performance of bio-treatment system by complete degradation and detoxification of contaminants from wastewater. Moreover, the satisfactory decolourization of dye could be achieved under optimized process and that too potentially higher when immobilized cell or enzymes are used than soluble state. Although numerous laboratory-scale studies related to degradation of dye solutions using pure enzymes either soluble or immobilized state have been reported in literature that indicates the lesser or no production of sludge than using whole cell, the feasibility of enzyme-based technique is still required to investigate and generate relative performance data for treatment of real textile effluent. The use of integrative techniques for both treatment and bio-energy generation could beneficially help to meet the growing demand of energy, promote the sustainable and profitable growth of industrialization, and reduce the overall operational cost. The existing bio-techniques could be improvised with the latest advances of genomics, transcriptomics, and proteomics to revolutionize the various aspects of biological sciences, which could explore the possibilities to enhance the performance of wastewater bio-treatment system.

References

Abraham CI (2013) Biodegradation of dyes. Ph.D. Thesis, School of Biosciences Mahatma Gandhi University, Kerela, India

Adnan LA, Sathishkumar P, Yusoff ARM, Hadibarata T, Ameen F (2017) Rapid bioremediation of Alizarin Red S and Quinizarine Green SS dyes using Trichoderma lixii F21 mediated by biosorption and enzymatic processes. Bioprocess Biosyst Eng 40:85–97

Almaguer MA, Carpio RR, Alves TLM, Bassin JP (2018) Experimental study and kinetic modelling of the enzymatic degradation of the azo dye Crystal Ponceau 6R by turnip (*Brassica rapa*) peroxidase. J Environ Chem Eng 6:610–615

Arabaci G, Usluoglu A (2014a) The enzymatic decolourization of textile dyes by the immobilized polyphenol oxidase from quince leaves. Sci World J. https://doi.org/10.1155/2014/685975

Arabaci G, Usluoglu A (2014b) The enzymatic decolourization of textile dyes by the immobilized polyphenol oxidase from quince leaves. Sci World J. https://doi.org/10.1155/2014/685975

Aragaw TA, Asmare AM (2018) Phycoremediation of textile wastewater using indigenous microalgae. Water PractTech 13(2). https://doi.org/10.2166/wpt.2018.037

Arteaga LC, Zavaleta MP, Eustaquio WM, Bobadilla JM (2018) Removal of aniline blue dye using live microalgae *Chlorella vulgaris*. J Energ Environ Sci 2(1)

Aruna B, Silviya LR, Kumar ES, Rani PR, Prasad DVR, VijayaLakshmi D (2015) Decolorization of acid blue 25 dye by individual and mixed bacterial consortium isolated from textile effluents. Int J Curr Microbiol App Sci 4(5):1015–1024

Ayed L, Bakir K, Achour S, Cheref A, Bakhrouf A (2017) Exploring bioaugmentation strategies for azo dye CI Reactive Violet 5 decolourization using bacterial mixture: dye response surface methodology. Water Environ J 3:80–89

Baldev E, Mubarakali D, Ilavarasi A, Pandiaraj D, Ishack KASS, Thajuddin N (2013) Degradation of synthetic dye, Rhodamine B to environmentally non-toxic products using microalgae. Colloids Surf B Biointerfaces 105:207–214

Bankole PO, Adekunle AA, Obidi OF, Chandanshive VV, Govindwar SP (2018a) Biodegradation and detoxification of Scarlet RR dye by a newly isolated filamentous fungus, Peyronellaea prosopidis. Sustain Environ Res 28:214–222

Bankole PO, Adekunleb AA, Govindwar SP (2018b) Enhanced decolorization and biodegradation of acid red 88 dye by newly isolated fungus, Achaetomium strumarium. J Environ Chem Eng 6:1589–1600

Bayramoglu G, Salih B, Akbulut A, Arica MY (2019) Biodegradation of Cibacron Blue 3GA by insolubilized laccase and identification of enzymatic byproduct using MALDI-ToF-MS: toxicity assessment studies by *Daphnia magna* and *Chlorella vulgaris*. Ecotoxic Environ Safe 170:453–460

Beg QK, Kapoor M, Mahajan L, Hoondal GS (2001) Microbial xylanases and their industrial applications: a review. Appl Microbiol Biotechnol 56:326–338

Behl K, Sinha S, Sharma M, Singh R, Joshi M, Bhatnagar A, Nigam S (2019) One-time cultivation of *Chlorella pyrenoidosa* in aqueous dye solution supplemented with biochar for microalgal growth, dye decolorization and lipid production. Chem Eng J 364:552–561

Bilal M, Rasheed T, Zhao Y, Iqbal HMN (2019) Agarose-chitosan hydrogel-immobilized horseradish peroxidase with sustainable bio-catalytic and dye degradation properties. Int J Biol Macromol 124:742–749

Blánquez A, Rodríguez J, Brissos V, Mendes S, Martins LO, Ball AS, Arias ME, Hernández M (2018) Decolorization and detoxification of textile dyes using a versatile Streptomyces laccase-natural mediator system. Saudi J Bio Sci. https://doi.org/10.1016/j.sjbs.2018.05.020

Britos CN, Gianolini JE, Portillo H, Trelles JA (2018) Biodegradation of industrial dyes by a solvent, metal and surfactant-stable extracellular bacterial laccase. Biocat Agri Biotech 14:221–227

Can OT, Bayramoglu M, Kobya M (2003) Decolorization of reactive dye solutions by electrocoagulation using aluminum electrodes. Ind Eng Chem Res 42:3391–3396

Cao X, Wang H, Zhang S, Nishimura O, Li X (2018) Azo dye degradation pathway and bacterial community structure in biofilm electrode reactors. Chemosphere 208:219–225

Chandanshive VV, Kadam SK, Khandare RV, Kurade MB, Jeon B, Jadhav JP, Govindwar SP (2018) In situ phytoremediation of dyes from textile wastewater using garden ornamental plants, effect on soil quality and plant growth. Chemosphere 210:968–976

Chandanshive VV, Rane NR, Gholave AR, Patil SM, Jeon BH, Govindwar SP (2016) Efficient decolourization and detoxification of textile industry effluent by *Salvinia molesta* in lagoon treatment. Environ Res 150:88–96

Chang JS, Chou C, Lin YC, Lin PJ, Ho JY, Hu TL (2001) Kinetic characteristics of bacterial azo-dye decolorization by *Pseudomonas luteola*. Wat Res 35(12):2841–2850

Chaudhari AU, Tapase SR, Markad VL, Kodam KM (2013) Simultaneous decolorization of reactive Orange M2R dye and reduction of chromate by *Lysinibacillus* sp KMK-A. J Hazard Mat 262:580–588

Chen SH, Cheow YL, Ng SL, Ting ASY (2019) Biodegradation of triphenylmethane dyes by non-white rot fungus *Penicillium simplicissimum*: enzymatic and toxicity studies. Int J Environ Res 13:273–282

Chen SH, Ting ASY (2015) Biodecolorization and biodegradation potential of recalcitrant triphenylmethane dyes by *Coriolopsis* sp. isolated from compost. J Environ Manage 150:274–280

Cheng S, Li N, Jiang L, Li Y, Xu B, Zhou W (2019) Biodegradation of metal complex Naphthol Green B and formation of iron–sulfur nanoparticles by marine bacterium *Pseudoalteromonas* sp CF10-13. Bioresour Tech 273:49–55

Chiong T, La SY, Lek ZH, Koh BY, Danquah MK (2016) Enzymatic treatment of methyl orange dye in synthetic wastewater by plant-based peroxidase enzymes. J EnvironChemical Eng 4:2500–2509

Dalakoti H, Mishra S, Chaudhary M, Singal SK (2018) Appraisal of water quality in the Lakes of Nainital District through numerical indices and multivariate statistics India. Int J River Basin Manag 16(2):219–229

Das A, Mishra S (2019) Complete biodegradation of azo dye in an integrated microbial fuel cell-aerobic system using novel bacterial consortium. Int J Environ Sci Technol 16:1069–1078

El-Rahima WMA, Moawad H, Azeiz AA, Sadowsky MJ (2017) Optimization of conditions for decolorization of azo-based textile dyes by multiple fungal species. J Biotech 260:11–17

Gioia L, Ovsejevi K, Manta C, Míguez D, Menéndez P (2018) Biodegradation of acid dyes by an immobilized laccase: an ecotoxicological approach. Environ Sci Water Res Technol 4:2125

Gita S, Shukla SP, Saharan N, Prakash C, Deshmukhe G (2019) Toxic effects of selected textile dyes on elemental composition, photosynthetic pigments, protein content and growth of a freshwater Chlorophycean Alga *Chlorella vulgaris*. Bull Environ Cont Toxic. https://doi.org/10.1007/s00128-019-02599-w

Goh SM, Chan MY, Ong LGA (2017) Degradation potential of basidiomycetes Trametes ljubarskyi on Reactive Violet 5 (RV 5) using urea as optimum nitrogen source. Biotech Biotechnol Equip 31(4):743–748

Gurav AA, Ghosh JS, Kulkarni GS (2011) Decolorization of anthroquinone based dye vat red 10 by Pseudomonas desmolyticum NCIM 2112 and Galactomyces geotrichum MTCC 1360. Int J Biotechnol Mol Biol Res 2(6):93–97

Haq I, Raj A (2018) Biodegradation of Azure-B dye by Serratia liquefaciens and its validation by phytotoxicity, genotoxicity and cytotoxicity studies. Chemosphere 196:58–68

He X, Song C, Li Y, Wang N, Xu L, Han X, Wei D (2018) Efficient degradation of Azo dyes by a newly isolated fungus Trichoderma tomentosum under non-sterile conditions. Ecotoxic Environ Safe 150:232–239

Holkar CR, Pandit AB, Pinjari DV (2014) Kinetics of biological decolorisation of anthraquinone based reactive blue 19 using an isolated strain of *Enterobacter* sp. F NCIM 5545. Bioresour Technol 173:342–351

Huang G, Wang W, Liu G (2015) Simultaneous chromate reduction and azo dye decolourization by *Lactobacillus paracase* CL1107 isolated from deep sea sediment. J Environ Manag 157:297–302

Hussain Z, Arslan M, Malik MH, Mohsine M, Iqbald S, Afzal M (2018) Integrated perspectives on the use of bacterial endophytes in horizontal flow constructed wetlands for the treatment of liquid textile effluent: phytoremediation advances in the field. J Environ Manag 224:387–395

Imran M, Arshad M, Negm F, Khalid A, Shaharoon B, Hussain S, Nadeem SM, Crowley DE (2016) Yeast extract promotes decolorization of azo dyes by stimulating azoreductase activity in *Shewanella* sp. strain IFN4. Ecotoxicol Environ Saf 124:42–49

Ji Q, Liu G, Zhou J, Wang J, Jin R, Lv H (2012) Removal of water-insoluble Sudan dyes by *Shewanella oneidensis* MR-1. Bioresour Technol 114:144–148

Jović M, Stanković D, Manojlović D, Anđelković I, Milić A, Dojčinović B, Roglić G (2013) Study of the electrochemical oxidation of reactive textile dyes using platinum electrode. Int J Electrochem Sci 8:168–183

Kagalkar AN, Khandare RV, Govindwar SP (2015) Textile dye degradation potential of plant laccase significantly enhances upon augmentation with redox mediators. RSC Adv 5:80505

Karim ME, Dhar K, Hossain MT (2018) Decolorization of textile reactive dyes by bacterial monoculture and consortium screened from textile dyeing effluent. J Gene Eng Biotech 16:375–380

Kaur B, Kumar B, Garg N, Kaur N (2015) Statistical optimization of conditions for decolorization of synthetic dyes by *Cordyceps militaris* MTCC 3936 Using RSM. BioMed Res Int. https://doi.org/10.1155/2015/536745

Khan S, Malik A (2018) Toxicity evaluation of textile effluents and role of native soil bacterium in biodegradation of a textile dye. Environ Sci Poll Res 25:4446–4458

Khan SS, Arunarani A, Chandran P (2015) Biodegradation of basic violet 3 and acid blue 93 by *Pseudomonas putida*. CLEAN Soil Air Water 43(1):67–72

Khandare RV, Govindwar SP (2015) Phytoremediation of textile dyes and effluents: current scenario and future prospects. Biotech Adv 33:1697–1714

Kooh MRR, Lim LBL, Lim L, Malik OA (2018) Phytoextraction potential of water fern (*Azollapinnata*) in the removal of a hazardous dye, methyl violet 2B: Artificial neural network modelling. Int J Phytoremed 20(5):424–431

Kulkarni AN, Watharkar AD, Rane NR, Jeon BH, Govindwar SP (2018) Decolorization and detoxification of dye mixture and textile effluent by lichen Dermatocarpon vellereceum in fixed bed upflow bioreactor with subsequent oxidative stress study. Ecotoxicol Environ Saf 148:17–25

Kuppusamy S, Sethurajan M, Kadarkarai M, Aruliah R (2017) Biodecolourization of textile dyes by novel, indigenous Pseudomonas stutzeri MN1 and Acinetobacter baumannii MN3. J Environ Chem Eng 5(1):716–724

Kurade MB, Waghmode T, Jadhav MU, Jeon B, Govindwar SP (2015) Bacterial-yeast consortium as an effective biocatalyst for biodegradation of sulphonated azo dye Reactive red 198. RSC Adv. https://doi.org/10.1039/C4RA15834B

Lade HS, Waghmode TR, Kadam AA, Govindwar SP (2012) Enhanced biodegradation and detoxification of disperse azo dye Rubine GFL and textile industry effluent by defined fungal-bacterial consortium. Int Biodeterior Biodegrad 72:94–107

Lalnunhlimi S, Krishnaswamy V (2016) Decolorization of azo dyes (Direct Blue 151 and Direct Red 31) by moderately alkaliphilic bacterial consortium. Brazil J Microbiol 47:39–46

Lebron YAR, Moreira VR, Santos LVS, Jacob RS (2018) Remediation of methylene blue from aqueous solution by *Chlorella pyrenoidosa* and Spirulina maxima biosorption: Equilibrium, kinetics, thermodynamics and optimization studies. J Environ Chem Eng 6:6680–6690

Liu W, Liu C, Liu L, You Y, Jiang J, Zhou Z, Dong Z (2017) Simultaneous decolorization of sulfonated azo dyes and reduction of hexavalent chromium under high salt condition by a newly isolated salt tolerant strain *Bacillus circulans* BWL1061. Ecotoxic Environ Safe 141:9–16

Ma L, Zhuo R, Liu H, Yu D, Jiang M, Zhang X, Yang Y (2014) Efficient decolorization and detoxification of the sulfonated azo dye Reactive Orange 16 and simulated textile wastewater containing Reactive Orange 16 by the white-rot fungus *Ganoderma* sp. En3 isolated from the forest of Tzu-chin Mountain in China. Biochem Eng J 82:1–9

Mahajan P, Kaushal J, Upmanyu A, Bhatti J (2019) Assessment of Phytoremediation potential of *Chara vulgaris* to treat toxic pollutants of textile effluent. J Toxicol. https://doi.org/10.1155/2019/8351272

Mahmood S, Khalid A, Arshad M, Mahmood T, Crowley DE (2015) Detoxification of azo dyes by bacterial oxidoreductase enzymes. Crit Rev Biotechnol. https://doi.org/10.3109/07388551.2015.1004518

Maiti A, Mishra S, Chaudhary M (2019) Nanoscale materials for arsenic removal from water. In: Nanoscale materials in water purification. 707–733. https://doi.org/10.1016/B978-0-12-813926-4.00032-X

Manai I, Miladi B, Mselmi AE, Smaali I, Hassen AB, Hamdi M, Bouallagui H (2016) Industrial textile effluent decolourization in stirred and static batch cultures of a new fungal strain *Chaetomium globosum* IMA1 KJ472923. J Environ Manag 170:8–14

Maqbool Z, Hussain S, Ahmad T, Nadeem H, Imran M, Khalid A, Abid M, Martin-Laurent F (2016) Use of RSM modeling for optimizing decolorization of simulated textile wastewater by *Pseudomonas aeruginosa* strain ZM130 capable of simultaneous removal of reactive dyes and hexavalent chromium. Environ Sci Pollut Res 23:11224–11239

Meerbergen K, Willems KA, Dewil R, Impe JV, Appels L, Lievens B (2018) Isolation and screening of bacterial isolates from wastewater treatment plants to decolorize azo dyes. J Biosci Bioeng 125:448–456

Mishra S, Maiti A (2018a) Process optimization for effective bio-discolouration of Reactive Orange 16 using Chemometric methods. Part A, Int J Environ Sci Health. https://doi.org/10.1080/10934529.2018.1541383

Mishra S, Maiti A (2018b) Optimization of process parameters to enhance the bio-decolourization of Reactive Red 21 by *Pseudomonas aeruginosa* 23N1. Int J Environ Sci Tech. https://doi.org/10.1007/s13762-018-2023-1

Mishra S, Maiti A (2018c) Process optimization for effective bio-discolouration of methyl orange by *Pseudomonas aeruginosa* 23N1 using chemometric methodology. The Canad J Chem Eng. https://doi.org/10.1002/cjce.23410

Mishra S, Maiti A (2018d) The efficacy of bacterial species to decolourise reactive azo, anthroquinone and triphenylmethane dyes from wastewater: a review. Environ Sci Poll Res 25:8286–8314

Mishra S, Maiti A (2019a) Applicability of enzymes produced from different biotic species for biodegradation of textile dyes. Clean Technol Environ Policy 21(4):763–81

Mishra S, Maiti A (2019b) Effectual bio-decolourization of anthraquinone dye reactive blue-19 containing wastewater by Bacillus cohnii LAP217: process optimization. Bioremed J. https://www.doi.org/10.1080/10889868.2019.1671793

Moghazy RM (2019) Activated biomass of the green microalga Chlamydomonas variabilis as an efficient biosorbent to remove methylene blue dye from aqueous solutions. Water SA. https://doi.org/10.4314/wsa.v45i1.03

Nabil GM, El-Mallah NM, Mahmoud ME (2014) Enhanced decolorization of reactive black 5 dye by active carbon sorbent-immobilized-cationic surfactant (AC-CS). J Indust Eng Chem 20:994–1002

Narayanan MP, Murugan S, Eva AS, Devina SU, Kalidass S (2015) Application of immobilized laccase from *Bacillus subtilis* MTCC 2414 on decolourization of synthetic dyes. Res J Microbiol 10:421–432

Naseer A, Nosheen S, Kiran S, Kamal S, Javaid MA, Mustafa M, Tahir A (2016) Degradation and detoxification of Navy Blue CBF dye by native bacterial communities: an environmental bioremedial approach. Desal. Water Treat. 57:24070–24082

Neetha JN, Sandesh K, Kumar KG, Chidananda B, Ujwal P (2019) Optimization of Direct Blue-14 dye degradation by Bacillus fermus (Kx898362) an alkaliphilic plant endophyte and assessment of degraded metabolite toxicity. J Hazard Mat 364:742–751

Nguyen TA, Fu CC, Juang RS (2016) Effective removal of sulfur dyes from water by biosorption and subsequent immobilized laccase degradation on crosslinked chitosan beads. Chem Eng J 304:313–324

Ning C, Qingyun L, Aixing T, Wei S, Youyan L (2018) Decolorization of a variety of dyes by *Aspergillus flavus* A5p1. Bioproc Biosyst Eng 41:511–518

Ogugbue CJ, Sawidis T (2011) Bioremediation and detoxification of synthetic wastewater containing triarylmethane dyes by *Aeromonas hydrophila* isolated from industrial effluent. Biotechnol Res Int 967925:11

Padmanaban VC, Geed SRR, Achary A, Singh RS (2016) Kinetic studies on degradation of reactive red 120 dye in immobilized packed bed reactor by *Bacillus cohnii* RAPT1. Bioresour Technol 213:39–43

Pandey A, Singh P, Iyengar L (2007) Bacterial decolorization and degradation of azo dyes. Int Biodeterior Biodegrad 59:73–84

Pandi A, Kuppuswami GM, Ramudu KN, Palanivel S (2019) A sustainable approach for degradation of leather dyes by a new fungal laccase. J Clean Prod 211:590–597

Parmar ND, Shukla SR (2017) Biodegradation of anthraquinone based dye using an isolated strain *Staphylococcus Hominis* Subsp. hominis DSM 20328. Environ Prog Sustain Energy 37:1. https://doi.org/10.1002/ep.12655

Qi J, Paul CE, Hollmann F, Tischler D (2017) Changing the electron donor improves azoreductase dye degrading activity at neutral pH. Enzym Microb Tech 100:17–19

Qi J, Schlömann M, Tischler D (2016) Biochemical characterization of an azoreductase from *Rhodococcusopacus* 1CP possessing methyl red degradation ability. J Mol Cat B: Enzym 130:9–17

Rane NR, Chandanshive VV, Watharkar AD, Khandare RV, Patil TS, Pawar PK, Govindwar SP (2015) Phytoremediation of sulfonated Remazol Red dye and textile effluents by *Alternanthera philoxeroides*: an anatomical, enzymatic and pilot scale study. Water Res 83:271–281

Rezaee A, Ghaneian MT, Khavanin A, Hashemian SJ, Maussavi G, Ghanizadeh G, Hajizadeh E (2008) Photochemical oxidation of reactive blue 19 dye (RB19) in textile wastewater by $UV/K_2S_2O_8$ process. Iran J Environ Sci Health Eng 5(2):95–100

Rosa ALD, Carissimi E, Dotto GL, Sander H, Feris LA (2018) Biosorption of rhodamine B dye from dyeing stones effluents using the green microalgae *Chlorella pyrenoidosa*. J Clean Produc 198:1302–1310

Roy U, Das P, Bhowal A, Datta S (2018) Biodegradation of Azo Dye Using the Isolated Novel Bacterial Species: Acinetobacter sp.S.K. Ghosh (ed.), Utilization and Management of Bioresources. https://doi.org/10.1007/978-981-10-5349-8_18

Saratale RG, Saratale GD, Chang JS, Govindwar SP (2011) Bacterial decolorization and degradation of azo dyes: a review. J Taiwan Inst Chem Eng 42:138–157

Sen SK, Raut S, Bandyopadhyay P, Raut S (2016) Fungal decolouration and degradation of azo dyes: a review. Fungal biol rev 30:112–133

Senthilkumar S, Perumalsamy M, Prabhu HJ (2014) Decolourization potential of white-rot fungus *Phanerochaete chrysosporium* on synthetic dye bath effluent containing Amido black 10B. J Saudi Chem Soc 18:845–853

Shang N, Ding M, Dai M, Si H, Li S, Zhao G (2019) Biodegradation of malachite green by an endophytic bacterium Klebsiella aerogenes S27 involving a novel oxidoreductase. Appl Microbiol Biotech 103:2141–2153

Sheekh MME, Abou-El-Souod GW, Asrag HAE (2018) Biodegradation of Some Dyes by The Green Alga *Chlorella vulgaris* and the Cyanobacterium *Aphanocapsa elachista*. Egypt J Bot 58:311–320

Sheth NT, Dave SR (2009) Optimisation for enhanced decolourization and degradation of reactive red BS C.I. 111 by *Pseudomonas aeruginosa* NGKCTS. Biodegradation 20(6):827–836

Singh AK, Majumder CB, Mishra S (2015a) Removal of Arsenic from contaminated wastewater using *Eichhornia crassipes*. Integr Res Adv 2(1):1–4

Singh RL, Singh PK, Singh RP (2015b) Enzymatic decolorization and degradation of azo dyes—A review. Int Biodeterior Biodegrad 104:21–31

Singh V, Sharma MP, Sharma S, Mishra S (2017) Bio-assessment of River Ujh using benthic macroinvertebrates as bioindicators. Int J River Basin Manag, India. https://doi.org/10.1080/15715124.2017.1394318

Sinha A, Lulu S, Vino S, Banerjee S, Acharjee S, Osborne WJ (2018) Degradation of reactive green dye and textile effluent by *Candida* sp. VITJASS isolated from wetland paddy rhizosphere soil. J Environ Chem Eng 6:5150–5159

Sinha A, Lulu S, Vino S, Osborne WJ (2019) Reactive green dye remediation by *Alternanthera philoxeroides* in association with plant growth promoting *Klebsiella* sp. VITAJ23: A pot culture study. Microbiol Res 220:42–52

Sinha A, Osborne WJ (2016) Biodegradation of reactive green dye (RGD) by indigenous fungal strain VITAF-1. Int Biodeterior Biodegrad 114:176–183

Song L, Shao Y, Ning S, Tan L (2017) Performance of a newly isolated salt-tolerant yeast strain *Pichia occidentalis* G1 for degrading and detoxifying azo dyes. Bioresour Technol 233:21–29

Sreedharan V, Rao KVB (2019) Biodegradation of textile azo dyes. Gothandam et al. (eds.), Nanoscience and biotechnology for environmental applications, environmental chemistry for a sustainable world 22. https://doi.org/10.1007/978-3-319-97922-9_5

Sudha M, Saranya A, Selvakumar G, Sivakumar N (2014) Microbial degradation of azo dyes: a review. Int J Curr Microbiol Appl Sci 3(2):670–690

Sun H, Yang H, Huang W, Zhang S (2015) Immobilization of laccase in a sponge-like hydrogel for enhanced durability in enzymatic degradation of dye pollutants. J Colloid Interface Sci 450:353–360

Talha MA, Goswami M, Giric BS, Sharma A, Rai BN, Singh RS (2018) Bioremediation of Congo red dye in immobilized batch and continuous packed bed bioreactor by *Brevibacillus parabrevis* using coconut shell biochar. Bioresour Technol 252:37–43

Tan L, He M, Song L, Fu X, Shi S (2016) Aerobic decolorization, degradation and detoxification of azo dyes by a newly isolated salt-tolerant yeast *Scheffersomyces spartinae* TLHS-SF1. Bioresour Technol 203:287–294

Telke AA, Kadam AA, Govindwar SP (2015) Bacterial Enzymes and their role in decolorization of azo dyes. Environ Sci Eng. https://doi.org/10.1007/978-3-319-10942-8_7

Thanavel M, Kadam SK, Biradar SP, Govindwar SP, Jeon BH, Sadasivam S (2018) Combined biological and advanced oxidation process for decolorization of textile dyes. SN Appl Sci 1:97. https://www.doi.org/10.1007/s42452-018-0111-y

The Federation of Indian Chambers of Commerce and Industry (FICCI), Knowledge Paper on Global Shifts in Textile Industry & India's Position, TAG 2016, September 2, 2016, Mumbai, India. https://www.usitc.gov/publications/332/PUB3401.pdf. Accessed on 25 Aug 2019

Uday USP, Bandyopadhyay TK, Bhunia B (2016) Bioremediation and detoxification technology for treatment of dye(s) from textile effluent. In: Akcakoca Kumbasar E (ed) Textile wastewater treatment. InTech. https://doi.org/10.5772/62309

Ventura-Camargo BC, Angelis DF, Marin-Morales MA (2016) Assessment of the cytotoxic, genotoxic and mutagenic effects of the commercial black dye in *Allium cepa* cells before and after bacterial biodegradation treatment. Chemosphere 161:325–332

Verma K, Saha G, Kundu LM, Dubey VK (2019) Biochemical characterization of a stable azoreductase enzyme from *Chromobacterium violaceum*: application in industrial effluent dye degradation. Int J Biol Macromol 121:1011–1018

Vigneshpriya D, Krishnaveni N, Renganathan S (2019) Untreated and Sargassum wightii-treated brilliant green dye toxicity impact on microflora and Allium cepa L. Appl Water Sci 9:16

Vigneswaran C, Ananthasubramanian M, Kandhavadivu P (2014) Enzymes in textile effluents. Bioprocessing of Textiles. Elsevier BV p 251–298. http://doi.org/10.1016/b978-93-80308-42-5.50005-6

Vijayaraghavan J, Basha SJS, Jegan J (2013) A review on efficacious methods to decolorize reactive azo dye. J Urban Environ Eng 7(1):30–47

Waghmode TR, Kurade MB, Sapkal RT, Bhosale CH, Jeon B, Govindwar SP (2019) Sequential photocatalysis and biological treatment for the enhanced degradation of the persistent azo dye methyl red. J Hazard Mat 371:115–122

Wang N, Chu Y, Wu F, Zhao Z, Xu X (2017) Decolorization and degradation of Congo red by a newly isolated white rot fungus, *Ceriporia lacerata*, from decayed mulberry branches. Int Biodeterior Biodegrad 117:236–244

Watharkar AD, Jadhav JP (2014) Detoxification and decolorization of a simulated textile dye mixture by phytoremediation using Petunia grandiflora and *Gailardia grandiflora*: A plant–plant consortial strategy. Ecotoxic Environ Safe 103:1–8

Wonyonyi WC, Onyari JM, Shiundu PM, Mulaa FJ (2019) Effective biotransformation of Reactive Black 5 Dye Using Crude Protease from *Bacillus Cereus* Strain KM201428. Energy proced 157:815–824

Wu BJ, Eiteman MA, SE Law (1998) Evaluation of membrane filtration and ozonation processes for treatment of reactive-dye wastewater.J Environ Eng 124(3): 272–277

Wu J, Doan H, Upreti S (2008) Decolorization of aqueous textile reactive dye by ozone. Chem Eng J 142:156–160

Xu H, Li M, Wang H, Miao J, Zou L (2012) Fenton reagent oxidation and decolorizing reaction kinetics of reactive red SBE. Energy Proced 16:58–64

Yang Q, Li C, Li H, Li Y, Yu N (2009) Degradation of synthetic reactive azo dyes and treatment of textile wastewater by a fungi consortium reactor. Biochem Eng J 43:225–230

Yang X, Zheng J, Lu Y, Jia R (2016) Degradation and detoxification of the triphenylmethane dye malachite green catalyzed by crude manganese peroxidase from Irpex lacteus F17. Environ Sci Pollut Res 23:9585–9597

You SJ, Damodar RA, Hou SC (2010) Degradation of Reactive Black 5 dye using anaerobic/aerobic membrane bioreactor (MBR) and photochemical membrane reactor. J Hazard Mat 177:1112–1118

Chapter 7
Impact of Biotechnology on the Climate Change

Saima Aslam, Shahid Ul Islam, Khurshid Ahmad Ganai and Epari Ritesh Patro

Abstract Increasing temperature, reducing ice and intensifying cyclones are the most common symptoms of climate change. Increasing CO_2 concentration in the environment is very dangerous for mankind. So, it is very essential to decrease or minimize the CO_2 concentration. To combat this situation, biotechnology offers the best solution to decrease greenhouse gases and mitigate climate change. Genetically modified crops can take more CO_2 from the atmosphere and convert it into oxygen. Various abiotic stress-tolerant varieties of crops or plants have been developed in order to response the climate change. This chapter reviews the impact of biotechnology on the climate change and describes the application of genetic engineering and molecular biology for the reduction in greenhouse gas emissions.

Keywords Climate change · Biotechnology · Biotic stresses · Drought

7.1 Introduction

Future technologies will provide global savings of carbon dioxide in the range of 500–1000 million tons per year. Industrial biotechnology therefore offers excellent opportunities to reduce greenhouse gas emissions and reduce dependence on fossil fuels (Hermann et al. 2007). Climate change plays an important role in determining

S. Aslam (✉)
Department of Biotechnology, Baba Ghulam Shah Badshah University, Rajouri, J&K, India
e-mail: saima.bt@gmail.com

S. Ul Islam
Department of Civil Engineering, Baba Ghulam Shah Badshah University, Rajouri, J&K, India
e-mail: shahidiitr@gmail.com

K. A. Ganai
Department of Botany, Islamia College of Science & Commerce, Srinagar, J&K, India
e-mail: khursheedtrali@yahoo.co.in

E. R. Patro
Department of Civil and Environmental Engineering, Politecnico di Milano, Milan, Italy
e-mail: epariritesh.patro@polimi.it

R. M. Singh et al. (eds.), *Environmental Processes and Management*,
Water Science and Technology Library 91,
https://doi.org/10.1007/978-3-030-38152-3_7

the distribution of biodiversity (Peterson et al. 2002). Simulations show that delays in the spread of modern research capabilities in biotechnology to developing countries will exacerbate the local food crisis. Similarly, global climate change will exacerbate these crises, stressing the importance of strengthening research capacities in developing countries (Evenson 1999). The impact of climate change on farmers varies by region, depending on the use of technology. Technology development determines farm productivity far more than its climatic and agricultural capabilities. Therefore, food insecurity is not only a result of "climate determinism" but can be addressed by improvements in economic, political and agricultural policy on a local and global scale. In areas where food insecurity prevails, agriculture is usually done manually, using a shovel and a low-input planting stick. The difference between the productivity of conventional farming and those that use petroleum-based fertilizers and pesticides, biotechnology-enhanced plant varieties, and mechanization is extreme (Thirtle et al. 2003). Climate change can increase food insecurity unless more frequent early warning systems and development programs are used (Brown and Funk 2008). The development of crops adapted to abiotic stress is now important in food production in many parts of the world. Expected changes in climate and variability, in particular, extreme temperature and rainfall changes are expected to make crop improvement more important for food production. Here, we discuss two important methods in biotechnology, molecular breeding and genetic engineering and their integration with traditional breeding to develop crops that are more resistant to abiotic stress (Varshney et al. 2011). There is an urgent need for more attention to the future adaptation of agriculture to climate change. Only few studies evaluated the possible effectiveness and degree of acceptance of possible response strategies (Howden et al. 2007). The impact of climate change over decades is critical for the future political environment of the poor. Free trade can help improve access to international supplies; investments in transportation infrastructure and planning will help ensure local deliveries as quickly as possible. Watering, promoting sustainable farming practices and ongoing technological advances can play a crucial role in providing stable national and international resources in times of climate change (Schmidhuber and Tubiello 2007). Biotechnology has changed the way drugs are discovered, designed and made often. New capabilities are spread across most of the world, sometimes amidst many controversies. Some proponents now also suggest that biotechnology will reform of the energy sector and that both the cultivated crops and the fuel they are made of change. Entrepreneurs have raised hundreds of millions of dollars for bio-energy companies working on "second-generation" fuel from crop residues, grasses or wood materials (Tollefson 2008). The scientific community agrees that the global climate is changing, surface temperatures are rising, snow and ice are melting, sea levels are rising, and climate change is intensifying. These changes are expected to have a major impact on public health (Frumkin et al. 2008). The challenge of climate change requires governments, citizens and businesses to work together to reduce greenhouse gas emissions, which requires information on risks, opportunities, strategies and emission levels carbon of the company (Reid and Toffel 2009).

7.2 Stress-Tolerant Varieties of Crops and Climate Change

Due to the adverse effects of high temperatures on plant growth, global warming is expected to have a generally negative impact on plant growth. The growing threat of extreme weather conditions, including extremely high temperatures, can lead to catastrophic loss of crop productivity and widespread famine (Bita and Gerats 2013). Environmental stresses such as drought, high temperatures, salinity and high CO_2 can affect plant growth and pose a growing threat to sustainable agriculture (Ahuja et al. 2016). Plants in nature are constantly exposed to many biotic and abiotic stresses. Among these stresses, drought stress is one of the most unfavorable factors for plant growth and productivity and poses a serious threat to sustainable crop production under climate change (Anjum et al. 2011).

7.2.1 Biotech Crops Adapted to Climate Change

Biotechnology makes it possible to modify crops faster than traditional crops, thus accelerating the implementation of strategies to cope with rapid and severe climate change. Resistant and biotech-resistant crops have been steadily growing to control pests and diseases, as new pests and diseases develop with climate change. Resistant varieties reduce the application of pesticides and therefore CO_2 emissions. Various crops have been developed which are adapted to the different abiotic stresses in response to climate change.

7.2.2 Salinity-Tolerant Crops

In general, when crop salinity reaches above threshold tolerance, then the yield decreases proportionately with increased salt concentration (Maas and Hoffman 1977). Soil salinity is a major environmental barrier to crop production and is estimated to affect 45 million hectares of irrigated land and is expected to increase due to global climate change and many irrigation methods (Rengasamy 2010; Munns and Tester 2008). The ultimate goal of salinity tolerance research is to increase the ability of plants to support saline soil growth and productivity relative to their growth in unsalted soils (Roy et al. 2014). Recent reports have shown that expression of the Arabidopsis ABF4 gene in potato increases tuber yield under normal and abiotic stress conditions, improves storage capability and processing quality of the tubers and enhances salt and drought tolerance. Potato is the third most important food crop in the world. Potato plants are susceptible to salinity and drought, which negatively affect crop yield, tuber quality and market value. The development of new varieties with higher yields and increased tolerance to adverse environmental conditions is a main objective in potato breeding (Muñiz García et al. 2018). Moreover, it is known that WRKY transcription factors (TFs) play a vital part in coping with different stresses. It is shown that overexpression of *DgWRKY2* in chrysanthemum enhanced tolerance to high-salt stress compared to the wild type (WT) (He et al. 2018).

7.2.3 Drought-Resistant Crops

Drought is one of the most important environmental stresses affecting the productivity of most field crops. The clarification of the complex mechanism underlying drought resistance in crops will promote the development of new varieties while improving drought resistance (Hu and Xiong 2014). Over the past decade, scientists have been studying the genetic and molecular mechanisms of drought resistance to improve the drought resistance of various crops and have made significant progress in avoiding drought and its tolerance (Fang and Xiong 2015). The development of drought-tolerant crops using modern biotechnology can contribute to global food security as drought-tolerant crops can be a factor in sustaining plant growth and productivity and increasing global arable land (Liang 2016). Recent studies have shown that overexpression of HSP90 family gene, OsHSP50.2 enhances drought stress tolerance in rice (Oryza sativa) (Xiang et al. 2018).

7.2.4 Cold-Tolerant Crops

Biotechnology offers new strategies for developing transgenic crops with improved resistance. The rapid development of recombinant DNA technology and the development of accurate and efficient gene transfer protocols have led to the efficient transformation and generation of transgenic lines of many crop types (Wani et al. 2008; Gosal et al. 2009; Wani and Gosal 2011). Low temperature inhibits the accumulation of chlorophyll in leaves that are actively developing. In rice, cold-tolerant lines such as japonica accumulate more chlorophyll under cold stresses than cold-sensitive lines (e.g., from indica rice) (Glaszmann et al. 1990). Low sensitivity of photosynthesis to cold observed in low temperature of maize inbreeds is partly related to specific enzyme portion of the process (Singh 2008). Cold is a major stress that limits the quality and productivity of economically important crops such as tomato (*Solanum lycopersicum* L.). Generating a cold-stress-tolerant tomato by expressing cold-inducible genes would increase agricultural strategies. Ectopic expression of *Arabidopsis Rare cold-inducible 2a(RCI2A)* in transgenic tomato enhances tolerance to cold stress without altering agronomic traits (Sivankalyani et al. 2015).

7.2.5 Heat-Resistant Crops

High-temperature tolerance has been genetically engineered in plants by modifying the protein levels of heat shock transcription factor, either directly or indirectly from the genes of heat shock proteins (Singh and Grover 2008). To better understand the plant responses to high temperatures, it is necessary to first determine the optimum temperature of the plant (Burke and Chen 2016). In genetically modified chrysanthemums containing the DREBIA gene from *Arabidopsis thaliana*, transgenic genes and other heat-sensitive genes such as HSP70 (thermal shock proteins) were strongly expressed when subjected to heat treatment. Genetically modified plants maintain high photosynthesis and high levels of photosynthesis enzymes (Hong et al. 2009). The negative effect of thermal stresses can be mitigated by developing plants with improved heat resistance using different genetic methods. For this purpose, however, it is imperative that the physiological responses of the plants to the high temperature, the temperature tolerance mechanisms and the possible strategies for improving the thermal heat of the plants are generally understood. Thermal stress affects plant growth throughout growth, but the threshold level varies greatly at different stages of growth (Wahid et al. 2007).

7.3 Biotechnology and Climate Change

7.3.1 Reduction in Greenhouse Gases Emission

Approximately, 25% of the greenhouse gases (carbon dioxide, methane and nitrous oxide) are dispersed into the atmosphere due to widespread agricultural practices such as artificial fertilizer use, overgrazing and deforestation. With the help of biotechnology, we can change the impact of climate change by reducing the use of energy-saving agriculture, carbon capture and synthetic fertilizers (Treasury 2009). GMO products in the last 16 years have contributed to the reduction of carbon dioxide emissions because GM plants do not require as much maintenance as regular products. It allows farmers to use less energy and fertilizers which are environmentally friendly (Fares 2014).

7.3.2 Use of Environmentally Friendly Fuels

Agricultural techniques should play a critical role in combating climate change. The production of biofuels from traditional products and from genetically modified crops such as sugarcane, oilseeds, rapeseed and jatropha will help to reduce the negative effects of carbon dioxide emissions in the transport sector (Treasury 2009; Sarin et al. 2007). Effective farming, therefore, will help to clean the atmosphere by growing non-edible oilseeds. Therefore, it was directly involved in the production of biodiesel for direct use in the energy sector. Then, it is mixed with fossil fuels and will help to reduce carbon dioxide emissions (Jain and Sharma 2010; Lua et al. 2009).

7.3.3 Carbon Sequestration

Absorption or capture of carbon-containing materials, in particular carbon dioxide (CO_2), is generally referred to as carbon sequestration. It is commonly used to describe any increase in soil organic carbon content caused by change of land management, with implication that the increased soil carbon storage mitigates climate change (Powlson et al. 2011). Therefore, it is important to enhance the war to reduce the concentration of carbon dioxide in the atmosphere by carbon sequestration in the soil. Reducing conventional tillage is one such way. Protection against soil erosion applications can help to prevent soil erosion and sequester soil carbon and methane (CH_4) consumption. Conservation practices that help prevent soil erosion may also sequester soil carbon and enhance methane (CH_4) consumption (West and Post 2002; Johnsona et al. 2007).

7.3.4 Reduced Synthetic Fertilizer

The adoption of agrochemicals to maintain productivity in marginal areas has led to global environmental pollution with toxins that alter biogeochemical cycles (Ogunseitan 2003). Low fertilizer use also means less nitrogen contamination of groundwater and surface waters. Inorganic nitrogen fertilizers such as ammonium sulfate, ammonium chloride, ammonium chloride, sodium nitrate and calcium nitrate are responsible for the formation and release of greenhouse gases (especially when interacting with common soil bacteria) (Brookes and Barfoot 2009). The use of bio-based fertilizers is encouraged to reduce the harmful effects of industrial fertilizers. The biotechnology option offers an advantage in reducing the use of synthetic fertilizers. Some properties of nitrogen-fixing bacteria have been developed by genetic engineering (Zahran 2001). It has also been reported that some non-leguminous crops (rice and wheat) can stabilize nitrogen in the soil (Yang et al. 2008; Saikia and Jain 2007). Another strategy is to grow the plants which efficiently use nitrogen. Examples of these crops are genetically modified canola, which has shown a significant reduction in the amount of nitrogen fertilizer lost to the atmosphere, increased profitability and an increased economy for farmers (Treasury 2009).

7.4 Adaptation to Biotic Stresses

7.4.1 Insect-Pest Resistance

India is the second-largest country in terms of population in the world. It is necessary to increase the production of agricultural products to meet the suffering of the population. As a result of growing populations and shrinking resources, there is an urgent need to enhance agricultural productivity. The most vital and viable idea is to inject new technology into Indian agriculture. Biotechnology crops are very promising to achieve this goal. The success of Bt-cotton, Bt-corn, Bt-Brinjal, Bt-maize and many other crops in many countries is clear evidence. Globally, developed countries are developing genetically modified crops such as maize, brinjal, maize and soya bean, but in Indian condition, only Bt-cotton is being successfully cultivated (Gangwar 2015). The first genetically modified cotton was developed in 1987 in the USA by Monsanto, Delta & Pine companies (Benedict and Altman 2001). At present, the resistance of insects and pests is generally insufficient for many crops. The use of chemical control measures is dangerous for consumers and also unsustainable for the environment. From a farmer's point of view, any genetic improvement that could reduce the cost of chemical application of pest control would be very beneficial. Bt (Cry) gene isolated from soil bacteria, Bacillus thuringiensis, has been shown to be highly effective in controlling many killer insects in a number of crops. Resistance to tomato was first reported using Bt. gene in 1987. Genetically modified Bt. Tomato plants showed resistance against *Spodoptera litura* and *Heliothis virescens* (Fischhoff et al. 1987). Brinjal is one of the most consumed vegetables in Asia and more particularly in the Indian subcontinent. However, it is severely damaged by *lepidopteran* insects, i.e., *Leucinodes orbonalis*. A synthetic cry1Ab gene coding for an insecticidal crystal protein (ICP) was transformed to brinjal (*Solanum melongena* cv. Pusa Purple Long) by co-cultivating cotyledons with *A. tumefaciens* (Kumar et al. 1998).

7.4.2 Disease Resistance

The next major obstacle to the production of horticultural crops is the various diseases caused by fungi, bacteria and pathogenic viruses. Traditional reproduction seems to be limited because of the lack of gene resistance gene(s) in gene poll in a given crop. One of the main goals of genetic transformation is to improve tolerance or to integrate plant resistance with different pathogens. Genetic engineering of disease resistance in crops has become popular and valuable in terms of cost and efficiency. In order to transfer resistance to bacterial and fungal diseases, different genes such as chitinase and osmotin are transferred to many horticultural crops around the world.

The various glycolic enzymes encoded by genes such as chitinase, gluconase and PR proteins within plant cells have cell wall-degrading capabilities that cause them to develop transgenic plants to incorporate resistance to fungal pathogens (Ceasar and Ignacimuthu 2012).

7.4.3 Resistance to Fungal Diseases

Chitin is a key component of cell walls for many fungal pathogens such as *Rhizoctonia solani* and can be analyzed by chitinase. On the other hand, β-1,3-glucanase is known to degrade glucans, which is also present in fungal cell walls. Chitinase and glucoside synthesis is known to occur in response to pathogens. When both enzymes are present at the same time, fungi are prevented from growing more effectively (Neuhaus 1999). Many laboratories have been able to transfer genes derived from plants or bacteria to plants and develop GM crops while promoting resistance to fungal diseases. These plants include grapevine (Yamamoto et al. 2000). peanut (Rohini et al. 2000) and cotton (Tohidfar et al. 2005, 2012). The combined expression of chitinase and glucanase in transgenic carrot, tomato and tobacco resulted in a much more effective prevention of fungal disease development (Melchers and Stuiver 2000).

7.4.4 Resistance to Bacterial and Viral Diseases

Bacterial blight is a destructive disease of domesticated rice (*Oryza sativa*) caused by the pathogen *Xanthomonas oryzae* pv. *oryzae*. The ethylene-responsive (ERF) transcription factors have been shown to play a role in regulating the expression of pathogenesis-related (PR) genes (Grennan 2008). The expression of cotton ERF in tobacco causes transgenic plants to exhibit greater level of resistance to *Xanthomonas* (Champion et al. 2009).

To date, more than 700 viruses have been identified from plants that are responsible for many diseases and significant crop losses. Due to lack of pesticides, viral diseases are traditionally controlled using virus-free certified plant material, the removal of infected plants and the spraying of chemicals against virus vectors. Coat protein-mediated resistance to viruses has been one of the successes of plant genetic engineering. Using this approach, many field crops have been designed to withstand important viral pathogens and eventually resistant plants have been commercialized, for example, potato event HLMT15-15 which is tolerant to potato Y virus (PYV) or potato event RBMT21-350 which is resistant to potato leaf roll virus (PLRV) (James 2013). Antibody technology has been developed as a new method of producing disease-resistant plants by expressing antibodies or rAb fragments capable of disrupting inhibitors or proteins involved in pathogenesis (Cardoso et al. 2014).

7.4.5 *Resistance to Herbicides*

The fourteen commercialized herbicide-tolerant crops have been introduced that comprise 222 events (James 2013). Glyphosate resistance in transgenic plants is obtained through the transfer of a mutated gene for EPSPS synthase that will better than its natural substrate enolpyruvate from glyphosate (Stalker et al. 1985). For example, in *Pseudomonas stutzeri* A1501, EPSPS gene was isolated that shows enhanced resistance to glyphosate up to 200 mM (Liang et al. 2008). *Amaranthus palmer*, another plant is under development for resistance to glyphosat (Gaines et al. 2010).

7.5 Conclusions

Climate change has profound implications for food security, health and safety, and approaches are needed to adapt to new climates. The effects of climate change are obvious, and we must now take steps to adapt quickly and avoid unexpected and undesirable consequences. The applications of recombinant DNA technology or genetic engineering to crop improvement contribute enormously to solving the problem of world hunger as the population grows day by day with the denial of sustainable intensification. Plant biotechnology and molecular biology can make a positive contribution to climate change adaptation and mitigation by reducing greenhouse gas emissions, carbon sequestration, fuel consumption, energy-saving agriculture and reduced artificial use. These measures help to improve agricultural productivity and protect the ecosystem from extreme weather events. Although biotechnology has so far had limited commercial success in horticultural crops, including fruits, vegetables, flowers and landscape plants, we cannot ignore the potential of this technique for promoting the genes of our crops. Horticultural crops to combat various production constraints such as biological stress or abiotic and fruit quality, genetic technology provides a potential technique for the promotion of genes using a characteristic of trait of interest to plants. Transcriptome sequence is now available in a generic database for different crops. This huge information will help identify the different genes that govern the various important traits and identify the target sites for genome editing and genetic transformation.

References

Ahuja I, de Vos RCH, Rohloff J, Stoopen GM, Halle KK, Ahmad SJN, Hoang L, Hall RD, Bones AM (2016) Arabidopsis myrosinases link the glucosinolate-myrosinase system and the cuticle. Sci Rep 6:38990. https://doi.org/10.1038/srep38990

Anjum SA, Xie X, Wang L, Saleem MF, Man C, Lei W (2011) Morphological, physiological and biochemical responses of plants to drought stress. Afr J Agric Res 6:2026–2032. https://doi.org/10.5897/ajar10.027

Benedict JH, Altman DW (2001) Commercialization of transgenic cotton expressing insecticidal crystal protein. In: Genetic improvement of cotton: emerging technologies. Oxf. & IBH Pub Co. Pvt. Ltd, New Delhi, pp 137–201

Bita C, Gerats T (2013) Plant tolerance to high temperature in a changing environment: scientific fundamentals and production of heat stress-tolerant crops. Frontiers Plant Sci 4:273

Brookes G, Barfoot P (2009) Global impact of biotech crops: Income and production effects, 1996–2007. J AgBio Forum 12(2):184–208

Brown ME, Funk CC (2008) Food security under climate change. NASA Publications, p 131. http://digitalcommons.unl.edu/nasapub/131

Burke JJ, Chen J (2016) Changes in cellular and molecular processes in plant adaptation to heat stress. Plant-environment interactions. CRC Press, pp 46–65

Cardoso FM, Ibanez LI, Van den Hoecke S, De Baets S, Smet A, Roose K, Schepens B, Descamps FJ, Fiers W, Muyldermans S, Depicker A, Saelens X (2014) Single-domain antibodies targeting neuraminidase protect against an H5N1 influenza virus challenge. J Virol 88:8278–8296. https://doi.org/10.1128/JVI.03178-13

Ceasar SA, Ignacimuthu S (2012) Genetic engineering of crop plants for fungal resistance: role of antifungal genes. Biotechnol Lett 34:995–1002

Champion A, Hebrard E, Parra B, Bournaud C, Marmey P, Tranchant C, Nicole M (2009) Molecular diversity and gene expression of cotton ERF transcription factors reveal that group IXa members are responsive to jasmonate, ethylene and Xanthomonas. Mol Plant Pathol 10:471–485. https://doi.org/10.1111/j.1364-3703.2009.00549.x

Evenson RE (1999) Global and local implications of biotechnology and climate change for future food supplies. Proc Natl Acad Sci 96(11):5921–5928

Fang Y, Xiong L (2015) General mechanisms of drought response and their application in drought resistance improvement in plants. Cell Mol Life Sci 72(4):673–689

Fares S (2014) Study: biotech crops return benefits to farmers, economy. Retrieved from: http://farmprogress.com/story-study-biotech-crops-return-benefits-farmers-economy-0-112173

Fischhoff DA, Bowdish KS, Perlak FJ, Marrone PG, McCormick SM, Niedermeyer JG, Dean DA, Kusano-Kretzmer K, Mayer EJ, Rochester DE, Rogers SG, Fraley RT (1987) Insect tolerant tomato plants. BioTechnol 5:807–813

Frumkin H, Hess J, Luber G, Malilay J, McGeehin M (2008) Climate change: the public health response. Am J Public Health 98:435–445. https://doi.org/10.2105/AJPH.2007.119362

Gaines TA, Zhang W, Wang D, Bukun B, Chisholm ST, Shaner DL, Nissen SJ, Patzoldt WL, Tranel PJ, Culpepper AS, Grey TL, Webster TM, Vencill WK, Sammons RD, Jiang J, Preston C, Leach JE, Westra P (2010) Gene amplification confers glyphosate resistance in Amaranth USPA Lmeri. Proc Natl Acad Sci U S A 107:1029–1034. https://doi.org/10.1073/pnas.0906649107

Gangwar RK (2015) Bt cotton—prospects and challenges—a review article 11. J Resour Dev Manage. https://www.iiste.org/Journals/index.php/JRDM/article/view/24897

Glaszmann JC, Kaw RN, Khush GS (1990) Genetic divergence among cold tolerant rices (*Oryza satva* L.). Euphytica 45:95–104

Gosal SS, Wani SH, Kang MS (2009) Biotechnology and drought tolerance. J Crop Improv 23:19–54

Grennan AK (2008) Ethylene response factors in jasmonate signaling and defense response. Plant Physiol 146(4):1457–1458

He L, Wu Y-H, Zhao Q, Wang B, Liu Q-L, Zhang L (2018) Chrysanthemum DgWRKY2 gene enhances tolerance to salt stress in transgenic chrysanthemum. Int J Mol Sci 19:2062. https://doi.org/10.3390/ijms19072062

Hermann BG, Blok K, Patel MK (2007) Producing bio-based bulk chemicals using industrial biotechnology saves energy and combats climate change. Environ Sci Technol 41(22):7915–7921

Hong B, Ma C, Yang Y, Wang T, Yamaguchi-Shinozaki K, Gao J (2009) Over-expression of AtDREB1A in chrysanthemum enhances tolerance to heat stress. Plant Mol Biol 70:231–240. https://doi.org/10.1007/s11103-009-9468-z

Howden SM, Soussana J-F, Tubiello FN, Chhetri N, Dunlop M, Meinke H (2007) Adapting agriculture to climate change. Proc Natl Acad Sci U S A 104:19691–19696. https://doi.org/10.1073/pnas.0701890104

Hu H, Xiong L (2014) Genetic engineering and breeding of drought-resistant crops. Annu Rev Plant Biol 65:715–741

Jain S, Sharma MP (2010) Prospects of biodiesel from Jatropha in India: a review. Renew Sustain Energy Rev 14(2):763–771. https://doi.org/10.1016/j.rser.2009.10.005

James C (2013) Global status of commercialized biotech/GM crops. Brief No. 46. ISAAA, Ithaca, NY

Johnsona JMF, Franzluebbersb AJ, Weyersa SL, Reicoskya DC (2007) Agricultural opportunities to mitigate greenhouse gas emissions. Environ Poll 150(1):107–124

Kumar PA, Mandaokar A, Sreenivasu K, Chakrabarti SK, Bisaria S, Sharma SR (1998) Insect-resistant transgenic brinjal plants. Mol Breed 4:3–37

Liang C (2016) Genetically modified crops with drought tolerance: achievements, challenges, and perspectives. In: Drought stress tolerance in plants, vol 2. Springer, Cham, pp 531–547

Liang A, Sha J, Lu W, Chen M, Li L, Jin D, Yan Y, Wang J, Ping S, Zhang W, Wang Y, Lin M (2008) A single residue mutation of 5-enolpyruvylshikimate-3-phosphate synthase in Pseudomonas stutzeri enhances resistance to the herbicide glyphosate. Biotechnol Lett. https://doi.org/10.1007/s10529-008-9703-8

Lua H, Liua Y, Zhoua H, Yanga Y, Chena M, Liang B (2009) Production of biodiesel from *Jatropha curcas* L. Oil Comp Chem Eng 33(5):1091–1096. https://doi.org/10.1016/j.compchemeng.2008.09.012

Maas Eugene V, Hoffman Glenn J (1977) Crop salt tolerance–current assessment. J Irrig Drainage Div 103(2):115–134

Melchers LS, Stuiver MH (2000) Novel genes for disease resistance breeding. Curr Opin Plant Biol 3:147–152

Muñiz García MN, Cortelezzi JI, Fumagalli M, Capiati DA (2018) Expression of the Arabidopsis ABF4 gene in potato increases tuber yield, improves tuber quality and enhances salt and drought tolerance. Plant Mol Biol 98:137–152. https://doi.org/10.1007/s11103-018-0769-y

Munns R, Tester M (2008) Mechanisms of salinity tolerance. Annu Rev Plant Biol 59:651–681

Neuhaus JM (1999) Plant chitinases (PR-3, PR-4, PR-8, PR-11). In: Datta SK, Muthukrishnan S (eds) Pathogenesis-related proteins in plants. CRC Press, Boca Raton, FL, pp 77–105

Ogunseitan OA (2003) Biotechnology and industrial ecology: new challenges for a changing global environment. Afr J Biotechnol 2(12):593–601

Peterson AT, Ortega-Huerta MA, Bartley J, Sánchez-Cordero V, Soberón J, Buddemeier RH, Stockwell DRB (2002) Future projections for Mexican faunas under global climate change scenarios. Nature 416:626–629

Powlson DS, Whitmore AP, Goulding KWT (2011) Soil carbon sequestration to mitigate climate change: a critical re-examination to identify the true and false. Eur J Soil Sci 62:42–55

Reid EM, Toffel MW (2009) Responding to public and private politics: Corporate disclosure of climate change strategies. Strateg Manag J 30(11):1157–1178

Rengasamy P (2010) Soil processes affecting crop production in salt-affected soils. Funct Plant Biol 37:613–620

Rohini V, Sankara K, Rao K (2000) Transformation of peanut (*Arachis hypogaea* L.) with tobacco chitinase gene: variable response of transformants to leaf spot disease. Plant Sci 160:889–898

Roy SJ, Negrão S, Tester M (2014) Salt resistant crop plants. Curr Opin Biotechnol 26:115–124

Saikia SP, Jain V (2007) Biological nitrogen fixation with non-legumes: an achievable target or a dogma? Curr Sci 93(3):317–322

Sarin R, Sharma M, Sinharay S, Malhotra RK (2007) Jatropha-palm biodiesel blends: an optimum mix for Asia. Fuel 86(10–11):1365–1371. https://doi.org/10.1016/j.fuel.2006.11.040

Schmidhuber J, Tubiello FN (2007) Global food security under climate change. Proc Natl Acad Sci 104(50):19703–19708

Singh BD (2008) Plant breeding. Kalyani Publishers, pp 443–460

Singh A, Grover A (2008) Genetic engineering for heat tolerance in plants. Physiol Mol Biol Plants 14(1-2):155

Sivankalyani V, Geetha M, Subramanyam K, Girija S (2015) Ectopic expression of Arabidopsis RCI2A gene contributes to cold tolerance in tomato. Transgenic Res 24:237–251. https://doi.org/10.1007/s11248-014-9840-x

Stalker DM, Hiatt WR, Comai L (1985) A single amino acid substitution in the enzyme 5-enolpyruvylshikimate-3-phosphate synthase confers resistance to the herbicide glyphosate. J Biol Chem 260:4724–4728

Thirtle C, Lin L, Piesse J (2003) The impact of research-led agricultural productivity growth on poverty reduction in Africa, Asia and Latin America. World Dev 31(12):1959–1975

Tohidfar M, Ghareyazie B, Mohammadi M, Abdmishani C (2005) *Agrobacterium*-mediated transformation of cotton using a chitinase gene. Plant Cell Tissue Organ Culture 83:83–96

Tohidfar M, Hossaini R, Shokhandan Bashir N, Tabatabaei M (2012) Enhanced resistance to *Verticillium dahliae* in transgenic cotton expressing an endochitinase gene from *Phaseolus vulgaris*. Czech J Genet Plant Breeding 4:345–355

Tollefson J (2008) Not your father's biofuels: if biofuels are to help the fight against climate change, they have to be made from more appropriate materials and in better ways. Jeff Tollefson asks what innovation can do to improve the outlook. Nature 451(7181):880–884

Treasury HM (2009) Green biotechnology and climate change. Euro Biol 12. Retrieved from http://www.docstoc.com/docs/15021072/GreenBiotechnology-andClimate-Change

Varshney RK, Bansal KC, Aggarwal PK, Datta SK, Craufurd PQ (2011) Agricultural biotechnology for crop improvement in a variable climate: hope or hype? Trends Plant Sci 16:363–371. https://doi.org/10.1016/j.tplants.2011.03.004

Wahid A, Gelani S, Ashraf M, Foolad MR (2007) Heat tolerance in plants: an overview. Environ Exp Bot 61:199–223. https://doi.org/10.1016/j.envexpbot.2007.05.011

Wani SH, Gosal SS (2011) Introduction of OsglyII gene into Indica rice through particle bombardment for increased salinity tolerance. Biol Plant

Wani SH, Sandhu JS, Gosal SS (2008) Genetic engineering of crop plants for abiotic stress tolerance. In: Malik CP, Kaur B, Wadhwani C (eds) Advanced topics in plant biotechnology and plant biology. MD Publications, New Delhi, pp 149–183

West TO, Post WM (2002) Soil organic carbon sequestration rates by tillage and crop rotation: a global analysis. Soil Sci Soc Am J 66:930–1046. https://doi.org/10.2136/sssaj2002.1930

Xiang J, Chen X, Hu W, Xiang Y, Yan M, Wang J (2018) Overexpressing heat-shock protein OsHSP50.2 improves drought tolerance in rice. Plant Cell Rep 37:1585–1595. https://doi.org/10.1007/s00299-018-2331-4

Yamamoto T, Iketani H, Ieki H, Nishizawa Y, Notsuka K, Hibi T, Hayashi T, Matsuta N (2000) Transgenic grapevine plants expressing a rice chitinase with enhanced resistance to fungal pathogens. Plant Cell Rep 19:639–646. https://doi.org/10.1007/s002999900174

Yan Y, Yang J, Dou Y, Chen M, Ping S, Peng J, Jin Q (2008) Nitrogen fixation island and rhizophere competence traits in the genome of root associated Pseudomonas stutzeri A1501. Proc Nat Acad Sci 105(21):7564–7569. https://doi.org/10.1073/pnas.0801093105

Zahran HH (2001) Rhizobia from wild legumes: diversity, taxonomy, ecology, nitrogen fixation and biotechnology. J Biotechnol 91:143–153

Chapter 8
Environmental Hazards of Limestone Mining and Adaptive Practices for Environment Management Plan

Harsh Ganapathi and Mayuri Phukan

Abstract Limestone is a fundamental raw material in various industrial sectors. It is formed due to biochemical precipitation of calcium carbonate, and further compaction over long periods of time. A high market for limestone products and its use in a growing number of industries has led to its widespread exploration and excavation. The most widely adopted method of limestone mining is through opencast pits with bench formation. Limestone mining causes widespread disturbance in the environment. Myriad impacts are observed as changes in land use pattern, habitat loss, higher noise levels, dust emissions and changes in aquifer regimes. These environmental concerns have brought about the need for sustainable Environment Management Plan in the mining sector, so as to reduce environmental degradation during operation as well as restoration of degraded lands after final mine closure. A well-formulated Environment Management Plan will help in mitigating the impacts of mining on the environment. The best practices adopted by industries around the world can be adapted as per site characteristics is to ensure sustainable mining along with the prevention of environmental degradation.

Keywords Limestone · Karst topography · Quarrying · Environmental pollution · Restoration

8.1 Introduction

Limestone is a carbonate sedimentary rock, consisting of calcium carbonate and in some cases magnesium carbonate as a secondary component. Limestone is most often mined from a quarry; however, underground limestone mines are found in some places of the world like central and eastern USA. Some of the biggest quarries in the world are in the state of Michigan in the USA, specifically near the Great

H. Ganapathi (✉) · M. Phukan (✉)
New Delhi, India
e-mail: harsh.ganapathi@gmail.com

M. Phukan
e-mail: mayuri.phukan1@gmail.com

R. M. Singh et al. (eds.), *Environmental Processes and Management*,
Water Science and Technology Library 91,
https://doi.org/10.1007/978-3-030-38152-3_8

Lakes' coastline (Critchfield 2017). Countries like China, the USA, Russia, Japan, India, Brazil, Germany, Mexico and Italy are some of the world's largest limestone producers today. It has numerous uses: as a major component in cement concrete, raw material in glass and metal-processing industries; as aggregate for the base of the road, building materials, chemical feedstock for the production of lime; as soil conditioner, industrial water treatment and many more. Due to the high demand for limestone and its products, it is mined in very large quantities around the world.

Impacts of limestone mining on the environment are manifold. Mining leads to landscape deformation and ecosystem destruction, which changes in groundwater regimes, dust and noise generation, during various mining operations like site excavation drilling, blasting and overburden. A major component of the mining activity includes mine closure and reclamation which gives a scope to alleviate the impacts caused due to the mining operations. Hence, adaptive and innovative measures are essential to incorporate sustainable limestone mining.

8.2 Limestone: Geological Formations and Mining Process

Limestone is a naturally formed mineral, primarily composed of calcium carbonate (Oates 2008). It forms commonly in shallow, calm and warm marine waters, as found in the Caribbean Sea, the Indian Ocean, the Persian Gulf and the Gulf of Mexico (King 2005). Another way of limestone that forms is through evaporation, with this type of limestone growing in caves around the world (Critchfield 2017).

8.2.1 Geological Formation

Limestone rocks are formed by the combination of dissolved calcium ions and carbon dioxide in sedimentary depositions. The sedimentation of calcium carbonate occurs with organic and inorganic mechanisms. The organic mechanism involves a variety of organisms that build shells, skeletons or secrete carbonate (e.g. bivalves, gastropods, brachiopods, corals, sponges, various algae), while the inorganic mechanism involves direct precipitation from sea and inland water, which has resulted in some commercially significant deposits of oolitic limestone and travertine. Some minor dolomite sediments are formed by direct precipitation from sea and lake waters (Oates 2008).

Some of the common structures of the limestone formations are the bedding planes which are formed due to tectonic activities, resulting in structures like anticlines, synclines, overthrust folds and faults. These bedding plane structures are formed as depositional structures over time and can also be formed after burial by the dissolution of limestone. Karstic dissolution of the limestone can also occur tens or hundreds of metres below the ground surface, producing cavity structures that varying in size from small cavities to large potholes and caves. Cavities near the surface are often filled with clay or overburden washed down from the surface (Oates 2008).

White coloured limestone is of high purity, while yellow, cream and red hues are indicative of iron and manganese impurities. Limestone exhibits an earthly odour due to the presence of carbonaceous matter (Lide 1998). Solubility of limestone increases in the presence of carbon dioxide forming calcium and magnesium bicarbonates. Limestone readily reacts with acid and is generally used for neutralization (Oates 2008).

8.2.2 Uses of Limestone

Limestone has myriad uses, depending on its chemical and physical specifications. It is applied to soil to combat soil acidity and to promote favourable conditions for the growth of essential microorganisms. Some of the Industrial applications of limestone include manufacturing of dye intermediates, in sugar production as a refining agent, as a flux in blast furnaces and in open hearth for steel production and as a raw material in cement production. It is also used as a neutralizing agent in the treatment of industrial wastes (Lamar 1961).

8.2.3 Limestone Mining Process: A Brief Review

The sum total of all activities that are undertaken during the lifetime of a mine can be categorized into four phases: mineral exploration, mine development, mining operation and mine closure (UNDP and UN Environment 2018).

8.2.3.1 Mineral Exploration

Mineral exploration phase consists of exploratory study and prospecting operations conducted through various geophysical and chemical analyses to identify ores from which minerals can be extracted. In India, the Ministry of Mines has recently initiated the National Mineral Exploration Policy, 2016, to promote research and exploratory study through the provision of baseline data in the public domain. The Ministry also undertakes geophysical survey throughout the country to find concealed mineral deposits (CUTS International 2018).

8.2.3.2 Mine Development

Mine development includes feasibility studies and environmental impact assessment studies. It is followed by land acquisition and displacement of population, if required.

Thus, this phase has major environmental and social impacts. Site preparation process involves land clearance of vegetation, construction of roads and other amenities (UNDP and UN Environment 2018).

8.2.3.3 Mining Operation

The activities involved in mining operation depend on the methodology adopted for mining. Method of mining largely depends on the characteristics of mineral deposits. For minerals that occur near the surface and larger deposits, like limestone, opencast mining process is used. Underground mining is conducted for deep-seated minerals like coal (UNDP and UN Environment 2018).

Opencast mining involves drilling and blasting to remove topsoil and overburden. The mineral-bearing ground is cut in the form of bench systems to excavate the ore. The extracted ore is hauled to the surface and subsequently transported using trucks or dumpers to the crusher, where the ore is converted into smaller pieces. This is followed by beneficiation of minerals, i.e. separation of the minerals from the ore. What remains the following milling of ores and separation of minerals is called tailings which need to be properly disposed. Minerals then undergo final processing (Haldar 2013).

8.2.3.4 Mine Closure

After the entire mineralized zone is excavated, mine closure phase commences. The ancillary facilities are decommissioned. Land reclamation including backfilling of pits and landscaping is an essential part of mine closure and is done to prevent further degradation of land and associated environment (UNDP and UN Environment 2018).

8.3 Environmental Hazards of Limestone Mining

Carbonate rock resources like limestone cannot be obtained from quarrying or mining activities without causing some environmental impacts although modern technologies have made it possible to reduce the associated environmental impacts. Mining operations induce change in geomorphology and land-use pattern with loss of habitat, dust, generation of high levels of noise, vibrations, erosion, subsidence and sedimentation. Limestone mining also causes changes in groundwater flow, groundwater quality and overall water quality. Many studies have described the potential hazards of mining limestone, some of which include Gatt (2001), Geomin Consultants (P) Ltd. (2009), Ochieng et al. (2010), Naja et al. (2010) and Barksdale (1991).

8.3.1 Impacts on Land and Soil

Opencast mining has been associated with a change in land use and land cover of a region. The process of clearing trees and vegetation in preparation for mineral excavation has huge impacts on prevailing ecosystems. In eco-sensitive areas like forests and hilly regions, loss of native and unique species is also coupled with habitat loss of the biodiversity of the region. The topsoil which is excavated is the most fertile component of the soil structure and takes thousands of years to form. Stripping of topsoil cannot be easily compensated, even with rigorous methods of land reclamation (Sharma and Ram 2014). For extraction of mineral, the layers of materials overlying the ore have to be removed. This layer is called the overburden. The quantity of overburden is usually very high and often consists of waste substances. These materials are dumped on open land at times and used for backfilling of pits after mineral extraction (ELAW 2010). Deterioration of soil quality and alteration in soil properties are known to occur due to prolonged dumping of lime mixed waste material (Lamare and Singh 2016).

Excavation results in the destruction of active caves, natural sinkholes and relicts. The extent of the geomorphic impact is a direct function of the size of the quarry, the number of quarries and the location of the quarry, with respect to the overall landscape and landform (Langer 2001). Gunn and Bailey (1993) report that crushed stone quarrying has removed an entire karst hill and large portions of other nearby karst hills in the Mendip Hills, UK.

8.3.2 Impacts Due to Blasting

Blasting is a technique where large amount of explosives is used in crushed site operations to obtain approximate size rubble of limestone. When an explosive is detonated, high amount of energy is released a part of which displaces the rocks from the quarry face while the rest is transmitted in the form of vibrations into the ground surface and through the air. Blast induced vibrations can cause naturally formed stalactites and stalagmites to break off and cause cave roofs to crack or collapse. It may also cause fracturing of quarry walls and increasing the permeability and drainage towards the quarry face (Langer 2001). Lolcama et al. (2002) describe a situation wherein blasting opened a conduit under a quarry floor which was connected to a local water storage basin and a river. Extensive grouting was required to stop the inflow of water from these sources (Fig. 8.1).

High noise levels are generated due to the use of machineries like excavators and during drilling and blasting; loading and unloading operations. The cumulative impact of all the noise created in the mining site can affect humans as well as the biodiversity present within the mine lease area (Ahmed et al. 2014). The distribution and impact of noise in the mine lease area depend on the sources of noise generation as well as on geographical attributes of a region. Meteorological factors like rain and

Fig. 8.1 Rock is drilled and blasted for use as crushed stone. In some isolated areas where people are not located nearby, larger amount of explosives may be used (Langer 2001)

storm add to the propagation of noise levels (Manwar et al. 2016). The mineworkers are the most affected due to the high noise levels, followed by residents of settlements surrounding the mine lease area.

Study conducted by (Manwar et al. 2016) on noise level mapping for an opencast mine in a hilly region identified high noise areas like drilling sites and crushers with increased noise levels during the night. With the results of the study, suitable locations for dumping overburden to act as a barrier to noise levels were also predicted.

8.3.3 Impacts on Air Quality

Emission of particulate matter is associated with various mining operations like excavations, construction of haul road and approach roads, drilling and blasting and transportation of minerals. Dust emissions are of great concern related to air quality surrounding mines. Air-blown particles from the stockpile of excavated material also raise the content of particulate matter in the air. Gaseous pollutants like sulphur dioxide (SO_2) and oxides of nitrogen (NO_x) are emitted from the Heavy Earth Moving Machineries (HEMM) like dumpers and excavators (Lamare and Singh 2016).

Studies have also suggested the impact of mining operations on global carbon budget. The loss of trees causes a loss of carbon dioxide uptake while heavy vehicles emit large quantities of carbon dioxide. Mining is a highly energy-intensive industry. The use of explosives for detonation and vehicles used for transportation consumes large amounts of fossil fuel and electricity. Thus, areas, where large-scale mining activities are concentrated, can contribute towards the worldwide phenomenon of climatic changes (ELAW 2010; Ruttinger and Sharma 2016).

8.3.4 Impacts on Water Resources

8.3.4.1 Impacts on Surface Water

Engineering activities associated with quarrying can directly change the course of surface water. Sinkholes created by quarrying can intercept surface water flow. Conversely, groundwater being pumped from quarries changes streams from gaining streams to losing streams and can drain other nearby surface water features such as ponds and wetlands. Similarly, blasting (see above) can modify groundwater flow, which ultimately can modify the surface water flow. Discharging quarry water into nearby streams can increase flood recurrence intervals (Langer 2001; Bhatnagar et al. 2014).

8.3.4.2 Impacts on Groundwater

Limestone mining is likely to result in relatively local impacts such as reduced water quality, rerouting of recharge water in aquifer, increased run-off and thereby leading to localized reduction in groundwater storage (Hobbs and Gunn 1998). The major impact of quarrying on water relates to mine dewatering and the associated decline of the water table. If a quarry intersects phreatic aquifer, it severely affects the transport of groundwater in such situations. The water may just flow out of the phreatic aquifer into the mine pit at very high velocity. The pit has to be dewatered to facilitate mineral excavation, and the water such pumped out is usually discharged to nearby areas. Thus, there is wastage of huge volume of water in the mining area while the water level in the downstream. Water pumped from a quarry is likely to be lost from the local groundwater system. Within the cone of depression, wells, springs and streams can go dry or have their flows significantly reduced, and the overall direction of groundwater flow may be changed (Hobbs and Gunn 1998). It is within this cone of depression that many human-induced sinkholes are formed.

Figure 8.2a, b shows how groundwater table is declined due to mining activities and the formation of sinkholes as a result of collapsing of the cavities in the karst system.

Karstic system (limestone) has a very low self-purification capacity (Kresic et al. 1992) which makes the water flowing through the karst system very susceptible to pollution. Mining activities can substantially change the direction of discharge and thereby change water quality (Bhatnagar et al. 2014). In aggregate mining, the target limestone, if unsaturated, may also act as a protective cover for the underlying aquifer. If this cover is removed due to mining, the hole created by the mining may direct the surface water to the groundwater, and if the surface water is contaminated, the quality of the groundwater can quickly degrade. Quarrying can also cause the formation of sinkholes that result in the capture of surface water. Dust can enter conduits and smaller openings and can be transported to the groundwater (Hobbs and Gunn 1998). Large amounts of silt and other effluents from quarries (oil, fuel

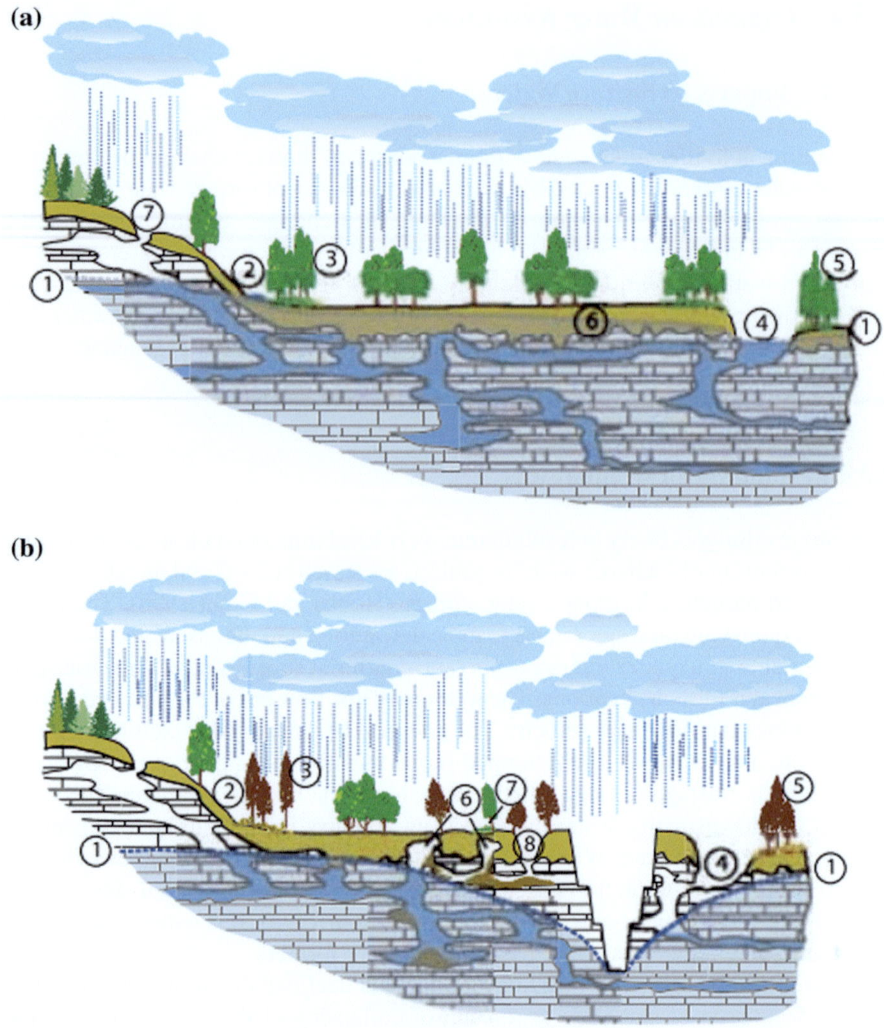

Fig. 8.2 **a** Hypothetical cross section showing karst area under conditions prior to quarry development. The water table (1) is generally above the soil/bedrock contact. Natural groundwater discharges to a spring (2), and a perennial stream (4), which supports wetland (3) and a riparian woodland (5). The surface of the bedrock is highly irregular (6) and is referred to as pinnacle bedrock. A natural sinkhole occurs where the water table is below the soil/bedrock contact (7) (Langer 2001). **b** Hypothetical cross section showing karst area under worst-case conditions after quarry development. Under actual conditions, none, some or all of these conditions may exist. Quarry dewatering has lowered the water table (1) below the soil/bedrock contact. Natural groundwater discharge to a spring (2) and perennial stream (4) has stopped, resulting in the destruction of the wetland (3), drying up of the stream (4) and destruction of the riparian woodland (5). Underground cavities formed in the soil in the area of the pinnacle bedrock due to the loss of buoyant support and piping (6). The ground above the cavity has subsided, resulting in the formation of a wet area, and the tilting of fence posts or trees (7). Ultimately, these cavities could collapse, creating a collapse sinkhole (8) (Langer 2001)

and waste) can pollute the nearby surface water and groundwater bodies within and far beyond the boundaries of limestone (Langer 2001).

Dissolution of calcium associated with limestone due to rain can also affect the water quality. Similarly, surface run-off from limestone can laden nearby seasonal streams with dissolved calcium, increasing the solid content in water and making it unfit for potable use (Sharma and Ram 2014). Based on the investigation conducted by (Eugene and Singh 2014) on water quality due to limestone mining in East Jaintia Hills, Meghalaya, India, it was concluded that in most cases water quality showed elevated levels of pH, electrical conductivity, total dissolved solids, hardness, alkalinity, calcium and sulphate concentrations due to the mining and processing of limestone. This was true for both the seasons of pre-monsoon and post-monsoon. The deterioration in water quality was concluded to have resulted due to additional accumulation, transportation, mixing and dispersion of pollutants (organic or inorganic) from the mining operations into the local water bodies.

8.4 Sustainable Mining for Environmental Hazard Management

Best practices for limestone mining are derived from a robust diagnosis of the environmental hazards, impacts of associated population and an accurate delimitation of the area of influence due to mining activity (Sanchez and Lobo 2018).

The goals mentioned for taking up adaptation measures are shown in Fig. 8.3.

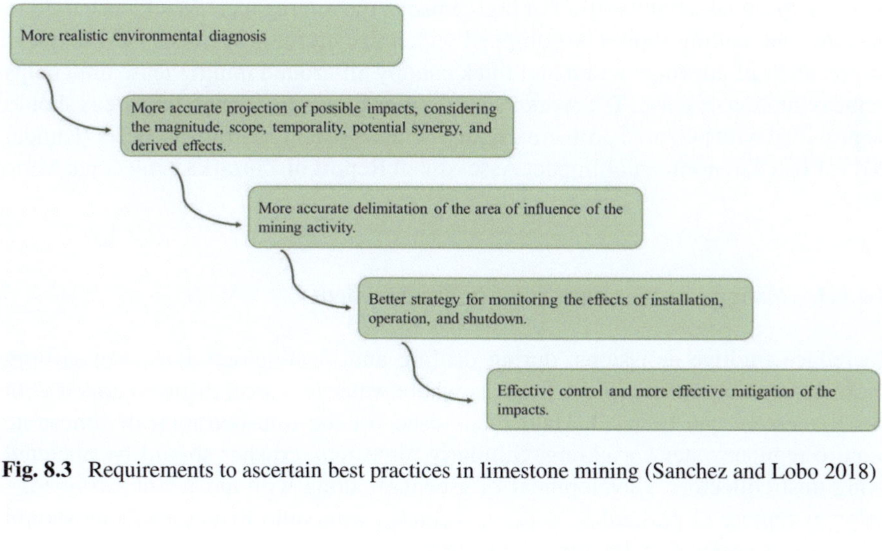

Fig. 8.3 Requirements to ascertain best practices in limestone mining (Sanchez and Lobo 2018)

8.4.1 Environment Management Plan

Environment Management Plan (EMP) is essential to be undertaken during active mining operations and post-completion of the project. It is important to re-establish the ecosystem lost due to the limestone mining activities. Involvement of stakeholders in post-closure monitoring of the mining area is equally essential as it is maintaining active care of the area (Sanchez and Lobo 2018). To achieve the same suitable policies and regulations for pre-mining, mining and post-mining activities need to be incorporated and proper EMP has to be done.

8.4.1.1 Mitigation Measures for Land and Soil Environment

During mining operations, the fertile topsoil requires to be stored separately from overburden that gets generated. It is necessary to reuse the topsoil within the shortest possible time, for plantations and greenbelt development. Construction of drains around the stockpiles prevents erosion and further wastage. Land reclamation activities like backfilling of exhausted pits, slope stabilization or conversion of pits to water reservoirs can be undertaken. Land rehabilitation is an activity with results that are demonstrated only after a long period of time (Kundal 2017).

8.4.1.2 Mitigation Measures for Noise Environment

Controlled blasting, between favourable hours (avoiding early morning or night time), helps to lessen the impact of high noise in the mining area. Machinery used in blasting and hauling should be equipped with noise muffling systems. Development of greenbelt of adequate width and thick canopy all-around mining lease area helps in the reduction of noise. The workers working in high noise generating areas should be provided with personal protective equipment to prevent hearing disorders (Kundal 2017; Final Environmental Impact Assessment Report of Zinzarka Limestone Mine 2016).

8.4.1.3 Mitigation Measures for Air Environment

To reduce fugitive emissions, during drilling and blasting activities, wet drilling should preferably be practised. In areas where water is scarce, drills equipped with dust collectors can be used. Haul roads used for the transportation of limestone require regular water sprinkling. Similarly, limestone crusher should be operated using dust collectors. Development of greenbelt along with the 7.5 m safety zone helps in capture of particulate matters. Vehicles with valid PUC Certificate should be allowed to operate in the mine lease area.

8.4.1.4 Mitigation Measures for Water Environment

As mine pits are dewatered, huge volume of water is generated which can be utilized in the mine office and ancillary facilities, thus reducing freshwater withdrawal. Effective network of drains with sedimentation pits should be developed to prevent the flow of eroded material to nearby drainage. Harvesting rainwater in mined-out pits will help towards maintaining the groundwater regime of the region (Kundal 2017).

8.4.1.5 Mitigation Measures for Biological Environment

The plantations around the mining lease area including the 7.5 m safety zone should include local species to maintain the ecological and biological environment of the site. Species survival rate should be recorded and accordingly, and plantation should be undertaken (Sagewill Limestone Quarry Environmental Management Plan & Reclamation Plan 2018).

8.4.2 Industry Best Practices from Around the World

8.4.2.1 Semi-Open Box Cut Mining by Siam Cement Kaeng Khoi Co. Ltd. (SKK), Thailand

SKK adopts semi-open-cut mining that combines open-cut mining and open-pit mining by leaving some forest area (about 40% of the mine lease area) as a buffer zone along with the quarry boundary line. This reduces noise and dust pollution, preserves significant forest cover and allows for mined land rehabilitation to be undertaken simultaneously with the production process (The Association of Southeast Asian Nations (ASEAN) 2017).

8.4.2.2 Use of Surface Miner in Tular Limestone Mine, Tamil Nadu

Fully mechanized opencast mining has been adopted by the Tular Limestone Mine of Madras Cement Works in Tamil Nadu, India. Excavators are used to remove the topsoil and surface miners were used to mine limestone. As no drilling and blasting are involved, there is a reduction in air pollution and high noise generation. The surface miners also have an inbuilt dust suppression system (Indian Bureau of Mines 2014).

8.4.2.3 Dump Stabilization at Halki Limestone Mine, Muddapur

Aloe vera saplings were planted over inactive overburden dumps by JK Cement Ltd. The process helps in the stabilization of the dump as well as prevents erosion due to wind or water (Indian Bureau of Mines 2015).

8.4.3 Recommendations

Some ways in which sustainable mining process can be achieved are mentioned below.

1. Use of the resistivity method to locate the precise location and size of underground caves to guide operations and anticipate with higher levels of safety procedures.
2. Improvements in blasting techniques such as the use of delaying fuse and air deck to reduce blasting materials. Observing the blasting schedules and finding the scope of improvement. Adoption of other methods of limestone size reduction such as surface miner, rock splitter, breaker and ripper to avoid secondary blasting and to reduce environmental impact of the blasting operations. Detonators and explosives used only during the day, while providing prior intimation to the nearby residents. Ensuring all drilling machines are equipped with dust filtration systems.
3. Storage of topsoil and overburden separately, in designated areas and provision of retaining walls. Provisions of drains along with dumps to arrest the flow of silt during the event of rainfall.
4. Retaining of native vegetation and development of green cover along with dust generation areas like haul roads.
5. Implementation of land reclamation plan after mineral excavation is complete. Reclamation work should be done after a thorough understanding of local geology and ecology. Landscaping will help recover the lost stability and contain immediate damages (The Association of Southeast Asian Nations (ASEAN) 2017; Natural Stone Council 2009).

References

Ahmed AF, Sharma HK, Ahmad RM, Rao RJ (2014) Impact of mining activities on various environmental attributes with specific reference to health impacts in Shatabdipuram, Gwalior, India. Int Res J Environ Sci 3(6)(2319–1414):81–87

Barksdale R (1991) The aggregate handbook. National Stone Association, Alexandria, USA

Bhatnagar D, Goyal S, Tignath S, Deolia D (2014) Impact of opencast limestone mining on groundwater in Katni river watershed, Madhya Pradesh, India—A geoinformatics approach. J Geomatics 8(1)

Critchfield K (2017) Worldatlas.com. Retrieved 12 Apr 2019, from https://www.worldatlas.com/articles/limestone-facts-geology-of-the-world.html
CUTS International (2018) Sustainable mining in India: overview of legal and regulatory framework, technologies and best process practises. Indian Chamber of Commerce
ELAW (2010) Guidebook for evaluating mining project EIAs. Environmental Law Alliance Worldwide, Eugene
Eugene LR, Singh O (2014) Degradation in water quality due to limestone mining in East Jaintia Hills, Meghalaya, India. Int Res J Environ Sci 3(5)(2319–1414):13–20
Final Environmental Impact Assessment Report of Zinzarka Limestone Mine (2016) Mantec Consultants Pvt Ltd., pp 148–153
Gatt P (2001) Limestone quarries and their environmental impacts. New York, USA
Geomin Consultants P Ltd. (2009) Environmental impact assessment and environmental management plan for Lumshong Limestone Mines Lumshong, Jaintia Hills District Meghalaya. Geomin Consultants (P) Ltd, Bhubaneswar
Gunn J, Bailey D (1993) Limestone quarrying and quarry reclamation in Britain. Environ Geol 21(3):167–172
Haldar S (2013) Elements of mining. In: Haldar S (ed) Mineral exploration principle and applications. Elsevier, pp 193–222
Hobbs S, Gunn J (1998) The hydrogeological effect of quarrying karstified limestone: options for prediction and mitigation. Q J Eng Geol Hydrogeol 31(2):147–157
Indian Bureau of Mines (2015) Innovation and noble work done by Halki Limestone Mine of M/S. JK Cement Works, Muddapur. Indian Bureau of Mines
Indian Bureau of Mines (2014) Tular Limestone and Alathiyur Limestone Mines of Madras Cement Limited, Tamil Nadu
King HM (2005) Geology.com. Retrieved 13 April 2019, from https://geology.com/rocks/limestone.shtml
Kresic N, Papic P, Golubovic R (1992) Elements of groundwater protection in a karst environment. Environ Geol Water Sci 20(3):157–164
Kundal M (2017) Environmental impact assessment report upgraded after public hearing for Malundha Limestone Mine. Excel Environ Tech, Ahmedabad, pp 201–228
Lamar J (1961) Uses of limestone and dolomite. Illinois State Geological Survey, Illinois
Lamare R, Singh OP (2016) Limestone mining and its environmental implications in Meghalaya, India. ENVIS Bull Himalayan Ecol:87–100
Langer WH (2001) Potential environmental impacts of quarrying stone in Karst- a literature review. U.S. Dept. of the Interior, U.S. Geological Survey, Denver, Colorado
Lide DR (1998) CRC handbook of chemistry and physics, 1998–1999. CRC Press, Boca Raton
Lolcama JL, Cohen HA, Tonkin MJ (2002) Deep karst conduits, flooding, and sinkholes: lessons for the aggregates industry. Eng Geol 65(2–3):151–157
Manwar VD, Mandal BB, Pal AK (2016) Environmental propagation of noise in mines and nearby villages: a study through noise mapping. Noise Health:185–193
Naja GM, Rivero R, Davis SE, Van Lent T (2010) Hydrochemical impacts of limestone rock mining. Water Air Soil Pollut 217(1–4):95–104
Natural Stone Council (2009) Best practices of the natural stone industry. Natural Stone Council; The University of Tennessee Center for Clean Products
Oates JA (2008) Lime and limestone. Wiley-VCH, Weinheim
Ochieng GM, Seanego ES, Nkwonta OI (2010) Impacts of mining on water resources in South Africa: a reviews. Sci Res Essays 5(22)(1992–2248):3351–3357
Ruttinger L, Sharma V (2016) Climate change and mining: a foreign policy perspective. Adelphi, Queensland
Sagewill Limestone Quarry Environmental Management Plan & Reclamation Plan (2018) New Brunswick: Sagewill Enterprises Ltd, pp 7–11
Sanchez L, Lobo H (2018) Guidebook of good environmental practices for the quarrying of limestone in Karst Areas, 1st edn. Brazilian Speleological Society, Campinas

Sharma A, Ram S (2014) Mining effects on forest area of Mussoorie division. Int J Sci Res Dev:250–255

The Association of Southeast Asian Nations (ASEAN) (2017) Sustainable minerals development: best practices in ASEAN. ASEAN, Jakarta

Transparency Market Research (2018) Limestone market global industry analysis, size, share, growth trends and forecasts 2019–2027. New York

UNDP and UN Environment (2018) Managing mining for sustainable development: a sourcebook. United Nations Development Programme, Bangkok

Chapter 9
Water Consumption Management for Thermal Power Plant

Seema A. Nihalani and Yogendra D. Mishra

Abstract A close nexus lies between water and energy production. Water is needed for energy production, and in turn, energy is utilized for pumping, conveying and treating water. Water is used to extract and produce energy, refine and process fuels, transportation and storage of electric power. With the rising scarcity of water, energy production too is facing a shrinking supply of freshwater due to the mounting cost of water along with growing pollution of water sources. Ministry of Environment and Forest has taken stringent steps to control environment and water pollution. MoEF has recently issued notification for the consumption of water usage and emission standards within thermal power plants. Based on the notification issued by MoEF, all the existing thermal power plants are expected to reduce specific water consumption up to 3.5 m^3/MWH (Mishra et al. in Paper on water resource management for 2 × 660 MW coal-based power plant and comparatives for wet and dry cooling system, New Delhi, India, 2016). All the new power plants that are proposed to be established after 1 January 2017 are required to use a maximum of 2.5 m^3/MWH as specific water consumption and achieve zero liquid discharge (ZLD). The current paper discusses water resource management for thermal power plants. It further evaluates various alternatives for water conservation in thermal power plants.

Keywords TPP—Thermal power plants · ACC—Air-cooled condenser · WCC—Water-cooled condenser · COC—Cycle of concentration

---

S. A. Nihalani (✉)
Civil Engineering Department, PIET, Parul University, Vadodara, Gujarat, India
e-mail: seema.nihalani@paruluniversity.ac.in

Y. D. Mishra
L&T-Sargent & Lundy Limited, Vadodara, Gujarat, India
e-mail: Yogendra.Mishra@lntsnl.com

R. M. Singh et al. (eds.), *Environmental Processes and Management*,
Water Science and Technology Library 91,
https://doi.org/10.1007/978-3-030-38152-3_9

9.1 Introduction

Water resources are mainly consumed by the irrigation sector, the domestic sector, power sector and other industrial sectors. With the increasing population and economic growth of a country, the demand for food grains also increases, which in turn increases the demand for water in the irrigation sector. The reason for growing water demand is rising population in addition to increased urbanization. It is projected that economic growth is fostered with the increase in urbanization.

Rapid economic growth and urbanization have necessitated the need for reliable round the clock electricity supply. Increasing demand for power in the country will make the power sector as one of the largest consumptive water users in India next to the irrigation sector.

Water is one of the unavoidable inputs needed for sustainable growth of the industrial sector of any country. As the need for power increases, the demand for freshwater supply for power generation will also grow. During the past years, there is a decrease in the availability of freshwater. Also, there is a rising realization in power plant managers to preserve freshwater and decrease the use of freshwater by employing various water conservation techniques including treatment, recycle and reuse. In addition, the need to discharge treated water in receiving stream to meet environmental regulations will further govern the future provision of freshwater supply. MoEF has recently issued notification for the consumption of water usage and emission standards within thermal power plants. This new environment norm compels the need for every power generation company to critically evaluate the water resource management within their premises. The first step in water resource management is to evaluate the water requirement of various consumers in the power plant and understand the quality of water required at various sources. This paper evaluates the possible alternatives to reduce the fresh makeup water requirement and discusses ways and means to achieve zero liquid discharge.

9.2 Zero Liquid Discharge

The thermal power industry is the largest contributor to wastewater generation and discharge. Both treatment and disposal of wastewater are crucial, and the treated water eventually gets mixed with various water sources, thus affecting the aquatic environment. Wastewater, discharged from thermal power plants, is considered to be the chief source of water pollution sources. On deriving cognizance form this, the Ministry of Environment and Forest has constricted water usage norms for thermal power plants. The new norms compel all the new power plants to bring down the water consumption from 3.5 to 2.5 m^3/MWH and achieve zero liquid discharge. This necessitates all the power plant owners, to treat, recycle and reuse the wastewater generated in their premises. It shall also be mandatory for all the power plants located within a radius of 50 km from a sewage treatment plant to reuse the treated sewage water.

Zero liquid discharge systems have several environmental benefits, reuse of treated water, and reduction in freshwater demand, less exploitation of water resources and supporting sustainability. Power plants with zero liquid discharge systems are expected to have wider public acceptance and acquire statutory approvals smoothly. Further, zero liquid discharge systems offer more options for site selection, as the power plant companies can select a site, away from the natural watercourse. In addition, wastewater recycling considerably reduces the freshwater demand of power companies. It further eliminates the costs linked with sewage disposal. Therefore, using innovative wastewater treatment technologies and optimization of effluent treatment plant operations serves to a viable option to bring down the freshwater demand.

9.3 Water Requirement in Thermal Power Plants

The main source of water for Indian thermal power plants is sea water or surface water sources being rivers, canals and ponds. In some cases, groundwater sources are also used for meeting the freshwater requirement of thermal power plants. The cooling water systems generally are of two types: direct cooling system and an indirect cooling system. In a direct cooling system, the water is returned to the source known as once-through cooling. In an indirect cooling system, cooling towers are a part of a closed-circuit system and only the makeup water is drawn from the source. Once-through cooling system is generally used for the coastal power plant as a huge amount of sea water source is available in the power plant vicinity. However, with the new environmental regulations, the once-through cooling needs to be replaced with the closed cooling system. Closed cooling system with the evaporative cooling tower is industrial practice for a power plant which usually receives water from river, bore well or canal water.

Currently, coal-based units contribute to 61% of the total power generation capacity in India. To meet the demand of the rapidly growing economy, coal is expected to be the main source of power generation in the future, although renewable energy will be the significant contributor due to the growing environmental concerns (Nihalani 2018). Water is an important requirement in the power plant for condenser cooling, cycle makeup, coal dust suppression, ash slurry preparation, HVAC makeup, cleaning and drinking. Rapid industrialization and the increasing water demand for water power and other industry sectors are turning water to be a scarce resource.

Freshwater availability for power sector has substantially reduced over the past few decades. This compels the power companies to bring down their freshwater intake and thereby contribute towards environmental protection. They employ various water conservation measures like wastewater treatment and reuse, increasing the cycle of concentration, use of high concentrated ash slurry disposal system, 100% dry fly ash utilization and the introduction of less water-intensive technology like air-cooled condensers.

Consumptive water requirement for the power sector is regulated several factors like raw water, condenser cooling system type, ash utilization, ash disposal system, wastewater treatment aspects, etc. In the past, power plants having generous water requirements were designed. Total makeup water requirement for a representative 2 × 660 MW coal-based supercritical plants has been estimated to be 4200 m^3/h which corresponds to the specific water consumption of around 3.2 m^3/MWH. This water requirement is established considering no recovery of water received from ash pond during the initial years of plant operation. Out of total water consumption value of 3.2, 2.2 m^3/MWH is being consumed by circulating cooling water system and 0.6 m^3/MWH is being consumed by an ash handling plant.

The following strategies can be used to optimize water consumption in thermal power plants:

- Minimize the withdrawal of freshwater and promoting recycling and reuse.
- Segregating the different wastewater streams to treat and reuse them differently.
- Recycling and reuse of plant service wastewater and ash dyke water.
- Recycling of clarifier sludge and filter backwash water.
- Using boiler blowdown for cooling tower makeup.
- Using treated wastewater for coal dust suppression and gardening.
- Treatment and reuse of sewage.
- Treatment and recycle of boiler blowdown by adopting membrane technology.
- Decentralized advanced effluent treatment plant.
- Using advanced chemical treatment for cooling water system to maintain high cycle of concentration.
- Adoption of higher cycles of concentration (4–5) to reduce blowdown water quantity.
- Adoption of dry air-cooled condensers to handle water scarcity.
- Use of dry cooling eliminates evaporation losses in the wet cooling system and also the need for CW makeup.

9.4 Water Resource Management in Thermal Power Plants

Water requirement in power generation is regulated by several factors like raw water quality, condenser cooling system, coal quality, utilization of ash, disposal of ash, wastewater handling, etc. Based on water quality and end-user concerns, the water utility in power plants can be divided into the following types:

1. Ash Handling Water: Water used for handling ash generated during the combustion process into slurry for disposal. It is ideal to use treated water for ash handling and coal dust suppression.
2. Cooling Tower Water: Water used in the cooling towers.

3. Makeup Water: Water used to compensate for the loss due to the evaporation of water in the cooling tower.
4. Demineralized Water (DM water): Water used in the boilers for generating steam must be demineralized. DM water comes at a higher cost as compared to cooling tower water, and its use is limited for a specific function.
5. Potable Water: Water used for drinking purpose within the plant premises and a nearby residential colony is known as potable water.
6. Service Water: Water used for purposes like coal dust suppression, fire fighting measures, toilets, and other utilities, and landscaping is termed as service water. It is recommended to use treated water for these purposes in place of freshwater.

9.4.1 Cooling Water System Water Requirements

Cooling water system mainly uses water to condense the steam coming out from turbine exhaust. The heat of condensed steam will again be rejected to the atmosphere with the help of wet cooling tower. In a closed cooling water system, cooling tower rejects the heat of hot water coming out from the condenser by evaporating some amount of water in the atmosphere. The process of evaporation inside cooling tower increases the total dissolved solid levels of circulating water which in turn increases scaling and corrosion problems in the system. To maintain dissolved solids levels in the closed cooling water system, some amount of water needs to be drained out from the system by making up freshwater to the system. The limit of dissolved solids levels in the cooling water system is dictated by a cycle of concentration. The cycle of concentration (COC) is how many times dissolved solids in water is getting concentrated without increasing scaling/corrosion issues and the loss of water to maintain COC of the system is termed as CW blowdown. In a closed cooling water system, water is required to be makeup to compensate for evaporation loss, blowdown loss and drift losses. Increasing COC of the system reduces the blowdown loss from the system; however, it will impact the circulating cooling water quality and affect the performance of the cooling water system. Generally, river water-based power plants are designed to maintain COC as 5 and coastal power plants using closed cooling water system are designed to maintain COC in the range of 1.2–1.3.

9.4.2 Ash Handling System Water Requirements

Ash is generated due to the burning of coal inside the boiler which needs to be disposed of to ash dyke. There are two types of ash generated from the boiler. Ash collected at bottom of boiler and economizer is termed as bottom ash, and ash collected at bottom of ESP is termed as fly ash. Generally, bottom ash and fly ash are generated in the ratio of 20 and 80% of the total ash generation during the burning of coal. In the ash handling system, water is required mainly to prepare ash slurry in specific concentration for evacuating bottom ash and fly ash to ash dyke with the help of slurry pumps (Nihalani et al. 2020). Water requirements in ash handling plant mainly depend on the type of ash handling system and quantity of ash available in the coal. For bottom ash disposal, mostly the practice is to use lean slurry disposal system in which 20–22% concentrated slurry is made with the help of water and transferred to ash dyke with the help of bottom ash slurry disposal system. However, a new trend of the dry bottom ash handling system is finding a place in recent approaches. For fly ash disposal, general practice is to use either of the following systems:

- The lean slurry disposal system
- High concentrated slurry disposal system
- The dry fly ash disposal system.

High concentrated slurry disposal system (HCSD) uses disposal of 50–55% concentrated fly ash slurry through HCSD pumps. Hence, water requirements of HCSD system are less as compared to the lean slurry fly ash disposal system. Further, due to MoEF norms for 100% utilization of fly ash from the fourth year of plant operation, all the power plants are planning to dispose of fly ash in wet mode during the initial operational period and then using dry fly ash disposal system for disposing fly ash in later years. Power plants are also equipped with ash water recirculation system to recover clear water from ash dyke area. With the installation of ash water recirculation system, approximately 70% of water can be recovered and utilized in ash handling system for slurry preparation after 1 year of plant operation.

With the help of water recovery from ash dyke after one year and 100% dry fly ash utilization after 4 years of plant operation, water requirements in the ash handling plant are expected to reduce to a considerable extent. However, during initial periods of plant operation, a considerable amount of water requirement is envisaged for wet bottom ash and fly ash disposal system with no recovery from ash dyke. For a typical 2 × 660 MW plant capacities with Indian coal quality that consists of high ash content, the total makeup water requirement for ash handling plant is estimated in the following range.

Table 9.1 shows that ash handling plant consumes 1.3–1.5 m^3 of water for the production of 1 MWH electricity. For the plant using lean slurry disposal system for the bottom and fly ash disposal, the above water requirement is reduced to 500–600 m^3/h after the first year of plant operation when recovery of water from ash dyke area is available. However, for a plant using the HCSD system for fly ash disposal, water requirement remains in the range of 700 m^3/h because no water could be

Table 9.1 Water requirement for ash handling

Description	Bottom and fly ash disposal system in lean slurry mode with no ash water recovery	Bottom ash disposal in lean slurry mode and fly ash disposal in HCSD mode with no ash water recovery
Makeup water requirement	1800–2000 m^3/h	1300–1400 m^3/h

recovered from fly ash HCSD system. However, during the fly ash utilization period, the water requirement will be in the range of 220–250 m^3/h approximately.

Total makeup water requirement of ash handling plant during the initial period of plant operation is generally being met by cooling tower blowdown water and remaining by raw water. During the ash water recovery scenario, water requirement can be met by cooling tower blowdown water and no additional raw water is needed. However, during less water requirement in case of fly ash utilization after the fourth year of plant operation, cooling tower blowdown water could not be fully utilized in ash handling plant for slurry making and this creates a situation of excess availability of cooling tower blowdown water. The scenario of existing power plants using dry fly ash utilization and recovery water from ash dyke creates concern for power companies on how zero liquid discharge could be achieved.

Thus, various recommendations for reduction in water consumption for ash handling include:

- Using optimum ash water ratio in the range of 25–30% ash by weight in the slurry.
- Fly ash handling system which is dry.
- Typically, 70–80% recovery of ash water from ash pond.
- Adopting new technology like high concentration slurry disposal system for ash handling because this system uses typically more than 60% solids by weight.
- 100% fly ash utilization by the fourth year as per MoEF notification.

9.4.3 Coal Handling System Water Requirements

In thermal power plants, coal is used as working fuel in the boiler for generating steam from water. The required amount of coal needs to be met from nearby coal mines through conveyer, railway wagons or trucks. In the process of making and transporting suitable size of coal in the boiler from the coal storage area, water is required to spray during unloading and transportation to avoid the emission of coal particles in the atmosphere. It is also required to spray water at the coal storage area to avoid spontaneous ignition of coal. Water requirement for coal handling plant varies from plant to plant based on the location of coal storage, numbers of coal stockpiles and numbers of transfer points. For typical 2 $\times$ 660 MW units, water requirement for coal handling plant is estimated in the range of 100–150 m^3/h.

9.4.4 Demineralized Water System Water Requirements

Demineralized water is mainly used as power cycle makeup in the condenser to compensate blowdown and other steam losses that occur in case of subcritical power plants. In supercritical power plants, cycle makeup is only required to compensate for the steam losses and other losses due to leakages. Power cycle makeup for supercritical power plants is generally considered as 0.5–1% of steam flow. Further consumers of demineralized water in power plant are condensate polishing units, demineralized water generation plant and closed auxiliary cooling water system. Condensate polishing plant is installed in all supercritical power plants to maintain the stringent quality of demineralized water in the power cycle as well as to reduce power cycle makeup water requirement. Condensate polishing plant is placed at condensate extraction pump discharge to treat the condensate coming from the condenser. In the condensate polishing plant, demineralized water is required to regenerate the resins of condensate polishing units. DM water is also used as makeup for the closed auxiliary cooling water system for water loss compensation due to leakages. For typical 2 × 660 MW supercritical power plant, total demineralized water requirement is estimated in the range of 50–60 m^3/h.

9.5 Water Balance for 2 × 660 MW Coal-Based Power Plants

The development of optimized water resource management for thermal power plants involves reviewing the present water balance scenarios within the plant premises. For a review of present water resource management, the water balance of 2 × 660 MW coal-based supercritical power plants has been used considering all modern practices followed by recent power plants. The following design considerations are adopted to develop typical water balance of 2 × 660 MW plant:

- River water is considered a source of fresh raw water.
- The water-cooled condenser is considered for circulating cooling water system, and the cycle of concentration for the system is considered as 5.
- Evaporation and drift losses inside the wet cooling tower are considered to be 1.6% and 0.005% of cooling tower flow, respectively.
- Cooling tower blowdown is waste having high dissolved solids. However, this waste will be utilized directly without treatment for ash slurry preparation in ash handling plant.
- Total demineralized water requirement is considered as 1.5% of flow considering regeneration water requirements, and conventional ion exchange resin-based demineralization (DM) plant is considered to generate the DM quality water.
- Demineralization and condensate polishing plant waste are generated during the regeneration of ion exchange resin. This waste is having high dissolved solids and will be utilized in ash handling and coal handling plant after neutralization of waste.
- Pretreatment plant sludge generated at bottom of clarifiers will further be treated using thickener and dewatering centrifuge for the generation of cake in semi-dry form.

- Floor wash along with oily waste generated will be treated in the clarifier and oily water separator to recover clear water. Clearwater will then be utilized to meet ash handling and coal handling plant water requirements.
- Generally, lean slurry disposal system is preferred for bottom ash disposal while high concentrated slurry disposal system is preferable for fly ash disposal during initial operation of the plant.
- Approximately 70% of the water from ash dyke is considered to be recovered after 1 year of plant operation and 100% fly ash will be utilized after fourth years of plant operation.
- Water requirement for future flue gas desulfurization (FGD) units is not considered while reviewing the present scenario of water balance.

Based on the above design consideration, net raw water makeup requirement for typical 2 × 660 MW coal-based supercritical plant has been estimated as 4230 m^3/h which corresponds to the specific water consumption of around 3.2 m^3/MWh. This water requirement is established considering no recovery water received from ash pond during the initial years of plant operation (Table 9.2).

The specific water consumption will further be reduced during ash water recovery scenario as well as 100% dry fly ash utilization period. However, during 100% dry fly ash utilization period there will be excess cooling tower blowdown in the range of 220–250 m^3/h within the plant and zero liquid discharge could not be achieved. Further, these values are estimated considering river water-based plant and COC of 5. For a plant which receives bore well water and sea water as raw water above specific water consumption values increases significantly. The specific water consumption values for current scenarios of river water and sea water-based plant using a closed cooling water system are shown in Table 9.3.

9.6 Analysis of New Environment Norms and Its Impacts

The study of new environment norms issued by MoEF reveals the following facts for coal-based power plants:

- Once-through cooling system is now discarded for all existing power plant in addition to the upcoming new power plants. Environment rules mandate to install a closed cooling water system for all existing once-through power plant. Further, rules also mandate specific water consumption of 3.5 m^3/MWH for all existing once-through power plant within 2 years.
- All new coal-based plants to be operational after 1 January 2017 should reduce specific water consumption to 2.5 m^3/MWH and achieve zero liquid discharge (Sharma and Mahapatra 2018).
- All existing, as well as new power plants, should meet the stringent requirements of SO_2 and NO_x emissions. To meet the new emission standards, most of the power plants should install flue gas desulfurization units.

Table 9.2 Consumptive water requirement of 2 × 660 MW power plants

S. No.	Systems	Water consumption (m^3/h)	Specific water consumption (m^3/MWH)
1	Makeup water requirement in the condenser cooling water system	3000	2.27
2	Total makeup water requirement for ash handling plant	1400	1.06
3	Total makeup water requirement for coal handling plant	150	0.11
4	DM water requirements	60	0.05
5	Miscellaneous service water and potable water requirements including HVAC, pump sealing	250	0.19
	Total water requirements	4860	3.68
6	CW blowdown water generated	600	–
7	Other waste generated including DM plant waste, CPU waste, floor washing waste and oily waste	30	–
	Total waste collected	630	–
9	Reduction in ash handling plant makeup water requirement due to the utilization of plant waste	480	–
10	Reduction in coal handling plant makeup water requirement due to the utilization of plant waste	150	–
11	Net raw water makeup requirement	4230	3.2

From above, it is observed that all the upcoming power plants irrespective of water sources available (i.e. sea water/river water/bore well) have to install closed cooling system and flue gas desulfurization units. Further, flue gas desulfurization system also uses water for the removal of SO_2 and other impurities from flue gases. There are various technologies available for SO_2 removal from the flue gas stream. Among various technologies including sea water-based FGD system, wet limestone FGD system is considered in this assessment for large coal-based power plants. Makeup water requirement for wet limestone FGD system has been estimated in the range of 220–250 m^3/h based on supplier's inputs for 2 × 660 MW units. The implementation

Table 9.3 Water consumption for river and sea water-based plant

S. No.	Description	Without ash water recovery	With ash water recovery	With ash water recovery and 100% fly ash utilization
A	Sea water-based plant considering COC of 1.3			
1	Specific water consumption	8.7 m^3/MWH	8.7 m^3/MWH	8.7 m^3/MWH
2	With ash water recovery and 100% fly ash utilization	7300 m^3/h	7800 m^3/h	8400 m^3/h
3	Is zero liquid discharge achieved	No	No	No
B	River water-based plant considering COC of 5			
1	Specific water consumption	3.2 m^3/MWH	2.65 m^3/MWH	2.5 m^3/MWH
2	Excess CW blowdown water available in plant	0 m^3/h	0 m^3/h	220–250 m^3/h
3	Is zero liquid discharge Achieved	Yes	Yes	No

of new environment norms must need to evaluate various alternatives for further optimization of water requirements and achievement of zero liquid discharge in coal-based power plants.

9.7 Adoption of Dry Cooling System

In this alternative, water resource management using a dry cooling system that is air-cooled condenser is evaluated with no change in ash handling system operation (Owen and Korger 2010). In the air-cooled condenser, heat is rejected to the atmosphere by sensible cooling only. It does not involve evaporative cooling (Mishra and Arya 2015). Hence, makeup water requirement for condenser cooling reduces to zero. The water resource management for a plant using an air-cooled condenser with fly ash HCSD system and plant using wet cooling tower with fly ash HCSD system can be correlated as below.

Table 9.4 shows that specific water consumption value of this alternative is almost 50% lower than the wet cooling system. With these alternatives, new environment norms related to constricted specific water consumption, as well as zero liquid discharge, can be achieved easily in all operating scenarios of plant, irrespective of water quality.

Table 9.4 Consumptive water requirement of 2 × 660 MW power plant with WCC and ACC

S. No.	Systems	Water consumption (m^3/h)	Specific water consumption (m^3/MWH)	Water consumption (m^3/h)	Specific water consumption (m^3/MWH)
		Water-cooled condenser		Air-cooled condenser	
1	Makeup water requirement in the condenser cooling water system	3000	2.27	0	0
2	Makeup water requirement in the auxiliary cooling system	–	–	200	0.15
3	Total makeup water requirement for ash handling plant	1400	1.06	1400	1.06
4	Total makeup water requirement for coal handling plant	150	0.11	150	0.11
5	DM water requirements	60	0.05	60	0.05
6	Miscellaneous service water and potable water requirements including HVAC, pump sealing	250	0.19	250	0.19
7	FGD system makeup water requirements	300	0.22	300	0.22
8	Total water requirements	5160	3.9	2360	1.8
9	CW blowdown water generated	600	–	70	–

(continued)

Table 9.4 (continued)

S. No.	Systems	Water consumption (m^3/h)	Specific water consumption (m^3/MWH)	Water consumption (m^3/h)	Specific water consumption (m^3/MWH)
		Water-cooled condenser		Air-cooled condenser	
10	Other wastes generated including DM plant waste, CPU waste, floor washing waste and oily waste	30	–	30	–
11	Total waste collected	630	–	100	–
12	Reduction in ash handling plant makeup water requirement due to the utilization of plant waste	480	–	100	–
13	Reduction in coal handling plant makeup water requirement due to the utilization of plant waste	150	–	0	–
14	Net raw water makeup requirement	4530	3.4	2260	1.7

9.8 Additional Water Conservation Techniques

The utility of tremendously large volume of water in the thermal power plants demands that the thermal power plant companies undertake numerous water conservation measures. Apart from the use of an air-cooled condenser, following water conservation techniques may also help to achieve new environmental norms of water consumption for many power plants.

9.8.1 Adoption of 100% Utilization of Fly Ash in Dry Mode

Present water balance of river water-based plant using HCSD system for fly ash disposal estimates specific water consumption in the range of 3.2 m^3/MWH during the initial period of plant operation. This meets the criteria of new environment norms specified for many existing power plants. These power plants could accomplish definite water consumption of 3.5 m^3/MWH by utilizing 100% dry fly ash disposal in place of the conventional lean slurry disposal system and using recovery water from ash dyke.

For river water-based plants to be installed after January 2017, new environment norms of 2.5 m^3/MWH could be achieved using 100% fly ash disposal in dry mode after 1 year of plant operation when recovery of water also starts from ash dyke area. However, during the above scenario, there was surplus CW blowdown water in the range of 220–250 m^3/h which could now be utilized to meet the makeup water requirement of recently added FGD system. Further, the wet FGD system also generates waste in the range of 10–15 m^3/h for 2 × 660 MW capacity plant. As waste from FGD system is less, this can be disposed to ash dyke after the removal of suspended solids and heavy metals through a suitable treatment process. With the above consideration, zero liquid discharge could be possible in the plant. This alternative helps to achieve new environment norms for river water-based plant after 1 year of plant operation without a substantial change in present systems of the plant. However, for plants where 5 COC of water could not be achieved due to the quality of raw water (in case of bore well and sea water), it is not possible to achieve new environmental norms of 2.5 m^3/MWH and zero liquid discharge with the adoption of 100% dry fly ash utilization.

9.8.2 Increasing Cycle of Concentration for Circulating Cooling Water System

The makeup water requirement in the CW system is a summation of cooling tower evaporation and drift losses and CW blowdown. The cooling tower evaporation and drift losses are a function of cooling tower performance and could not be reduced. CW blowdown loss is a function of the cycle of concentration being maintained in the CW system. The cycle of concentration for CW system will depend on the quality of makeup water. Generally, river or canal water having low dissolved solids may achieve a higher cycle of concentration in the range of 6–7. Most of the recent river water-based plant has been designed to achieve 5 COC. The increasing cycle of concentration will reduce the amount of CW blowdown waste which in turn will reduce the net CW makeup water requirement. For typical 2 × 660 MW plant, increasing COC from 5 to 6 will reduce CW blowdown quantity in the range of 60–80 m^3/h.

9.8.3 Installation of Ash Water Recovery System from Ash Dyke

Ash dyke receives the ash slurry from bottom ash and fly ash disposal system. In ash dyke, heavy particles of ash will get settled at the bottom. Over a period of time, clear water from ash slurry will overflow from the ash dyke area. To reuse clear water received from the ash dyke area, recovery system shall be installed. In ash water recovery system, clear water overflow from ash dyke area will be passed through clarifiers for further clarification. Clarified water can then be recycled or reused for ash slurry preparation. Generally, overflow from ash dyke area starts after one year of plant operation based on ash dyke volume. Installation of ash water recovery system recovers approximately 70% of water. Hence, ash handling plant water requirement during ash water recovery scenario will reduce to 30% of water being supplied at initial plant operation.

9.8.4 Recycling of Cooling Water Blowdown to Other Systems

The cooling water (CW) blowdown is waste generated from circulating cooling water system to maintain dissolved solids levels. CW blowdown can directly be reused for ash slurry preparation as well as for coal dust suppression system in power plant. Reusing of CW blowdown in ash handling plant and coal handling plant also reduces fresh makeup water requirements. However, dry fly ash utilization, as well as ash water recovery scenario, creates a situation of excess CW blowdown waste available within the plant. This excess CW blowdown water needs to be treated and recycled suitably to meet the zero liquid discharge condition.

9.8.5 Optimization of the Ash–Water Ratio

Out of the total water used in thermal power plants, around 45–50% is utilized for ash handling. Using treated wastewater or outlet from the condenser reduces freshwater demand significantly. Bottom and fly ash are flushed using high-pressure water while water with low pressure is utilized for ash hopper filling. Bottom and fly ash can be managed individually or collectively in a common pit. The ash slurry is then drained to ash dyke with the help of ash slurry disposal pumps. Typical ash water ratios considered for design are 1:8 for bottom ash and 1:5 for fly ash. However, combined ash water ratios found in the field are around 1:20. It is presumed that with one percent reduction of ash water ratio, saving water potential is about 60 m^3/h of water.

9.8.6 Reducing Leakage and Overflows

Various sources of water loss in plant premises can be leakages from valves, taps, firefighting hoses, firefighting lines, cooling tower basin, gardening hoses, etc. Overflows from overhead tanks as well as cooling towers of AC plants due to non-functional float systems are a common feature in many thermal power plants. This water loss can be significantly curtailed by plugging the leakages and checking the overflows.

9.8.7 Recycling of Treated Wastewater

Various power plants have shown per capita availability of water to be in the range of 600–700 litres per capita per day against the WHO norm of 110 litres per capita per day. The per capita demand can be significantly reduced by reducing the running hours for the water pumps by about 25% which will further reduce the overuse of water. Higher water demand can also be attributed to water losses due to lower operating efficiencies of pumps as compared to their design values. This high-water demand can be brought down significantly by employing an effective wastewater treatment system, to enable treatment, recycle and reuse of wastewater within the plant premises. The installation of such wastewater treatment plants shall enable recycling and reuse of about 60–80% of the wastewater generated which can be reused for gardening, dust suppression and firefighting.

9.9 Approaches for Zero Liquid Discharge

Zero liquid discharge means zero discharge of wastewater from the thermal power plant. A ZLD system includes various advanced wastewater treatment technologies to recycle, recover and reuse "treated" wastewater to ensure there is no wastewater discharge to the environment.

A typical ZLD system consists of the following components:

1. Pretreatment
2. Reverse osmosis (membrane process)
3. Evaporator and crystallizer (thermal processes).

Various wastewater sources in power plants include cooling tower blowdown and wet flue gas desulphurization systems. The wastewater contains salts of sodium sulphate, sodium chloride, calcium, magnesium and bicarbonates. Effluent treatment plants used in thermal power plants usually comprise evaporators, crystallizers and dewatering units to remove these salts and produce clean water for recycling within plant premises.

Cooling tower blowdown, generally contains total dissolved solids less than 10,000 mg/L hence, reverse osmosis used to pre-concentrate this stream before concentrating it in an evaporator and the remainder is reduced to solids using a crystallizer. The salts present in cooling tower blowdown are usually composed of sodium sulphate, sodium chloride, and small quantities of calcium, magnesium and bicarbonate. All these salts can be crystallized readily by involving evaporation method.

Wastewater from wet flue gas desulphurization systems consists of as calcium and ammonium chlorides in soluble form in addition to heavy metal salts; hence, it is difficult to crystallize this stream by using evaporation. Such wastewater shall be given pretreatment with chemicals to remove the crystalline solids produced. For this kind of pretreatment, wastewater is taken to reactor tanks, where caustic soda, lime, sodium sulphide or organosulphide may be added to precipitating heavy metals as insoluble hydroxides and sulphide salts. Then, ferric chloride, alum and polymers are added carrying out coagulation, flocculation and sedimentation of remaining solids. This treatment method is adopted for removing suspended solids, metals and acidity from the wastewater, but it is ineffective in removal of soluble salts like calcium, magnesium, sodium and ammonium combined with chloride, nitrate or organic compounds. These soluble salts can be further removed by employing biological treatment process. Thus, a ZLD system including evaporators and crystallizers, when integrated with the high recovery RO plant, shall prove to be more effective and economical.

Storm Water Separation or Rainwater Harvesting

Separate RCC drains for collection of rainwater or storm water from plant and township shall be provided. Storm water drains from plant and township shall be connected to storm water pump houses separately. The water from storm water pump houses shall be sent to reservoirs.

Effluent Treatment Plant

The quality of wastewater regulates the overall design of an effluent treatment system. Depending on wastewater quality, wastewater may be subjected to a suitable treatment system to remove or reduce the following:

(i) Suspended solids
(ii) Oil and grease
(iii) Biochemical oxygen demand and chemical oxygen demand
(iv) Biodegradability of organics in the wastewater.

During these treatment operations, it may be required to adjust the pH, add necessary chemicals or maintain dissolved oxygen concentration for getting desired results.

Sewage Treatment Plant

The sewage treatment plant will treat the raw sewage of township generated from water closets, kitchen and bathrooms and include the waste from other non-residential buildings, i.e. club restaurant, guesthouse and canteen. The treated sewage quality will meet permissible standards as prescribed by the pollution control board. After treatment, BOD shall be reduced from to 30 mg/L and TSS shall be reduced to 100 mg/L depending on the disposal norms. The treated sewage shall be disinfected before being used for horticulture or gardening in the township. There shall be no treated sewage discharge outside the plant premises.

9.10 Discussions and Conclusion

New environment norms for water reduction and zero liquid discharge were reviewed along with existing plant, water resource management and possible alternative options are discussed and compared to optimize the water requirements for existing as well as future power plants. The outcome of the various alternative options discussed in this paper is summarized below:

- For Indian conditions, adopting a dry cooling system like ACC shall reduce the plant output by about 7% and increase the gross heat rate by about 7%. Hence, the capital cost of the plant per MW shall increase by about 10%[7].
- For future, river water-based plants, installation of wet limestone FGD system, HCSD system for fly ash disposal and achieving 100% utilization of dry fly ash from the first year could attain explicit water consumption of 2.5 m^3/MWH with zero liquid discharge after 1 year of plant operation when recovery of water from ash dyke starts.
- For existing river, water-based plants, installation of wet limestone FGD system and achievement of 100% dry fly ash utilization could meet the new environmental norms.
- For seawater based once through cooling thermal power plants, Installation of the closed cooling system, as well as FGD system, is now the necessity of new environmental norms. Since sea water FGD plant requires a huge amount of sea water, wet limestone FGD system is the probable solution to reduce the FGD water requirements in sea water-based plants. Further, installation of the wet cooling tower for sea water-based plants also end up with a huge amount of excess CW blowdown water in the range of 7300–8400 m^3/h. Achievement of zero liquid discharge with such huge amount of excess blowdown does not seem practical with the help of available water treatment technologies in the market.
- Specific water consumption of 3.5 m^3/MWH could also not be achieved for the existing sea water-based plant with the installation of wet cooling towers. At present, the dry cooling system (air-cooled condenser) is only probable alternatives to meet the new environment norms for sea water-based plants. However, installation of the air-cooled condenser to existing sea water-based plants using a water-cooled condenser is challenging in terms of meeting rated plant output as well as retrofitting of the air-cooled condenser in place of the water-cooled condenser.
- For new as well as existing plants using bore well water which operates on low COC of 3–4, specific water consumption with zero liquid discharge could be

achieved by utilizing 100% fly ash and installing wastewater treatment plants for excess blowdown water. However, in some of the cases, costly thermal desalination technology may be required to achieve zero liquid discharge.
- Industry and academia shall work towards reducing the operating performance and capital cost gap between dry cooling and the wet cooling. The stringent environmental norms and scarcity of water favour the use of the dry cooling system as it shall be the prominent technology in the future.

Acknowledgements The study is part of Industry Defined Research Project titled "To Reduce Water Consumption and Optimize Efficiency of Thermal Power Plants by Using Air Cool Condensers" funded by Royal Academy of Engineering under Newton Bhabha Fund in collaboration of L&T Power with Parul University.

References

Mishra P, Arya M (2015) A review of literature on air-cooled steam condenser. Int J Res Aeronaut Mech Eng 3(10)

Mishra YD, Belligundu K, Parekh N (2016) Paper on water resource management for 2 × 660 MW coal-based power plant and comparatives for wet and dry cooling system. New Delhi, India

Nihalani SA (2018) Emission control technologies for thermal power plants. IOP Conf Ser: Mater Sci Eng

Nihalani SA, Mishra YD, Meeruty AR (2020) Handling and utilisation of fly ash from thermal power plants. In: Ghosh S, Kumar V (eds) Circular economy and fly ash management. Springer, Singapore

Owen MTF, Kröger DG (2010) The effect of screens on air-cooled steam condenser performance under windy conditions. Department of Mechanical and Mechatronic Engineering, University of Stellenbosch, Private Bag X1, Stellenbosch 7599, South Africa

Sharma N, Mahapatra SS (2018) A preliminary analysis of increase in water use with carbon capture and storage for Indian coal-fired power plants. Environ Technol Innovation

Chapter 10
Geoengineering Structures of Crabs and Their Role in Nutrient Cycling in Mangrove Ecosystem of Mahanadi Delta, Odisha, India

Kakoli Banerjee, Nihar Ranjan Sahoo and Gopal Raj Khemundu

Abstract Ecological function of bioturbation and nutrition profiling by the crabs are considered to be the best ecological engineering works affecting the physical and chemical processes of the sediments. In the present survey programme, an attempt has been made to study the burrow structure of two families of brachyuran crabs, viz. Ocypodidae and Grapsidae along with their role in nutrient cycling (with respect to organic carbon and nitrogen). Burrow structure was studied through plaster of paris cast and L-, J-, Y-, I- and S-shaped burrows were identified and recorded. Seasonal analysis of surface sediment temperature, pH, salinity, sediment texture, organic carbon and nitrogen was performed at all the four sampling sites. Depth analysis of sediment nutrients was performed at four selected sites namely Sanatubi, Batighara, North Jamboo and South Jamboo where a significant variation of sediment nutrient has been proved at all depths between 0–5 cm, 5–10 cm and 10–15 cm, respectively. Significant relationship ($p < 0.05$) between sediment temperature, pH and salinity along with sediment nutrients was pronounced seasonally. Similar relationship ($p < 0.05$) was also shown between density sediment nutrients which proves that the density of crabs plays a vital role in maintaining the natural biogeochemical cycling process in the environment.

Keywords Carbon · Crabs burrows · Bioturbation · Brachyuran crabs · Ecosystem engineer · Mahanadi delta · Mangrove sediments · Nitrogen

10.1 Introduction

Mangroves are large and widespread on the east coast of India due to the rich alluvial soil formed by the sedimentation of fine-grain silt or clay by the long and wide rivers—Ganga, Brahmaputra, Mahanadi, Godavari, Krishna and Cauvery due to its perennial supply of fresh water along the deltaic and estuarine coasts (Kathiresan and Qasim 2005). Mahanadi is the third-largest delta of east coast of India which

K. Banerjee (✉) · N. R. Sahoo · G. R. Khemundu
Department of Biodiversity and Conservation of Natural Resources, Central University of Orissa, Landiguda, Koraput 764020, India

R. M. Singh et al. (eds.), *Environmental Processes and Management*, Water Science and Technology Library 91,
https://doi.org/10.1007/978-3-030-38152-3_10

is claimed as shrinking. It is the ecological and socio-economic hub of the state of Odisha. The river is a complex morphologic and hydrodynamic system. Its 1400 arcuate-shaped delta is an amalgamation of three river deltas (the rivers Brahmani, Baitarani and Mahanadi). The delta's landforms are mainly denuded hills and erosion plains, whose depressions contain water bodies. The central part of the delta is distinct for its extensive plains, levees and paleochannels. The coastal parts contain spits, bars, lakes, creeks, swamps, beaches, tidal flat and mangroves. The mangrove sediment of Mahanadi delta was Pleistocene deposits comprised of clay, sand, silt and 'kankar', with reddish-brown cemented pebbles and gravels due to high degree of oxidation which little bit resemble with Bhitarkanika mangrove sediments. Mangrove largely constitutes fertile estuary with rich source of nutrients having detritus-based food web. The vegetation forms a direct source of food for insects, crustaceans, invertebrate and vertebrates. Most of the productions from mangroves get transferred to other trophic levels by means of litterfall and detritus pathway. So, the mangrove forests are highly productive ecosystems in terms of primary production, and in contrast to other tropical forests, they allocate a large amount of biomass below ground as root system (Komiyama et al. 2008). Based on their high productivity and slow sediment carbon degradation rate, due to predominant anaerobic conditions, they function as potential carbon sinks (Alongi et al. 2001; Bouillon et al. 2011; Donato et al. 2011; Sanders et al. 2010). Mangrove sediments are known to be the world's most fertile soils as they are rich in nutrients which are deposited from the mangrove litter and detritus, out-welling of adjacent areas, and upwelling of the seawater. So, mangrove sediments are potential habitat of all invertebrate species. Mangrove forests are typically dominated by various burrowing decapods, such as ocypodid, sesarmid and portunidid crabs. They are herbivores that retain, bury, macerate and ingest litter and microalgal mats (Kristensen and Alongi 2006). By doing so, they prevent loss of nutrients and promote decomposition processes.

Bioturbation is defined as biological reworking of soils and sediments through animal activities like burrowing and feeding. Invertebrate feeding on resources in the sediments evidently affects key processes, such as organic carbon mineralization (Otani et al. 2010), nutrient dynamics (Karlson et al. 2007; McHenga and Tsuchiya 2008), sulphur and iron cycling (Gribsholt et al. 2003; Nielsen et al. 2003), sediment texture modification and particle mixing (Paarlberg et al. 2005). Such altered soil characteristics might further impact microbial activities, zooplankton recruitment and other biotic components or might generate niches for smaller organisms). One type of dominant bioturbations in coastal ecosystems is crab burrowing which can transport sediments and modify sediment texture, accelerating ecosystem nutrient cycling.

The availability of oxygen to the deeper sediments (anoxic-zone) creates a favourable environment for certain bacteria responsible for nutrient cycling. The benthic nitrogen cycle and the mineralization of organic matter are among the microbial processes that are stimulated by bioturbation (Kristensen et al. 2000). Nitrification processes can be stimulated in burrow walls that are periodically aerated by ventilation and exposure to the NH_4^+ excreted by the inhabiting animal (Mayer et al. 1995).

On the other hand, denitrification can be promoted by the facilitated NO_3^- penetration into deep sediment layers that become anoxic periodically. Crab burrows, in particular, create ideal conditions for both nitrification and denitrification processes. Through these interactions, crab burrows can effectively remove nitrogen loads from aquatic ecosystem.

Ecosystem engineers are organisms that modulate resource flows and in doing so modify, maintain and create habitats for other organisms (Odling-Smee et al. 2013). Consequently, ecosystem engineers have direct and indirect positive and negative influences on the ecological interactions and fitness of organisms that share their environment (Erwin 2008; Odling-Smee et al. 2013). Engineering impacts are often greatest when abiotic factors such as soils are modified because they integrate many resources (living space, nutrients, prey, etc.) within one locale. Organisms with small individual impacts can also have huge ecological effects, providing they occur at sufficiently high densities over large areas (Jones et al. 1994). Since their influence on an ecosystem is disproportional to their biomass, ecosystem engineers may be considered as keystone position organisms (Bond 2001).

There are two major types of crab found in mangrove vegetation, which play active role in sediment processing, bioturbation and mangrove ecosystem functioning. Firstly, ecological roles of the burrowing activities of fiddler crabs of the genus *Uca* have been widely studied in temperate regions of the world. Burrowing activities have been reported to increase drainage and oxidation of the substrate, increase the rate of decomposition of plant debris within the sediments and enhance growth of substrate microorganisms. Burrows are important to the crabs, because they serve as a refuge from predators and environmental extremes, provide water for the crabs' physiological needs and are also sites for moulting and reproduction. The architecture of these burrows is related to the characteristics (i.e. sex, size, reproductive stage, etc.) of the resident crab fauna (Lim and Diong 2003; Lim 2006). Moreover, burrow architecture is also dependent on some abiotic factors, i.e. sediment composition, vegetation, root cover, human disturbance, etc. (Penha-Lopes et al. 2009).

Secondly, leaf-eating sesarmid crabs play a significant role in the detritus food chain and energy flow in most mangrove environments (Robertson 1986; Poovachiranon and Tantichodok 1991; Lee 1998; Skov and Hartnoll 2002). The crab is known to process 30–70% of the total leaf production by eating both attached leaves and fallen leaf litter. By feeding on leaf litter and other debris that has fallen from trees, sesarmid crabs prevent tidal export of valuable nutrients from the mangrove environment. The crabs also initiate and enhance the breakdown of mangrove detritus and recycling of nutrients. After maceration and passage through crab guts, the litter is incompletely digested and returns to the environment as faecal pellets, which contain readily available substrates for bacterial colonization and ultimately for other organisms of the detritus food chain (Lee 1997; Kristensen and Pilgaard 2001), thereby maintaining the C:N balance in sediments which affects the mangrove development as well as the survival of crabs.

Within mangrove forests, deposit-feeding fiddler crabs (*Uca*, Ocypodidae) and sesarmids (Grapsidae, Sesarmidae) affect availability of resources for other living organisms in mangroves through burrow construction and maintenance, processing of plant litter and forcing of the microbial distribution and activity in sediment (Thongtham and Kristensen 2005; Kristensen and Alongi 2006; Kristensen 2008). While sesarmids have long been known to play a significant role in promoting nutrient recycling and shaping mangrove structure (Lee 1998, 2008; Dahdouh-Guebas and Koedam 2002; Thongtham and Kristensen 2005; Kristensen and Alongi 2006), it is only recently that the importance of fiddler crabs has been widely recognized (Cannicci et al. 2008; Kristensen et al. 2008; Penha-Lopes et al. 2009). Fiddler crabs and sesarmids are considered engineers of mangrove ecosystems because they have a considerable impact on ecosystem functionality. Since the engineering effects are different for sesarmid and fiddler crabs, presumably due to foraging differences, comparative work on the distribution and abundances of sesarmids and fiddler crabs is required (Kristensen 2008).

Quantifying fiddler crab and sesarmid abundance allows scaling of their engineering impact, which is important for understanding the structure and function of mangrove ecosystems and for providing essential information for the role they play in global carbon budgets. Identifying the environmental conditions influencing the spatial and temporal distribution and abundance of these crabs would allow us to make predictions of engineering impacts over a wide variety of habitats, which is essential given the inherent environmental variability of mangrove systems (Kristensen et al. 2008).

Biotic factors, such as the abundance and distribution of vegetation, are thought to substantially affect the distribution of mangrove fauna (McGuinness 1994; Kristensen and Alongi 2006; Cannicci et al. 2008; Nagelkerken et al. 2008). For instance, sesarmids are abundant in vegetated mangrove areas and rarely seen in canopy gaps. Fiddler crabs, on the other hand, may prefer vegetated or non-vegetated mangroves areas depending on the species (Hagen and Jones 1989; Skov et al. 2002; Nobbs 2003).

Studies have emphasized the importance of burrowing and leaf litter feeding crabs for functioning of mangrove ecosystems. Crabs rework the sediment by constructing and maintaining burrows and process leaf litter, which is otherwise exported by the tide. Thus, crabs can influence sediment characteristics, energy and nutrient cycling and, consequently, other ecosystem components such as tree growth. However, almost no studies have documented the nutrient cycling by crabs and geo-engineering structures of ocypodid and sesasrmid crabs in sediment nutrient cycling in Mahanadi delta.

With the above-mentioned background, the present research programme aims at studying the microhabitat of selected crab species in the intertidal zone of Mahanadi delta mangrove ecosystem with respect to sediment composition, soil salinity and soil pH, burrow structure with respect to depth, nutrient profile (carbon and nitrogen) of soil with respect to depth and the variation of selected parameters with respect to depth.

10.2 Materials and Methods

10.2.1 Study Area

The field studies were conducted in mangrove ecosystem of Mahanadi delta area (Fig. 10.1 and 10.2) extending from 20°18′ to 20°31′ N latitude to 86°41′ to 86°46′ E longitudes and are located between Kanhupur in the north and river Mahanadi in the south, Bay of Bengal in the east and Kendrapara town in the west. The selection of sites was based on their relationship to the open coast and the level of anthropogenic

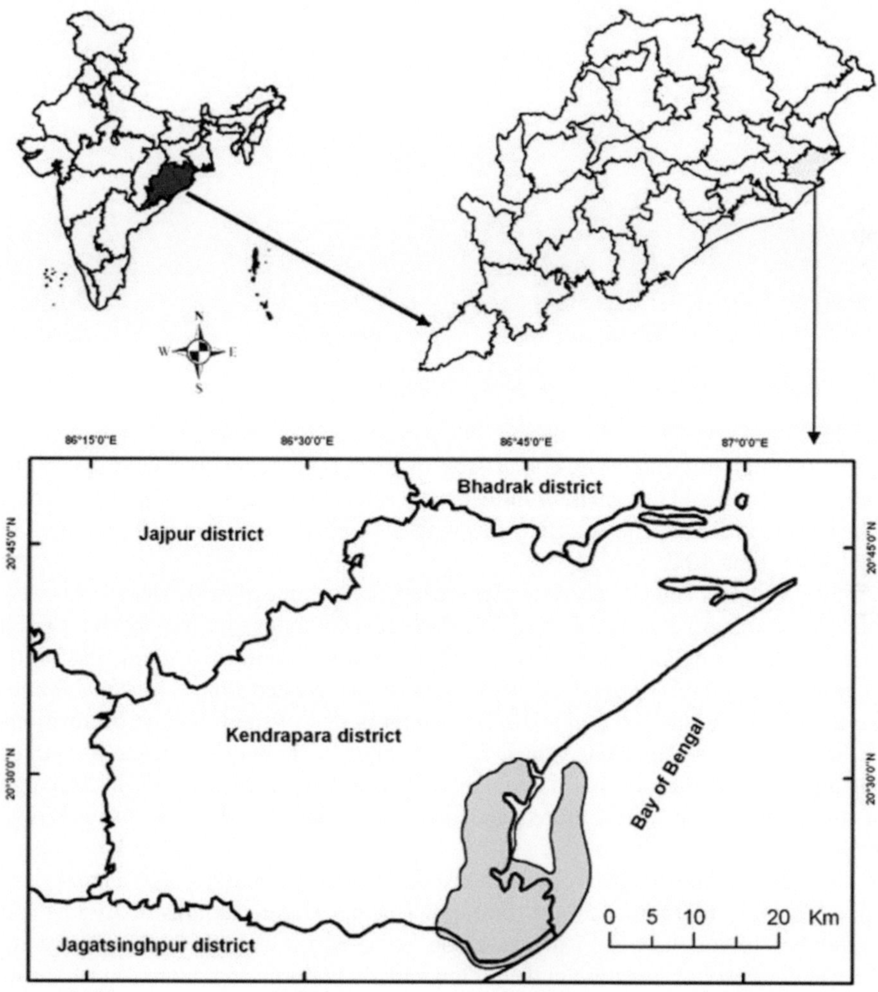

Fig. 10.1 View of study area in Mahanadi delta mangrove ecosystem

Fig. 10.2 Satellite image of Mahanadi delta showing sampling

pressure. Four different sites comprising of mangrove vegetation with dissimilar inundation level and human interference were selected along the mudflats of Gobari and Mahanadi River. The selected sites with their characteristics are represented in Tables 10.1 and 10.2.

River Mahanadi and its distributaries are the main source of fresh water to this wetland. Total mangrove cover is 45 sq. km. 26 species of mangrove have been reported from Mahanadi delta (Agarwal et al. 2017). Among them, *Avicennia officinalis*, *A. marina*, *Sonneratia apetala*, *Exocoecaria agallocha* and *Rhizophora mucronata* are the dominant species (Table 10.3). Fishing is one of the principle occupations of the people living in villages near forest fringes. Continuous increase in anthropogenic pressure has caused degradation of mangrove ecosystem. The discharge of wastewaters from aquaculture ponds has caused mass destruction of mangrove habitats.

The delta region enjoys tropical monsoon climate with an average annual rainfall of around 1800 mm. 75% of the rainfall occurs during months of August and September, although in 2017 the monsoon extended till end of October. There are three main seasons prevailing in the region namely premonsoon, monsoon and postmonsoon. The data thus collected were grouped to represent different seasons of a

Table 10.1 Selection of study sites with their characteristics

Sl. No.	Name of the sites	Latitude	Longitude	Characteristics
I.	Stations under Kharanasi block			
1	Site-1 (Sanatubi)	20°21′26.35″ N	86°42′41.32″ E	It is situated 6 km away from Bay of Bengal and west of Mahanadi delta. It is one of the important fishing stations and under anthropogenic pressure. Some small patches of *Avicennia* sp. are found in this area along with *Exocoecaria agallocha*
2	Site-2 (Batighara)	20°19′57.41″ N	86°45′06.37″ E	It is situated 4 km from sea and south of Mahanadi delta. Some villagers inhabit near the mangrove forest and fishing is the most important means of livelihood for local people. This area is under anthropogenic pressure. Degraded mangrove patches of *Avicennia sp., E. agallocha, Rhizophora sp.* are found in this area
II.	Stations under Jambu block			
1	Site-3 (South Jambu)	20°24′33.37″ N	86°42′26.14″ E	It is situated near the bank of Gobari River, north-west of Mahanadi delta. It is 4 km from sea. The area is well connected through roadways to nearest town of Kendrapara. Fishing and human habitation are a major problem. Scarce patches of *Avicennia* spp., *E. agallocha* are found in this area
2	Site-4 (North Jambu)	20°26′34.19″ N	86°43′39.40″ E	It is situated north of Mahanadi delta, mostly inhabited by local villagers. It is situated 3 km away from sea. Dense mangrove patches of *Avicennia* spp., *E. agallocha*, *Ceriops decandra* are predominantly found

Table 10.2 Brachyuran crabs found in four sites

Family	Name of the species	Stations			
		Site-1	Site-2	Site-3	Site-4
Ocypodidae	*Uca rosea*	p	p	p	p
	Uca accuta	a	p	a	p
	Uca annulipes	p	p	a	p
	Uca paradussumieri	p	p	a	a
	Uca triangularis	p	p	p	p
Grapsidae	*Episesarma bidens*	p	p	a	a
	Episesarma versicolor	p	p	p	p
	Parasesarma plicatum	p	p	a	a
	Metasesarma obesum	p	p	p	a
	Metopograpsus messor	p	p	p	p
	M. maculatus	p	p	a	a
	Metalpax longipes	p	p	p	p
	Neosarmatium meinerti	p	p	p	p

'a' means absence and 'p' means presence

Table 10.3 Dominant mangrove species of the sampling sites with respect to $10 \times 10\ m^2$ quadrates

Serial Number	Species	Site-1	Site-2	Site-3	Site-4
1	*Avicennia marina*	++	+++	+++	+++
2	*Avicennia officinalis*	+++	++	+	+
3	*Avicennia alba*	++	+	++	++
4	*Bruguiera sexangula*	–	–	+	+
5	*Bruguiera cylindrica*	–	–	–	+
6	*Ceriops decandra*	+	–	+	+
7	*Excoecaria agallocha*	++	+	++	+++
8	*Heritieria fomes*	+	++	+++	++
9	*Kandelia candel*	–	–	+	+
10	*Rhizophora mucronata*	–	+	–	++
11	*Rhizophora apiculata*	–	+	+	–
12	*Sonneratia apetala*	–	–	+	–
13	*Xylocarpus granatum*	–	–	–	+

'+' means <5; '++' means 5–10; '+++' means >10; '–' means absent)

calendar year 2016–2017 viz., monsoon 2016, postmonsoon 2016 and premonsoon 2017, respectively.

10.2.2 Sampling Technique

The identification of the burrow is an important issue in the study of burrow morphology of *Uca* species with distribution, and it was quite difficult to determine which burrow belonged to which crab species without identifying a pure distribution patch of each species. Before the inception of this study, individual representatives of the available different fiddler crab species were selected from the study area (Plate 10.1). The crabs were identified following Ng et al. (2008) and World Register of Marine Species (WORMS) according to their major morphological characters, whose identification was later confirmed by the Zoological Survey of India (ZSI), Kolkata. A $(1 \times 1)\ m^2$ quadrant was made in the intertidal zone for studying the crab microhabitat and sediment composition. To identify individual burrow of fiddler crab, each quadrate was closely monitored for about 2 h from a distance of about 1 m immediately after the tide receded. Burrows inhabited by different species were marked with wooden stems. Care was taken to ensure that the stems were away from the burrow entrances so that movement of the crabs in and out of the burrows remained uninterrupted. The total numbers of each species were counted for each quadrate until no more crabs emerged from the freshly opened burrows. The sex of each crab was visually determined (since male fiddler crab can easily be distinguished from the female by the presence of an oversized cheliped).

10.2.3 Creation of Plaster Casts to Study the Burrow Structure

The plaster casts were found to be perfect for measuring the diameter, total length and shape of the burrows (Plate 10.2). The total numbers of open burrows within each quadrate were counted. In each quadrate, three or four burrows were randomly selected for casts. Liquid solution of plaster of Paris was made by mixing up two parts of powder and one part of water. It was then poured into the selected crab burrows with the help of a syringe until the burrows were completely filled, then allowed to dry for 30–60 min (Warburg and Shuchman 1978). If the crab emerged during this process, it was handpicked, washed properly and placed in adjacent mudflat. The casts were then carefully removed from sediments by hand, or with a spade in the case of hard substratum, then cleaned to remove as much sediment as possible from the surface of the cast. Each cast was separately placed in pre-marked poly bags and brought back to the laboratory for further analyses of the burrow shape and burrow length. After the casts were dug out, the area within each quadrate was excavated to a depth of 15 cm and the crabs were counted to calculate the relationship between the density of fiddler crabs and crab burrows.

Plate 10.1 Dominant brachyuran crab species documented in study sites

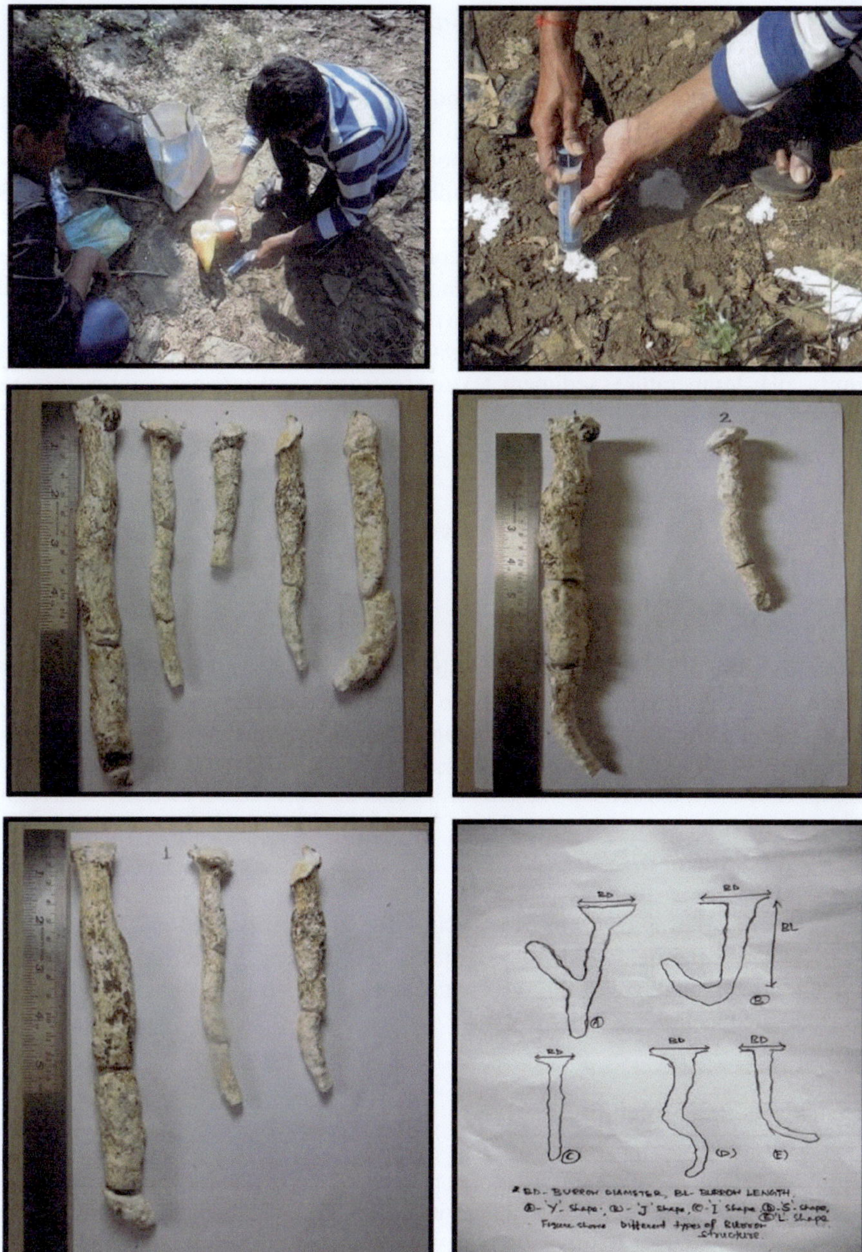

Plate 10.2 Creation of plaster cast and study of burrow structure

10.2.4 Collection of Sediment Samples

The sediment samples were collected randomly from selected quadrant of the mangrove ecosystem of Mahanadi delta in different seasons. Sediments were carefully scraped off from the opening shafts to bottom, i.e. 0–15 cm hole depth using sterilized spatula. The exact location of sampling stations was taken with Garmin GPS etrex10 (GPS). Collected sediments were kept in plastic bag with proper labelling. The sediment samples were kept in cool condition (i.e. icebox) till the samples arrive up to laboratory. In laboratory, the sediment samples were air-dried, homogenized and passed through a 2-mm-mesh sieve to remove the stone pieces and large root particles. The composite sediment sample was used for detail analysis of the sediment salinity, pH, temperature, sediment texture, nitrogen and organic carbon through standard methods.

10.2.5 Analysis of Sediment Physio-Chemical Parameters

10.2.5.1 Sediment Temperature

The sediment temperature was measured by using multi-thermometer (Model: Mextech; St-9283B). During the field study, the digital thermometer was dipped randomly in the sediment directly and five readings were taken and the average sediment temperature reading was calculated as final.

10.2.5.2 Sediment pH

The sediment pH was determined by using digital pH meter (Model: Systronics; DM15). During the field study, the digital pH meter was dipped randomly in the sediment directly and five readings were taken and the average sediment pH reading was calculated as final.

10.2.5.3 Sediment Salinity

Salinity of sediment saturation extract was determined by measuring the electrical conductivity with the help of a conductivity meter. 20 gm of sediment was taken in 100 ml beaker, and then 40 ml of distilled water was added and intermittently shaken for about 30 min. Then the electrical conductivity of saturation extract was used for salinity determination.

10.2.6 Sediment Texture Analysis

Sediment texture was measured by the hydrometer method (Gee and Bauder 1986). Fifty grams of air-dried sediment was weighed into a 600 ml container and 100 ml of 5% sodium hexametaphosphate (Calgon) was added. Five hundred millilitres of distilled water was added to the mixture and put on an automatic rotational shaker overnight. The mixture was transferred into a 1000 ml measuring cylinder and distilled water was added to the 1000 ml mark. Using a plunger, the suspension was thoroughly mixed by moving it up and down for 1 min. The hydrometer was gently placed into the cylinder, and its reading noted at 40 s and 6.52 h after plunging. The temperature (°C) of the solution was also noted. The amount of silt and clay suspended was recorded from the hydrometer reading. The following equations were used to calculation of silt and clay.

$$\text{Percentage of clay} = \left(\frac{\text{Corrected Hydrometer reading at 6.52 h}}{\text{Weight of sample}}\right) \times 100 \quad (10.1)$$

$$\text{Percentage of silt} = \left(\frac{\text{Corrected Hydrometer reading at 40 s}}{\text{Weight of sample} - \text{percentage of clay}}\right) \times 100 \quad (10.2)$$

$$\text{Percentage of sand} = 100\% - \%\text{ of silt} - \%\text{ of clay} \quad (10.3)$$

10.2.7 Analysis of Sediment Organic Carbon

Sediment organic carbon content of the sediment samples was analyzed by the standard method (Walkley and Black 1934). For organic carbon estimation, 1 gm of dried sediment was taken in 500 ml conical flask; 1 ml of phosphoric acid (H_3PO_4) and 1 ml distilled water were added. Then the mixture was heated for 10 min 100^0–110 °C. Then 10 ml 1 N $K_2Cr_2O_7$ and 20 ml concentrated H_2SO_4 with Ag_2SO_4 were added, mixed allow to stand for 30 min. Then the mixture was diluted to 200 ml, and 10 ml phosphoric acid and 1 ml indicator (Diphenylanine) were added, the colour of the mixture changes to bluish purple. Then the mixture was titrated by the help of Mohr salt solution (Ammonium iron (II) sulphate $[(NH_4)_2Fe(SO)_2 \cdot 6H2O]$ until the colour of the solution modified to brilliant green. The same titration was repeated without taking sediment and the volume of $K_2Cr_2O_7$ required to oxidize organic carbon was calculated from the difference. The calculation of sediment organic carbon was done by the formula.

$$\text{Sediment organic carbon }(\%) = \frac{3.951}{g}\left(1 - \frac{S}{B}\right) \quad (10.4)$$

where

S Mohr salt solution consumed by sample
B Mohr salt solution consumed by blank and
G weight of sediment in grams.

10.2.8 Statistical Analysis

The data collected were expressed as mean ± standard deviation. ANOVA was performed to find the difference of nutrients (carbon and nitrogen) with respect to depth. Correlation coefficient was calculated to find the relationship between crab burrow and density of crab, burrow shape, diameter and depth, respectively.

10.3 Results

India has a variety of national coastal ecosystems with beaches, marshes, lagoons and deltas, particularly in the eastern coast of India. The exposed mudflat of the east coast is dominated by almost 2600 species of marine crabs which includes *Scylla serrata*, *S. tranquebarica*, *Portunus pelagicus* and *P. sanguinolentus* among the leading species. Crustaceans comprising of prawns, lobsters and crabs accounted for 4.4% of this, which are the most highly valuable commodities, both in seafood industries and in ecosystem functioning.

In the survey of the four selected sites of Mahanadi delta, about five species of *Uca* belonging to family Ocypodidae and two species of Episesarma, one species of Parasesarma, three species of Metasesarma, one each of Metalpax and Neosesarmatium of family Grapsidae were documented (Table 10.2). Mangroves provide a suitable habitat (Table 10.3) for the physiological and metabolic processes of crabs along with the ecological abiotic interactions of the crab species with that of the ambient environment. The effect of tidal variations in the study sites (characterized by semidiurnal tide with tidal amplitudes of 3–3.5 m) plays a pivotal role in sediment nutrient interactions and in controlling the marine food web. The order of dominance of mangrove species in the study sites was *Avicennia marina* > *Avicennia officinalis* > *Avicennia alba* > *Heritieria fomes* > *Excoecaria agallocha*.

10.3.1 Crab Burrows Structure

A total of 13 crab species belonging to family Ocypodidae and Grapsidae were recorded from study sites (Table 10.2) out of which *Uca rosea* and *Uca triangularis* belonging to family *Ocypodidae* and *Episesarma versicular*, and *Neosarmtium meinerti* belonging to family Grapsidae were the most dominant. Burrow construction and maintenance are not only the engineering effect of mangrove crabs but also handling and ingestion of food substances that have a substantial effect on other organisms. Many sesarmid crab species are remarkable in their ability to consume mangrove litter (Dye and Lasiak 1987; Robertson et al. 1992). The engineering effects of fiddler crab foraging are generally confined to the upper few millimetres of the sediment (Dye and Lasiak 1986). They feed on fine particles by picking sediment from the surface using the minor chelae and placing it on mouth cavity. Here, the edible particles including microalgal cells, nematodes and bacteria are sorted and ingested. The non-edible particles are shaped into small irregular balls that are deposited at the sediment surface at regular intervals by the minor chelae (France 1998).

Most sesarmid and fiddler crabs construct and maintain burrow structures in mangrove sediment with significant engineering effect on the distribution and activities of associated organisms. The burrows are used by the crabs as a retreat or refuse when the environmental conditions at the sediment surface are intolerable, for example, during high tide or when conditions are too dry and hot (De la Iglesia et al. 1994; Botto and Iribarne 2000; Thongtham and Kristensen 2003).

The two groups of crabs generally occupy different habitats of different mangrove ecosystems. Sesarmids are most abundant on the tree canopies where their food source, leaf litter, is abundant, while the network of pneumatophores and cable roots provide protection against predators and extreme temperatures (Thongtham and Kristensen 2003). Fiddler crabs, on the other hand, prefer open areas, particularly near creeks banks, where strong sunlight stimulates growth of microphytobenthos, which is a primary food source for these crabs (Nobbs 2003). Both types of crabs clearly modify their habitat and create much of the visible topographic structures, e.g. mounds and depressions that are typical for mangrove sediments (Warren and Underwood 1986).

The morphology of fiddler crab burrows is quite simple and similar among species and usually consists of more or less permanent vertical shaft extending 10 to 40 cm into the sediment (Kristensen 2007). In the present study, and in all seasons, the fiddler crabs *Uca rosea* and *Uca triangularis* have shown in bent L or J shape ending with the chamber in all the four selected sites. In monsoon season, 2016, the number of burrows in all selected sites ranged from 4 ± 1.67 for *Uca rosea* at Site-2 to 7 ± 2.70 at Site-4 for *Uca triangularis.* The population of crabs, which has been calculated in term of density (No./m^2) of crabs, ranged from 3 ± 1.12 (No./m^2) in *Uca rosea* at Site-2 to 5 ± 1.58 (No./m^2) in *Uca triangularis* at Site-4. The burrow diameter ranged from 1.10 ± 0.29 cm for *Uca rosea* at Site-1 to 1.54 ± 0.25 cm for *Uca triangularis* at Site-2. Burrow length also varied from 9.26 ± 0.50 cm for *Uca rosea* at Site-2 to 8.34 ± 0.67 cm for *Uca triangularis* at Site-2, respectively (Tables 10.4, 10.5 and 10.6).

Burrows of sesarmid crabs varied considerably in morphological structure and some species do not construct burrow, but utilize crevices or abandoned burrows of other species (Gillikin and Kamanu 2005). Burrow shape of sesarmid crabs usually recorded in the study sites varied as per Y- or J-shaped for *Episesarma versicular* and I- or S-shaped for *Neosarmatium meinerti*. The number of burrows ranged from 8 ± 2.79 both in the case of *Episesarma versicular* and *Neosarmatium meinerti* at

Table 10.4 Average data of burrow structure found at the study sites during monsoon 2016

Site-1					Site-2			
Parameters	Ocypodidae		Grapsidae		Ocypodidae		Grapsidae	
	Uca rosea	*Uca triangularis*	*Episesarma versicolor*	*Neosarmatiummeinerti*	*Uca rosea*	*Uca triangularis*	*Episesarma versicolor*	*Neosarmatiummeinerti*
Number of Burrows	4 ± 1.70	7 ± 1.74	8 ± 2.79	8 ± 2.79	4 ± 1.67	6 ± 1.72	10 ± 2.10	11 ± 2.21
Density of crabs (No./m^2)	3 ± 1.38	5 ± 1.34	3 ± 1.48	4 ± 1.78	3 ± 1.12	5 ± 0.98	4 ± 1.09	5 ± 1.43
Burrow diameter (cm)	1.10 ± 0.29	1.50 ± 0.25	5 ± 0.49	5.5 ± 2.34	1.10 ± 0.76	1.54 ± 0.25	5.10 ± 0.65	5.8 ± 2.34
Burrow length (cm)	9.36 ± 0.50	7.21 ± 0.67	25 ± 4.94	34 ± 2.45	9.26 ± 0.50	7.67 ± 0.68	26 ± 4.94	34.65 ± 2.46
Burrow shape	L	J	Y or J	I or S	L	J	Y or J	I or S
Site-3					Site-4			
Parameters	Ocypodidae		Grapsidae		Ocypodidae		Grapsidae	
	Uca rosea	*Uca triangularis*	*Episesarma versicolor*	*Neosarmatiummeinerti*	*Uca rosea*	*Uca triangularis*	*Episesarma versicolor*	*Neosarmatiummeinerti*
Number of burrows	6 ± 2.33	7 ± 2.43	13 ± 2.56	12 ± 2.79	6 ± 2.70	7 ± 2.70	13 ± 2.79	14 ± 2.79
Density of crabs (No./m^2)	3 ± 1.38	4 ± 1.34	5 ± 1.48	4 ± 1.78	4 ± 1.58	5 ± 1.58	5 ± 1.48	6 ± 1.78

(continued)

Table 10.4 (continued)

Site-3					Site-4			
Parameters	Ocypodidae		Grapsidae		Ocypodidae		Grapsidae	
	Uca rosea	*Uca triangularis*	*Episesarma versicolor*	*Neosarmatiummeinerti*	*Uca rosea*	*Uca triangularis*	*Episesarma versicolor*	*Neosarmatiummeinerti*
Burrow diameter (cm)	1.19 ± 0.23	1.51 ± 0.26	5 ± 0.54	5.5 ± 2.11	1.20 ± 0.29	1.5 ± 0.25	5.04 ± 0.47	5.6 ± 2.79
Burrow length (cm)	9.54 ± 0.50	8.21 ± 0.67	26 ± 4.94	35 ± 2.45	9.56 ± 0.50	8.34 ± 0.67	27 ± 4.94	36 ± 2.45
Burrow shape	L	J	Y or J	I or S	L	J	Y or J	I or S

Table 10.5 Average data of burrow structure found at the study sites during postmonsoon 2017

Site-1					Site-2			
Parameters	Ocypodidae		Grapsidae		Ocypodidae		Grapsidae	
	Uca rosea	*Uca triangularis*	*Episesarma versicolor*	*Neosarmatium meinerti*	*Uca rosea*	*Uca triangularis*	*Episesarma versicolor*	*Neosarmatium meinerti*
Number of burrows	8 ± 1.34	10 ± 1.79	12 ± 2.54	15 ± 2.21	7 ± 1.62	8 ± 1.71	17 ± 2.50	16 ± 2.33
Density of crabs (No./m^2)	3 ± 1.32	5 ± 1.29	5 ± 1.56	6 ± 1.79	3 ± 1.26	3 ± 0.56	6 ± 1.54	7 ± 1.98
Burrow diameter (cm)	1.6 ± 0.65	1.66 ± 0.75	5.5 ± 0.75	5.7 ± 2.62	1.06 ± 0.76	1.54 ± 0.75	5.00 ± 0.65	5.21 ± 2.31
Burrow length (cm)	9.56 ± 0.50	7.78 ± 0.67	25 ± 3.20	34 ± 2.24	9.12 ± 0.50	7.69 ± 0.68	29 ± 4.94	34 ± 2.63
Burrow shape	L	J	Y or J	I or S	L	J	Y or J	I or S
Site-3					Site-4			
Parameters	Ocypodidae		Grapsidae		Ocypodidae		Grapsidae	
	Uca rosea	*Uca triangularis*	*Episesarma versicolor*	*Neosarmatium meinerti*	*Uca rosea*	*Uca triangularis*	*Episesarma versicolor*	*Neosarmatium meinerti*
Number of burrows	12 ± 2.66	13 ± 2.43	11 ± 2.56	18 ± 2.62	16 ± 2.78	15 ± 2.74	19 ± 2.56	17 ± 2.28
Density of crabs (No./m^2)	4 ± 1.38	3 ± 1.37	4 ± 1.48	7 ± 1.78	7 ± 1.45	5 ± 1.58	6 ± 1.88	8 ± 1.79
Burrow diameter (cm)	1.56 ± 0.77	1.78 ± 0.26	5.7 ± 0.54	6 ± 2.34	1.23 ± 0.29	1.5 ± 0.25	5.12 ± 0.47	5.8 ± 2.79
Burrow length (cm)	10 ± 0.56	8.34 ± 0.87	26 ± 4.94	37 ± 2.45	9.67 ± 0.50	8.45 ± 0.67	28 ± 4.94	36 ± 2.45
Burrow shape	L	J	Y or J	I or S	L	J	Y or J	I or S

Table 10.6 Average data of burrow structure found at the study site during premonsoon 2017

Site-1					Site-2			
Parameters	Ocypodidae		Grapsidae		Ocypodidae		Grapsidae	
	Uca rosea	*Uca triangularis*	*Episesarma versicolor*	*Neosarmatium meinerti*	*Uca rosea*	*Uca triangularis*	*Episesarma versicolor*	*Neosarmatium meinerti*
Number of burrows	10 ± 1.73	14 ± 1.79	19 ± 2.79	21 ± 2.79	9 ± 1.67	10 ± 1.78	18 ± 2.10	17 ± 2.21
Density of crabs (No./m^2)	2 ± 1.32	4 ± 1.24	3 ± 1.42	5 ± 1.79	2 ± 1.16	2 ± 0.98	4 ± 1.09	5 ± 1.43
Burrow diameter (cm)	1.8 ± 0.25	1.76 ± 0.25	5.9 ± 0.65	6 ± 2.32	1.10 ± 0.76	1.54 ± 0.25	5.10 ± 0.65	5.8 ± 2.31
Burrow length (cm)	9.60 ± 0.50	7.87 ± 0.65	25 ± 4.94	34 ± 2.34	9.26 ± 0.50	7.67 ± 0.68	29 ± 4.94	34 ± 2.23
Burrow shape	L	J	Y or J	I or S	L	J	Y or J	I or S
Site-3					Site-4			
Parameters	Ocypodidae		Grapsidae		Ocypodidae		Grapsidae	
	Uca rosea	*Uca triangularis*	*Episesarma versicolor*	*Neosarmatium meinerti*	*Uca rosea*	*Uca triangularis*	*Episesarma versicolor*	*Neosarmatium meinerti*
Number of burrows	14 ± 2.33	17 ± 2.43	13 ± 2.56	20 ± 2.72	17 ± 2.70	18 ± 2.70	19 ± 2.56	19 ± 2.22
Density of crabs (No./m^2)	5 ± 1.38	4 ± 1.31	5 ± 1.48	4 ± 1.78	4 ± 1.58	5 ± 1.58	5 ± 1.48	7 ± 1.79
Burrow diameter (cm)	1.56 ± 0.23	1.54 ± 0.26	5 ± 0.54	6 ± 2.11	1.20 ± 0.29	1.5 ± 0.25	5.04 ± 0.47	5.6 ± 2.79
Burrow length (cm)	9.78 ± 0.56	8.78 ± 0.67	26 ± 4.94	35 ± 2.45	9.56 ± 0.50	8.34 ± 0.67	27 ± 4.94	36 ± 2.45
Burrow shape	L	J	Y or J	I or S	L	J	Y or J	I or S

Site-1 to 14 ± 2.79 for *Neosarmatium meinerti* at Site-4. The population of crabs varied from 3 ± 1.48 (No./m^2) for *Episesarma versicular* at Site-1 to 6 ± 1.78 (No./m^2) in the case of *Neosarmatium meinerti* at Site-4. Burrow diameter, which was found in the case of sesarmid crabs, ranged from 5 ± 0.49 cm for *Episesarma versicular* at Site-1 to 5.8 ± 2.34 cm for *Neosarmatium meinerti* at Site-2. The burrow length varied from 25 ± 4.94 cm for *Episesarma versicular* at Site-1 to 36 ± 2.49 cm for *Neosarmatium meinerti* at Site-4, respectively (Tables 10.4, 10.5 and 10.6).

In postmonsoon season 2017, the number of burrows in all selected sites ranged from 8 ± 1.34 at Site-1 to 16 ± 2.78 at Site-4 for *Uca rosea*. The population of crabs, which has been calculated in term of density (No./m^2) of crabs, ranged from 3 ± 1.26 (No./m^2) at Site-2 to 7 ± 1.45 (No./m^2) at Site-4 for *Uca rosea*. The burrow diameter ranged from 1.06 ± 0.76 cm for *Uca rosea* at Site-2 to 1.78 ± 0.26 cm for *Uca triangularis* at Site-3. Burrow length also varied from 7.69 ± 0.68 cm for *Uca triangularis* at Site-2 to 10 ± 0.56 cm for *Uca rosea* at Site-3, respectively. The burrow shapes for *Uca rosea* were L and *Uca triangularis* were J (Tables 10.4, 10.5, 10.6 and Plate 10.2).

For sesarmid, the number of burrows ranged from 11 ± 2.56 at Site-3 to 19 ± 2.56 for *Episesarma versicular* at Site-4. Density of crabs varied from 4 ± 1.48 (No./m^2) for *Episesarma versicular* at Site-3 to 8 ± 1.79 (No./m^2) in the case of *Neosarmatium meinerti* at Site-4. Burrow diameter ranged from 5 ± 0.65 cm for *Episesarma versicular* at Site-2 to 6 ± 2.24 cm for *Neosarmatium meinerti* at Site-3. The burrow length varied from 25 ± 3.20 cm for *Episesarma versicular* at Site-1 to 37 ± 2.50 cm for *Neosarmatium meinerti* at Site-3. Burrows of sesarmid crabs usually varied as Y- or J-shaped in *Episesarma versicular* and I- or S-shaped in the case of *Neosarmatium meinerti* in all the sites, respectively (Tables 10.4, 10.5, 10.6 and Plate 10.2).

In premonsoon season 2017, the number of burrows for fiddler crabs varied from 9 ± 1.67 for *Uca rosea* at Site-2 to 18 ± 2.70 for *Uca triangularis* at Site-4. The density of crabs, which has been calculated in term of density No./m^2 of crabs, ranged from 2 ± 1.16 No./m^2 in *Uca rosea* at Site-2 to 5 ± 1.58 No./m^2 for *Uca triangularis* at Site-4. The burrow diameter ranged from 1.10 ± 0.76 cm at Site-2 to 1.8 ± 0.25 cm for *Uca rosea*. Burrow length also varied from 7.67 ± 0.68 cm for *Uca triangularis* at Site-2 to 9.78 ± 0.56 cm for *Uca rosea* at Site-3. The burrow shapes were similar to that of postmonsoon and monsoon seasons, respectively (Tables 10.4, 10.5, 10.6 and Plate 10.2).

For sesarmid, the number of burrows ranged from 17 ± 2.21 at Site-2 for *Neosarmatium meinerti* to 21 ± 2.79 for *Neosarmatium meinerti* at Site-1. Density of crabs varied from 3 ± 1.42 No./m^2 for *Episesarma versicular* at Site-1 to 7 ± 1.79 No./m^2 in the case of *Neosarmatium meinerti* at Site-4. Burrow diameter ranged from 5 ± 0.54 cm for *Episesarma versicular* at Site-3 to 6 ± 2.32 cm for *Neosarmatium meinerti* at Site-1. The burrow length varied from 25 ± 4.94 cm for *Episesarma versicular* at Site-1 to 36 ± 2.25 cm for *Neosarmatium meinerti* at Site-4. Burrow shapes were similar to that of monsoon 2016 and postmonsoon 2017, respectively (Tables 10.4, 10.5, 10.6 and Plate 10.2).

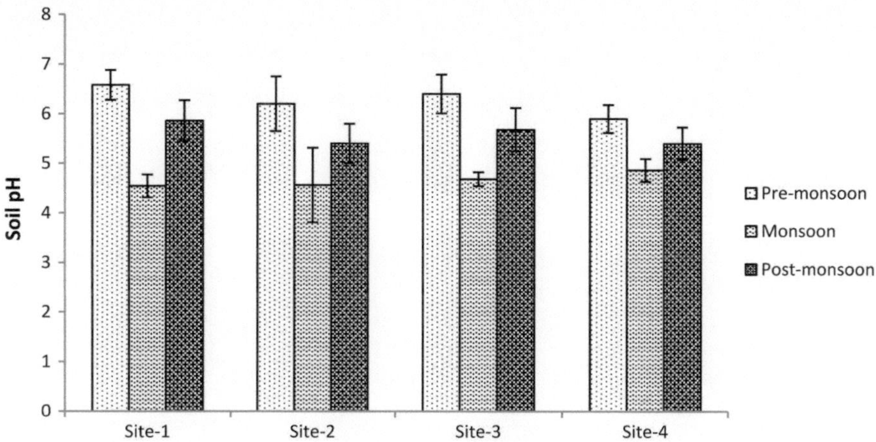

Fig. 10.3 Seasonal variations of soil pH at four selected sites

10.3.2 Physico-Chemical Parameters of Sediment

Sediment biogeochemical processes play important role in metabolism and nutrient cycling of mangrove ecosystems (Webb and Eyre 2004). Burrowing of the crabs significantly maintains a nutrient balance between the coastal vegetation and estuarine waters by transporting soil nutrients to marshy areas. Therefore, crabs directly or indirectly affect ecosystem processes and functioning of the mangrove ecosystems (Wang et al. 2010). In the present research programme, an attempt was made to study the nutrient of surface sediments along with the changes in the depth from 0 to 15 cm. The physico-chemical parameters of the sediments with respect to sediment temperature, sediment pH, sediment salinity and sediment texture (sand, silt and clay) were studied simultaneously in order to pinpoint the microhabitat of the crab species in the intertidal zone of the Mahanadi delta.

The surface sediment temperature in the study sites ranged from 22.76 ± 0.51 °C during postmonsoon 2017 at Site-1 to 32.46 ± 0.41 °C during premonsoon 2017 at Site-2, soil pH ranged from 5.4 ± 0.43 during postmonsoon 2017 at Site-4 to 6.58 ± 0.30 during premonsoon 2017 at Site-1, and soil salinity ranged from 2.15 ± 0.13 psu during postmonsoon 2017 at Site-2 to 7.6 ± 0.36 psu during premonsoon 2017 at Site-4, respectively (Figs. 10.3, 10.4 and 10.5).

Soil organic carbon ranged from 0.64 ± 0.049% during premonsoon 2017 at Site-1 to 0.94 ± 0.036% during postmonsoon 2017 at Site-4 and soil nitrogen varied from 218.40 ± 5.27 (kg/ha) during monsoon 2016 at Site-1 to 392 ± 6.89 (kg/ha) during postmonsoon 2017 at Site-4, respectively (Figs. 10.6 and 10.7).

Soil texture which represented in the form of sand, silt and clay (%) varied with respect to seasons and sites. Sand % varied from 16.20 ± 2.16% during postmonsoon 2017 at Site-1 to 19.80 ± 1.92% during monsoon 2016 at Site-3, silt % varied from 22.40 ± 2.88% during premonsoon 2017 at Site-1 to 26.2 ± 1.92% during monsoon

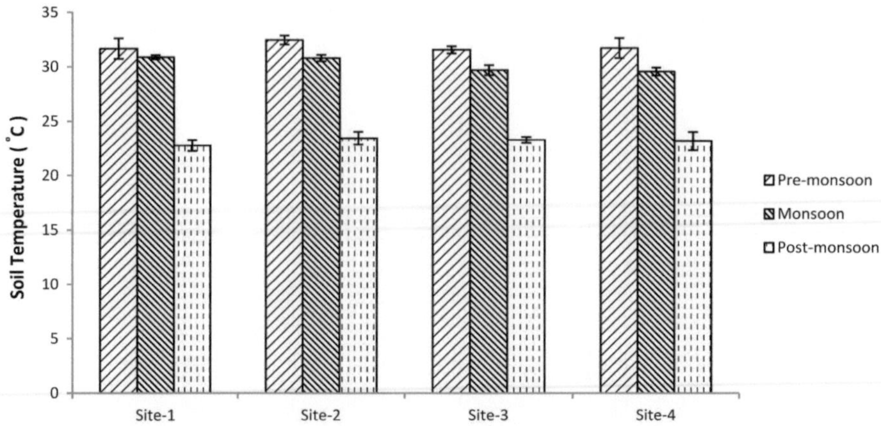

Fig. 10.4 Seasonal variations of soil temperature at four selected sites

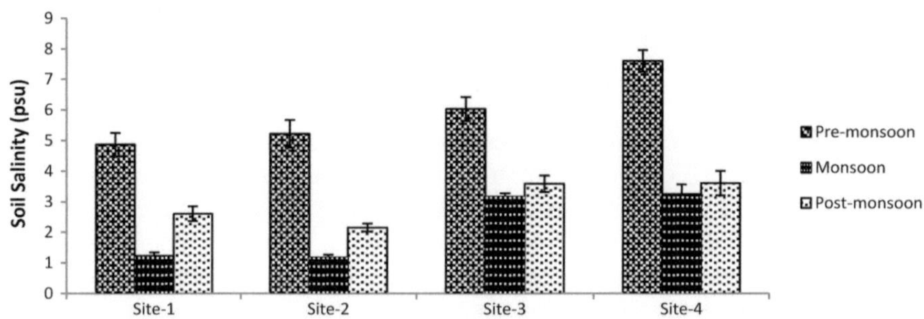

Fig. 10.5 Seasonal variations of soil salinity at four selected sites

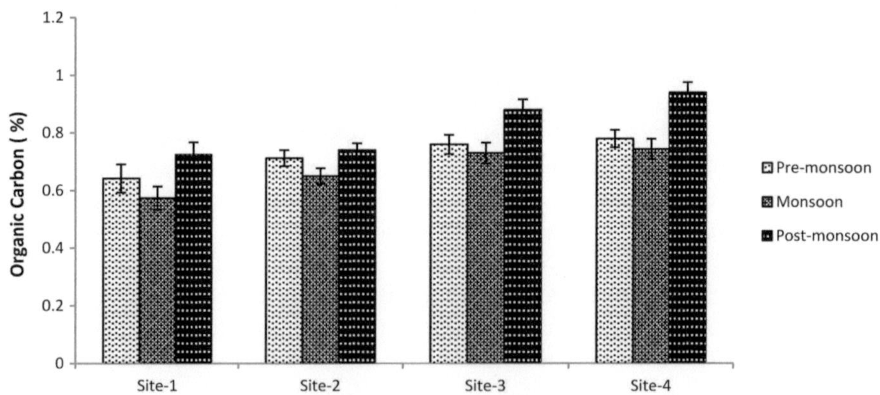

Fig. 10.6 Seasonal variations of soil organic carbon at four selected sites

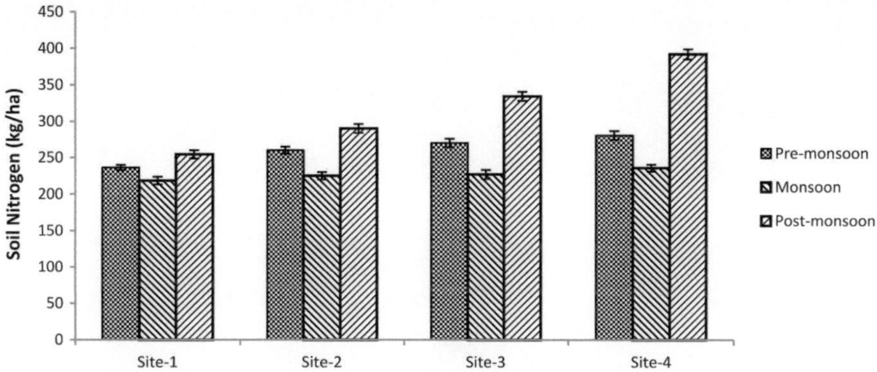

Fig. 10.7 Seasonal variations in soil nitrogen at four selected sites

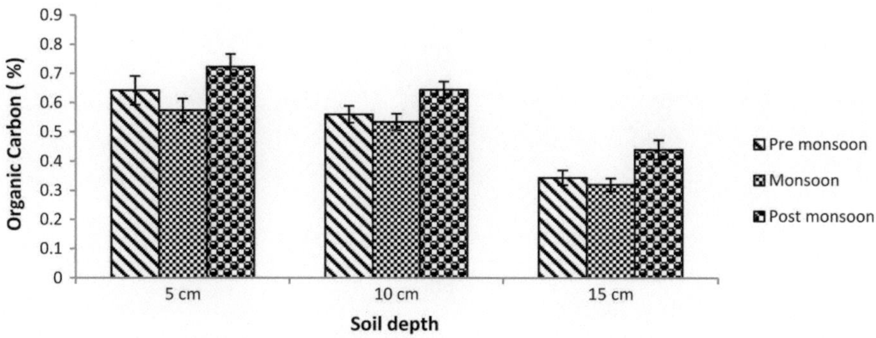

Fig. 10.8 Depth-wise variations of soil organic carbon at Site-1

2016 at Site-2 and clay % varied from 36.4 ± 2.07% during premonsoon 2017 at Site-2 to 41.80 ± 2.48% during postmonsoon 2017 at Site-1, respectively (Figs. 10.16, 10.17 and 10.18).

Depth-wise variation of sediment organic carbon and sediment nitrogen was studied at four sites with depth varying from 0–5 cm, 5–10 cm and 10–15 cm, respectively. In all the sites, both the parameters (sediments organic carbon and sediment nitrogen) exhibited higher values in 0–5 cm depth which proves the fertility of topsoil of mangrove ecosystem. The depth-wise variation of sediment organic carbon varied from 0.34 ± 0.025% at 10-15 cm depth at Site-1 to 0.78 ± 0.030% at depth 0–5 cm depth at Site-4 during premonsoon 2017 0.32 ± 0.022% at 10-15 cm depth at Site-1 to 0.74 ± 0.034% at 0–5 cm depth at Site-4 during monsoon 2016 and 0.44 ± 0.033% at 10–15 cm depth at Site-1 to 0.94 ± 0.036% at 0–5 cm depth at Site-4 during postmonsoon 2017, respectively (Figs. 10.8, 10.9, 10.10 and 10.11).

Similar variation of sediment nitrogen with respect to depth was also studied at four selected sites where values ranged from 121 ± 6.48 (kg/ha) at 10–15 cm depth at Site-1 to 280 ± 6.14 (kg/ha) at 0–5 cm depth at Site-4 during premonsoon, 2017,

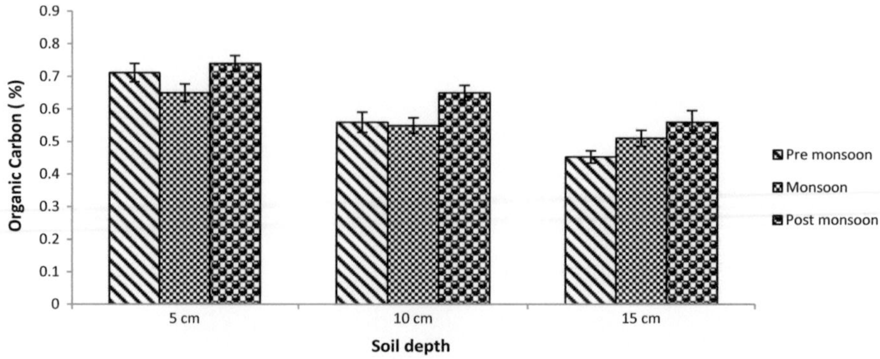

Fig. 10.9 Depth-wise variations of soil organic carbon at Site-2

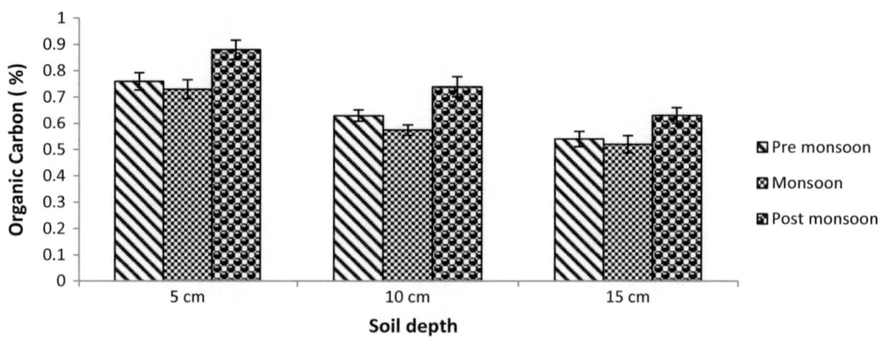

Fig. 10.10 Depth-wise variations of soil organic carbon at Site-3

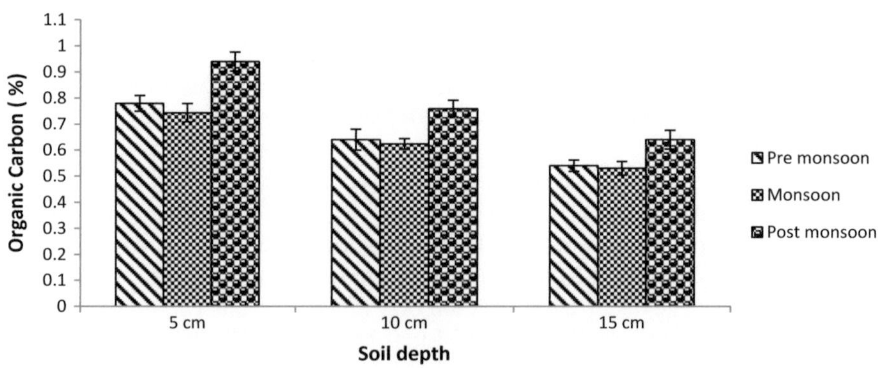

Fig. 10.11 Depth-wise variations of soil organic carbon at Site-4

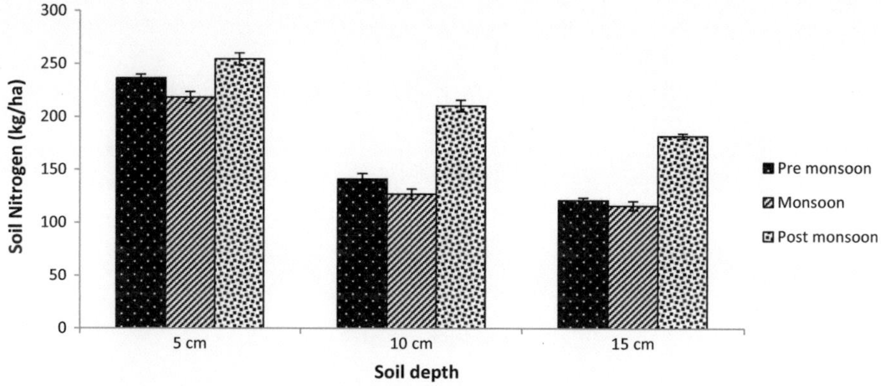

Fig. 10.12 Depth-wise variations of soil nitrogen at Site-1

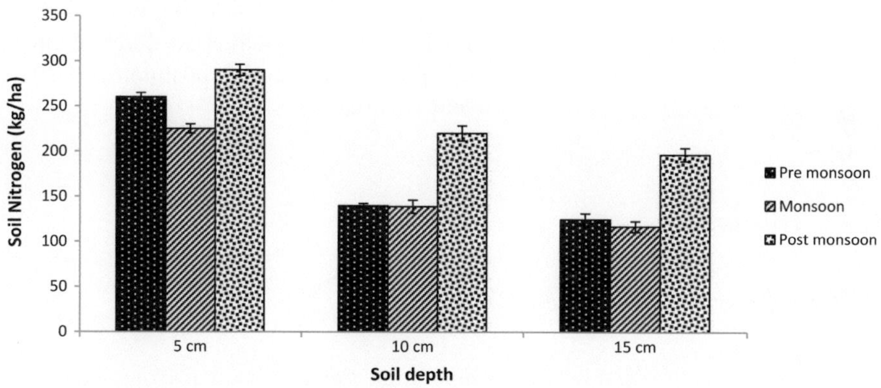

Fig. 10.13 Depth-wise variations of soil nitrogen at Site-2

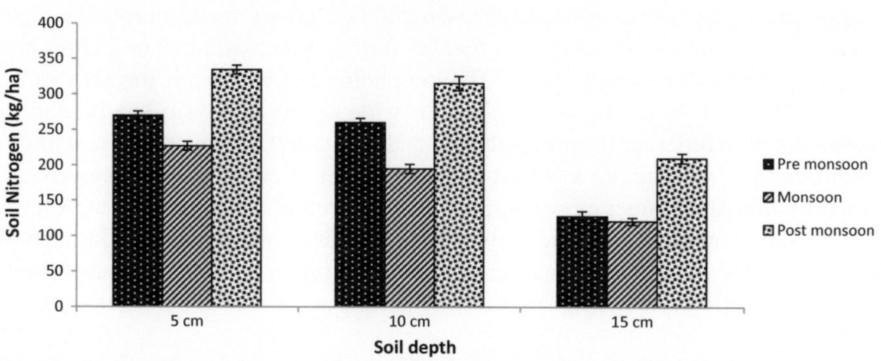

Fig. 10.14 Depth-wise variations of soil nitrogen at Site-3

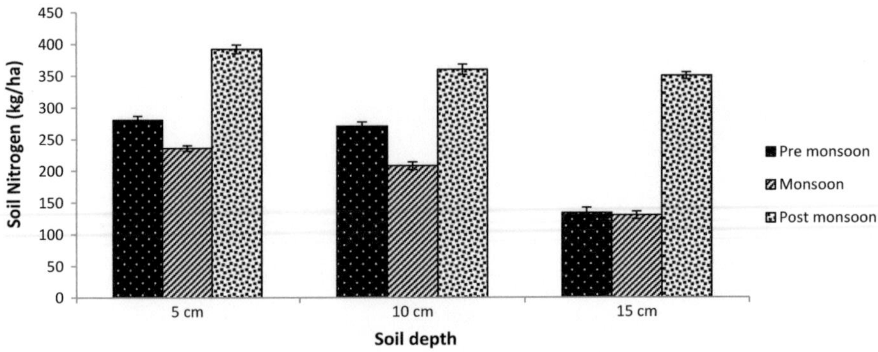

Fig. 10.15 Depth wise variations of soil nitrogen at Site-4

161 ± 4.30 (kg/ha) at 10–15 cm depth at Site-1 to 235.8 ± 4.65 (kg/ha) at 0–5 cm depth at Site-4 during monsoon 2016 and 182 ± 2.44 (kg/ha) at 10–15 cm depth at Site-1 to 392 ± 6.89 (kg/ha) at 0–5 cm depth at Site-4 during postmonsoon 2017, respectively (Figs. 10.12, 10.13, 10.14 and 10.15).

10.4 Discussion

10.4.1 Burrow Structure

Burrowing is a widespread behaviour among invertebrates in mangrove sediments. The resulting biogenic structure and the engineering processes involved in constructing and maintaining them generate complex interactions that coexist alongside trophic webs (Reise 2002; Kristensen 2008). Burrow depth morphology and the irrigation intensity by the burrow dweller further affect sediment properties and biogeochemistry (Kristensen 2008). The morphology of burrows is mostly species specific (Wolfrath 1992). However, given the wide variety of physical and chemical factors between different types of sediment and vegetation, burrowing species might modify burrow architecture to adjust to a specific set of environmental parameters. Crabs are among the larger more conspicuous burrowers and some species are known to have significant intra-specific variations in their burrow morphology in relation to environmental factors such as sediment composition, surface hardness and root mat density of the surrounding vegetation (Bertness and Miller 1984; Morrisey et al. 1999).

In the present study, the density of fiddler crabs was more towards the lower littoral zone proving that microbes being dominant diet of the fiddler crabs, whereas sesarmids crabs were dominant under the mangrove tree trunk and pneumatophores proving that mangrove litter to be the favourite diet for them. Owing to the potential difference of burrows of fiddler crabs and sesarmid crabs, the burrows of sesarmid crabs are long-lived which may actually be the result of continued construction of several generation of occupancies. On contrary, there is frequent collapse of near-shore part of burrows of fiddler crabs, and hence, they must continually maintain, rearrange and often reposition their entrance (Micheli et al. 1991).

Considering all the three seasons and four sites, it has been recorded that the burrows of Grapsidae are longer in size in comparison with Ocypodidae. In order to find out the relationship between burrow number and density of crabs during monsoon 2016, postmonsoon 2017 and premonsoon 2017, it was observed that for family Ocypodidae, there was significant positive correlation at 1% level of significance for all the seasons except Site-1 during monsoon 2016, Site-3 during postmonsoon 2017 and Site-3 during premonsoon 2017, respectively (Tables 10.7, 10.9 and 10.11). In the case of family Grapsidae, it also showed 1% level of significance during monsoon 2016, postmonsoon 2017 and premonsoon 2017, respectively except Site-2

Table 10.7 Inter-relationship between burrow number, crab density, burrow diameter and burrow length for Ocypodidae in four sites during monsoon 2016

Site-1					Site-2			
Parameters	Number of burrows	Density of crabs (No./m^2)	Burrow diameter (cm)	Burrow length (cm)	Number of burrows	Density of crabs (No./m^2)	Burrow diameter (cm)	Burrow length (cm)
Number of burrows	1				1			
Density of crabs (No./m^2)	0.260	1			0.993	1		
Burrow diameter (cm)	0.655	−0.201	1		0.630	0.570	1	
Burrow length (cm)	−0.729	−0.607	0.030	1	−0.356	−0.418	0.485	1
Site-3					Site-4			
Parameters	Number of burrows	Density of crabs (No./m^2)	Burrow diameter (cm)	Burrow length (cm)	Number of burrows	Density of crabs (No./m^2)	Burrow diameter (cm)	Burrow length (cm)
Number of burrows	1				1			
Density of crabs (No./m^2)	0.967	1			0.997	1		
Burrow diameter (cm)	0.897	0.856	1		0.799	0.753	1	
Burrow length (cm)	0.005	−0.246	0.387	1	0.843	0.851	0.536	1

during postmonsoon 2017 and Sites 2, 3 and 4 during premonsoon 2017, respectively (Tables 10.8, 10.10 and 10.12).

Relationships were also drawn between burrow diameter and burrow length for family Ocypodidae, which revealed significant positive relationship at 5% level of significance for Sites 2 and 4 during monsoon 2016, Site-4 during postmonsoon 2017 and Sites 1 and 3 during premonsoon 2017 (Tables 10.13, 10.14, 10.15, 10.16, 10.17, 10.15, 10.16, 10.17, 10.18, 10.19, 10.20, 10.21, 10.22, 10.23 and 10.24). For family Grapsidae, significant positive relationship at 5% level of significance was shown in Site-2 during monsoon 2016, Sites 3 and 4 during postmonsoon 2017 and 1% level of significance at Sites 2 and 4, respectively (Tables 10.13, 10.14, 10.15, 10.16, 10.17, 10.15, 10.16, 10.17, 10.18, 10.19, 10.20, 10.21, 10.22, 10.23 and 10.24). The probable reason for such relationship may probably be due to better substratum characteristics at Site-4 excepting family Grapsidae in monsoon 2016 and family Ocypodidae during premonsoon 2017. The probable reason for insignificant

Table 10.8 Inter-relationship between burrow number, crab density, burrow diameter and burrow length for Grapsidae in four sites during monsoon 2016

Site-1					Site-2			
Parameters	Number of burrows	Density of crabs (No./m^2)	Burrow diameter (cm)	Burrow length (cm)	Number of burrows	Density of crabs (No./m^2)	Burrow diameter (cm)	Burrow length (cm)
Number of burrows	1				1			
Density of crabs (No./m^2)	0.694	1			0.996	1		
Burrow diameter (cm)	0.801	0.595	1		0.945	0.926	1	
Burrow length (cm)	0.209	0.844	0.171	1	0.742	0.756	0.530	1
Site-3					Site-4			
Parameters	Number of burrows	Density of crabs (No./m^2)	Burrow diameter (cm)	Burrow length (cm)	Number of burrows	Density of crabs (No./m^2)	Burrow diameter (cm)	Burrow length (cm)
Number of burrows	1				1			
Density of crabs (No./m^2)	0.995	1			0.965	1		
Burrow diameter (cm)	0.600	0.529	1		0.878	0.851	1	
Burrow length (cm)	−0.701	−0.754	0.057	1	−0.122	−0.289	0.232	1

Table 10.9 Inter-relationship between burrow number, crab density, burrow diameter and burrow length for Ocypodidae in four sites during postmonsoon 2017

Site-1					Site-2			
Parameters	Number of burrows	Density of crabs (No./m^2)	Burrow diameter (cm)	Burrow length (cm)	Number of burrows	Density of crabs (No./m^2)	Burrow diameter (cm)	Burrow length (cm)
Number of burrows	1				1			
Density of crabs (No./m^2)	0.984	1			0.702	1		
Burrow diameter (cm)	0.527	0.389	1		0.831	0.924	1	
Burrow length (cm)	−0.581	−0.683	0.375	1	−0.381	0.355	0.183	1
Site-3					**Site-4**			
Parameters	Number of burrows	Density of crabs (No./m^2)	Burrow diameter (cm)	Burrow length (cm)	Number of burrows	Density of crabs (No./m^2)	Burrow diameter (cm)	Burrow length (cm)
Number of burrows	1				1			
Density of crabs (No./m^2)	−0.007	1			0.899	1		
Burrow diameter (cm)	0.961	0.229	1		0.508	0.111	1	
Burrow length (cm)	−0.022	0.996	0.226	1	0.982	0.913	0.502	1

relationship between crab density and number of crab burrows, in case for both families, may be due to the fact that the semidiurnal tide in the study sites might have devastated the burrow structure from time to time, and hence during sampling of the crabs, they might be seen more with less number of burrows and vice versa. Similar studies on tidal action on the burrow structures have also been studied by Skov and Hartnoll (2002). Similar studies on burrow diameter and crab density have also been studied by earlier workers (Genoni 1991).

The aqueous solution of plaster of Paris was poured into the selected crab burrows during premonsoon 2017. Owing to the fact that this season is favourable for drying of the Paris within the burrow, the moisture content of the sediment is comparatively less in this season. About 30 casts per m^2 were poured; however, only 10–15 casts could be dugged out with a shape and cleaned to remove sediment. Our results of burrow structure have clearly revealed L or J for family Ocypodidae and Y or J and I or S for family Grapsidae. Similar structure on burrow shape for *Uca* species

Table 10.10 Inter-relationship between burrow number, crab density, burrow diameter and burrow length for Grapsidae in four sites during postmonsoon 2017

Site-1					Site-2			
Parameters	Number of burrows	Density of crabs (No./m^2)	Burrow diameter (cm)	Burrow length (cm)	Number of burrows	Density of crabs (No./m^2)	Burrow diameter (cm)	Burrow length (cm)
Number of burrows	1				1			
Density of crabs (No./m^2)	0.810	1			−0.048	1		
Burrow diameter (cm)	0.451	0.888	1		0.423	0.869	1	
Burrow length (cm)	0.985	0.702	0.302	1	−0.864	0.510	0.032	1
Site-3					Site-4			
Parameters	Number of burrows	Density of crabs (No./m^2)	Burrow diameter (cm)	Burrow length (cm)	Number of burrows	Density of crabs (No./m^2)	Burrow diameter (cm)	Burrow length (cm)
Number of burrows	1				1			
Density of crabs (No./m^2)	0.996	1			0.107	1		
Burrow diameter (cm)	0.599	0.656	1		0.749	0.728	1	
Burrow length (cm)	1.000	0.995	0.598	1	−0.216	0.945	0.482	1

(Family Ocypodidae) has also been described by Katz (1980) and Genoni (1991) and for family Grapsidae by Thongtham and Kristensen (2003), Qureshi and Saher (2012).

10.4.2 Nutrient Profiling by Crabs

Mangrove ecosystems are highly dynamic and productive ecosystems, which provide suitable habitat for benthic flora and fauna being situated in the intertidal zone of the coastal ecosystems. They are adapted to growth in intertidal sediment saturated with sea water. The pneumatophores and top roots provide physical support to soft mud (Kitaya et al. 2002). Benthic fauna associated with mangrove forest is typically dominated by burrowing decapods, such as fiddler crabs and sesarmid crabs. They

Table 10.11 Inter-relationship between burrow number, crab density, burrow diameter and burrow length for Ocypodidae in four sites during premonsoon 2017

Site-1					Site-2			
Parameters	Number of burrows	Density of crabs (No./m^2)	Burrow diameter (cm)	Burrow length (cm)	Number of burrows	Density of crabs (No./m^2)	Burrow diameter (cm)	Burrow length (cm)
Number of burrows	1				1			
Density of crabs (No./m^2)	0.992	1			0.707	1		
Burrow diameter (cm)	0.248	0.365	1		0.961	0.872	1	
Burrow length (cm)	−0.753	−0.664	0.441	1	−0.373	0.384	−0.104	1
Site-3					Site-4			
Parameters	Number of burrows	Density of crabs (No./m^2)	Burrow diameter (cm)	Burrow length (cm)	Number of burrows	Density of crabs (No./m^2)	Burrow diameter (cm)	Burrow length (cm)
Number of burrows	1				1			
Density of crabs No./m^2)	−0.674	1			0.995	1		
Burrow diameter (cm)	0.257	0.489	1		0.914	0.945	1	
Burrow length (cm)	−0.515	0.931	0.562	1	−0.508	−0.454	−0.223	1

are typically herbivores that retain, bury, macerate, ingest litter and microalgal mats (Kristensen and Alongi 2006).

The process of ingestion promotes decomposition processes and helps in prevention of loss of nutrients. In the process of burrowing, otherwise called ‘ecosystem engineering’, these decapods engulf the nutrient from the surface down below through which there is a constant regulation of sediment nutrients. On the other hand, the faecal pellets, which are dropped at the sediment surface, then form mud balls containing sediment nutrient mainly carbon and nitrogen. Since the mangrove litter is accumulated on the topsoil, its degradation is governed by sediment temperature, pH and salinity. Moreover, our results have also shown higher rate of carbon and nitrogen values at 0–5 cm depths providing topsoil to be the most fertile sediment for the habitation of crab species (Figs. 10.8, 10.9, 10.10, 10.11, 10.12, 10.13, 10.14 and 10.15).

Table 10.12 Inter-relationship between burrow number, crab density, burrow diameter and burrow length for Grapsidae in four sites during premonsoon 2017

Site-1					Site-2			
Parameters	Number of burrows	Density of crabs (No./m^2)	Burrow diameter (cm)	Burrow length (cm)	Number of burrows	Density of crabs (No./m^2)	Burrow diameter (cm)	Burrow length (cm)
Number of burrows	1				1			
Density of crabs (No./m^2)	0.998	1			−0.102	1		
Burrow diameter (cm)	0.663	0.706	1		0.238	0.939	1	
Burrow length (cm)	0.938	0.915	0.370	1	−0.600	0.836	0.603	1
Site-3					Site-4			
Parameters	Number of burrows	Density of crabs (No./m^2)	Burrow diameter (cm)	Burrow length (cm)	Number of burrows	Density of crabs (No./m^2)	Burrow diameter (cm)	Burrow length (cm)
Number of burrows	1				1			
Density of crabs (No./m^2)	−0.616	1			0.304	1		
Burrow diameter (cm)	0.829	−0.089	1		0.785	0.790	1	
Burrow length (cm)	0.353	−0.611	0.029	1	−0.059	0.927	0.550	1

In our study, we try to draw a relationship between density of crab with surface sediment temperature, pH, salinity, organic carbon and nitrogen. Our results have shown the relationship with temperature only during monsoon 2016 at Sites 2 and 3 and premonsoon 2017 at Sites 1 and 2, respectively, proving that temperature did not play a significance role on the density of crabs being tropical country. In the case of pH, a significant positive relationship with density of crabs only exists during monsoon 2016 at Sites 1, 3 and 4, excepting Site-2. On the other hand, during postmonsoon and premonsoon 2017, pH has not shown any role with respect to density of crabs, proving that monsoon season is the best season for low pH in sediment because of bacterial degradation. Surface sediment salinity has shown significant positive relationship with respect to density of crabs during monsoon season 2016 than the other sites. The density of crabs has also shown significant positive relationship with soil salinity at Sites 1, 3 and 4 during monsoon 2016, Sites 2 and 4 during postmonsoon 2017 at Sites 1 and 4 during premonsoon 2017

Table 10.13 Inter-relationship between burrow number, crab density and sediment physico-chemical parameters in Site-1 during monsoon 2016

Parameters	Number of burrows	Density of crabs	Sediment temperature	Sediment pH	Sediment salinity	Sediment organic carbon	Sediment nitrogen
Number of burrows	1						
Density of crabs	0.177	1					
Sediment temperature	0.891	0.053	1				
Sediment pH	−0.086	0.753	0.053	1			
Sediment salinity	0.246	0.632	0.412	0.895	1		
Sediment organic carbon	0.423	0.515	0.553	0.763	0.917	1	
Sediment nitrogen	0.425	0.582	0.575	0.798	0.959	0.959	1

Table 10.14 Inter-relationship between burrow number, crab density and sediment physico-chemical parameters in Site-2 during monsoon 2016

Parameters	Number of burrows	Density of crabs	Sediment temperature	Sediment pH	Sediment salinity	Sediment Organic carbon	Sediment nitrogen
Number of burrows	1						
Density of crabs	0.071	1					
Sediment temperature	0.125	0.987	1				
Sediment pH	0.885	0.112	0.187	1			
Sediment salinity	0.232	0.383	0.480	0.563	1		
Sediment organic carbon	0.417	0.390	0.496	0.692	0.864	1	
Sediment nitrogen	0.344	0.606	0.664	0.635	0.892	0.878	1

Table 10.15 Inter-relationship between burrow number, crab density and sediment physico-chemical parameters in Site-3 during monsoon 2016

Parameters	Number of burrows	Density of crabs	Sediment temperature	Sediment pH	Sediment salinity	Sediment Organic carbon	Sediment nitrogen
Number of burrows	1						
Density of crabs	0.016	1					
Sediment temperature	0.790	0.579	1				
Sediment pH	0.700	0.670	0.947	1			
Sediment salinity	−0.030	0.448	0.240	0.295	1		
Sediment organic carbon	0.329	0.617	0.538	0.607	0.705	1	
Sediment nitrogen	0.382	0.622	0.567	0.715	0.694	0.907	1

Table 10.16 Inter-relationship between burrow number, crab density and sediment physico-chemical parameters in Site-4 during monsoon 2016

Parameters	Number of burrows	Density of crabs	Sediment temperature	Sediment pH	Sediment salinity	Sediment organic carbon	Sediment nitrogen
Number of burrows	1						
Density of crabs	0.041	1					
Sediment temperature	0.323	0.311	1				
Sediment pH	0.253	0.891	0.117	1			
Sediment salinity	0.213	0.759	0.714	0.659	1		
Sediment organic carbon	0.598	0.573	0.591	0.612	0.749	1	
Sediment nitrogen	0.675	0.544	0.532	0.621	0.777	0.874	1

Table 10.17 Inter-relationship between burrow number, crab density and sediment physico-chemical parameters in Site-1 during postmonsoon 2017

Parameters	Number of burrows	Density of crabs	Sediment temperature	Sediment pH	Sediment salinity	Sediment organic carbon	Sediment nitrogen
Number of burrows	1						
Density of crabs	−0.518	1					
Sediment temperature	0.786	−0.519	1				
Sediment pH	0.553	−0.164	0.248	1			
Sediment salinity	0.659	0.232	0.491	0.575	1		
Sediment organic carbon	0.437	0.344	0.323	0.476	0.927	1	
Sediment nitrogen	0.090	0.453	−0.026	0.317	0.666	0.824	1

Table 10.18 Inter-relationship between burrow number, crab density and sediment physico-chemical parameters in Site-2 during postmonsoon 2017

Parameters	Number of burrows	Density of crabs	Sediment temperature	Sediment pH	Sediment salinity	Sediment organic carbon	Sediment nitrogen
Number of burrows	1						
Density of crabs	−0.057	1					
Sediment temperature	0.402	0.059	1				
Sediment pH	−0.340	0.302	0.208	1			
Sediment salinity	−0.067	0.436	0.242	0.647	1		
Sediment organic carbon	0.101	0.457	0.176	0.362	0.781	1	
Sediment nitrogen	0.248	0.185	0.083	0.198	0.655	0.885	1

Table 10.19 Inter-relationship between burrow number, crab density and sediment physico-chemical parameters in Site-3 during postmonsoon 2017

Parameters	Number of burrows	Density of crabs	Sediment temperature	Sediment pH	Sediment salinity	Sediment Organic carbon	Sediment Nitrogen
Number of burrows	1						
Density of crabs	0.007	1					
Sediment temperature	−0.297	−0.339	1				
Sediment pH	0.338	−0.086	0.193	1			
Sediment salinity	0.058	0.318	0.288	0.521	1		
Sediment organic carbon	−0.005	0.248	0.287	0.551	0.970	1	
Sediment nitrogen	−0.023	0.522	0.158	0.286	0.923	0.853	1

Table 10.20 Inter-relationship between burrow number, crab density and sediment physico-chemical parameters in Site-4 during postmonsoon 2017

Parameters	Number of burrows	Density of crabs	Sediment temperature	Sediment pH	Sediment salinity	Sediment organic carbon	Sediment nitrogen
Number of burrows	1						
Density of crabs	−0.169	1					
Sediment temperature	0.789	0.023	1				
Sediment pH	0.495	0.095	0.317	1			
Sediment salinity	0.440	0.562	0.578	0.607	1		
Sediment organic carbon	0.108	0.713	0.480	0.124	0.694	1	
Sediment nitrogen	0.220	0.794	0.390	0.340	0.885	0.721	1

Table 10.21 Inter-relationship between burrow number, crab density and sediment physico-chemical parameters in Site-1 during premonsoon 2017

Parameters	Number of burrows	Density of crabs	Sediment temperature	Sediment pH	Sediment salinity	Sediment organic carbon	Sediment nitrogen
Number of burrows	1						
Density of crabs	−0.359	1					
Sediment temperature	−0.537	0.530	1				
Sediment pH	0.577	0.289	0.284	1			
Sediment salinity	0.365	0.560	0.398	0.932	1		
Sediment organic carbon	−0.282	0.801	0.638	0.485	0.671	1	
Sediment nitrogen	0.223	0.682	0.406	0.821	0.914	0.789	1

Table 10.22 Inter-relationship between burrow number, crab density and sediment physico-chemical parameters in Site-2 during premonsoon 2017

Parameters	Number of burrows	Density of crabs	Sediment temperature	Sediment pH	Sediment salinity	Sediment Organic carbon	Sediment Nitrogen
Number of burrows	1						
Density of crabs	−0.348	1					
Sediment temperature	0.385	0.498	1				
Sediment pH	0.398	0.312	0.288	1			
Sediment salinity	0.234	0.534	0.290	0.894	1		
Sediment organic carbon	0.674	0.181	0.446	0.839	0.745	1	
Sediment nitrogen	0.014	0.481	0.308	0.843	0.860	0.576	1

Table 10.23 Inter-relationship between burrow number, crab density and sediment physico-chemical parameters in Site-3 during premonsoon 2017

Parameters	Number of burrows	Density of crabs	Sediment temperature	Sediment pH	Sediment salinity	Sediment Organic carbon	Sediment nitrogen
Number of burrows	1						
Density of crabs	−0.006	1					
Sediment temperature	−0.537	−0.188	1				
Sediment pH	0.800	−0.167	−0.011	1			
Sediment salinity	0.216	0.096	0.629	0.603	1		
Sediment Organic carbon	0.351	−0.010	0.470	0.682	0.919	1	
Sediment nitrogen	0.457	−0.011	0.329	0.818	0.745	0.785	1

Table 10.24 Inter-relationship between burrow number, crab density and sediment physico-chemical parameters in Site-4 during premonsoon 2017

Parameters	Number of burrows	Density of crabs	Sediment temperature	Sediment pH	Sediment salinity	Sediment organic carbon	Sediment nitrogen
Number of burrows	1						
Density of crabs	−0.639	1					
Sediment temperature	0.256	−0.254	1				
Sediment pH	0.483	0.024	0.629	1			
Sediment salinity	0.014	0.234	0.390	0.721	1		
Sediment organic carbon	−0.079	0.516	0.533	0.762	0.761	1	
Sediment nitrogen	0.198	−0.069	0.302	0.648	0.692	0.575	1

(Tables 10.7, 10.8, 10.9, 10.10, 10.11 and 10.12). The density of crabs has also shown significant positive relationship with soil organic carbon at Sites 1, 3 and 4 during monsoon 2016, Sites 2 and 4 during postmonsoon 2017 and Sites 2 and 4 during premonsoon 2017. This might probably be due to comparatively dense mangrove forest at Site-4, which has increased the sediment organic carbon load. Similar relationship was also found in case of surface nitrogen at Sites 1, 3 and 4 during postmonsoon 2017 and Sites 1 and 2 during premonsoon 2017, respectively (Tables 10.7, 10.8, 10.9, 10.10, 10.11 and 10.12).

The higher significance of sediment nutrients (carbon and nitrogen) was found during monsoon 2016 which might probably be due to the fact that monsoon season provides a better ecological condition for higher degradation of litter and high run-off from the adjacent land masses. Significant positive relationship of sediment nutrient (organic carbon and nitrogen) with sediment temperature, pH and salinity was observed during monsoon 2016 at all the four sites, proving monsoon season to be the best season for surface sediment nutrient enrichment. During postmonsoon 2017, significant positive relationship of sediment salinity with nutrients was observed at all the sites. However with carbon, only sediment pH has shown positive relationship at Sites 1 and 3, respectively. This proves that at Sites 1 and 3, there is lesser degradation of litter carbon as these sites being located far away from the sea. During premonsoon 2017, significant positive relationship was observed for sediment nutrients (organic carbon and nitrogen) with sediment temperature, pH and salinity at all the four sites excepting with that of temperature at Sites 2, 3 and 4. This might be probably due to the fact that temperature did not play significant role in case of Sites 2, 3 and 4 which are located nearby to each other and having the same tropical characteristics (Tables 10.13, 10.14, 10.15, 10.16, 10.17, 10.15, 10.16, 10.17, 10.18, 10.19, 10.20, 10.21, 10.22, 10.23 and 10.24). Similar studies are also been highlighted by Saravanakumar et al. (2007).

ANOVA was performed for sediments parameters between sites and seasons which showed significance difference between seasons with respect to all the selected parameters (Tables 10.25 and 10.26). However, significant variation between sites was observed with respect to sediment organic carbon only, which proves that the density of mangroves has played a very significant role in contribution to organic carbon in the sediment (Table 10.25). ANOVA was performed for sediment texture between sites and seasons with respect to sand, silt and clay which has shown significant difference between seasons proving that there is significant contribution of tide and land run-off at the study sites (Table 10.26; Figs. 10.16, 10.17 and 10.18).

ANOVA was also performed for understanding the depth-wise variation for organic carbon and nitrogen (Tables 10.27 and 10.28). The data revealed that there is significant variation between depth and seasons with respect to sediment nutrients (organic carbon and nitrogen) at all the four selected sites, proving that the selected sites are prominently differing from each other with respect to nutrient load. This might be probably due to the location of the sites from the Bay of Bengal. Since burrowing activity of crabs has let to engagement/utilization of sediment nutrient by crabs, the depth-wise variation of nutrient was recorded in the present study. Such works on sediment organic matter with respect to organic carbon and nitrogen have

Table 10.25 ANOVA for sediment parameters between sites and seasons

Parameters	Variables	F calculated value	F critical value
Sediment pH	Between sites	0.618718	1.867332
	Between seasons	39.93016	3.244818
Sediment temp.	Between sites	0.88798	1.867332
	Between seasons	908.2751	3.244818
Sediment salinity	Between sites	10.2811	1.867332
	Between seasons	329.4298	3.244818
Sediment OC	Between sites	9.530203	1.867332
	Between seasons	64.69105	3.244818
Sediment N	Between sites	3.025844	1.867332
	Between seasons	66.79355	3.244818

Table 10.26 ANOVA for sediment texture between sites and seasons

Sediment texture	Variables	F calculated value	F critical value
Sand	Between sites	1.732906	1.867332
	Between seasons	7.163343	3.244818
Slit	Between sites	2.448521	1.867332
	Between seasons	4.211742	3.244818
Clay	Between sites	1.321564	1.867332
	Between seasons	13.88462	3.244818

also been studied by Kristensen et al. (1995), Twilley and Chen (1997), Dittmar and Lara (2001), Thongtham and Kristensen (2003), Chatterjee et al. (2014) and Wang et al. (2010).

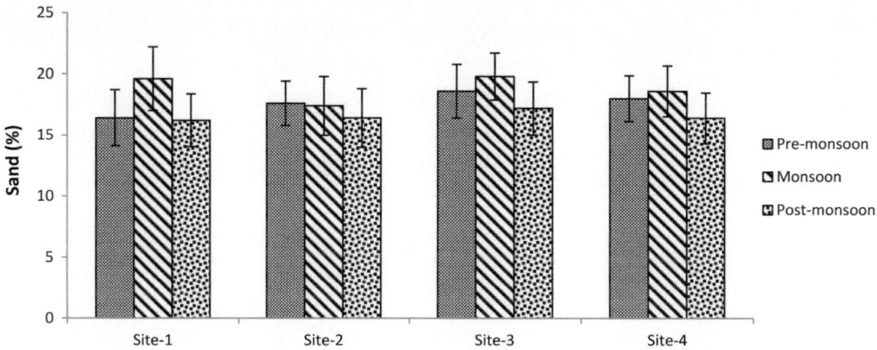

Fig. 10.16 Seasonal variations in sediment texture (sand) at four selected sites

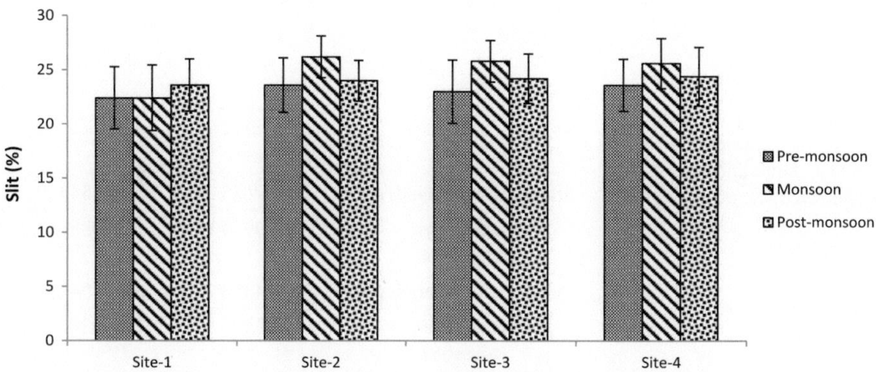

Fig. 10.17 Seasonal variations in sediment texture (silt) at four selected sites

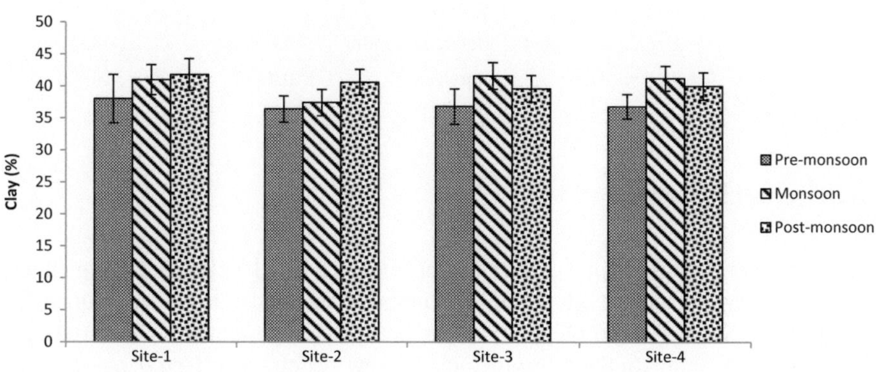

Fig. 10.18 Seasonal variations in sediment texture (clay) at four selected sites

Table 10.27 ANOVA for depth-wise sediment organic carbon between depth and seasons among selected sites

Sites	Variables	F calculated value	F critical value
Site-1	Between depth	53.42811	2.063541
	Between seasons	71.83831	3.340386
Site-2	Between depth	17.09497	2.063541
	Between seasons	24.48606	3.340386
Site-3	Between depth	22.42563	2.063541
	Between seasons	64.38673	3.340386
Site-4	Between depth	26.35997	2.063541
	Between seasons	72.15191	3.340386

Table 10.28 ANOVA for depth-wise sediment N in between depth and seasons at the selected sites

Sites	Variables	F calculated value	F critical value
Site-1	Between depth	31.87186	2.063541
	Between seasons	91.49974	3.340386
Site-2	Between depth	46.94348	2.063541
	Between seasons	144.8353	3.340386
Site-3	Between depth	50.45682	2.063541
	Between seasons	226.4396	3.340386
Site-4	Between depth	6.63858	2.063541
	Between seasons	150.4475	3.340386

10.5 Conclusion

Mangrove ecosystems are one of the most privileged coastal wetlands among all natural ecosystems, which provide suitable habitats for decapod crustaceans. Burrowing brachyuran (Grapsidae and Ocypodidae) is the most important macroinvertebrates of marine ecosystems. In the present study, burrow structure for *Uca rosea, Uca triangularis, Episesarma versicolor* and *Neosarmatium meinerti* revealed L, Y, J, I and S patterns in most of the selected sites. Depth-wise variation of organic carbon and nitrogen was proved through ANOVA, where 0–5 cm depth has shown the maximum amount of organic carbon and nitrogen, respectively. The crab burrows and density of crab have also shown significant relationship with respect to site and season, which proved that number of burrows and crab density varied as per sediment texture and the physico-chemical parameters of the soil. A long-term study on nutrient profiling by crabs and mangrove assemblage can pinpoint the role of crabs in biogeochemical processes of nutrients in mangrove ecosystem.

Acknowledgements The authors are grateful to Ministry of Earth Sciences, Govt. of India project (Sanction No. MoES/36/OOIS/EXTRA/44/2015 dated 29/11/2016) for financial assistance.

References

Agarwal S, Banerjee K, Pal N et al (2017) Carbon sequestration by mangrove vegetations: a case study from Mahanadi mangrove wetland. J Environ Sci Comput Sci Eng Technol 7(1):16–29

Alongi et al (2001) Organic carbon accumulation and metabolic pathways of mangrove forests in Southern Thailand. Mar Geol 179:85–103

Bertness MD, Miller T (1984) The distribution and dynamics of *Uca pugnax* (Smith) burrows in a New England saltmarsh. J Exp Mar Biol Ecol 83:211–237

Bond W (2001) Keystone species—hunting the snark. Science 292(5514):63–64

Botto F, Iribarne O (2000) Contrasting effects of two burrowing crabs (*Chasmagnathus granulata* and *Uca uruguayensis*) on sediment composition and transport in estuarine environments. Estuar Coast Shelf Sci 51:141–151

Bouillon S, Connolly RM, Gillikin DP et al (2011) Use of stable isotopes to understand food webs and ecosystem functioning in estuaries. In: Wolanski E, McLusky D (eds) Treatise on estuarine and coastal science. Academic Press, Waltham, Massachusetts, USA, pp 143–173

Cannicci S, Burrows D, Fratini S et al (2008) Faunal impact on vegetation structure and ecosystem function in mangrove forests: a review. Aquat Bot 89:186–200

Chatterjee S, Mazumdar D, Chakraborty SK (2014) Ecological role of fiddler crabs (*Uca* spp.) through bioturbatory activities in the coastal belt of East Midnapore, West Bengal, India. J Mar Biol Assoc India 56(1):1–10

Dahdouh-Guebas F, Koedam N (2002) A synthesis of existent and potential mangrove vegetation structure dynamics from Kenyan, Sri Lankan and Mauritanian case-studies. Food Agric Organ United Nations 48:487–511

De la Iglesia HO, Rodriguez E, Dezi RE (1994) Burrow plugging in the crab *Uca uruguayensis* and its synchronization with photoperiod and tides. Physiol Behav 55(5):913–919

Dittmar T, Lara RJ (2001) Driving forces behind nutrient and organic matter dynamics in a mangrove tidal creek in North Brazil. Estuar Coast Shelf Sci 52:249–259

Donato CD, Kauffman JB, Murdiarso D et al (2011) Mangroves among the most carbon-rich forests in the tropics. Nat Geosci 4:293–297

Dye AH, Lasiak TA (1986) Assimilation efficiencies of fiddler crabs and deposit-feeding gastropods from tropical mangrove sediments. Comp Biochem Physiol 87A(2):341–344

Dye AH, Lasiak TA (1987) Microbethos, meiobenthos and fiddles crabs: trophic interactions in typical mangrove sediment. Mar Ecol Progr Res 32:259–264

Erwin D (2008) Macroevolution of ecosystem engineering, niche construction and diversity. Trends Ecol Evol 23(6):304–310

France R (1998) Estimation the assimilation of mangrove detritus by fiddler crabs in Laguna Joyuda, Puerto Rico, using dual stable isotopes. J Trop Ecol 14:413–425

Gee GW, Bauder JW (1986) Particle-size analysis. In: Klute A (ed) Methods of soil analysis, Part 1. Physical and mineralogical methods. Agronomy monograph, no 9, 2nd edn. American Society of Agronomy/Soil Science Society of America, Madison, WI, pp 383–411

Genoni GP (1991) Increased burrowing by fiddler crabs *Uca rapax* (Smith) (Decapoda: Ocypodidae) in response to low food supply. J Exp Mar Biol Ecol 147:267–285

Gillikin DP, Kamanu CP (2005) Burrowing in the East African mangrove crab, *Chiromantes ortmanni* (Crosnier, 1965) (Decapoda, Brachyura, Sesarmidae). Crustaceana 78:1273–1275

Gribsholt B, Kostka JE, Kristensen E (2003) Impact of fiddler crabs and plant roots on sediment biogeochemistry in a Georgia saltmarsh. Mar Ecol Prog Ser 259:237–251

Hagen VHO, Jones DS (1989) The fiddler crabs (Ocypodidae*: Uca*) of Darwin, Northern Territory, Australia. Beagle, Rec Northern Territory Mus Arts Sci 6(1):55–68

Jones CG, Lawton JH, Shachak M (1994) Organisms as ecosystem engineers. Oikos 69:373–386

Karlson K, Bonsdorff E, Rosenberg R (2007) The impact of benthic macrofauna for nutrient fluxes from Baltic Sea sediments. Ambio 36:161–167

Kathiresan K, Qasim SZ (2005) Biodiversity of mangrove ecosystem. Hindustan Publishing Corporation, India, p 251

Katz LC (1980) Effects of burrowing by the fiddler crab, *Uca pugnax* (Smith). Estuar Coast Mar Sci 11:233–237

Kitaya Y, Yabuki K, Kiyota M et al (2002) Gas exchange and oxygen concentration in pneumatophores and prop roots of four mangrove species. Trees 16:155–158

Komiyama A, Ong JE, Poungparn S (2008) Allometry, biomass, and productivity of mangrove forest: a review. Aquat Bot 89:128–137

Kristensen E (2007) Carbon balance in mangrove sediments: the driving processes and their controls. Greenhouse Gas and Carbon Balances in Mangrove Coastal Ecosystems. Gendai Tosho, Kanagawa, Japan, pp 61–78

Kristensen E (2008) Mangrove crabs as ecosystem engineers; with emphasis on sediment processes. J Sea Res 59:30–43

Kristensen E, Alongi DM (2006) Control by fiddler crabs (*Uca vocans*) and plant roots (*Avicennia marina*) on carbon, iron, and sulfur biogeochemistry in mangrove sediment. Limnol Ocenography 51(4):1557–1571

Kristensen E, Pilgaard R (2001) The role of fecal pellet deposition by leaf-eating sesarmid crabs on mineralization processes in a mangrove sediment (Phuket, Thailand). In: Aller JY, Woodin SA, Aller RC (eds) Organism-sediment interactions. University of South Carolina Press, Columbia, pp 369–384

Kristensen E, Holmer M, Banta GT et al (1995) Carbon, nitrogen and sulfur cycling in sediments of the Ao Nam Bor mangrove forest, Phuket, Thailand: a review. Phuket Mar Biol Centre Res Bull 60:37–64

Kristensen E, Anderson FO, Holmboe N et al (2000) Carbon and nitrogen mineralization in sediments of the Bangrong mangrove area, Phuket, Thailand. Aquat Microb Ecol 22:199–213

Lee SY (1997) Potential trophic importance of the faecal material of the mangrove Sesarmine crab *Sesarma messa*. Mar Ecol Prog Ser 159:275–284

Lee SY (1998) Ecological role of grapsid crabs in mangrove ecosystems: a review. Mar Freshw Res 49:335–343

Lim SSL (2006) Fiddler crab burrow morphology: how do burrow dimensions and bioturbative activities compare in sympatric populations of *Uca vocan* (Linnaeus, 1758) and *U. annulipes* (H. Milne Edwards, 1837). Crustaceana 79(5):525–540

Lim SSL, Diong CH (2003) Burrow-morphological characters of the fiddler crab, *Uca annulipes* (H. Milne Edwards, 1837), and ecological correlates in a lagoonal beach on Pulau Hantu, Singapore. Crustaceana 76:1055–1069

Mayer RC, Davis JH, Schoorman FD (1995) An integrative model of organization trust. Acad Manag Rev 20(3):709–734

McGuinness KA (1994) The climbing behaviour of *Cerithidea anticipata* (Mollusca: Gastropoda): the roles of physical and biological factors. J Ecol South Hemisphere 19(3):283–289

Mchenga ISS, Tsuchiya M (2008) Nutrient dynamics in mangrove crab burrow sediments subjected to anthropogenic input. J Sea Res 59:103–113

Micheli F, Gherardi F, Vanini M (1991) Feeding and burrowing ecology of two East African mangrove crabs. Mar Biol 111:247–254

Morrisey DJ, DeWitt TH, Roper DS et al (1999) Variation in the depth and morphology of burrows of the mud crab *Helice crassa* among different types of intertidal sediments in New Zealand. Mar Ecol Prog Ser 182:231–242

Nagelkerken I, Blaber SJM, Bouillon S et al (2008) The habitat function of mangroves for terrestrial and marine fauna: a review. Aquat Bot 89:155–185

Ng PKL, Guinot D, Davie PJF (2008) Systema Brachyurorum: part I. An annotated checklist of extant brachyuran crabs of the world. Raffles Bull Zool 17:1–286

Nielsen OI, Kristensen E, Macintosh DJ (2003) Impact of fiddler crabs (*Uca* spp.) on rates and pathways of benthic mineralization in deposited mangrove shrimp pond waste. J Exp Mar Biol Ecol 289:59–81

Nobbs M (2003) Effects of vegetation differ among three species of fiddler crabs (*Uca* spp.). J Exp Mar Biol Ecol 284(1–2):41–50

Odling-Smee J, Erwin DH, Palkovacs EP et al (2013) Niche construction theory: a practical guide for ecologists. Q Rev Biol 88(1):3–28

Otani S, Kozuki Y, Yamanaka R et al (2010) The role of crabs (*Macrophthalmus japonicus*) burrows on organic carbon cycle in estuarine tidal flat, Japan. Estuar Coast. Estuar Coast Shelf Sci 86:434–440

Paarlberg AJ, Knaapen MAF, de Vries MB et al (2005) Biological influences on morphology and bed composition of an intertidal flat. Estuar Coast Shelf Sci 64:577–590

Penha-Lopes GP, Bouillon S, Mangion P et al (2009) Population structure, density and food sources of Terebralia palustris (Potamididae: Gastropoda) in a low intertidal Avicennia marina mangrove stand (Inhaca Island, Mozambique). Estuar Coast Shelf Sci 84(3):318–325

Poovachiranon S, Tantichodok P (1991) The role of sesarmid crabs in the mineralization of leaf litter of *Rhizophora apiculata* in a mangrove, Southern Thailand. Phuket Mar Biol Centre Res Bull 56:63–74

Qureshi NA, Saher NU (2012) Burrow morphology of three species of fiddler crab (*Uca*) along the coast of Pakistan. Belg J Zool 142(2):114–126

Reise K (2002) Sediment mediated species interactions in coastal waters. J Sea Res 48:127–141

Robertson AL (1986) Leaf-burying crabs: their influence on energy flow and export from mixed mangrove forests (*Rhizophora* spp.) in northeastern Australia. J Exp Mar Biol Ecol 102:237–248

Robertson A, Alongi DM, Boto KG (1992) Food chains and carbon fluxes. In: Robertson A, Alongi DM (eds) Tropical mangrove ecosystems. AGU, Washington, DC, pp 293–326

Sanders CJ, Smoak JM, Naidu AS et al (2010) Organic carbon burial in a mangrove forest, margin and intertidal mud flat. Estuar Coast Shelf Sci 90(3):168–172

Saravanakumar D, Vijayakumar C, Kumar NR (2007) PGPR-induced defense responses in the tea plant against blister blight disease. Crop Protection 26:556–565

Skov MW, Hartnoll RG (2002) Paradoxical selective feeding on a low-nutrient diet: why do mangrove crabs eat leaves? Oecologia 131(1):1–7

Skov MW, Vannini M, Shunula JP et al (2002) Quantifying the density of mangrove crabs: Ocypodidae and Grapsidae. Mar Biol 141:725–732

Thongtham N, Kristensen E (2003) Physical and chemical characteristics of mangrove crab (*Neoepisesarma versicolor*) burrows in the Bangrong mangrove forest, Phuket, Thailand; with emphasis on behavioural response to changing environmental conditions. Life and Environment 53:141–151

Thongtham N, Kristensen E (2005) Carbon and nitrogen balance of leaf-eating sesarmid crabs (*Neoepisesarma versicolor*) offered different food sources. Estuar Coast Shelf Sci 65:213–222

Twilley RR, Chen R (1997) A water budget and hydrology model of a basin mangrove forest in Rookery Bay, Florida. Mar Freshw Res 49:309–323

Walkley A, Black IA (1934) An examination of Degtjareff method for determining soil organic matter and a proposed modification of the chromic acid titration method. Soil Sci 37:29–37

Wang JQ, Zhang XD, Jiang LF et al (2010) Bioturbation of burrowing crabs promotes sediment turnover and carbon and nitrogen movements in an Estuarine Salt Marsh. Ecosystems 13:586–599

Warburg MR, Schuchman E (1978) Dispersal, population structure and burrow shape of *Ocypode cursor*. J Mar Biol 49(3):255–263

Warren JH, Underwood AJ (1986) Effects of burrowing crabs on the topography of mangrove swamps in New South Wales. J Exp Mar Biol Ecol 102(2–3):223–235

Webb AP, Eyre BD (2004) The effect of natural populations of the borrowing and grazing soldier crab (*Mictyris longicarpus*) on sediment irrigation, benthic metabolism and nitrogen fluxes. J Exp Mar Biol Ecol 309(1):1–19

Wolfrath B (1992) Burrowing of the fiddler crab *Uca tangeri* in the Ria Formosa in Portugal and its influence on sediment structure. Mar Ecol Prog Ser 85(3):237–243

Chapter 11
Application of a Low-Cost Technology to Treat Domestic Sewage and to Improve Fertility of a Barren Lateritic Soil

Kruti Jethwa, Samir Bajpai and P. K. Chaudhari

Abstract Selecting low-cost, less power-intensive or power-independent, with less mechanization, and efficient alternative technologies for wastewater treatment is essential to improve treatment capacity, especially in developing regions. Widespread application of constructed wetland (CW) is limited due to land cost. Use of locally available lands having infertile, barren soils nearby townships will be economical, because these types of soils are not in use and also cost of transportation of such land is minimum. The shallow soil media will also help in maintaining aerobic conditions, overall improving the performance of CW. Another advantage of use of infertile, barren soil as substrate of CWs can be improvement in fertility of such soil, thereby resulting in overall benefits for the area. Performance of six different aquatic macrophytes—*Typha latifolia*, *Colocasia esculenta*, *Alternanthera sessilis*, *Polygonum*, *Canna indica* and *Ocimum americanum* L—grown in mono-culture and poly-culture for treatment of domestic sewage and retention of major nutrients nitrogen (N) and phosphorus (P) in lateritic soil substrate (barren or infertile) was studied in sixteen laboratory-scale CWs. Fourteen numbers of laboratory-scale CWs, having dimensions 1.0 m × 0.36 m × 0.35 m (L × W × H) for mono-culture and a pair of CWs having dimension 2.5 m × 1.0 m × 0.35 m for poly-culture, were used. In every CW, the depth of lateritic soil substrate was kept shallow at 0.30 m to maintain aerobic conditions. Depth of 0.05 m was kept as free board. CWs were operated in batch mode with sewage loading for six hours to maintain aerobic conditions. Sewage residence time was twenty-four hours, and applied hydraulic loading was 0.028 m^3/m^2/day. Good removal of BOD and COD was observed for all the CWs.

K. Jethwa (✉) · S. Bajpai
Department of Civil Engineering, National Institute of Technology, Raipur, Chhattisgarh 492010, India
e-mail: krutijethwa171@gmail.com

S. Bajpai
e-mail: sb@nitrr.ac.in

P. K. Chaudhari
Department of Chemical Engineering, National Institute of Technology, Raipur, Chhattisgarh 492010, India
e-mail: pkchaudhari.che@nitrr.ac.in

R. M. Singh et al. (eds.), *Environmental Processes and Management*,
Water Science and Technology Library 91,
https://doi.org/10.1007/978-3-030-38152-3_11

BOD removal of 36.0–80.6% for the CWs with mono-culture of plants and 71.3–79.8% for poly-culture of plants was observed. Similarly, COD removal of 32.2–72.6% for the CWs with mono-culture of plants and 34.0–63.5% for poly-culture of plants was observed. For nutrient removal, best results were obtained for CW operated with *T. latifolia* with average removals of 69% for total Kjeldahl nitrogen (TKN) and 89% soluble reactive phosphorus (SRP) at plant density of 4 plants/CW. Nutrient removal rate for the plants followed the pattern—*T. latifolia* > *C. indica* > *A. sessilis* > *Polygonum* > *C. esculenta* > *O. americanum* L. > *Pistia stratiotes.* Highest amount of average mineralizable phosphorus retention in the lateritic soil substrate was 85%, and nitrogen retention was 46.2% for *P. stratiotes* plant with the density of 4 plants/CW. Plant uptake was the major nutrient removal pathway in the wetland microcosms. Nutrient removal from domestic wastewater in CWs with poly-culture of plants was not greater than that in CWs with mono-culture plants, due to interspecies competition. CWs prove to be a viable option for simultaneous treatment of sewage and enhancement of fertility of barren lateritic lands.

Keywords Available/mineralizable nitrogen · Available/mineralizable phosphorus · Domestic sewage · Lateritic soil

11.1 Introduction

Constructed wetlands (CWs) are engineered systems that utilize the processes in natural wetlands, mimicking the role of wetland plants, soils and their microbial assemblages to treat wastewater (Brix 1994; USEPA 2000). CWs provide simple, sustainable, low-energy and cost-effective alternative treatment technology to conventional wastewater treatment systems, especially for small communities and remote areas.

However, use of CW in India remains largely unexplored. Widespread application of CWs is limited due to land cost. To overcome the land cost barrier, use of lands with infertile, barren soils, available locally, nearby townships, will be economical, due to the low cost of such lands. Further, if such locally available, infertile and barren soils can be used in shallow thickness in CWs, the cost of construction may further be reduced. The shallow soil media will also help in maintaining aerobic conditions, overall improving the performance of CW. Another advantage of use of local infertile, barren soil as substrate of CWs can be subsequent improvement in fertility of such soil, thereby resulting in overall benefits for the area.

Conventional wastewater treatment methods based on activated sludge process (ASP), stabilization pond and trickling filter (TF), though good in removal of organic load, are not competent in removing nitrogen (N) and phosphorus (P) (Sunarsih et al. 2015). The main reason for poor nitrogen removal is incomplete nitrification due to availability of limited oxygen and low retention time. Also, the precipitation and adsorption opportunities for phosphorus removal are not available in common treatment systems based on ASP and TF. The P removal in treatment systems is strongly associated with the physicochemical and hydrological properties of the reactor system (Bunce et al. 2018).

CWs have been reported to remove N and P (Vohla et al. 2007; Vymazal 2007). Ilyas and Masih (2018) demonstrated higher P removal by aerated CWs than the non-aerated systems. However, artificial aerated CWs seem to be costly for developing countries.

Lateritic soils are found in the tropical environment in which there are severe chemical weathering and leaching of soluble minerals. Laterites are highly weathered and altered residual soils formed by the in situ weathering and decomposition of rocks under tropical conditions. These soils occupy an area of about 49,000 sq. miles in India (Raychaudhuri 1980). The lateritic soil has a deficiency of mineralizable phosphorus, mineralizable nitrogen and soil organic carbon (Shrivastav et al. 2015). Lateritic soils are widely distributed in the tropical humid areas across the globe and form major portion of fertility-challenged land area. There is need to evaluate the utility of CWs in improving fertility of lateritic soils while treating domestic wastewater.

Limited information is available for pollutant removal in lateritic soil media using CWs. For example, Wood and McAtamney (1996) used typical thumb rule of 5 m^2 area per population equivalent for removal of ortho-phosphorus and heavy metal from the raw leachate from the landfill site using crushed laterite and gravel with *P. australis* and *Phalaris arundinacea* as macrophytes. Sekiranda and Kiwanuka (1998) used laboratory-scale 40 L experimental buckets with lateritic gravel rooted *Phragmites* and floating *Phragmites* for removal of N and P from domestic sewage.

Successful treatment of wastewater using shallow soil substrate depth in the CWs has been reported by several researchers (Huang et al. 2000; García et al. 2005; Song et al. 2009; Ávila et al. 2014). Thus, CWs need to be studied to provide a solution for not only effective treatment of wastewater but also as an economical method to improve the fertility of barren soils like laterites by removal and retention of carbon and important nutrients, N and P.

The objective was to study nutrient, N and P, retention by barren lateritic soil (locally known as 'Bhata' in Chhattisgarh) as substrate in CWs using native wetland plants.

11.2 Materials and Methods

Sixteen numbers of laboratory-scale CWs, with the lateritic soil media depth of 0.3 m, were tested for treatment of domestic sewage (DS) and nutrient (N and P) removal. Out of fourteen CWs of size (1.0 m × 0.36 m × 0.35 m), two CWs were operated as blank (without any plants, only with soil media), six pair of CWs (12 in numbers) were operated with mono-culture of plants with the respective plant density of 4 plants/CW and 2 plants/CW and two CWs of size 2.5 m × 1.0 m × 0.35 m were operated with poly-culture of plants having plant density of, respectively, 24 plants/CW and 12 plants/CW. Plants used in mono-culture and poly-culture were:

T. latifolia, *C. esculenta*, *A. sessilis*, *Polygonum*, *C. indicia*, *O. americanum* L and *P. stratiotes*. Details of CWs, plant species and density are given in Table 11.1.

Table 11.1 Details of CWs with plant densities

S. No.	Nomenclature for CWs	Name of plants	Common name of plants	Nomenclature for plants	Density of plants
1	B1	–	–	–	Zero
2	B2	–	–	–	Zero
3	M1T	*Typha latifolia*	Cattail	T	4 plants/CW
4	M2T	*Typha latifolia*	Cattail	T	2 plants/CW
5	M1Co	*Colocasia esculenta*	Elephant ears	Co	4 plants/CW
6	M2Co	*Colocasia esculenta*	Elephant ears	Co	2 plants/CW
7	M1C	*Canna indica*	Canna lily	C	4 plants/CW
8	M2C	*Canna indica*	Canna lily	C	2 plants/CW
9	M1O	*Ocimum americanum* L	American basil	O	4 plants/CW
10	M2O	*Ocimum americanum* L	American basil	O	2 plants/CW
11	M1A	*Alternanthera sessilis and Polygonum*	Alligator weed	A	4 plants/CW
12	M2A	*Alternanthera sessilis and Polygonum*	Alligator weed	A	2 plants/CW
13	M1P	*Pistia stratiotes*	Water lettuce	P	4 plants/CW
14	M2P	*Pistia stratiotes*	Water lettuce	P	2 plants/CW
15	P1	*Typha latifolia, Colocasia esculenta, Alternanthera sessilis, Polygonum, Canna indicia, Ocimum americanum* L, *Pistia stratiotes*	–		24 plants/CW

(continued)

Table 11.1 (continued)

S. No.	Nomenclature for CWs	Name of plants	Common name of plants	Nomenclature for plants	Density of plants
16	P2	*Typha latifolia, Colocasia esculenta, Alternanthera sessilis, Polygonum, Canna indicia, Ocimum americanum* L, *Pistia stratiotes*	–		12 plants/CW

Note M stands for mix, and P1 and P2 stand for poly-culture of plant species

The experiments were conducted approximately for a year from December 2016 to December 2017 at the Environmental Engineering Laboratory of National Institute of Technology, Raipur (21.2497° N and 81.6050° E), India.

11.2.1 Lateritic Soil Substrate

Lateritic soil containing 51% of SiO_2, 15% of $A1_20_3$, 7% of CaO, 3% of MgO, 26% of Fe_20_3 and 1% of TiO_2 of minerals, initial pH 6.2, available nitrogen 148.50 kg/ha and available phosphorus 5.2 kg/ha was used as soil media.

11.2.2 CW Plants

Plants of similar size (20–30-cm height) were selected from their local native habitat and washed with tap water in order to remove soil and dead tissues from their roots. Plant densities of 2 plants/CW and 4 plants/CW were selected based on the previous work to study the effects of plant density on wastewater treatment (Jethwa and Bajpai 2016).

Figure 11.1a, b represents general layout, horizontal and vertical cross section of CWs units, depicting the substrate layers of each CW.

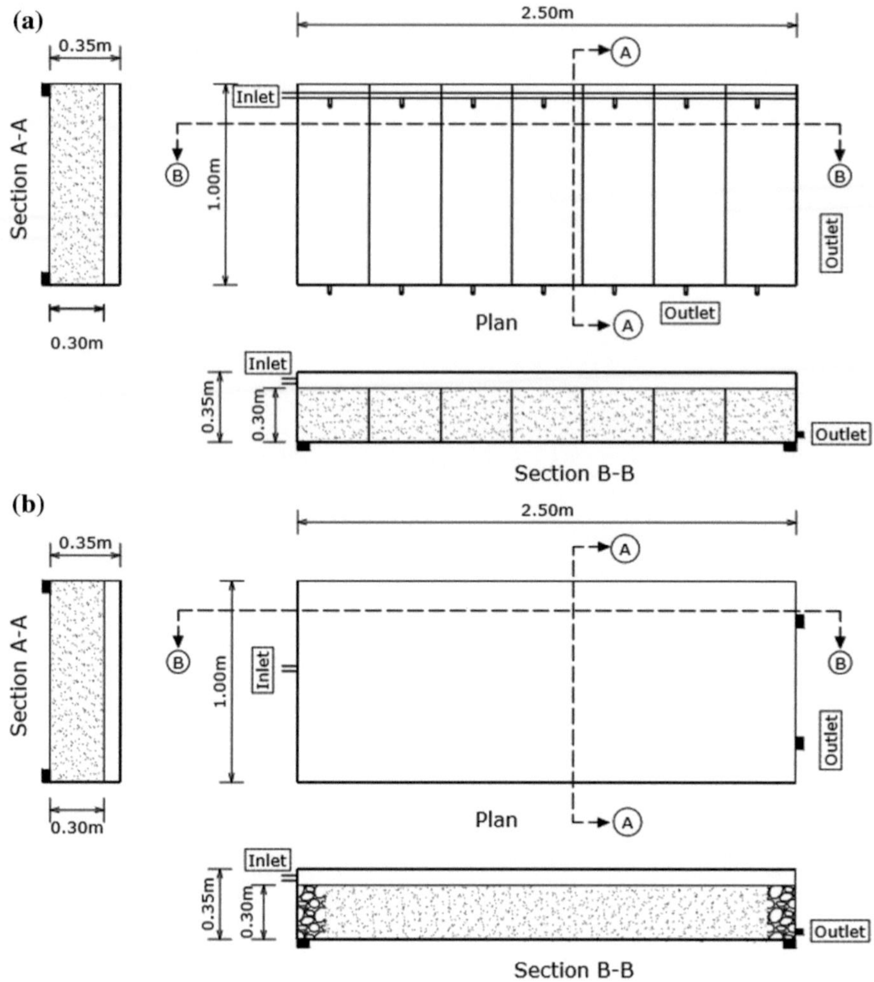

Fig. 11.1 **a** Typical plan and cross section of CW using mono-culture. **b** Typical plan and cross section of CW using poly-culture

11.2.3 Experimental Protocol

DS was collected from staff quarters and manhole and allowed to settle for two hours in two tanks of 250 L capacity. Surface loading rate of 0.0285 m^3/m^2/day was used from December to April. As the rate of evaporation was very high during May and June months, average ambient temperature was above 45 °C so increased surface loading rate of 0.045 m^3/m^2/day was used in the months of May and June. July onwards, the rate was again reduced to 0.0285 m^3/m^2/day. To maintain aerobic conditions, DS was applied on alternate days, for six hours a day only, maintaining alternate wetting and drying cycles. Twenty-four hour retention to the applied wastewater was provided in the CWs by closing the drainage valve located at the opposite side-bottom end of the reactor. Later, the bottom drainage valve was opened to drain out the treated wastewater. Initially, different parameters were analysed more frequently (twice a week), and later analyses were done twice a month. In the field application, alternate day dosing of domestic sewage on two plots of CWs can be adopted.

11.2.4 Analysis of Parameters

Biochemical oxygen demand (BOD), chemical oxygen demand (COD), total Kjeldahl nitrogen (TKN) and soluble reactive phosphorus (SRP) were analysed according to the standard methods (Eaton et al. 1995). BOD (5-day BOD_{20} test) and COD (closed reflux, titrimetric method) using HACH DRB 200.TKN of sewage were measured by digestion and distillation using KEL PLUS KES 06iL/12I and Kjeloplus-EAS VA instrument. SRP of sewage was measured by ascorbic acid method using UV spectrophotometer (Make: Lab India, Model: UV3000).

11.2.5 Media and Plant Analysis

Soil samples were air dried and mixed in equal proportion to obtain a homogenous sample. Air-dried soil was then passed through 1.18-mm sieve, and various physicochemical parameters were analysed—available nitrogen in soil was measured by Subbiah and Asija (1956) method, available phosphorus was measured by Olsen (1954) method; plant P was measured by Banton method (Cresser and Parsons 1979), and plant's TKN was measured by the micro-Kjeldahl method (Kalra 1999).

11.3 Results and Discussion

Plants were fully established within 2 months of plantation and showed luxuriant growth during the next 12 months of operation of the wetlands. After the initial 2 months of start-up period, the inlet and outlet sewage samples were collected by composite sampling, during the next 12 months of the experiment, at regular intervals, and analysed.

11.3.1 Nitrogen Dynamic

11.3.1.1 TKN Removal

The average TKN concentration in influent was 27.3 mg/L (5.6–49.0 mg/L), and average TKN in the effluent from different CWs was 18.25 mg/L in B1 and 18.13 mg/L in B2; 8.3 mg/L in M1T and 11.7 mg/L in M2T; 9.7 mg/L in M1Co and 12.34 mg/L in M2Co; 9.0 mg/L in M1C and 11 mg/L in M2C; 10.7 mg/L in M1O and 13.6 mg/L in M2O; 9.7 mg/L in M1A and 13.3 mg/L in M2A; 14.20 mg/L in M1P and 16.16 mg/L in M2P; and 10.20 mg/L in P1 and 14.10 mg/L in P2. Average influent concentration of TKN was consistently less than 100 mg/L, below the effluent quality standards for TKN (100 mg/L) prescribed by Central Pollution Control Board (CPCB) (1986), India. However, mono-culture CWs with plant species *T. latifolia* and *C. indica* performed better than the poly-culture plant species CWs. The effluent TKN concentration in the blank CWs with lateritic soil substrate media was higher in comparison with the CWs with plants (Fig. 11.2).

Overall M1T and M1C are removing more TKN as compared to both poly-culture CWs. This result also supports observation of Brisson and Chazarenc (2009) that *T. latifolia* is more efficient in nitrogen removal compared to other tropical wetland plants.

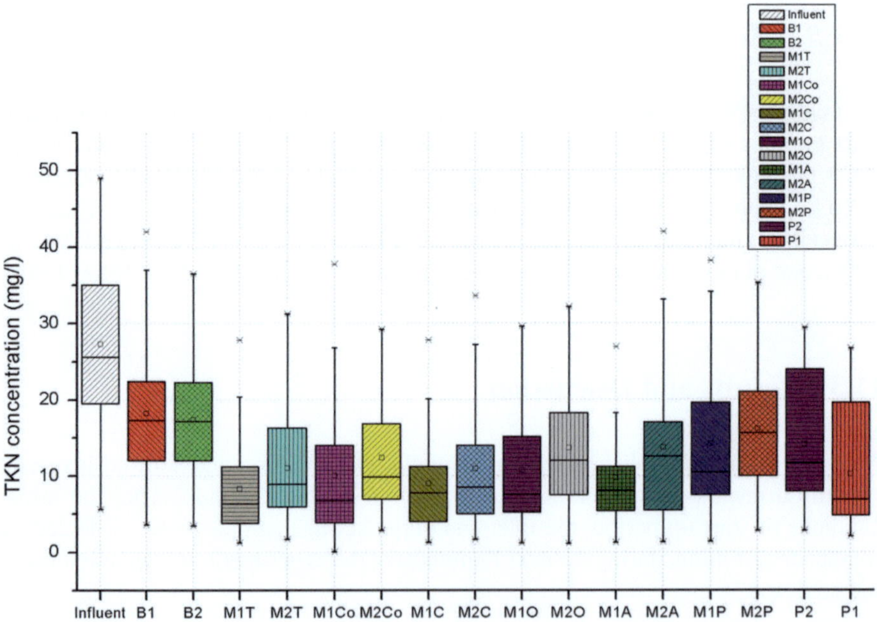

Fig. 11.2 TKN removal in CWs

Leung et al. (2016) suggested that cultural arrangement (*Acorus calamus, C. indica* and *Phragmites australis*) using mono-culture vs. poly-culture had no significant effect on the treatment efficiency. Coleman et al. (2001) reported species mixture (*Juncus effusus* L., *Scirpus validus* L. and *T. latifolia L*) outperformed species mono-cultures. Fraser et al. (2004) also found that the four-species mixture (*S. validus*, *Carex lacustris*, *P. arundinacea* and *T. latifolia*) was generally more effective in nutrient removal; however, the results were not significantly different from the mono-cultures. Zhu et al. (2018) and Wang et al. (2013) showed combination of plant species (*C. indica* + *L. salicaria*) and (grasses, forbs and legumes) reported superior performance of poly-culture than mono-culture plant species.

The difference with the findings of other researchers between performance of mono- and poly-culture is due to the specific mono-culture plant species (*T. latifolia* and *C. indicia*) being dominant in nitrogen removal than other plant species forming the poly-culture mix (*O. americanum* L, *P. stratiotes*, *A. sessilis* and Polygonum), this may be leading to interspecies completion for resources and in turn affecting the nutrient uptake, some of the plants being fragile (*O. americanum* L and *P. stratiotes*) died and decomposed and became part of the soil, releasing the nutrients to the soil and also to the effluent, similarly leaf shading by some plants (*P. stratiotes*) was higher than others, and these leaves decomposed and contributed nutrients to the soil and effluent.

In CWs, adsorption, ionic exchange, volatilization, plant absorption and uptake, and nitrification–denitrification by micro-organisms are the most important nitrogen removal pathways from wastewater (Billore et al. 1999).

Plants were analysed for total nitrogen content before and after the treatment. Dominant removal mechanisms for TKN in order of ranking are: (i) plant absorption, uptake and retention (shown by sizeable amounts of N in the plant tissue during 12-month biomass growth, system being quite fertile); (ii) ammonia volatilization due to favourable pH, influent wastewater pH varied between 7 and 9 (above 7) and higher ambient temperature (yearly average +30 °C); and (iii) nitrification, under favourable aerobic conditions due to the plant species and shallow soil depth, and higher contact time, leading subsequently to nitrate fixation, due to microbial immobilization and sorption in lateritic biofilm in surface litter and standing dead biomass of the plant species and denitrification in anaerobic micro-sites of the bed in the CWs.

Observed TKN removal pattern was: M1T > M1C > M1Co > M1A > P1 > M1O > M2T > M2C > M2Co > M2A > M2O > P2 > M1P > M2P > B2 > B1. The results presented in Fig. 11.2 show significant removal efficiency for TKN ($p < 0.05$), with better performance by the higher plant density of 4 plants/CW in comparison with the lower plant density of 2 plants/CW. Hence, higher density of plant mix resulted in removal of more nitrogen.

11.3.1.2 Nitrogen Retention in Soil

Depending on soil characteristics, organic carbon availability and pH, and climate conditions such as moisture and temperature, the mineral nitrogen may be metabolized by micro-organisms and may return to the organic form in a process called immobilization (Bengtsson et al. 2003).

Initial available nitrogen in the soil was 14.85 g/kg (6.0 g/m^2/year or 148.5 kg/ha) in the CWs. The average nitrogen concentration in the influent was 27.26 mg/L (278.50 g/m^2/year), and after application of sewage increase in the average available nitrogen concentration in soil is 36.5% in B1 and 37.9% in B2; 37.0% in M1T and 41.3% in M2T; 35.1% in M1Co and 40.9% in M2Co; 41.1% in M1C and 36.0% in M2C; 37.0% in M1O and 45.0% in M2O; 36.4% in M1A and 42.7% in M2A; 46.2% in M1P and 43.1% in M2P; 34.5% in P1 and 36.3% in P2, respectively (Fig. 11.3).

Figure 11.3 shows that available N retention in lateritic soil used in the CWS increased from 148.5 kg/ha (<108 kg/ha is very less, resulting in infertile soil) to 234–261 kg/ha (<108–280 kg/ha is medium range) indicating improvement in fertility of soil.

In the present study, after application of DS on the lateritic soil, pH increase from 6.2 to 7.1 was observed. Increased soil pH affects the N fixation in soil as wet and alkaline soil shows a higher conversion of NH_4^+ to NH_3, thereby increasing N losses by volatilization (Brandy and Raymond 2012). Also, when ambient temperature is above 30 °C there could be volatilization of NH_3 (Jones et al. 2013). In Raipur, located in tropical climate, except winter (3 months), average ambient temperature is above 30 °C in most of the months.

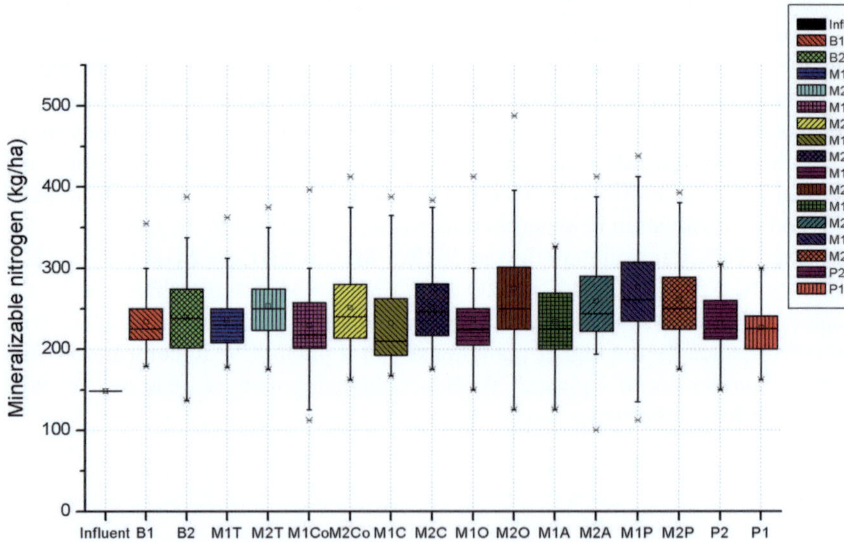

Fig. 11.3 Nitrogen retention in the CW soil media

11.3.1.3 TKN Mass Balance

The mass balance approach was used by researchers to quantify the contributions of different removal pathways in removing N and evaluating N transformations in wetland microcosms. The factors considered were—(1) amounts imported and exported from the microcosm systems; (2) amounts assimilated by plants; (3) amounts stored (e.g. adsorbed and precipitated) in the sediment; (4) N_2O emission amounts due to nitrification–denitrification; and (5) other losses involving N_2 emission via nitrification–denitrification process, ammonia volatilization (Reddy and Delaune 2008).

The nutrient removal efficiency was not increased by mixed culture between *T. latifolia, C. esculenta, A. sessilis, Polygonum, C. indicia, O. americanum* L and *P. stratiotes* compared with mono-cultures. The results agreed with the finding of Fraser et al. (2004) and Zhang et al. (2009). The present study might suggest that there was a significant interspecies competition between mono-culture and poly-culture, with *T. latifolia* being the superior competitor. In the present study, the major removals of N in CWs were through gasification. In poly-culture CWs, P1 and P2, respectively, plants with 55% and 46% N uptake of the influent, were the major nutrient sinks for N. Loss due to gasification was negligible in poly-culture in comparison with mono-culture plant species. *T. latifolia* plant species showed the least N retention of 12 and 14% in M1T and M2T CWs.

In mono-culture CWs, N removal in effluent was 30–60% and N loss due to volatilization was 35–60%, while in poly-culture CWs, 38 and 52% of N removed in effluent, for P1 and P2 CWs, respectively. Highest amount of N was passed in effluent for *P. stratiotes* (53% and 60% of influent N concentration in M1P and M2P CWs, respectively), while highest N removal in the effluent was observed in *T. latifolia* (effluent concentration of 31% and 41% in M1T and M2T).

CWs' soil with *P. stratiotes* retains the highest amount of N 4.2% and 3.9% in M1P and M2P, respectively, in comparison with other plant species; however, difference is almost negligible. Both the poly-culture CWs, P1 and P2, respectively, retain almost same amount of N in soil media (3.3% and 3.4%).

It seems that retention of available N in soil media is the highest in CWs planted with *P. stratiotes* plant species, followed by *O. americanum* L. This may be because *P. stratiotes* species assimilates N in plant tissues and after some time it disintegrates and becomes humus in the CWs, while *O. americanum* L species had a low survival rate in CW continuous wet conditions and quickly assimilated by the soil media releasing N stored in plant tissues to the soil. Other plant species consumed more TKN in comparison with *P. stratiotes* and *O. americanum* L. Alligator weed leafy species *A. sessilis* and *Polygonum* having good amount of protein and microbial infestation have been categorized into 4 cover grades (Grade I–IV) from lowest (no/negligible: <10% cover) to highest: >60% cover), helping in N retention in soil (Chatterjee and Dewanji 2014). The present study showed that plant species *Pistia stratiotes* and *O. americanum* L consumed higher amount of TKN per unit plant weight and contributed to soil N due to low survival rate and assimilation by soil.

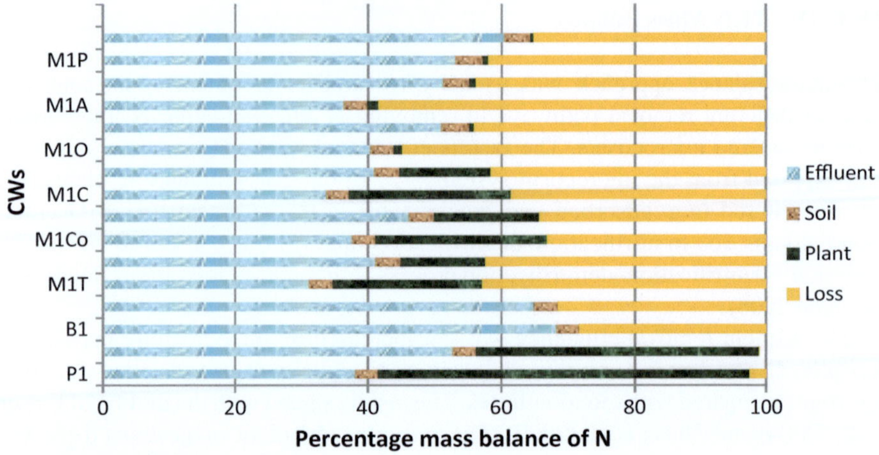

Fig. 11.4 Proportion of N removed by different pathways among different wetlands during the experimental period

According to researchers, *T. latifolia* (Polomski 2009), *C. esculenta* (Lakhanpaul et al. 2003), *C. indica* (Zhang et al. 2007), *O. americanum L* (Sharafzadeh and Alizadeh 2011) and *P. stratiotes* (Lieberei et al. 2002) have rhizome type of root structure which helps in N retention in soil.

Proportion of N removed by different pathways among different wetlands during the experimental period was calculated (Fig. 11.4).

Limited information is available for N and P mass balance using domestic sewage, so the present study in comparison with different types of wastewater was conducted (Tables 11.2 and 11.3). The removal efficiencies of N were comparable to reported studies on other plant species (Table 11.2). For example, Zhang et al. (2007) reported that plant uptake (86% of influent) was the major nutrient removal pathway in the wetland microcosms. Lee et al. (2014) attributed 45% N in effluent and N loss of 34%, to less plant coverage area (<3%). Chung et al. (2007) and Sekiranda and Kiwanuka (1998) showed that nitrification and denitrification remained high in CWs with shallow depth and therefore ammonium-N and nitrate-N could be effectively removed from the wastewater. Wu et al. (2013a, b) treating slightly polluted river water in surface CW reported plant uptake and sediment storage as the key factors limiting nitrogen removal besides microbial processes.

The results agreed with the observations of Coleman et al. (2001) that *T. latifolia* was the superior competitor among the three-species (*J. effusus, T. latifolia* and *S. cyperinus*) mixture in small-scale CWs. However, no research literature could be located for N uptake using poly-culture of *T. latifolia, C. esculenta, A. sessilis, Polygonum, C. indicia, O. americanum* L, *P. stratiotes.*

Increase in soil pH value, ambient temperature above 30 °C and anaerobic conditions (wetting condition) may result in lower retention of nitrogen in soil. Thus,

Table 11.2 Mass balance of N in comparison with the result of other researchers

References	Present study	Zhang et al. (2007)	Lee et al. (2014)	Wu et al. (2013a, b)	Sekiranda and Kiwanuka (1998)		Chung et al. (2007)	
Wastewater type	Domestic sewage	Secondary municipal wastewater	Piggery wastewater	Slightly polluted river	Domestic sewage		Municipal wastewater	
Name of plant species	T, C, Co, A, O, P[a] Mono-culture and Poly-culture	Mix culture of *C. Indica* and *Svalidus*	—	*Typha orientalis*	*Phragmites*	*Phragmites*	T with two HRTs 10 days and 5 days	
Media type	Laterite	Sand	—	River sand	Laterite		Sandy loamy soil with compost (2:1—v/v)	
Effluent (%)	31–68	6	45	40	–	–	41	59
Plant (%)	0.5–58	86	1	10	12	5	<5	<5
Media (%)	3.3–4.2	16	7	10	19	29	–	–
Gasification (%)	–	–	–	–	–	–	58	39
Loss (%)	2–58	14	34	25	70	66	–	–

[a]T: *Typha latifolia*, Co: *Colocasia esculenta*, A: *Alternanthera sessilis*, Pa: *Polygonum*, C: *Canna indicia*, O: *Ocimum americanum* L, P: *Pistia stratiotes*

Table 11.3 Mass balance of P in comparison with the results of other researchers

References	Present study	Zhang et al. (2007)	Lee et al. (2014)	Wu et al. (2013a, b)	Sekiranda and Kiwanuka (1998)		Chung et al. (2007)		Menon and Holland (2013)
Wastewater type	Domestic sewage	Secondary municipal wastewater	Piggery wastewater	Slightly polluted river	Domestic sewage		Municipal wastewater		2.5 ppm of phosphorus solution
Name of plant species	T, Co, C, O, A, P[a]	Mix culture of *C. Indica* and *Svalidus*	–	*Typha orientalis*	*Phragmites*		T		*J. effusus*, *C. lurida*, and *D. acuminatum*
	Mono-culture and Poly-culture	–	–	–	Floating Phragmites	Laterite-gravel rooted Phragmites	10 days HRT	5 days HRT	–
Media type	Laterite	Sand	–	River sand	Laterite		Sandy loamy soil with compost (2:1—v/v)		Sandy and clay mixture
Effluent (%)	11–55	32	–	–	25	8	41	59	11
Media (%)	37–55	16	34	36–49	–	–	<1	<1	7
Plant (%)	0.06–25	50	9	4.8–22	37	37	–	–	63
Loss (%)	8–46	14	30	–	38	38	58	39	–
Microbial uptake (%)	–	–	–	0.26–4.13	–	–	–	–	–

[a]T: *Typha latifolia*, Co: *Colocasia esculenta*, A: *Alternanthera sessilis*, Po: *Polygonum*, C: *Canna indicia*, O: *Ocimum americanum* L, P: *Pistia stratiotes*

a dynamic system involving organic and inorganic forms of N mediated by biological activity promotes a series of changes in the element transformations, such as mineralization, immobilization, nitrification and denitrification.

11.3.2 Phosphorus Dynamics

11.3.2.1 Phosphorus Removal from Domestic Sewage

The average P concentration in influent was 7 mg/L as PO_4^- (0.19–17.15 mg/L), and average concentration in effluent from the mono-culture CWs was 4 mg/L in B1 and 4 mg/L in B2; 0.8 mg/L in M1T and 1.3 mg/L in M2T; 1.0 mg/L in M1Co and 1.3 mg/L in M2Co; 1.0 mg/L in M1C and 1.3 mg/L in M2C; 2.0 mg/L in M1O and 2.0 mg/L in M2O; 1.7 mg/L in M1A and 2.0 mg/L in M2A; and 2.3 mg/L in M1P and 2.3 mg/L M2P, while in poly-culture CWs effluent phosphorus concentrations as PO_4^- were 1.3 mg/L and 1.4 mg/L in P1 and P2, respectively (Fig. 11.5).

Effluents from all the CW types—mono-culture, poly-culture and blank—satisfied permissible discharge limit for disposal of treated sewage to natural stream ($\leq$5 mg/L). CWs with mono-culture performed marginally better than CW with poly-culture of plants in SRP removal. Lateritic soil substrate media shows good phosphorus adsorption capacity. The results presented in Fig. 11.5 show significant removal efficiency for SRP ($p < 0.05$) for planted CWs than the CWs without plants

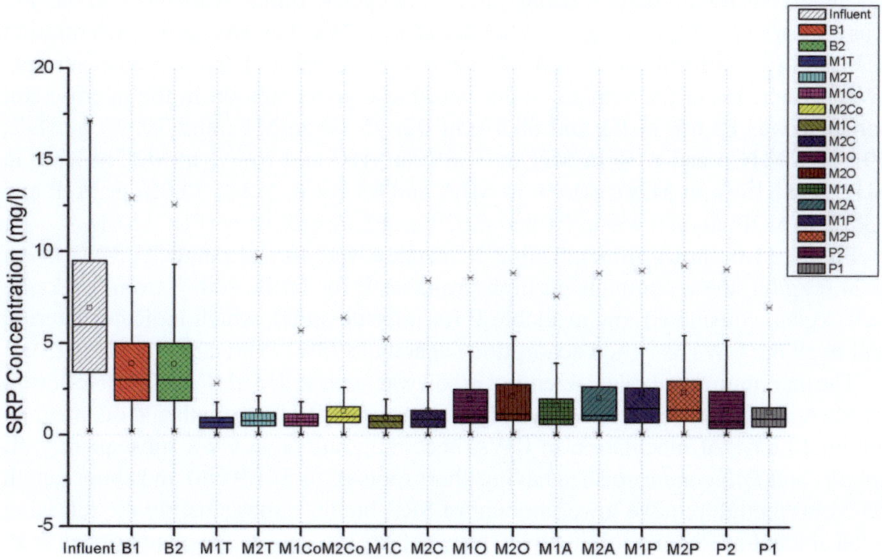

Fig. 11.5 SRP removal in CWs

(B1 and B2). CWs with higher plant density of 4 plants/CW showed better performance in comparison with the CWs with lower plant density of 2 plants/CW, though the increase in removal efficiency is not in direct proportion to the plant density.

Also, wetting and drying, and reduction and oxidation cycles may lead to changes in the form of various constituents (such as iron), resulting in their association with phosphate and subsequent adsorption and removal. However, there may be a distinct change in iron phosphate association due to possible reduction of Fe^{3+} into Fe^{2+} due to water-logged conditions, resulting in a decrease in iron phosphate in soil substrate. The released ferrous phosphate is likely to get distributed into other forms (Mishra and Gupta 1971).

Sartori et al. (2015) observed only 55% phosphorus removal in a gravel-based SFS to SF to oxidation pond system to treat wastewater, while Yu et al. (2012) using pilot-scale rural wastewater treatment system removed 86.3% total phosphorus (TP) with the average influent concentration of only 1.1 mg/L. In the work of Zhu et al. and Coleman et al. (2001), relatively higher removal of TP by poly-culture (with plant species of *J. effusus* L., *S. validus* L. and *T. latifolia* L.) in comparison with mono-culture of plants was reported, because *J. effusus* L., *S. validus* L. and *T. latifolia* L. are good at P removal.

11.3.2.2 Phosphorus Retention in the Soil Media

Dry and wet conditions in CW help in P fixation in the soil media (Laura 1976). Throughout the study, the CWs were loaded only twice a week with pre-settled sewage to ensure alternate wetting and drying cycles. Initial available P in the soil was 0.59 g/kg as PO_4^- (5.9 g/m^2/year), for all the CWs. The average P concentration as PO_4^- in the influent was 7 mg/L (71.4 g/m^2/year), and after application of sewage, average increase in the available/mineralizable P concentration in the lateritic soil substrate was 81.6% in B1 and 80.9% in B2; 75.7% in M1T and 79.5% in M2T; 79.2% in M1Co and 81% in M2Co; 77.8% in M1C and 79.7% in M2C; 81.7% in M1O and 81.4% in M2O; 80.4% in M1A and 82.6% in M2A; 85.0% in M1P and 83.8% in M2P; and 80.4% in P1 and 82.2% in P2, respectively (Fig. 11.6).

Figure 11.6 shows mineralizable P increase was in the range 25.5–40 kg/ha (>24.6 kg/ha, the recommended high available P for fertile soils), from 5.9 kg/ha (<10 kg/ha, considered low available P for infertile soils), which indicates lateritic soil in all the CWs has good adsorption capacity (>75%) (Shrivastav et al. 2015).

The maximum available P retention in soil was achieved in M2P and M1P followed by M2A and M1A. *Pistia stratiotes* plants consume SRP for growth and disintegrate within 15 days of inoculation in CWs, becoming humus in CWs subsequently. *A. sessilis* and *Polygonum* species having short roots of (1.5–1.9 cm) and shoot length (2–5 cm) though remove lower amount of SRP, but they subsequently die releasing good amount of P to the soil media in terms of dry biomass of the plant (Abbas et al. 2016; Walter et al. 2014). However, soil in blank CWs retained 43% of P indicating good PO_4^- adsorption capacity of lateritic soil.

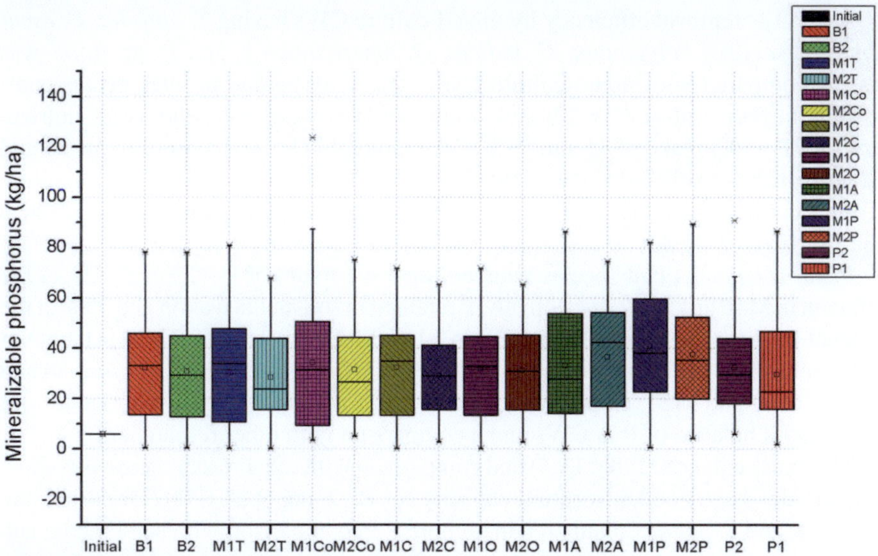

Fig. 11.6 Phosphorus retention in the CW soil media

11.3.2.3 P Mass Balance

A mass balance model for PO_4^- was applied in order to quantify the contributions of different PO_4^- removal pathways for each wetland. The factors considered were (i) amounts imported and exported from the microcosm systems; (ii) amounts assimilated by plants; (iii) amounts stored by the media; and (iv) other losses (Fig. 11.7).

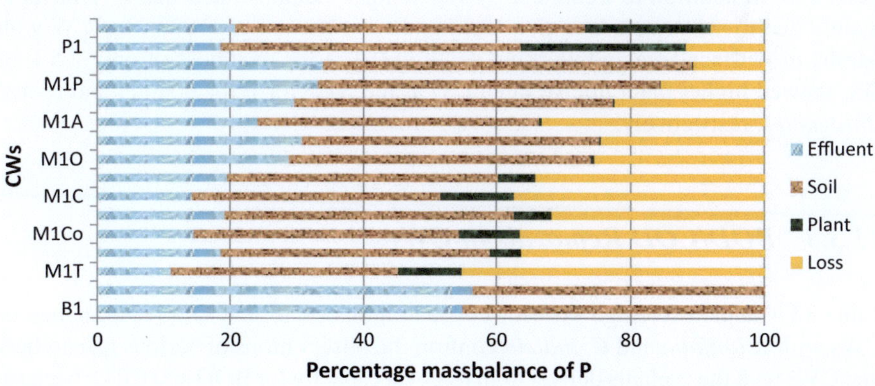

Fig. 11.7 Proportion of P removed by different pathways among different wetlands during the experimental period

The PO_4^- removal efficiency by mixed-culture CWs having *T. latifolia, C. esculenta, A. sessilis, Polygonum, C. indicia, O. americanum* L and *P. stratiotes* was less as compared with mono-culture CWs. This may be due to interspecies competition in poly-culture CWs. In poly-culture CWs, plants were the major nutrient sinks for P with uptake of 45 and 53% of the influent PO_4^- concentration for P1 and P2 CWs, respectively, also having negligible loss in comparison with mono-culture CWs. However, for mono-culture CWs phosphate passage in effluent was 11–34% and loss was 10–45%.

Pistia stratiotes plant species retained highest amount of P—55 and 51% of the influent, in M1P and M2P, respectively. P retained in soil media was 45 and 55% of the influent for both the poly-culture CWs in P1 and P2, respectively. The least amount of P retention in the lateritic soil media was for CWs planted with *T. latifolia* 34 and 40% in M1T and M2T, respectively, due to maximum uptake by plants. Table 11.3 shows mass balance of P in CWs and a comparison with other researchers.

This mass balance (Table 11.3) and comparison with other researcher works show that laterites have good adsorption capacity for P. Zhang et al. (2007) reported that *C. indica* and *S. validus* plants retained more P in comparison with sand media and wastewater effluent. Menon and Holland (2013) showed that the CWs with poly-culture of plant species consumed more P in comparison with sand and clay mixture and effluent, and suggested that proper selection of plant species is important for achieving the full functioning efficiency of a CW. Dzakpasu et al. (2015) found that the sediment storage (60%) was the major P removal pathway in the integrated constructed wetland (ICW). The authors used surface area of 27.58 m^2/PE which is quite larger than the present study (2.5 m^2/PE). Chung et al. (2007) reported higher adsorption of P on soil substrate (mixture of sandy loamy soil with compost) in comparison with plants and effluent. Lee et al. (2014), using a CW consisting of six cells—each cell performing a specific treatment function; Cell 1 for sedimentation of particulates, aeration in Cell 2 and sedimentation of organics in the subsequent Cells 3–6, in addition to treatment by vegetation—demonstrated that P removal is mainly due to substrate adsorption. Sekiranda and Kiwanuka (1998), for CW with 0.0962 m^2 surface area and working depth of 0.3 m for treatment of only 2.5 L of DS, showed higher plant uptake rate of 37% and 54% for P for floating and rooted *Phragmites*, respectively.

11.3.3 BOD/COD Removals in CWs

Table 11.4 depicts average effluent concentrations of BOD/COD performance of CWs with *T. latifolia* and *C. indicia* confirm the observations of various researchers that CWs with these plants deliver high removal capacity for BOD and COD, because of high oxygen transfer due to the higher green foliage (large leave area and green stem) resulting in higher photosynthetic activity ensuring transport of higher concentration of oxygen to the fine-fibrous and bulbous type of root structure of the plants (Billore et al. 1999; Rai et al. 2013), ensuring greater surface area for growth

Table 11.4 Average BOD/COD removal of CWs

CWs	BOD (mg/L)	COD (mg/L)
B1	109	230
B2	113	236
M1T	29	95
M2T	36	141
M1Co	46	131
M2Co	59	167
M1C	34	120
M2C	40	160
M1O	52	145
M2O	75	194
M1A	56	143
M2A	68	174
M1P	78	188
M2P	91	218
P1	36	127
P2	51	200

of micro-organisms and for greater contact surface and hence higher contact time for various physico-biochemical reactions.

Pistia stratiotes, having a little stem and shallow root system, has been reported to have the least oxygen transfer rate, resulting in the least pollutant removal as compared to other plants (Reddy et al. 1996), while the least BOD removal was observed for *P. stratiotes* with 2 plants/CW. BOD removal is slightly better in CWs with mono-culture of plant species than poly-culture; however, the difference is not that significant. Certain plants (such as *T. latifolia* and *C. indica*) play dominant role in BOD/COD removal and should be the plants of choice for any CW system. The density played important role in BOD/COD removal for plants such as *O. americanum* L, *A. sessilis* and *Polygonum*. For other plants, such as *T. latifolia* and *C. indica*, the plant density did not play a significant role. This difference may be due to root structure of such plants. In case of *O. americanum* L, *A. sessilis* and *Polygonum*, the root structure is short, thin tailed and hairy, while for *T. latifolia* and *C. indicia*, the root structure is fine, fibrous and bulbous.

Role of plant species for pollutant removal using vegetated wetland in comparison with non-vegetated was also studied by Huett et al. (2005). The presence of plants improves uptake of carbon, nitrogen and phosphorus. Plant roots provide sites for variety of micro-organisms that actively decompose organic matter for their metabolism, effectively reducing the BOD. Plant roots also supply oxygen to the soil system, helping in maintaining the aerobic conditions resulting in more efficient reduction of BOD/COD.

This study showed significant removal efficiency for COD ($p < 0.05$) with plant density 4/CW, with the best performance by *T. latifolia* (average 72.6% removal) followed with *C. indica* (65.7% removal).

Overall, BOD/COD removal pattern in CWs was M1T > M1C > P1 > M2T > M2C > M1Co > P2 > M1O > M1A > M1P > M2Co > M2O > M2P > M2P > B1 > B2. Poly-culture CWs with higher plant density always performed better than the lower plant density CWs. Overall, mono-culture CWs with higher density of *T. latifolia* and *C. indica* gave best performance in BOD/COD removal.

11.4 Conclusion

Sixteen laboratory-scale CWs, with shallow lateritic soil media, were operated for 365 days with various densities of six plant species to deal with settled domestic sewage. Vegetation type played an important role in CW performance. Specifically, *T. latifolia* and *C. indica* contributed to higher pollutant removal. The domestic sewage surface loading rate of 0.0285 m^3/m^2/day resulted in 70–85% of BOD removal with *T. latifolia* (83.4%) and *C. indica* (80.6%) plants with density of 4 plants/CW. Mono-culture CWs remove more BOD/COD and TKN compared to poly-culture of plant species. Higher density of plants for the thin and short-root plant species resulted in more BOD/COD, SRP and TKN removal, while for plants with fibrous and bulbous type of root structures density of plants did not affect the performance of CWs. CWs with poly-culture were a combination of individual plant species with good performance and poor performance, and the overall effect reduced pollutant removal rate. Hence, section of plant species plays an important role. With good plant species, poly-culture CWs will give better performance. Average TKN removal was 18% in B1 and 18% in B2; 8% in M1T and 11% in M2T; 10% in M1Co and 12% in M2Co; 9% in M1C and 10% in M2C; 10% in M1O and 14% in M2O; 10% in M1A and 13% in M2A; 14% in M1P and 16% in M2P; and 10% in P1 and 14% in P2, respectively, whereas average SRP removal was 44% in B1 and 44% in B2; 89% in M1T and 82% in M2T; 86% in M1Co and 81% in M2Co; 86% in M1C and 81% in M2C; 71% in M1O and 69% in M2O; 76% in M1A and 71% in M2A; 67% in M1P and 66% in M2P; and 82% in P1 and 79% in P2, respectively.

CWs with higher plant density performed well in the removal of BOD/COD, TKN and SRP and retention of available phosphorus and available nitrogen in lateritic soil, resulting in substantial improvement in the fertility value of the soil. Higher plant density resulted in better performance, but the improvement was not in direct proportion to the plant density. Alternate wetting and drying cycles help to maintain aeration in CWs which helps in oxidizing the BOD/COD, SRP and TKN.

Average available nitrogen concentration in soil is 36.5% in B1 and 37.9% in B2; 37.0% in M1T and 41.3% in M2T; 35.1% in M1Co and 40.9% in M2Co; 41.1% in M1C and 36.0% in M2C; 37.0% in M1O and 45.0% in M2O; 36.4% in M1A and 42.7% in M2A; 46.2% in M1P and 43.1% in M2P; and 34.5% in P1 and 36.3% in P2, respectively. Average increase in the available/mineralizable P concentration in the lateritic soil substrate was 81.6% in B1 and 80.9% in B2; 75.7% in M1T and 79.5% in M2T; 79.2% in M1Co and 81% in M2Co; 77.8% in M1C and 79.7% in M2C; 81.7% in M1O and 81.4% in M2O; 80.4% in M1A and 82.6% in M2A; 85.0% in M1P and 83.8% in M2P; and 80.4% in P1 and 82.2% in P2, respectively.

These values were obtained in the soil substrate after one year of study, indicating shifting of nutrient values of the lateritic soil towards that of more fertile soils of the region. Overall, for the objective of removal of BOD/COD, TKN and SRP, *T. latifolia, C. esculenta*, and *C. indicia* plant species should be preferred, while for retention of nutrients in soil *P. Stratiotes*, *A. sessilis* and *Polygonum* plant species should be preferred.

Volatilization of ammonia due to higher ambient temperature played a major role in TKN removal. Adsorption of P by lateritic soil played a major role in SRP removal.

In the present work, the successful use of barren, infertile lateritic soil media for the satisfactory treatment of domestic wastewater using readily available plants, and improving its fertility, may open up opportunity for adoption of this eco-friendly solution by the developing countries which are resources challenged for installation of the costly high-tech treatment plants.

References

Abbas T, Tanveer A, Khaliq A, Safdar ME (2016) Comparative allelopathic potential of native and invasive weeds in rice ecosystem. Pak J Weed Sci Res 22(2):269–283

Avila C, Matamoros V, Reyes-Contreras C, Piña B, Casado M, Mita L, Rivetti C, Barata C, García J, Bayona JM (2014) Attenuation of emerging organic contaminants in a hybrid constructed wetland system under different hydraulic loading rates and their associated toxicological effects in wastewater. Sci Total Environ 470–471:1272–1280

Bengtsson G, Bengtson P, Mansson KF (2003) Gross nitrogen mineralization-, immobilization-, and nitrification rates as a function of soil C/N Ratio and microbial activity. Soil Biol Biochem 35:143–154

Billore SK, Singh N, Sharma JK, Dass P, Nelson RM (1999) Horizontal subsurface flow gravel bed constructed wetland with *Phragmites karka* in central India. Water Sci Technol 40(3):163–171. https://doi.org/10.1016/S0273-1223(99)00461-8

Brandy N, Raymond W (2012) The nature and properties of soils. Pearson Education. ISBN: 978-0133254488:1-1052

Brisson J, Chazarenc F (2009) Maximizing pollutant removal in constructed wetlands: should we pay more attention to macrophyte species selection? Sci Total Environ 407(13):3923–3930. https://doi.org/10.1016/j.scitotenv.2008.05.047

Brix H (1994) Function of macrophytes in constructed wetlands. Water Sci Technol 29(4):71–78

Bunce JT, Ndam E, Ofiteru ID (2018) A review of phosphorus removal technologies and their applicability to small-scale domestic wastewater treatment systems. Front Environ Sci 6:1–15

Central Pollution Control Board (1986) General standards for discharge of environmental pollutants. Environ (Prot) Rules 174(2):145–160

Chatterjee A, Dewanji A (2014) Effect of varying *Alternanthera philoxeroides* (alligator weed) cover on the macrophyte species diversity of pond ecosystems: a quadrat-based study. Aquat Invasions 9(3):343–355

Chung AKC, Wu Y, Tam NFY, Wong MH (2007) Nitrogen and phosphate mass balance in a sub-surface flow constructed wetland for treating municipal wastewater. Ecol Eng 2(2000):81–89

Coleman J, Hench K, Garbutt K, Sexstone A, Bissonnette G, Skousen J (2001) Treatment of domestic wastewater by three plant species in constructed wetlands. Water Air Soil Pollut 128:283–295
Cresser MS, Parsons JW (1979) Sulfuric-perchloric acid digestion of plant material for the determination of nitrogen, phosphorus, potassium, calcium. Anal Chim Acta 109:431–436
Dzakpasu M, Miklas S, Mccarthy V, Jordan S (2015) Phosphorus retention and mass balance in an integrated constructed wetland treating domestic wastewater. Water Environ J 29(2):298–306
Eaton AD, Clesceri LS, Greenberg AE (1995) Standard methods for the examination of water and wastewater, 19th edn. Washington, DC
Fraser LH, Carty SM, Steer D (2004) A test of four plant species to reduce total nitrogen and total phosphorus from soil leachate in subsurface wetland microcosms. Biomed Res Int 94:185–192
García J, Aguirre A, Barragán J, Mujeriego R, Matamoros M, Bayona JM (2005) Effect of key design parameters on the efficiency of horizontal subsurface flow constructed wetlands. Ecol Eng 25:405–418
Gliński J, Horabik J, Lipiec J (2003) Encyclopedia of earth sciences series encyclopedia of agrophysics, 1st edn. Springer, Netherland, pp 1–900
Huang J, Reneau RB, Hagedorn C (2000) Nitrogen removal in constructed wetlands employed to treat domestic wastewater. Water Res 34(9):2582–2588
Huett DO, Morris SG, Smith G et al (2005) Nitrogen and phosphorus removal from plant nursery runoff in vegetated and unvegetated subsurface flow wetlands. Water Res 39(14):3259–3272
Ilyas H, Masih I (2018) The effects of different aeration strategies on the performance of constructed wetlands for phosphorus removal. Environ Sci Pollut Res 25(6):5318–5335
Jethwa K, Bajpai S (2016) Role of plants in constructed wetlands (CWS): a review. J Chem Pharm Sci 2:4–10
Jones C, Brown BD, Engel R, Horneck D, Olson-Rutz K (2013) Factors affecting nitrogen fertilizer volatilization. http://landresources.montana.edu/soilfertility/. Accessed 20 Dec 2019
Kalra YP (1999) Handbook of reference methods for plant analysis. CRC Press, Taylor and Francis Group, Boca Raton, Boston, London, New York, Washington, D.C., pp 1–287
Lakhanpaul S, Velayudhan KC, Bhat KV (2003). Analysis of genetic diversity in Indian taro using random amplified polymorphic DNA (RAPD) markers. Genet Resour Crop Evol, 603–609
Laura RD (1976) On the stimulating effect of drying a soil and the retarding effect of drying a plant material. Plant Soil 44(2):463–465
Lee S, Maniquiz-Redillas MC, Choi J, Kim L-H (2014) Nitrogen mass balance in a constructed wetland treating piggery wastewater effluent. J Environ Sci (China) 26(6):1260–1266
Leung JYS, Cai Q, Tam NFY (2016) Comparing subsurface flow constructed wetlands with mangrove plants and freshwater wetland plants for removing nutrients and toxic pollutants. Ecol Eng 95:129–137. https://doi.org/10.1016/j.ecoleng.2016.06.016
Lieberei R, Bianchi HK, Boehm V, Reisdorff C (2002) Neotropical ecosystems, proceedings of the German-Brazilian workshop, Hamburg 2000. GKSS-Geesthacht
Menon R, Holland MM (2013) Phosphorus retention in constructed wetlands vegetated with *Juncus effusus*, *Carex lurida*, and Dichanthelium Acuminatum Var. Acuminatum, Water, Air, Soil Pollut 224(7):1602
Mishra SG, Gupta BP (1971) Effect of alternate wetting and drying on laterite and their engineering behaviour. Technology 8(2):154–157
Olsen SR (1954) Estimation of available phosphorus in soils by extraction with sodium bicarbonate. U. S. Department of Agriculture Circular No. 939. Banderis, A. D., D. H. Barter and K. Anderson. Agricultural and Advisor
Polomski RF (2009) Nitrogen and phosphorus remediation of aquatic garden plants in laboratory-scale constructed wetlands. PhD thesis, Graduate School of Clemson University, Department of Plant and Environmental Sciences
Rai UN, Tripathi RD, Singh NK, Upadhyay AK, Dwivedi S, Shukla MK, Mallick S, Singh SN, Nautiyal CS (2013) Constructed wetland as an ecotechnological tool for pollution treatment for conservation of the Ganga river. Bioresour Technol 148:535–541. http://dx.doi.org/10.1016/j.biortech.2013.09.005

Raychaudhuri SP (1980) The occurrence, distribution, classification and management of laterite and lateritic soil. J Coh ORSTOM Sér P Mol 18(3–4):249–252

Reddy KR, Delaune RD (2008) Biochemistry of wetland science and application. CRC Press, Taylor and Francis Group, Boca Raton Boston London New York Washington, DC

Reddy KR, Flaig EG, Graetz DA (1996) Phosphorus storage capacity of uplands, wetlands and streams of the Lake Okeechobee Watershed. Fla Agric Ecosyst Environ 59:203–216

Sartori L, Canobbio S, Riccardo F, Riccardo C, Francesca M, Valeria M (2015) COD, nutrient removal and disinfection efficiency of a combined subsurface and surface flow constructed wetland: a case study. Int J Phytorem 18(4):416–422

Sekiranda SBK, Kiwanuka S (1998) A Study of nutrient removal efficiency of *Phragmites mauritianus* in experimental reactors in Uganda. Hydrobiologia 364(1984):83–91

Sharafzadeh S, Alizadeh O (2011) Nutrient supply and fertilization of basil. Adv Env Biol 5(5):956–960

Shrivastav LK, Mishra VN, Banvashi R, Jatav GK, Pradip D (2015) Soil testing manual. Indira Gandhi Krishi Vishwa Vidhyalaya, Raipur

Song H, Nakano K, Taniguchi T, Nomura M, Nishimura O (2009) Estrogen removal from treated municipal effluent in small-scale constructed wetland with different depth. Bioresour Technol 100(12):2945–2951

Subbiah BV, Asija GL (1956) A rapid procedure for the estimation of available nitrogen in soils. Curr Sci 25:256–260

Sunarsih, Purwanto, Budi WS (2015) Modeling of domestic wastewater treatment facultative stabilization ponds. Int J Technol 4:689–698

USEPA (2000) Manual constructed wetlands treatment of municipal wastewaters. Risk Management (EPA-625-R (September):166. Retrieved http://www.epa.gov/ORD/NRMRL/Pubs/2001/wetlands/625r99010.pdf

Vohla C, Alas R, Nurk K, Baatz S, Mander Ü (2007) Dynamics of phosphorus, nitrogen and carbon removal in a horizontal subsurface flow constructed wetland. Sci Total Environ 380:66–74

Vymazal J (2007) Removal of nutrients in various types of constructed wetlands. Sci Total Environ 380(1–3):48–65

Walter T, Merish S, Tamizhamuthu M (2014) Review of *Alternanthera sessilis* with reference to traditional Siddha medicine. Int J Pharmacognosy Phytochem Res 6(2):249–254

Wang H, Chen Z-X, Zhang X-Y, Zhu S-X, Ge Y, Chang S-X, Zhang C-B, Huang C-C, Chang J (2013) Plant species richness increased belowground plant biomass and substrate nitrogen removal in a constructed wetland. Clean—Soil, Air, Water 41(7):657–664

Wood RB, McAtamney CF (1996) Constructed wetlands for waste water treatment: the use of laterite in the bed medium in phosphorus and heavy metal removal. Hydrobiologia 340(1–3):323–331

Wu H, Zhang J, Wei R, Liang S, Li C, Xie H (2013a) Nitrogen transformations and balance in constructed wetlands for slightly polluted river water treatment using different macrophytes. Environ Sci Pollut Res 20(1):443–451

Wu H, Zhang J, Li C et al (2013b) Mass balance study on phosphorus removal in constructed wetland microcosms treating polluted river water. Clean—Soil, Air, Water 41(9):844–850

Yu R, Wu Q, Lu X (2012) Constructed wetland in a compact rural domestic wastewater treatment system for nutrient removal. Environ Eng Sci 29:751–757

Zhang Z, Rengel Z, Meney K (2007) Nutrient removal from simulated wastewater Using *Canna indica* and *Schoenoplectus validus* in mono-culture and mixed-culture in wetland microcosms. Water Air Soil Pollut 183(1–4):95–105

Zhang HM, Xu MG, Zhang WJ, He XH (2009) Factors affecting potassium fixation in seven soils under 15-year long-term fertilization. Chin Sci Bull 54(10):1773–1780. http://link.springer.com/10.1007/s11434-009-0164-9

Zhu H, Zhou QW, Yan BX, Liang YX, Yu XF, Gerchman Y, Cheng XW (2018) Influence of vegetation type and temperature on the performance of constructed wetlands for nutrient removal. Water Sci Technol 77(3):829–837

Chapter 12
Partial Replacement of Fine Aggregates with Defluoridation Sludge in Cement Mortars Manufacturing: A Critical Review

Swati Dubey, Madhu Agarwal and A. B. Gupta

Abstract Generation of an excessive amount of sludge with various water treatment plants has become a major problem worldwide. Due to the scarcity of land and environmental concern, management of sludge is the prime concern nowadays. In fluoride removal process via coagulation, aluminium sulphate (alum) and poly aluminium chloride (PACl) are the chief coagulants, and the sludge is generated in massive amount from the defluoridation plants. Previous researchers pointed towards the health concerns related to its disposal. In Rajasthan state, India, Bisalpur dam supplies 16.2 TMC per day out of which 5.1 TMC is supplied to Ajmer district and 11.1 to Jaipur district. About 16.5 metric tonne sludge is generated per day which is a huge quantity. As aluminium sulphate is used as coagulant in water treatment plant, this sludge is highly toxic as it contains metals like aluminium, so its disposal is a major problem for Public Health Engineering Department (PHED). According to Central Pollution Control Board (CPCB), India, India is producing huge amount of inevitable waste everyday at their water treatment plants which requires proper handling and disposal. Therefore, there is an urgent need to look out for a proper management of the sludge. Moreover, the use of a large volume of sludge as a construction material can solve disposal problems and makes an approach towards eco-friendly construction. This paper presents an overview on the behaviour of cement mortars with different proportions of sludge. The compressive strength and microstructure analysis in comparison with the control mix are the vital factors for the cement mortars. The description about the performance of the both sludge: alum and PACl has

S. Dubey
Department of Chemical Engineering, Banasthali Vidyapeeth, Tonk, Rajasthan 304022, India
e-mail: swatidubey2021@gmail.com

M. Agarwal (✉)
Department of Chemical Engineering, MNIT, Jaipur, Rajasthan 302017, India
e-mail: magarwal.chem@mnit.ac.in

A. B. Gupta
Department of Civil Engineering, MNIT, Jaipur, Rajasthan 302017, India
e-mail: abgupta.ce@mnit.ac.in

R. M. Singh et al. (eds.), *Environmental Processes and Management*,
Water Science and Technology Library 91,
https://doi.org/10.1007/978-3-030-38152-3_12

been reviewed which suggests that PACl sludge is better in handling and does not have a negative effect on compressive strength of the mortars. However, alum sludge has sulphate content, and it retards the compressive strength of the mortars after certain proportion of sludge.

Keywords Defluoridation sludge · Mortar · Compressive strength · Microstructural analysis

12.1 Introduction

The treatment of drinking water to remove fluoride and other harmful ions is increasing nowadays. The generation of considerable amount of sludge generated is the major problem of the water treatment plants. The sludge generated in water treatment plants consists of organic and inorganic matter in solid, liquid, and gaseous states and varies in terms of physical, chemical, and biological characteristics (Bourgeois et al. 2004). The volume of discarded waste depends on the characteristics of the operational units involved and the quality of the raw water.

Sludge from water treatment plants may also contain other heavy metals from raw water or from contaminants resulting from the addition of coagulants (Sotero-Santos et al. 2005). Alum and PACl are chief coagulants used for the fluoride removal process through coagulation. The sludge generated is one of the major concerns of this process and therefore demands attention for its disposal and handling costs. As a result, sludge with high amount of water is discarded into adjacent aquatic systems (de Azevedo et al. 2018). Water purifying organizations have major environmental responsibilities since the land for throwing the sludge is limited (de Oliveira Andrade et al. 2018). In addition to this, an extra burden by the Pollution Control Board has been given to the water industries regarding legal and constitutional necessities for disposing of the sludge (Ahmad et al. 2016). In Rajasthan state, India, Bisalpur dam supplies 16.2 TMC per day out of which 5.1 TMC is supplied to Ajmer district and 11.1 to Jaipur district (Stearns 2009). About 16.5 metric tonne sludge is generated per day which is a huge quantity. As aluminium sulphate is used as coagulant in water treatment plant, this sludge is highly toxic as it contains metals like aluminium, so its disposal is a major problem for Public Health Engineering Department (PHED) (CPCB n.d.). According to Central Pollution Control Board (CPCB), India, India is producing huge amount of inevitable waste everyday at their water treatment plants which requires proper handling and disposal. But, due to lack of sludge management strategies, most of the water treatment plants in India discharge their filter backwash water and sludge into nearby drains which ultimately meet the water source. Some of the water treatment plants dispose the clarifier sludge on nearby open lands (CPCB n.d.). Therefore, it points to an urgent need to use this metal-containing sludge in a gainful manner to reduce environmental pollution and development of clean technology with zero-waste generation. The use of this sludge can improve the efficiency of primary sewage treatment and rationalize the treatment of water in sludge deposition (Vijayalakshmi et al. 2013).

To comply of disposal waste standards set by the local/federal authority makes researchers looking for alternative construction materials as a substitute to traditional materials likes cement, ceramic, bricks, tiles and aggregates in manner of reducing the impact of these waste on environment. Series of researches aimed to beneficial reuse in an effort to close the gap between enormous amounts of alum sludge and relieve pollution. However, the most common methods of disposal still depend on land application, reuse for agricultural purposes, and attempts to reuse it as a coagulant in the primary treatment of sewage. One of the possibilities for the defluoridation sludge is reuse it in the construction sector. The construction sector consumes huge volume of materials every year which gives construction sector potential to reuse alum sludge in making constructional material and concrete works. Thus, there is a need to do more laboratory experiments to determine maximum percentage that could be used as substitution on construction material. Thereby, the growing problem of sludge disposal can be alleviated if new disposal options other than of landfill can be found. This paper critically reviews the use of defluoridation sludge in construction materials and the performance of various concrete mixes on the basis of compressive strength, etc.

12.2 Impact of Defluoridation Sludge on Health and Environment

Increased environmental awareness among people has resulted in extensive pressure on the water production industry to develop and implement safe disposal techniques for the residues generated in water treatment plants (WTP) (Vouk et al. 2016). Although various chemicals are available in the market, alum is widely used as a coagulant due to economics and availability (Dassanayake et al. 2015). Thus, alum sludge remains an inescapable by-product of the water treatment process.

Influences of alum sludge resulting from the water treatment on environment can also lead to damage to human health. There are some studies that confirm this correlation presence of sludge with Alzheimer's disease (Pokhara 2015). Some of these effects may occur immediately, others may take some time to impact (with a cumulative effect), and therefore, health effects generally associated with environmental pollution that includes the use of chemicals should be examined and tested for the potential contamination of the environment (Jagtap et al. 2012). The leachability of Al from alum sludge is an operational consideration for management of a water treatment plant. Disposal of alum sludge is a recognized concern in various parts of the world. Since alum sludge contains Al, there may be a concern that land application of sludge will increase extractable Al and may increase the potential for Al phytotoxicity and Al movement into surface waters, which may damage aquatic ecosystems (Codling et al. 2014). Nevertheless, at low soil pH values ($pH < 5.2$)

where extractable Al becomes more available and can be toxic to plants (Sparks 2003). Aluminium phytotoxicity is directly connected to environmental conditions that control Al solubility in the soil. If the soil pH is greater than five, Al would not likely create any toxic problem (Mahdy et al. 2009). In Malaysia, more than 2.0 million tons of water treatment sludge is produced annually by the water operators throughout the country. The increasing cost of landfill (limited available land), the needs for sustainable best practices, and the increasing demand for high quality of drinking water causes the daily vast amount production of alum sludge that may remain unavoidable. Thus, the research that involves using alum sludge in construction and concrete works contributes to reduce the environmental impact. Increasing amount of alum sludge as a waste is not only difficult to dispose, but they also cause serious health hazards. Therefore, efforts are to be made for controlling pollution arising out of the disposal of alum sludge by conversion of these unwanted wastes into utilizable raw materials in construction sector in beneficial uses.

12.3 Use of Defluoridation Sludge for Making Bricks

Several experiments have been performed to investigate the use of defluoridation sludge in brick-making, concrete- and mortar-making processes. Table 12.1 describes the different researches performed to use alum sludge for brick making and compressive strength obtained in each case.

Table 12.1 Compressive strength of brick produced with alum sludge

S. No.	% Replacement of alum sludge	Temperature °C)	Compressive strength (N/mm^2)	Reference
1.	10 20 30 40 50	850	1.88 1.79 0.35 0.11 0.11	(Elangovan and Subramanian 2011)
2.	10	985	8.9	(Liew et al. 2004)
	20		5.4	
	30		3.1	
	40		2.0	
3.	5	850	5	(Victoria 2013)
	10		3.8	
	15		1.3	
	20		0.97	
4.	10	850	3	(Hii et al. 2013)
	20		2	

In a study done by Elangovan and Subramanian (2011), the alum sludge generated during water treatment was used as a partial substitute for clay in a clay brick manufacturing process. Alum sludge and commercial local clay were blended in various proportions and sintered at different temperatures to produce clay-sludge bricks. Physical and mechanical properties of clay-sludge bricks, such as loss on ignition, water absorption, and compressive strength, were investigated. The results from this study indicated that alum sludge could be used as a partial substitute in commercial clay bricks to a maximum of 20% without compromising the strength of the bricks. The lighter weight of clay-sludge bricks was due to the presence of the lightweight element, aluminium in the sludge. The clay-sludge bricks with 20 wt% sludge content sintered at a temperature of 850 °C were the most suitable for load-bearing wall structures.

Torres et al. (2012) investigated the use of alumina sludge for the manufacture of ceramic bricks. The results showed that it is feasible to use these sludges in partial replacement of one of the constituent materials of brick, in this case, the sand in percentages above 10%; however, to avoid compromising the compression resistance, it should be optimized the previous sludge dewatering to increase the potential waste-to-energy scheme. The brick obtained had appropriate characteristics for non-structural use.

Vicenzi et al. (2019) evaluated the feasibility of using alum sludge, a by-product from potable water plants, as raw material for ceramic production, and 10, 20, and 30 wt% of alum sludge were added to a clay used for the production of building materials.

In a study by Liew et al. (2004), bricks were produced from sewage sludge in different compositions were investigated. The physical, mechanical, and chemical properties of the bricks that were supplemented with various proportions of dried sludge from 10 to 40 wt% generally complied with the General Specification for Brick as per the Malaysian Standard MS 7.6:1972, which dictated the requirements for clay bricks used in walling in general. A standard leaching test method also showed that the leaching of metals from the bricks was very low.

Kevin et al. investigated the potential for reusing desalination sludge by using it as a partial replacement material in clay bricks. The bricks were made by mixing incinerated sludge ash into clay at different ratios corresponding to 0, 10, 20, 30, and 40% dried desalination sludge content by weight. The results showed that the compressive strength decreased with increasing dried desalination sludge could be produced, but the produced bricks were very fragile.

Victoria (2013) performed experiments to analyse the performance of sludge generated from purification process. The results demonstrated that the sludge could be used as a colourant and clay supplement in brick making. The percentage of replacement of clay by sludge was 0, 5, 10, 15, and 20 of the total weight of sludge. Brick was fired in a heat controlled furnace at evaluated temperature of 850, 900, 950, 1000, and 1050 °C. The percentage up to 20% can be applied into brick without losing its plastic behaviour and environmental sustainability.

12.4 Use of Alum Sludge for Making Concrete

Table 12.2 shows the different works done on use of alum sludge for making cement concrete. (Owaid et al. (2013) examined the influence of alum sludge powder as partial Portland cement type I replacement on the mechanical properties of high-performance concrete. The percentages of the alum sludge by weight of cement were 0, 6, 9, 12, and 15%. Slump tests were performed on fresh concrete to measure workability. The mechanical properties, compressive strengths, and splitting tensile strengths of the concrete samples were investigated at the end of 3, 7, and 28 days. Specimens without alum sludge were compared with those that contained alum sludge. The results revealed that the workability of the concrete consistently increased as the amount of cement replaced with alum sludge increased. It was found that the concrete with 6% alum sludge–cement replacement demonstrated improved compressive strength and splitting tensile strength at all ages, compared with the control concrete.

In a recent study by Quraatu et al. (2015), alum sludge (AS) was utilized as partial replacements (0, 5, and 10%) of natural granite coarse aggregate (by mass) to form a lightweight concrete. The water/cement ratio is 0.65. The water absorption of the alum sludge is 22.06%. The slump, density, compressive strength, and split

Table 12.2 Compressive strength of concrete and mortars produced with alum sludge

S. No.	Type of cement mix	% Replacement of alum sludge	Compressive strength (N/mm^2)		Reference
			7 days curing	28 days curing	
1.	Concrete	5	64	76	(Owaid et al. 2013)
		10	58	70	
		15	56	65	
2.	Concrete	5	14.5	16.7	(Quraatu et al. 2015)
		10	12.7	14.2	
3.	Concrete	10	8.21	8.38	(Kaosol 2010)
		20	4.79	5.2	
		30	3.8	4.2	
		40	3.5	3.7	
		50	2.5	2.8	
4.	Concrete				(Owaid et al. 2016)
5.	Concrete	2	7.5	25	(Lee et al. 2012)
		4	6.5	24	
		6	8	27	
6.	Mortar	10	–	15	(Ramirez Zamora et al. 2008)

tensile strength of the lightweight alum sludge aggregate concrete (LASAC) reduced as the alum sludge aggregate content increased. The compressive strength reduced from 25.6 MPa to 16.7 MPa and 14.2 MPa at 0, 5, and 10% replacement of alum sludge aggregate, respectively. Results show that strength of alum sludge lightweight aggregate concrete was better than lightweight crumb tyre aggregate concrete.

Alqam et al. (2011) investigated the use of water treatment sludge to replace cement in the production of paving tiles for external use. The study utilized sludge–cement replacement percentages of 10, 20, 30, 40, and 50%. Produced tiles were tested for water absorption and breaking (bending) strength. Leaching of sludge metals from tiles was assessed using TCLP. The study showed that all produced tiles exhibited a water absorption ratio of around 10%. The study concluded that produced tiles, except for 50% sludge–cement replacement, comply with the breaking strength requirements of 2.8 MPa for tiles for external use. The TCLP results indicated that metal leaching from tiles is negligible.

Breesem et al. (2014) carried out study on behaviour of self-compacting concrete using different sludge and waste material. The alum sludge from water treatment can be used in the production of cement and as ingredient in concrete mixes. The alum sludge was taken as 10, 20, 30, 40, and 50% proportions to weight of cement. The concrete blocks were cured for 28 days, strength was checked, and strength obtained was 805 higher than control concrete block. The reason behind it is due to pozzolanic reaction of alum sludge with $Ca(OH)_2$ that produces calcium silicate hydrate (C–S–H), so the strength increases.

Kaosol (2010) presented potential applications of the water treatment sludge for beneficial uses. The results in showed that the water treatment sludge mixtures can be used to produce hollow non-load-bearing concrete blocks, while 10 and 20% water treatment sludge mixtures can be used to produce the hollow load-bearing concrete blocks. Economically, the 10 and 20% water treatment sludge mixtures can reduce the cost at 0.64 and 1.05 Thai baht per block, respectively. The 50% of water treatment sludge ratio in mixture to make a hollow non-load-bearing concrete block can reduce the maximum cost at 2.35 baht per block.

Owaid et al. (2016) studied on high-performance concrete using alum sludge in concrete mixes. They investigated using alum sludge from 0 to 15% by weight of cement. Viscocrete-2044 as supper-plasticizer was used to improve the workability at constant w/c ratio. Compressive strength of concrete with 6% alum sludge increases with all ages. Density of alum sludge concrete mix decreases as the replacement levels increase, but it was opposite with workability.

12.5 Use of Water Treatment Sludge for Making Cement Mortars

Ramirez Zamora et al. (2008) evaluated drinking water treatment sludge as a supplementary cementations and sand substitute. The best binary (sludge–cement and sludge–mortar 90–10%) and ternary (90% sludge 5% lime 5% cement) formulations, prepared to produce cube mortar and concrete specimens, showed compressive strength values (ranging from 130 to 150 kg cm^{-2}) higher than the proposed values for equivalent standard specimens, according to the Mexican complementary technical criteria for the design and construction of masonry.

Rodríguez et al. (2010) investigated the reuse of drinking water treatment plant sludge as an addition for the cement industry. They found that the drinking water treatment plant sludge has a chemical composition and a particle size similar to Portland cement. The mortars that were made with 10–30% atomized sludge showed lower mechanical strength than the control cement and decline in slump. The results indicated that the properties of drinking water treatment in majority depend on chemical compositions that are important for its potential reuse.

Lee et al. (2012) prepared mortar specimens with various amounts of the sludge and a solidification agent, and their properties were examined. Addition of sludge reduced the mechanical strength of the concrete; on the other hand, addition of the solidification agent increased the mechanical strength of the concrete to an acceptable range. The strength of the specimen was higher than the typically required strength of 210 kgf/cm^2. The concrete specimens could meet the compressive strength requirement with as low as 2% solidification agent and up to 30% sludge as a fine aggregate replacement. The results of the slump test indicate that the workability of most sludge/solidification agent/cement mixtures also falls into the acceptable range of 14–26 cm. The low toxicity characteristic leaching procedure (TCLP) leachate concentrations, good mechanical property, appropriate workability, and a lower disposal cost make the use of water purification plant sludge in concrete mix a viable alternative for beneficial use.

12.5.1 Partial Replacement of Fine Aggregates with Defluoridation Sludge in Cement Mortars

Researchers have performed solidification/stabilization for activated alumina sludge, but there was a decrement in the strength of the bricks and acid resistance prepared having 0–40% replacement of fine aggregates with sludge (Pokhara 2015). The reason behind the reduction in strength was the application of high amount of acids and alkalis in activated alumina fluoride removal technique that makes the sludge chemically offensive. On the contrary, sludge generated from the coagulation process may provide impressive outcomes because it works at neutral pH (Agarwal et al. 2017). In a recent review, it has been inferred that coagulation is the most cost-effective technique in developing countries like India and Africa, where the communities cannot afford purchasing and operating RO for drinking water due to its high initial cost (Dubey et al. 2018). The fouling and energy costs mentioned for groundwater.

Many researchers have investigated the use of water treatment plant sludge with the partial replacement of fine aggregates. Although, the utilization of the sludge from defluoridation plant which treats groundwater has not been investigated. Also, aluminium leaching through mortars has not been studied earlier. Thus, the partial replacement of fine aggregates with defluoridation sludge in making cement mortars can be done. Studies have been performed with fluoride cement the results of the research work provide a good basis for the wider adoption of fluoride cement as an alternative to ordinary Portland cement, especially in developing economies (Malatachirwa 2012; Tran 2011). In fact, disposal of such sludges poses hazards to the environment, which has been reduced here substantially. This is an attempt to bring circularity in the system achieving net-zero wastage.

12.6 Conclusion

On the basis of literature, it could be concluded that alum sludge has the potential application in construction area, and it can be an efficient solution for the sludge disposal problem. Therefore, there are more experiments needed to be performed to determine the optimum percentage replacement of sludge to satisfy all the properties. According to papers, alum sludge can effectively use as a cement replacement because alum sludge has similar composition as cement and importantly contains silica, alumina, and ferric oxide which gives higher strength. The use of alum sludge in manufacturing of brick has been explored and in studies reviewed, and it was seen that utilization of this waste material in building purposes (brick and tiles) will contribute in minimizing the cost of disposal and the existing problem with disposing of waste in the environment. The use of alum sludge in concrete has been studied (used in cement production and replacement of cement in concrete, used in concrete as coarse aggregate replacement) but not fully explored as replacement of fine aggregate in concrete. The past literature lacks information on the condition of the sludge before used as replacement of fine aggregate in mortar or concrete. Finally, recent researches strive and are directed towards finding environmentally friendly solutions ensuring a more sustainable development.

References

Agarwal M, Dubey S, Gupta AB (2017) Coagulation process for fluoride removal by comparative evaluation of Alum & PACl coagulants with subsequent membrane microfiltration. Int J Environ Technol Manage 20(3–4):200–224

Ahmad T, Ahmad K, Ahad A, Alam M (2016) Characterization of water treatment sludge and its reuse as coagulant. J Environ Manage 182:606–611

Alqam M, Jamrah A, Daghlas H (2011) Utilization of cement incorporated with water treatment sludge. Jordan J Civil Eng 5(2):268–277

Bourgeois JC, Walsh ME, Gagnon GA (2004) Treatment of drinking water residuals: comparing sedimentation and dissolved air flotation performance with optimal cation ratios. Water Res 38(5):1173–1182

Breesem KM, Faris FG, Abdel-magid IM (2014) Behavior of self-compacting concrete using different sludge and waste materials—a general overview. Int J Chem Environ Biol Sci 2(3):3–8
Codling EE, Mulchi CL, Chaney RL (2014) Grain yield and mineral element composition of maize grown on high phosphorus soils amended with water treatment residual. J Plant Nutr 30(225–240)
CPCB 2008 (n.d.) Status of water treatment plants in India. Central Pollution Control Board
CPCB 2011 (n.d.) Central Pollution Control Board Ministry of Environment & Forest
Dassanayake KB, Jayasinghe GY, Surapaneni A, Hetherington C (2015) A review on alum sludge reuse with special reference to agricultural applications and future challenges. Waste Manage:1–16
de Azevedo ARG, Alexandre J, de Xavier GC, Pedroti LG (2018) Recycling paper industry effluent sludge for use in mortars: a sustainability perspective. J Cleaner Prod 192:335–346
de Oliveira Andrade JJ, Wenzel MC, da Rocha GH, da Silva SR (2018) Performance of rendering mortars containing sludge from water treatment plants as fine recycled aggregate. J Clean Prod 192:159–168
Dubey S, Agrawal M, Gupta AB (2018) Advances in coagulation technique for treatment of fluoride-contaminated water: a critical review. Rev Chem Eng 35(2):1–29
Elangovan C, Subramanian K (2011) Reuse of alum sludge in clay brick manufacturing. Water Sci Technol Water Supply 11(3):333–341
Hii K, Mohajerani A, Slatter P, Eshtiaghi N (2013) Reuse of desalination sludge for brick making. In: Hirschfeld K (ed) Chemeca 2013: Challenging tomorrow, Brisbane, Australia, 29 September–2 October 2013, pp 1–8
Jagtap S, Yenkie MK, Labhsetwar N, Rayalu S (2012) Fluoride in drinking water and defluoridation of water. Chem Rev 112(4):2454–2466
Kaosol T (2010) Reuse water treatment sludge for hollow concrete block manufacture. Energy Res J 1(2):131–134
Lee Y, Lo S, Kuo J, Tsai C-C (2012) Beneficial uses of sludge from water purification plants in concrete mix. Environ Eng Sci 29(4):284–289
Liew AG, Idris A, Samad AA, Wong CHK, Jaafar MS, Baki AM (2004) Reusability of sewage sludge in clay bricks 41–47
Mahdy AM, Elkhatib EA, Fathi NO, Lin Z-Q (2009) Effects of co-application of biosolids and water treatment residuals on corn growth and bioavailable phosphorus and aluminum in alkaline soils in egypt. J Environ Qual 38:1501–1510
Malata-chirwa CD (2012) Manufacture and properties of fluoride cement. Iowa State University
Owaid HM, Hamid R, Majeed ZH, Jawad IT (2016) Utilization of alum sludge as pozzolanic material in sustainable high performance concrete 1–4
Owaid HM, Hamid R, Sheikh Abdullah SR, Tan Kofli N, Taha MR (2013) Physical and mechanical properties of high performance concrete with alum sludge as partial cement replacement. Jurnal Teknologi 65(2):105–112
Pokhara P (2015) Activated alumina sludge from water defluoridation plants as partial substitute for fine aggregates in brick making. Indian Institute of Technology, Kanpur, India
Quraatu N, Mohd A, Hamid R (2015) Mechanical properties of lightweight alum sludge aggregate concrete 413–416
Ramirez Zamora RM, Alfaro OC, Cabirol N, Ayala FE, Moreno AD (2008) Open access. Am J Environ Sci 4(3):223–228
Rodríguez NH, Ramírez SM, Varela MTB, Guillem M, Puig J, Larrotcha E, Flores J (2010) Cement and concrete research re-use of drinking water treatment plant (DWTP) sludge: characterization and technological behaviour of cement mortars with atomized sludge additions. Cem Concr Res 40(5):778–786 (Elsevier Ltd)
Sotero-Santos RB, Rocha O, Povinelli J (2005) Evaluation of water treatment sludge toxicity using the Daphnia bioassay. Water Res 39(16):3909–3917
Sparks DL (2003) Environmental Soil Chemistry 1–350
Stearns J (2009) Supplying clean water to Jaipur: a study of two current government projects. Water Supply

Torres P, Hernández D, Paredes D (2012) Productive use of sludge from a drinking water treatment plant for manufacturing ceramic bricks. Revista ingeniería de Construcción 27(3):145–154

Tran TT (2011) Fluoride mineralization of Portland cement. Ph.D. Thesis, Arhus University, Denmark

Vicenzi J, Bernardes AM, Alegre P (2019) Evaluation of alum sludge as raw material for ceramic products. Ind Ceram 25:3–5

Victoria AN (2013) Characterisation and performance evaluation of water works sludge as bricks material. Int J Eng Appl Sci 3(3):69–79

Vijayalakshmi M, Sekar ASS, Ganesh Prabhu G (2013) Strength and durability properties of concrete made with granite industry waste. Constr Build Mater 46:1–7

Vouk D, Serdar M, Nakic D, Aleksandra A-V (2016) Use of sludge generated at WWTP in the production of cement mortar and concrete. Gradevinar 68:199–210

Chapter 13
Impact of Genetically Modified Crops on Environment

Saima Aslam and Nadia Gul

Abstract The remarkable increase in population day by day has turned the world into food hunting ground due to global food crises. Millions of people suffer from hunger and malnutrition. This claims the lives of people every year since fixed amount of land is available for farming. The realistic solution to this problem lies to utilize this available arable land to its fullest so as to meet ever increasing demand of food due to population growth. Although agricultural yield of crops should also increase as by 2050 food demands will get doubled. Change in climate is quite evident which is perceived by crops. It can further worsen this problem as temperature fluctuations cause the decline in yield of crops. So, stabilizing the population turns out to be the very daunting task. Plant with desired characters should be raised that should have the capability to withstand biotic and abiotic stresses which can be attained by using conventional breeding techniques but it is very time-consuming process. One way to circumvent all the hurdles is to make use of genetically modified crops. This technology assists the crops to grow much more at local level thereby helping to deal with global food crisis. These genetically modified crops tend to grow at fast rate than indigenous varieties. Moreover, it mitigates the food crisis along with poverty via increase in income of poor farmers. As per reports, genetically modified organisms (GMO) reduce the emission of greenhouse gases and pesticide usage. Both greenhouse gases and pesticides have very negative effect on environment. So, by reducing these harmful entities it paves the way for overall sustainable development. These genetically modified crops turn out to be not less than a magical wand having capability to tackle problems that world faces due to climate change and food crisis.

Keywords GMO · Phytoremediation · Pesticide · Biotechnology · Herbicide

---

S. Aslam (✉) · N. Gul
Department of Biotechnology, Baba Ghulam Shah Badshah University, Rajouri, J&K, India

R. M. Singh et al. (eds.), *Environmental Processes and Management*,
Water Science and Technology Library 91,
https://doi.org/10.1007/978-3-030-38152-3_13

13.1 Introduction

Worldwide population is presently around 7 billion and it is estimated that by 2050 it will exceed to 9.5 billion (United Nations 2015). Millions of mouths are meant to fed yearly, but cultivable land for farming being constant around 1.5 billion hectares (United Nations 2015; James 2014). Currently, 795 million people being under nourished, threatening the level of 3.1 billion children every year (Von Grebmer et al. 2015; Black et al. 2013). So, to tackle the problem of world hunger high yielding crop varieties should be grown so as to make use of the available cultivable land at its highest potential.

Conventional breeding methods are in use since era of decades for producing the plant with characteristic features which involves selecting, combing and propagation of favourable trait. It takes about 15 years to produce new progeny. But with advancement in science and technology via the opening of genetic engineering field, the processes of producing new variety have been geared up in a highly targeted manner. It also circumvents the hurdles involved in conventional breeding like sexual incompatibility between species and profound increase in available gene pool (Southgateetal 1995). Introduction of genetically modified (GM) crops in field of agriculture has turned breakthrough in global agriculture (Qaim and Zilberman 2003; Moschini 2008; Kathage and, Qaim 2012; Xu et al. 2013; Barrows 2014). Genetically modified crops were first introduced in 1996 in market. At global level, the genetically modified plants occupied 52.6 million hectares in 2001 (James 2001). Recombinant deoxyribonucleic acid (rDNA) technology involves the expression of desired gene in non-native plant. The protein expressed by this transgene confers particular character or trait to this plant. The technology involves in producing varieties that are resistant to biotic and abiotic stress. It can be also manipulated to produce plant with improved nutritional status. Moreover, recombinant medicines, industrial products like vaccines, monoclonal antibodies etc. are also produced via recombinant DNA technology (Sticklen 2005; Conrad 2005; Ma et al. 2003). There is 12% growth rate of global area of genetically modified crops planted since 2007 (Executive summary of Global Status 2007). Soyabean, maize, canola, cotton and rice are included in this global increase rate (Brandt 2003).

With passage of time, there has been development of keen interest pertaining to impact of genetically modified crops on environment (Dale et al. 2002). Conventional mode of cropping method requires treatment of chemicals like pesticides in order to generate crop yield at its highest potential. But these pesticides involving weedicides, fungicides and herbicides have disastrous effect on environment (EPA 2013). Moreover, there is global decrease in emission of greenhouse gases by using genetically modified crops (Intergovernmental Panel on Climate Change 2006).

13.2 Genetic Modification of a Plant

Genetically modified plants are produced using different techniques. Most commonly used techniques include Agrobacterium mediated transfer which is having natural capability of DNA transfer to plants and gene gun technology, which involves shooting of nano or microscopic particle bound DNA particles to plants cells (Southgate et al. 1995). Targeted plant cells develop into new plant via tissue culture techniques. Further, selectable markers like that of kanamycin resistance marker conferred by exgene so as to demarcate genetically modified plant from non-genetically modified plants (Gay and Gillespie 2005; Bennett et al 2004). Agrobacterium mediated gene transfer is most widely used among the high gene transfer methods or strategies (Conner et al. 2003). Many other mechanisms other than transduction, conjugation and transformation are also involved (Burmeister 2015).

13.3 Pest Resistance

Crop production worldwide is severely damaged by lepidopteron pests (Gresssel et al. 2004). Insecticidal spray proved to be ineffective and non-eco-friendly (Kush 2001). So, development of insecticide resistant variety of crops using *Bacillus thuringiensis* (Bt) turned out to be the alternative way of tackling the damage via pests (Tang et al. 2006). Bt acts as source of wide range of insecticidal proteins called cry proteins, having 100 holotypes and each holotype in turn has many subgroups with narrower host range (Bravo et al. 1998). Some of these Bt toxins were incorporated into few crop plants like root crop, cereals, forage crops and vegetables (Shelton et al. 2002).

European corn borer is potent pest which affects maize grain and is responsible for 0.75–7.5 million tonnes loss per year in USA. With advancement in genetic engineering, insect resistant variety of maize has been released and 6.8-million-hectare area is under this cultivation (James 2000). It has led to decline in pesticide usage. There was about 84% decrease in pesticide usage in 2013 that accounts for 8.2 million kg of pesticide. From 1996 to 2013, overall 72 million kg insecticide were saved (Brookes and Barfoot 2015).

Cotton crop is mainly affected by pest like bollworm. So, it requires surplus pesticide treatment. Cotton accounts for 16% global insecticide usage (Environmental Justice Foundation 2007). During 2013, Bt cotton cultivation saved the 21.3 million kg of insecticide which accounts for 48.3% reduction (Brookes and Barfoot 2015). China is among the major pesticide users and cotton producer worldwide. But the use of Bt cotton in china changed the scenario of pesticide usage, it tremendously decreased from 55 to 16 kg/Ha and spray reduced from 20 to 7 (Haung et al. 2001). The Arizona cotton research and protection (2000) has revealed that Bt cotton has reduced the insecticide usage in Arizona over past 20 years. Area under Bt. cotton cultivation has been seen to increase over last few years covering an area around 25 million acres in 2007 (James 2007). Moreover, the cotton advisory board of India recorded 5.27 million tons of cotton crops with increasing area under cultivation to 23.8 million acres in 2007–2008, due to which India has turned out to be second largest exporter of cotton worldwide after USA in the year 2007–2008, which is on whole attributed to the introduction and cultivation of Bt cotton (Khadi et al. 2007; CAI 2008). However, it was in 2006 when genetic engineering approved committee (GEAC) approved introduction of broader spectrum of pest control via two Bt gene incorporation. In India, around 131 Bt variants were grown in 2007–2008 (James 2007).

Brinjal is mainly produced by Bangladesh, and over 124,526 acres is under brinjal cultivation (BBS 2016). This crop is susceptible to various species of pests. *Leucinodesorbonalis* is most common among pests that attack brinjal (Alam et al. 2003). To tackle this problem, farmers rely on spraying pesticide about every day or frequently per growing season (Tacio 2013). So Bt brinjal carrying an inbuilt cry1AC gene to resist the pests (Gosh et al. 2003). Dependence on pesticide is circumvented by using Bt brinjal as per evidence found in Bangladesh (http://bteggplant.cornell.edu).

13.4 Herbicide Resistance

Weeds are major threat associated with agriculture. So, to sequester them chemical treatments are given. About 275 million kg of herbicide are consumed by agriculture crops of USA every year. But these tend to damage crop as well (Pimentel 1995; Pimentel et al. 1992). Resistant crops help to save crop damage due to herbicide treatment (Mannion 1995). The herbicide tolerant (HT) crops have led to switch over to less toxic herbicide. These tend to have replaced more toxic and more persistent herbicide with least toxic and low persistent in soil (Roland 2011). Moreover, no till farming practices decreased soil erosion and lessen greenhouse gas emission (Qaim and Traxler 2005; Brookes and Barfoot 2008). During period of 1996–2012, 18.3 million kg of net decrease in herbicide use was estimated using HT cotton. Alone in 2012, there was 2 million kg reduction of herbicide-related active constituent. The environmental profile has been improved by much better than conventional alternative (Peter and Gram 2014). In 2011, out of 160 million ha 126 million ha had herbicide tolerant traits. Out of 126 million, 94 million ha had herbicide tolerant trait and 42 million ha comprises insect tolerant and herbicide tolerant. However, there is also 10% increase in adoption of herbicide resistant crops globally (James 2011). Moreover, herbicide resistant crops tend to decrease overall herbicide use (Frisvold et al. 2009).

13.5 Transition to Conservation Tillage and Reduced Green House Gas Emission

Crops that are broad spectrum herbicide tolerant enabled no tillage or conservative tillage adoption (Brookes and Barfoot 2015). Tillage is a mechanical process of weed control that results in soil erosion, wind erosion, carbon dioxide emission from soil. It is the reason behind nitrification and dead zone formation in the Gulf of Mexico due to farm or field run off (Brookes and Barfoot 2011; National Research Council 2010). On the contrary, conservation tillage accounts only loosening of soil without turning. Only minor grooves can be made in the soil and seeds can be sown directly avoiding any kind of soil disturbance (Brookes and Barfoot 2015). Conservation tillage leads to decreased soil erosion, wind erosion, nutrient water pollution, reduced CO_2 emission and fossil fuel consumption (Fawcett and Towery 2002; Cerdeira and Duke 2006). Glyphosate tolerant crop variety turned out to be the vital factor favouring crop formation via no-tillage. Moreover, no-tillage land increased from 45 to 111 million ha from 1999 to 2009 globally (Bonny 2011; Derpsch et al. 2010).

Also, no tillage area in USA doubled while in Argentina five-fold inclement was seen. With decline in tillage process, fuel consumption and greenhouse gas emission reduced, thirty-two percent fuel savings was estimated in China using insect resistant variety of cotton alone. While insect resistant corn leads to 4 and 18% fuel savings, different estimates depicted 30–73% fuel savings (Jasa 2000; Sanders 2000; Mitchell et al. 2006; USDA-NRCS 2008). According to assumption made by Barfoot and Brookes 2014, 2.7 kg of CO_2 is produced by 1 litre of fuel. Total fuel savings from 1996 to 2012 accounts for 6268 million litres of fuel that in turn accounts for 16,730 million kg of CO_2 production. However, CO_2 sequestration in soil increased via no-tillage causing decline in CO_2 production or emission providing substantial environmental benefits.

13.6 Phytoremediation

Pollution is one of the global environmental problem, having disastrous effect on the flora and fauna by affecting their health, productivity and economic interests (Conesa et al. 2012). Increased anthropogenic activities tend to release harmful entities into the environment thereby increasing level of environmental pollution (Jaak et al. 2015). As per European environment agency, clean up of polluted area in Europe costs about 59–109 billion EUR (Conesa et al. 2012). Biotechnology mediated phytoremediation offers way to tackle this menace in economical way. It involves degradation of harmful and recalcitrant compounds present in the environment transgenic technology (Salt et al.1998; Sonali 2014). This technique involves increasing the remedial tendency of plant system up to much significant level using gene manipulation and transformation technological aspects (Kraomer 2005). Inorganic compounds like fertilizers, metals, metalloids, etc., can be sequestered using phytoremediation process.

Synergistic expression of γ-glutamylcysteine (γECS), phytochelatins (PCS) gene in Arabidopsis and two bacterial genes in transgenic plants imparts greater accumulation and tolerance of arsenic (As) (Guo et al. 2008; Dhankher et al. 2002). While incorporation of two bacterial genes mer A and mer B in different plant species like tobacco, rice, cottonwood and *Arabidopsis thaliana* delivers ten times greater capability to sequester mercury (Heaton et al. 2005, 2003; Che et al. 2003; Rugh et al. 1996). Over expression of ATP sulfurylase in *Brassica juncea* showed rapid selenate accumulation, three-fold more than wild type (Pilon et al. 1999).

Genetically engineered tobacco offers tolerance to trinitrotoluene (TNT) using bacterial gene like NADPH-dependent nitroreductase (Crockert et al. 2006). While transgenic *Arabdopsis* with flavodoxin cytochrome P450 enzyme can be used to degrade hexa-hydro-1,3,5-trinitro-1,3,5-triazine (RDX). (Jackson et al. 2007). Incorporation of atz A gene in tobacco, alfa alfa imparts atrazine pollution control (Kawahigaashi et al. 2006). Solvents like trichloroethylene (TCE) present in shallow water bodies is phytoremediated using P450 CYP2E1 in poplar plants and transgenic tobacco (Parkash et al. 2012). Also, transgenic plants with cytochrome P450 2E1 and CYP2E1 genes helps to detoxify toluene and benzene, respectively (James et al. 2008; Guo et al. 2008; Schnoor et al. 1995; Suresh and Ravishanker 2004). So, plant biotechnology has paved the way forward for reducing environmental pollution using advanced phyto-remedial measures.

13.7 Sustainable Environment via Biotechnological Intervention

Current industrial sector consumed health, environment to vast extent. So, it has become pivotal and need to adapt sustainable way. The eco-friendly technologies tend to decrease energy consumption, minimize hazardous chemical use and least waste production and utilize easily available renewable starting material. Industrial biotechnology is an emerging field in the way of sustainable development. It circumvents the problems associated with current industrial system. Industrial biotechnology or white biotechnology is aimed to focus on green biotechnology and medical sector (Wim and Erick 2006). One of the aspects of biotechnology involves green chemistry that is focused to use renewable resources for production of wide variety of substances like vitamins, colourants, solvents, biofuels, bioplastics, etc. (Dale 2003). Nowadays bioplastic is produced using transgenic crops. *Alcaligenes eutropus* posses polyhydroxybutyrate genes. These genes were incorporated in corn. So, as to produce biodegradable plastic in very affordable and competitive in market (Choudhury et al. 2008). Moreover, plant genetic engineering is used to produce vaccines, antibodies and other pharmacologically important compounds at low cost. Various plant-based therapies have been approved against dental caries, Gauchers disease, Vitamin B12 deficiency, anti-infection and anti-inflammatory, etc. (Obembe et al. 2011; Fox 2012).

13.8 Waste Management

Waste is generated at an alarming rate globally (Deepak and Archana 2017). Advancement in field of biotechnology has led to introduction of new branch known as molecular microbial ecology. The genome dependent tools have led to discovery of new micro-organisms with new metabolic tendencies (Rittmann et al. 2006, 2008). Different biotechnological approaches are involved in treating and disposal of wastewater in collaboration with environmental engineering. Wastewater treatment involves trickling filter, bio-filters, oxidation ponds, etc., while solid waste management involves bio-sorption, bio-trickling etc. (Hanife and Levent 2009). In all treatment processes, micro-organisms have potent role. Set-up is made in a way that micro-organisms derive their food from waste (Weiner and Mathews 2002). So, biotechnological waste disposal paves an important way to dispose off waste (Deepak and Archana 2017).

13.9 Disadvantages of Genetically Engineered Crops on Environment

The impact of genetically modified crops on environment fluctuates itself based on adapted agricultural practice (GM science review panel 2003). The biotechnology or genetic engineering is an illustrative field that enhances genetic diversity among crops via introduction of novel or beneficial gene that tend to overcome or suffice the short comings of the traditional crops. The abiotic stress tolerance and post-harvest reduction of crops has been attained via the genetic engineering tools (Slabbert et al. 2004; Gressel et al. 2004). GMOs are aimed to protect crops, disease control via pest and herbicide management (Agapito-Tenfen et al. 2014). But development of resistance in some cases like that in animal rye grass and horse weed against glyphosate has been reported (Dale et al. 2002). The development of these resistant might be due to over expression of target gene, decreased translocation rate and sensitivity to glyphosate (Wakelin et al. 2004). Also, use of pest resistant pesticides has developed resistance in diamond month population (Tabashnik et al. 2013). Moreover, development of cross resistance has also been reported in Bt cotton, Bt corn etc. (Zhoa et al. 2001; Burd et al. 2003).

13.10 Conclusion

The ecological and environmental implications associated with the genetically modified crops ought to be assessed before their release. The level of risk associated with them tend to be variable. However, area under GM crops cultivation increased from last 3 decades. Stringent screening is to be implemented so as to avoid any negative effect associated with GMO. The side effects associated with them can be circumvented via using tissue specific promoters so as to avoid gene transfer, resistant species development. Next generation sequencing based artificial promoters could be developed to minimize the risk. With the passage of time, population increases globally so demand for food also increases. In order to feed the existing population, adoption of GMO has become necessary. GMOs also led to upliftment of the sustainable environment via pesticide and herbicide resistant varieties. Million kgs of pesticides and herbicides that were used earlier do have very determinable effects on environment. Moreover, GMOs have reduced greenhouse gas emission and also helps to curve the menace of pollution via phytoremediation aided biotechnological tools and no tillage practices. With advancement in genetic engineering, biodegradable plastic is also being produced at large scale. Also, environmental engineering in collaboration with biotechnological approach wastewater management via different treatments percolates the maximum amount of pollutants present in it.

So, field of genetic engineering or biotechnology turned out to be a scientific way that helps to eradicate unfriendly environmental ailments thereby ensuing the way for sustainable development and conservation of ecosystem. So, this field aims to balance conservation and technology in realistic way by reducing its negative impact on the environment via risk assessment management.

References

Agapito-Tenfen SZ, Vilperte V, Benevenuto RF, Rover CM, Traavik TI, Nodari RO (2014) Effect of stacking insecticidal cry and herbicide tolerance epsps transgenes on transgenic maize proteome. BMC Plant Biol 14(1):346

Alam SN, Rashid MA, Rouf FMA, Jhala RC, Patel JR., Satpathy S., Shivalingaswamy TM., Rai S, Wahundeniya I, Cork A, Ammaranan C, Talekar NS (2003) Development of an integrated pest management strategy for eggplant fruit and shoot borer in South Asia. AVRDC—the World Vegetable Center, Technical Bulletin No. 28, Shanhua, Taiwan

Bangladesh Bureau of Statistics (BBS) (2016) Yearbook of agricultural statistics of Bangladesh, pp 3–15

Barfoot Peter, Brookes G (2014) Key global environmental impacts of genetically modified (GM) crop use 1996–2012. GM Crops Food: Biotechnol Agric Food Chain 5(2):149–160. https://doi.org/10.4161/gmcr.28449

Barrows G, Sexton S, Zilberman D (2014) Agricultural biotechnology: The promise and prospects of genetically modified crops. J Econ Perspect 28:99–120

Bennett PM, Livesey CT, Nathwani D, Reeves DS, Saunders JR, Wise R (2004) An assessment of the risks associated with the use of antibiotic resistance genes in genetically modified plants: report of the Working Party of the British Society for Antimicrobial Chemotherapy. J Antimicrob Chemother 53:418–431

Black RE, Victora CG, Walker SP, Bhutta ZA, Christian P, de Onis M, Ezzati M, Uauy SR (2013) Maternal and child under-nutrition and overweight in low-income and middle-income countries. Lancet 832:427–451. https://doi.org/10.1016/s0140-6736(13)60937-x

Bonny S (2011) Herbicide-tolerant transgenic soybean over 15 years of cultivation: pesticide use, weed resistance, and some economic issues. case USA Sustain 3:1302–1322

Brandt P (2003) Overview of the current status of genetically modified plants in Europe as compared to the USA. J Plant Physiol 160:735–740

Bravo A, Sarabia S, Lopez L, Ontiveros H, Abarca C, Ortiz A, Ortiz M, Lina L, Villalobos FJ, Pena G, Nunes-Valdez ME, Soberon M, Quintero R (1998) Characterization of cry genes in a mexican*Bacillusthuringiensis* strain collection. Appl Environ Microb 64:4965–4972

Brookes G, Barfoot P (2008) GM crops: global socioeconomic and environmental impacts 1996–2008. PG Econ, Dorchester

Brookes G, Barfoot P (2011) Global impact of biotech crops: environmental effects 1996–2009. GM Crops 2(1):34–49, PG Econ, Dorchester

Brookes G, Barfoot P (2015) Environmental impacts of genetically modified (GM) crop use 1996–2013: impacts of pesticide use and carbon emissions. GM crops and food

Burd AD, Gould F, Bradley JR, Van Duyn JW, Moar WJ (2003) Estimated frequency on non-recessive Bt resistance genes in bollworm, Helioverpazea (Boddie) (Lepidoptera: Noctuidae) in Eastern North Carolina. J Econ Entomol 96:137–142

Burmeister A (2015) Horizontal gene transfer. Evol Med Public Health 1:193–194. https://doi.org/10.1093/emph/eov018

Buyukgungor H, Gurel L (2009) The role of biotechnology on the treatment of wastes. Afr J Biotechnol 8(25):7253–7262

CAI (Cotton Association of India) (2008) Cotton statistics and news. (Issue No. 42, 15 January 2008). Available on the World Wide Web: http://www.eica.in

Cerdeira AL, Duke SO (2006) The current status and environmental impacts of glyphosate-resistant crops: a review. J Environ Qual 35:1633–1658

Che D, Meagher RB, Heaton AC, Lima A, Rugh CL, Merkle SA (2003) Expression of mercuric ion reductase in Eastern cottonwood (Populusdeltoides) confers mercuric ion reduction and resistance. Plant Biotechnol J 1:311–319

Choudhury H, Bhattacharjee A, Upaddhaya K (2008) Environmental biotechnology: bioplastics and their environmental importance. In: Proceedings of national seminar on toxicity of chemicals and their hazards with special reference to heavy metals, pp 121–125

Conesa HM, Evangelou MW, Robinson BH, Schulin R (2012) A critical view of current state of phytotechnologies to remediate soils: still a promising tool? Sci World J 173–829

Conner AJ, Glare TR, Nap J (2003) The release of genetically modified crops into the environment. Part II. Overview of ecological risk assessment. Plant J 33:19–46

Conrad U (2005) Polymers from plants to develop biodegradable plastics. Trends Plant Sci 10:511–512

Crocker FH, Indest KJ, Fredrickson HL (2006) Biodegradation of the cyclic nitramine explosives RDX, HMX, and CL-20. Appl Microbiol Biotechnol 73:274–290

Dale BE (2003) "Greening" the chemical industry: research and development priorities for biobased industrial products. J Chem Technol Biotechnol 78:1093–1103

Dale PJ, Clarke B, Fontes EMG (2002) Potential for the environmental impact of transgenic crops. Nat Biotechol 20:567–574

Derpsch R, Friedrich T, Kassam A, Li H (2010) Current status of adoption of no-till farming in the world and some of its main benefits. Int J Agric Biol Eng 3:1–25

Dhankher OP, Li Y, Rosen BP, Shi J, Salt D, Senecoff JF et al (2002) Engineered tolerance and hyperaccumulation of arsenic in plants by combining arsenate reductase and γ-glutamylcysteine synthetase expression. Nat Biotechnol 20:1140–1145

Dille D, Archana K (2017) Biotechnology for solid waste management—a critical review. Open Access Int J Sci Technol 2:1

Environmental Justice Foundation (2007) The deadly chemicals in cotton. Available through http://efoundation.org/sites/default/files/public/the_deadly_chemicals_in_cotton.pdf

Environmental Protection Agency (2013) What are pesticides and how do they work? NSW Government. Retrieved from http://www.epa.nsw.gov.au/pesticides/pestwhatrhow.htm

Executive summary of Global Status of Commercialised Biotech/GM crops: 2007. ISAAA Briefs No. 37. Ithaca, NY ISAAA, 2007

Fawcett R, Towery D (2002) Conservation tillage and plant biotechnology: how new technologies can improve the environment by reducing the need to plow. Conservation Technology Information Center, West Lafayette, IN
Fox JL (2012) First plant-made biologic approved. Nat Biotechnol 30(6):472
Frisvold GB, Hurley TM, Mitchell PD (2009) Overview: herbicide resistant crops—diffusion, benefits, pricing, and resistance management. Ag Bio Forum 12:244–248
Gay PB, Gillespie SH (2005) Antibiotic resistance markers in genetically modified plants; a risk to human health. LancetInfect Dis 5:637–646
Ghosh SK, Laskar N, Senapati SK (2003) Estimation of loss in yield of brinjal due to pest complex under Terai region of West Bengal. Environ Ecol 21:764–769
GM Science Review First Report (2003) GM science review panel
Gressel J, Hanafi A, Head G, Marasas W, Obilana AB, Ochanda J, Souissi T, Tzotzos G (2004) Major heretofore intractable biotic constraints to African food security that may be amenable to novel biotechnological solutions. Crop Prot 23:661–689
Guo J, Dai X, Xu W, Ma M (2008) Overexpressing GSH1 and AsPCS1 simultaneously increases the tolerance and accumulation of cadmium and arsenic in Arabidopsis thaliana. Chemosphere 72:1020–1026
Heaton AC, Rugh CC, Kim T, Meagher RB (2003) Toward detoxifying mercury polluted aquatic sediments with rice genetically engineered for mercury resistance. Environ Toxicol Chem 22:2940–2947
Heaton ACP, Rugh CL, Wang NJ, Meagher RB (2005) Physiological responses of transgenic merAtobacco (Nicotiana tabacum) to foliar and root mercury exposure. Water Air Soil Pollut 161:137–155
Huang J, Hu R, Pray C, Quiao F, Rozelle S (2001) Biotechnology as an alternative to chemical pesticides: a case study of Bt cotton in China. In: Proceedings of 5th international consortium agriculture, biotechnology research. Ravello (Italy), pp 109–110 (Abstr.)
Intergovernmental Panel on Climate Change. Chapter 2: generic methodologies applicable to multiple land-use categories. Guidelines for national greenhouse gas inventories volume 4. Agriculture, forestry and other land use (2006). Available at: http://www.ipccnggip.iges.or.jp/public/2006gl/pdf/4_Volume4/V4_02_Ch2_Generic.pdf
Jaak T, Marika T, Mikk E, Hiie N, Jaanis J (2015) Phytoremediation and plant-assisted bioremediation in soil and treatment wetlands: a review. Open Biotechnol J 9:85–92
Jackson RG, Rylott EL, Fournier D, Hawari J, Bruce NC (2007) Exploring the biochemical properties and remediation applications of the unusual explosivedegrading P450 system XplA/B. Proc Natl Acad Sci USA 104:16822–16827
James C (2000) Global review of commercialised transgenic crops. Publ. International Service for the Acquisition of Agri-biotech Applications, Ithaca, NY
James CA (2001) Global review of commercialized transgenic crops, 2001. ISAAA Briefs No. 24. Preview ISAAA, Ithaca, NY. http://www.isaaa.org/publications/briefs/Brief_24.htm
James C (2007) Global status of commercialized biotech/GM crops: 2007 (ISAAA Brief No. 37). International Service for the Acquisition of Agri-Biotech Applications, Ithaca, NY
James C (2011) Global status of commercialized biotech/GM crops: 2011. ISAAA Brief No. 43, ISAAA, Ithaca, NY
James C (2014) Global status of commercialized biotech/GM crops: 2014. ISAAA Brief 49. Retrieved from http://www.isaaa.org/resources/publications/briefs/49/
James CA, Xin G, Doty SL, Strand SE (2008) Degradation of low molecular weight volatile organic compounds by plants genetically modified with mammalian cytochrome P450 2E1. Environ Sci Technol 42:289–293
Jasa PJ (2000) Conservation tillage systems. Paper presented at the International Symposium on Conservation Tillage, in Mazatlan, Mexico, 24–27 Jan. http://agecon.okstate.edu/isct/labranza/jasa/tillagesys.doc
Kathage J, Qaim M (2012) Economic impacts and impact dynamics of Bt (Bacillus thuringiensis) cotton in India. Proc Natl Acad Sci U.S.A. 109:11652–11656

Kawahigashi H, Hirose S, Ohkawa H, Ohkawa Y (2006) Phytoremediation of the herbicides atrazine and metolachlor by transgenic rice plants expressing human CYP1A1, CYP2B6, and CYP2C19. J Agric Food Chem 54:2985–2991

Khadi BM, Rao MRK, Singh M (2007) Potential to improve lives of ryots, (The Hindu survey of Indian agriculture). The Hindu, Chennai

Khush GS (2001) Green revolution: the way forward. Nat Rev Genet 2:815–822

Kraomer U (2005) Phytoremediation: novel approaches to cleaning up polluted soils. Curr Opin Biotechnol 16:133–141

Ma JKC, Drake PMW, Christou P (2003) The production of recombinant pharmaceutical proteins in plants. Nature 4:794–805

Mannion AM (1995) Agriculture and environ-mental change. Wiley, Chichester (UK)

Mitchell J, Daniel P, Munk S, Prys B, Klonsky K, Wroble J, De Moura R (2006) Conservation tillage production systems compared in San Joaquin Valley Cotton. Cali Agric 60(3):140–45

Moschini G (2008) Biotechnology and the development of food markets: Retrospect and prospects. Eur Rev Agric Econ 35:331–355

National Research Council (2010) The impact of genetically engineered crops on farm sustainability in the United States. Natl Acad Press

Obembe OO, Popoola JO, Leelavathi S, Reddy SV (2011) Advances in plant molecular farming. Biotechnol Adv 29:210–222

Parkash D, Elizabeth AH, Pilon S, Richard BM, Doty S (2012) Biotechnological approaches for phytoremediation (Plant Biotechnology and Agriculture). Elsevier Inc

Pilon-Smits EAH, Hwang S, Lytle CM, Zhu Y, Tai JC et al (1999) Overexpression of ATP sulfurylase in Indian mustard leads to increased selenate uptake, reduction, and tolerance. Plant Physiol 119:123–132

Pimentel D, Stachow U, Takacs DA, Brubaker HW, Dumas AR, Meaney JJ, O'Neil J, Onsi DE, Corzilius CB (1992) Conserving biological diversity in agricultural/forestry systems. Bioscience 42:354–362

Pimentel D et al (1995) Environmental and economic costs of soil erosion and conservation benefits. Science 267:1117–1123

Qaim M, Traxler G (2005) Roundup ready soybeans in Argentina: farm level and aggregate welfare effects. Agric Econ 32:73–86

Qaim M, Zilberman D (2003) Yield effects of genetically modified crops in developing countries. Science 299:900–902

Rittmann BE et al (2006) A vista for microbial ecology and environmental biotechnology. Environ Sci Technol 40:1096–1103

Rittmann BE, Krajmalnik-Brown R, Halden RU (2008) Pregenomic, genomic, and post-genomic study of microbial communities involved in bioenergy. Nat Rev Microbiol 6:604–612

Roland P (2011) Plant genetics, sustainable agriculture and global food security. Genetics 188:11–20

Rugh CC, Summers AO, Meagher RB (1996) Mercuric ion reductase and resistance in transgenic Arabidopsis thaliana expressing modified bacterial merA gene. Proc Natl Acad Sci USA 93:3182–3187

Salt DE, Smith RD, Raskin I (1998) Phytoremediation. Annu Rev Plant Physiol Plant Mol Biol 49:643–668

Sanders LD (2000) The economics of conservation and conservation tillage. Paper presented at the international symposium on conservation tillage, Mazatlan, Mexico, 24–27 Jan. http://agecon.okstate.edu/isct/labranza/sanders/mazecon00.doc

Schnoor JL, Licht LA, McCutcheon SC, Wolfe NL, Carreira LH (1995) Phytoremediation of organic and nutrient contaminants. Environ Sci Technol 29:318A–323A

Sharma P, Pandey S (2014) Status of phytoremediation in world scenario. Int J Environ Bioremediat Biodegradation 2(4):178–191

Shelton AM, Zhao JZ, Roush RT (2002) Economic, ecological, food safety, and social consequences of the deployment of Bt transgenic plants. Annu Rev Entomol 47:845–881

Slabbert R et al (2004) Drought tolerance, traditional crops and biotechnology: breeding towards sustainable development. South African J Botany 70:116–123

Soetaert W, Erick (2006) Impact of industrial biotechnology. Biotechnol J 756–769. https://doi.org/10.1002/bot.200600066

Southgate EM, Davey MR, Power JB, Merchant R (1995) Factors affecting the genetic engineering of plants by microprojectile bombardment. Biotechnol Adv 13:631–657

Sticklen M (2005) Plant genetic engineering to improve biomass characteristics for biofuels. Curr Opin Biotechnol 17:315–319

Suresh B, Ravishankar GA (2004) Phytoremediation–a novel and promising approach for environmental clean-up. Crit Rev Biotechnol 24:97–124

Tabashnik BE, Brevault T, Carriere Y (2013) Insect resistance to Bt crops: lessons from the first billion acres. Nat Biotechnol 31:510–521. https://doi.org/10.1038/nbt.2597

Tacio HD (2013) Is genetically-modified talong safe to eat?GMA News Online. Available at: http://www.gmanetwork.com/news/story/33983/scitech/science/isgeneticallymodified-talong-safe-to-eat

Tang W, Chen H, Xu C, Li X, Lin Y, Zhang Q (2006) Development of insect-resistant transgenic indica rice with a synthetic *cry1C** gene. Mol Breed 18:1–10

United Nations, Department of Economic and Social Affairs, Population Division (2015) World population prospects: the 2015 revision, key findings and advance tables. Population Dev Rev 41(3):557–561. https://doi.org/10.1111/j.1728-4457.2015.00082.x

US Department of Agriculture–National Resources Conservation Service (USDA-NRCS) (2008) Energy consumption awareness tool: tillage. Washington, DC. http://ecat.sc.egov.usda.gov/

Von Grebmer K, Bernstein J, de Waal A, Prasai N, Yin S, Yohannes Y (2015) 2015 Global hunger index: Armed conflict and the challenge of hunger. Global Hunger Index. https://doi.org/10.2499/9780896299641

Wakelin AM, Lorraine-Colwill DF, Preston C (2004) Glyphosate resistance in four different populations of Loliumrigidum is associated with reduced translocation of glyphosate to meristematic zones. Weed Res 44:453–459

Weiner RF, Matthews RA (2002) Environmental engineering, 4th edn. Butterworth-Heinemann, USA

Xu Z, Hennessy DA, Sardana K, Moschini G (2013) The realized yield effect of genetically engineered crops: US maize and soybean. Crop Sci 53:735–745

Zhao JZ, Li YX, Collins HL, Cao J, Earle ED, Shelton AM (2001) Different cross resistance patterns in the diamondback moth (Lepidoptera: Plutellidae) resistant to Bacillus thuringiensis toxin Cry1C. J Econ Entomol 94:1547–1552

Chapter 14
Source Apportionment of Particulate Matter—A Critical Review for Indian Scenario

Seema A. Nihalani, Anjali K. Khambete and Namrata D. Jhariwala

Abstract Particulate matter pollution is an area of major interest in the urban regions on this planet. Ambient particulate matter is increasingly becoming a concern around the globe due to its negative effects on human beings. Major health hazards associated with it are respiratory problems and health effects leading to premature mortality. The fine and ultra-fine particulates which carry along with them metals such as iron, zinc, copper, cadmium, and lead have toxicological and carcinogenic effects. The key aspect of air quality control lies in identifying the correct sources of the pollutant. Therefore, an increased emphasis on the particulate matter analysis is required. Source apportionment of particulate matter is a significant method for examining processes of particulate formation and transformation. It also helps in assessing air pollution control strategies and reasons for non-compliance with air quality standards. Receptor model is a widely used tool of source apportionment which is based on the study of chemical composition of aerosols measured at specific sites. The current paper presents a review of various receptor model studies undertaken in India. An exhaustive evaluation for particulate matter studies and their signature sources has been made in the Indian context, bearing in mind most appropriate sources and their variations.

Keywords Particulate matter · Receptor model · Source apportionment · Principal component analysis · Positive mass factorisation

S. A. Nihalani (✉) · A. K. Khambete · N. D. Jhariwala
SVNIT, Surat, India
e-mail: seema.nihalani@paruluniveristy.ac.in

A. K. Khambete
e-mail: anjali_khambete3@yahoo.co.in

N. D. Jhariwala
e-mail: nd_jariwala@yahoo.co.in

S. A. Nihalani
Parul University, Vadodara, Gujarat, India

R. M. Singh et al. (eds.), *Environmental Processes and Management*, Water Science and Technology Library 91,
https://doi.org/10.1007/978-3-030-38152-3_14

14.1 Introduction

Air quality all over the world has reached an alarming situation due to concentrations of certain pollutants exceeding applicable norms, particularly in developing countries. Particulate matter is regarded as one of the key pollutants due to its harmful influence on human beings. In order to control the particulate matter concentration in cities, several norms have been proposed. However, to devise an effective abatement program and formulate policies for reducing particulate matter concentration in ambient air, it is vital to have precise data about air pollution origin and meteorological factors.

There is a positive correlation between particulate matter concentration and atmospheric pressure; however, the correlation between particulate matter concentration and atmospheric temperature is considerably negative. The correlation of particulates concentration with relative humidity and wind speed is considerably poor. Precipitation has larger impact on coarse particulate concentration as compared to fine particulate matter.

a. **Air Temperature**: The correlation between particulate matter concentration and atmospheric temperature is negative, which means that with the increase in air temperature, the particulate matter concentration reduces. The reason behind this is heating of the earth's underlying surface by sun's radiation during summers. Due to this, the turbulence strengthens as the lower atmosphere is not very stable, which promotes diffusion of particulates. Hence, the concentration of particulates reduces with the increase in temperature.
b. **Atmospheric Pressure**: When low pressure exists near the earth's surface, high-pressure air mass is formed moving it away from the earth's surface, which drifts the air pollutants upwards and reduces the particulate concentration near earth's surface. Contrary, to this when high pressure exists near the earth's surface, an air mass of low pressure develops above it resulting in low wind speed condition and formation of thermal inversion layer. This leads to stable atmospheric condition, reducing the diffusion of particulate pollutants and aggravating the air pollution condition. Hence, a positive co-relation exists between air pressure and particulate concentration.
c. **Precipitation**: Precipitation or rainfall has a larger impact on coarser particulates as compared to fines ones. It is believed that heavy rainfall can reduce dust significantly by making it settle on surrounding ground or earth's surface. Hence, precipitation largely gets rid of coarser particulates as compared to finer particulates. Therefore, it is assumed that concentration PM10 reduces significantly with heavier rainfall, while concentration of PM2.5 declines less as compared to PM10.
d. **Wind Speed**: There exists a negative co-relation between wind speed and concentrations particulate matter. As the wind speed increases, more diffusion of pollutants occurs, and hence, the concentration of particulates reduces. However, the influence of wind speed on finer particulates is more obvious than coarser particulates.

14.2 Source Apportionment

Source apportionment (SA) implies methods that are used to compute different source contributions to atmospheric PM concentration. SA of ambient particulate matter computes the influence of discrete sources to particulate matter contribution depending on the characteristics of the receptor as well as source and type of pollutants. This can be achieved by examining explicit tracer markers by using filter analysis or by numerical analysis of a particular marker with prevalent weather condition or using emission record data along with dispersal models. The basic SA methods existing are (a) chemical conveyance models, (b) receptor-oriented models, and (c) emission data with dispersal models. There is extensive literature available on SA involving dispersal models and monitoring data. However, a substantial amount of research work has been undertaken using receptor models. Such models are explicitly termed as receptor models because variable information about the composition of particulates is taken from the receptor site.

The significant output of RM is the percentage contribution of an individual source to the concentration of pollutant. Such models are predominantly useful at places where comprehensive emissions data is not accessible. RMs are utilised for identifying the origin and individual source contributions of ambient particulates.

The chief clusters of source apportionment methods are:

(a) Chemical transport models working on pollutant speciation depend on meteorological variables and assessment of observed data. Handling of elementary information is required to recognise the origin. Examples are: (1) identifying source locations by using association of wind pathway with measured pollutant concentration, (2) identifying source association by using relationship between gaseous contaminants and particulate constituents, (3) identifying the contributions from regional/city background by deducting local backdrop pollutants from those existing in urban cities or at roadsides, and (4) deducting particulate concentration at local backdrop site from those at urban backdrop sites to quantify natural PM contributions. The key benefit of this method is simplicity and subsequent small influence of mathematical equations owing to the proper handling of data.
(b) Emission data and dispersal models are used for simulation of pollutant discharge, creation, conveyance, and dumping. Such prototypes need exhaustive discharge data which might not be easily accessible and might be restricted by emission data precision.
(c) Receptor models are based on statistical assessment and analysis of PM data collected from receptor sites.

Several SA studies for various atmospheric pollutants are available in India, and majority of them have used receptor type of models working on an observed concentration of particulate matter and its origin contour. Depending on the type of particle attributes used for simulation, receptor models are generally categorised as microscopic receptor models and chemical receptor models. Microscopic RM analyses structural aspects of ambient particulate matter using an electron microscope or automated SEM that are adequately competent to characterise pollutants in mixing states. However, microscopic RM limits are not applicable on a large scale since it does not yield quantitative results in most cases and produces only qualitative results, while chemical receptor models classify and recognise explicit PM sources by carrying out the chemical configuration of particulates. Several models used in this study are chemical mass balance, enrichment factor, factor analysis including principal component analysis, UNMIX, positive matrix factorisation, and multilinear engine.

14.3 Receptor Models

The essential proposition for receptor model is to assume conservation of matter and use mass equilibrium theory to classify and recognise of particulate matter sources. A mass balance equation is used to consider all chemical types (m) in samples (n) as involvement from individual sources (p).

$$X_{ij} = \sum_{p=1}^{P} g_{ip} f_{ip} + e_{ij} \tag{14.1}$$

where X_{ij} is observed the quantity of jth type in the ith sample, f_{jp} is the quantity of jth type in substance discharged from origin p, g_{ip} is the quantity of pth origin to ith sample, and e_{ij} is measurement share that the model cannot identify. In F matrix, RMs utilise experimental data of receptor species quantification in source discharge as input; this is usually termed to be source profile. Also, RMs can obtain receptor species concentrations in the matrix (F) by iteration method, which is usually referred to as a factor profile.

The basic presumptions behind the mass balance analysis equation are as follows:

(1) Source profiles do not alter considerably with time or they alter in a replicable way so as the system is pseudo-stationary. (2) Receptor species do not undergo chemical reactions or phase separation (solid or gas and solid or liquid) from origin to receptor transport (they augment linearly). The additional inherent presumptions being (i) information represents geographical area under study and depends on the abstract model and (ii) corresponding investigative methods can be applied for receptor sites for the entire study and source profile classification.

14.4 Source Profiles Known—Chemical Mass Balance

When the numbers and nature of sources are known, the only unidentified variable is the contribution of the matter of individual origin to the individual sample. Such values are calculated by applying regression analysis. Winchester and Nifong initially proposed Chemical Mass Balance (CMB) model in 1971 followed by Miller in 1972. This method uses the least square method of variance to solve a problem.

The basis of the CMB method is conservation of matter which assumes that signature markers do not transform chemically while travelling to the receptors from the source. It considers the concentration of receptor to be a direct summation of a source profile and may be calculated when suitable indeterminate estimations are existing. CMB uses profile of source and receptor along with a suitable indeterminate estimate to be input to generate discrete source involvement with rational indeterminacy using multiple linear least square regression algorithms. However, source profile choice shall evade linearity and probability of identical geographical areas; or else the model relevance shall reduce considerably. To decrease the linearity of source profiles, it is recommended combining like a group of sources together. CMB is one of the unconventional receptor model techniques and is suitable when limited monitoring data is obtainable. Accessibility of comprehensive data about source profile reduces requisite samples; however, a smaller data set probably escalates the level of uncertainty. CMB is represented by Eq. 14.2 shown below:

$$\mathrm{C}_{ik} = \sum_{n=1}^{N} F_{in} S_{kn} \tag{14.2}$$

where C_{ik} is the airborne quantity of ith species from kth source observed at receptor location and S_{kn} is influence from the kth source. F_{in} is identified source profile, and S_{nk} is source influence of measured quantity (C_{ik}) at the receptor site. The linear equation shown by the above equation can be solved using the least square method of weighted variance.

CMB method uses certain presumptions that certainly cannot be accomplished totally because as particulate elements react with one another, source configuration cannot be constant and may vary based on prevailing conditions. Since particulate while travelling to receptor from source is likely to endure chemical alteration, Winchester and Nifong (1971) presented a coefficient of fractionation to Eq. (14.3)

$$\mathrm{C}_{ik} = \sum_{n=1}^{N} a_{iN} F'_{in} S_{nk} \tag{14.3}$$

where F'_{in} denotes configuration of pollutant at source location while F_{in} is receptor configuration. The calculation of a_{iN} is particularly intricate practically. The indeterminacy related to the source is evaluated using variance weights and is shown by Eq. 14.4

$$(\omega_e)_{ii} = \frac{1}{\sigma_i^2} + \sum_{n=1}^{N} \sigma_m^2 s_n^2 \tag{14.4}$$

where σ_i is observed indeterminacy of contaminant quantity, x_i, and σ_{in} is the of observed indeterminacy of i species released by n source.

14.5 Unidentified Source Profile

Owing to the absence of native origin-related discharge data and alteration in source discharge fundamental profiles, development of the new method is vigorous when the profile of source is unknown. Such methods are termed as factor analysis, where the problem is extended to source profiles resolution and contribution for a sample set. The elementary equation is expressed as

$$X = GF' \tag{14.5}$$

where $G =$ the matrix of source contribution, $F' =$ source profile transpose matrix.

14.5.1 Enrichment Factor

Enrichment factor (EF) method is used for source apportionment of particulates along with know-how of source profile for suggesting emission source. EF is utilised to evaluate predictable species origin and magnitude of human-induced events linked with the discharge of particulates. It is one of the elementary receptor models that suggest occurrence or absenteeism of specific markers. It also gives elementary information associated with the creation of secondary particulates. EF associates the comparative ratio of the basic configuration of predictable elements with a reference element in the sample to the equivalent ratio in the normal backdrop configuration.

$$\text{Enrichment factor} = \frac{(C_x/C_b)\text{sample}}{(C_x/C_b)\text{background}} \tag{14.6}$$

If EF is greater than one, natural or human-induced sources are expected to be predominant. If a particular source is governing, EF analysis is done by direct regression analysis or constituent configuration ratio. EF offers partial data regarding discrete sources and is incapable to enumerate discrete influence from a source for the composite group. EF analysis uses (1) variable regression algorithmic rule, (2) boundaries of a 2D dispersion plot, and (3) influence ratios of the intended marker and tracer element in airborne particulates monitored for a specific time interval when a sole source is predominant. EF analysis provides the influence of individual source species measured in the analysis. If elementary presumptions of mass balance are satisfied; EF method may be used for selection of information or backing presumptions for receptor sources if inadequate data is accessible. Thus, results from EF analysis should always be inferred with care.

14.5.2 Factor Analysis

Primary applicability of this method is to decrease variables for a definite set of data. Furthermore, factor analysis (FA) is also utilised for eliminating repetitiveness for a data set of interrelated variables. It is regularly utilised for investigating data structure, reducing data groups to a suitable numeral, and inferring results from the initial data group. FA is an imperative numerical technique for finding accurate results by utilising the flexible system and isolates the origin of particulates based on observation at a receptor location. SA studies involving FA are beneficial as data associated with source profile is redundant but helps in differentiating like sources. Hence, the outcome requires a subjective explanation before finally drawing conclusions. These limitations can be overcome by using confirmative methods of factor analysis such as UNMIX and PMF, wherein the modeller predetermines the explicit constraints based on hypothetical expectancy.

A. **Principal component analysis**

Amid various factor analysis methods, principal component analysis (PCA) is regularly utilised to be an investigative technique that syndicates factor analysis with multilinear regression to compute source involvement of particulates. The basic equation governing PCA is the conventional mass balance equation. The chief purpose of PCA is to transform several possibly interrelated variants into a group of linearly uncorrelated variants, termed as principal components (PCs). These components are successively inferred by the modeller as the probable source profiles. This method utilises orthogonal breakdown in such a way that the first principal component accounts maximum variability in the data. Further every succeeding component, next higher variance is possible so that it is uncorrelated with the previous component. Thus, PC with maximum variance is inferred as utmost dominant sources while each successive PC is next higher variance. A greater correlation occurs between individual set of components, whereas no or minimal correlation exists between individual PCs. Thus, the loading factors connect discrete variables with different components using orthogonal rotations like varimax.

The principal components are uncentred by subtracting a zero-valued pseudo-sample and carrying out regression against entire particulate matter mass, to overcome the effects of mean-centring of PCA score. This method is known as absolute principal components analysis (APCA). This method results in quantitative apportionments. The first PCA outcomes are initially rotated, and then, the component scores are uncentred with reference to zero-pollution case. This results in scaling coefficients for the component scores associated with source impacts and also for component loadings linked to source profiles. Hence, optimisation of results is done to describe influence information variability without limiting adequate constituent of matter. This result reduces independent rotations since it utilises orthogonal rotation. The equation for APCA is given as:

$$X_i = \sum_{k=1}^{p} z_k \text{APCS}_{Ki} \tag{14.7}$$

where X_i is mass of PM observed for ith observation, APCS_{ki} is rotated absolute score for k component for ith observation, and z_k is regression coefficient between PCs and pollutant mass.

B. **Confirmatory factor analysis (CFA)**

Confirmatory factor analysis is like the PCA method, but has the hypothesis prototype as its basis and may produce a solution that can be physically inferred and assessed. Confirmatory analysis methods include UNMIX and positive matrix factorisation (PMF). The confirmatory analysis puts a limitation on potential source influence results and necessitates them to follow some physical constraints like non-negative source impacts.

C. **Positive matrix factorisation (PMF)**

PMF is one of the complex factor analysis tools which used for source apportionment of atmospheric particulates. It is a multivariable analysis method similar to principal component analysis; however, it eliminates total non-positive items. Moreover, in spite of complete reliance on the arithmetic relationship of data group, it practises minimum square reduction to associate input factors. PMF differentiates particulate species data group to a diverse matrix such as numeral of factors, the influence of factor, and their profiles. This data straightaway relates to sources if a severe group of presumptions is satisfied. Furthermore, it deals every data point discretely to control discrete contribution based on confidence interval measurement. Investigational indeterminacy is utilised as input for resolving weighted factorisation and permitting discrete handling of elements.

Various factor analysis methods utilise eigenvalue analysis depending on sole value disintegration (SVD). PMF uses a dissimilar method to solve a problem. X matrix is expressed as

$$X = USV' = \overline{USV} + E \tag{14.8}$$

where U and V are initial columns of matrices U and V matrix, matrix U and V are derived using eigenvector analysis of matrix $X'X$ and XX'. As shown on the right-hand side of Eq. (14.9), the next term estimates X by the least square method to produce least probable value.

$$\sum_{i=1}^{m}\sum_{j=1}^{n} e_{ij}^2 = \sum_{i=1}^{m}\sum_{j=1}^{n}\left[x_{ij} - \sum_{p=1}^{P} g_{ip} f_{jp}\right]^2 \tag{14.9}$$

Thus, eigenvalue investigation is an inherent least square study which minimises the total of squared residuals. For PCA, data is scaled by row/column for normalising it, and data scaling causes misrepresentations of the study. Furthermore, an optimal ranking of information shall be to rank every data point discretely to generate accurate data with greater influence on results rather than data points having larger indeterminacy. However, exhaustive proportion provides a proportionate data matrix which might not be replicated by a regular factor analysis method. Thus, PMF uses the methodology of a clear least square method that decreases the objective function:

$$Q = \sum_{j=1}^{n} \sum_{i=1}^{m} \left| \frac{X_{ij} - \sum_{p=1}^{P} g_{ip} f_{jp}}{S_{ij}} \right|^2 \tag{14.10}$$

where s_{ij} is indeterminacy assessment of *j*th variable observed for *i*th sample. The problem of FA then reduces Q (E) in relation to G and F with a constriction that every G and F element shall be non-negative. PMF holds similar benefits like PCA; however, PMF has the surplus benefit of treating missing data or data which is at a level below detection. But this method necessitates a big data group, rather greater than the number of associated elements and a weighted factor related to every observation must be allotted.

D. **UNMIX**

The basic philosophy for this method is to enforce minimum presumptions, thus allowing information to communicate itself. This model has an innovative and arithmetically exhaustive program to assess the numeral of sources. For a given amount of sources, UNMIX utilises PCA to decrease the dimensions of data sets. UNMIX model solves mass balance equivalence by applying eigenvector analysis to decrease dimensions of the data group by not centring the initial data. UNMIX produces a group of principal components, having covariance receptor classes, that are consequently inferred by modeller as probable origin profiles.

Some features of this model are the ability to substitute lacking information and capability to approximate huge numeral of sources by utilising dual perception for RMs. It assesses indeterminacy in origin configuration by applying a blocked bootstrap method which considers the sequential interrelation of data. This model does not integrate inaccuracies in the study; however, restriction of UNMIX is that it does not generate a precise answer for the mass balance equation. UNMIX is capable to solve maximum powerful sources, but the fragile sources generally display low compliance of anticipated and assessed source involvement. Additionally, UNMIX shall be simply utilised for locations for which exhaustive origin profile information is not accessible. The below equation is used for this model,

$$C_{\overline{ij}} = \sum_{i=1}^{p} \left(\sum_{k=1}^{p} U_{ik} D_{ki} \right) V_{ij} + E_{ij} \tag{14.11}$$

where U, D, and V are $n\ x\ p$, $p\ x\ p$ diagonal, and $p\ x\ m$ matrix, respectively. E_{ij} is inaccuracy comprising total variance in C_{ij} not calculated by initial principal component (p). The utility of this model necessitates great numeral of variants and data set larger than 100, to generate an articulate result.

14.6 Hybrid Models

The theory of hybrid model combines CMB and positive factor analysis and offers superior regulation of concluding result. Two groups of hybrid models are used: (i) constrained receptor models and (ii) trajectory receptor models. The trajectory model utilises contaminant concentration, wind direction, and wind speed observed near the receptor location. This method utilises multivariable factor analysis and uses a clear overview of supplementary data (along with wind and its paths) to decrease the rotation uncertainty of the result.

14.6.1 Constrained Physical Receptor Model (COPREM)

COPREM combines the aspects of CMB and FA. The model can be resolved by decreasing a C2 function through a dual-stage iteration process. The modeller can utilise the backdrop know-how to deviate the reiteration towards a coherent solution. For instance, selecting trajectories that are relative to acknowledged source profiles by picking up restraints that keep a portion or entire profile as constant and thereby preventing undesirable mingling of source trajectories. Model output is a source power matrix, a profile matrix, function C2, and degree of freedom (n). A single-factor study is done on residuals to divulge a probable overlooked source. The indeterminacy in the source profiles is assessed using a precise model by linear regression. The estimated indeterminacy, however, denotes lower bound values, since rotation indeterminacy and independent variable indeterminacies are ignored.

It is a multivariable receptor model using linear model and weighted data deduction rule with restraints which permit to offer improved origin separation with measured indeterminacy. It primarily generates a matrix for origin profile employing origin trajectories and consequently augments extra restraints to decrease mingling of origin profiles. Further, an overview of restraints assists to control not-physical result such as non-positive source profile, and it provides profile constituents in continual ratio. Hence, the choice of suitable constraint is particularly crucial and can be selected based on appropriate data of initial source configuration. This model offers the feasibility of integrating backdrop know-how to control the undesirable mingling of sources. COPREM needs huge data with comprehensive information about source profile which occasionally restricts its applicability.

14.6.2 Extended Factor Analysis

The conventional linear factor analysis implemented on a two-dimension sample matrix by receptor elements is expanded to resolve additional composite multilinear equivalence by utilising multilinear engine (ME) platform. ME produces a table which stipulates the study and resolves it by Conjugate Grade Program. In this type of model, the rotation uncertainty of FA is decreased by incorporating supplementary constraints like identified origin profile, acknowledged source influence, and meteorology variants. The flexibility of ME is subjugated to produce data set explicit models and to develop diverse information such as particulate configuration and its size classification.

Extended factor analysis model solves the different multilinear and pseudo-multilinear problems by including constraints. ME is backdrop algorithm utilised for running PMF. The problem is denoted by a group of equivalence, where every equivalence estimates sole data signifying diverse indeterminate. Indeterminates are mentioned for diverse data as per the model configuration, and positive restraints are involved to decrease the probability of non-positive source distribution. It gives a general set-up for numerous ME models together with bilinear or trilinear and multilinear model that improvises its flexibility. In sums-of-products form, this model is shown below

$$\mathrm{x}_i = y_i + e_i = \sum_{k=1}^{Kf} \pi f_i + e_i (i = 1, 2 \ldots M) \tag{14.12}$$

where i index indicates the equivalence for the model, every equivalence relates to a single observed value (X_i), M denotes numeral of equivalences which is total of numeral of observations and ancillary equivalences, if any, built-in value y_i for every data point x_i is designated as total product for all factors, and K_i specifies the numeral of product terms in every equivalence. The enclosure of diverse data groups, particulate classification, particle size distributions, meteorology variants, and indeterminacy with the flexibility to alter the input as per requirement are some of the benefits of the ME model.

14.7 Features of Various Receptor Models

In India, receptor models have been used for source apportionment study of airborne particulates in various cities. Most of the studies have applied multivariate statistical methods to generate factors that are characterised by permutations of elemental and ionic constituents. The study began with the use of enrichment factor analysis, then shifting to use of principal component analysis, and finally culminating in the use of positive matrix factorisation model. The features of each model are summarised below:

CMB:

- The results of the CMB model analysis are influenced intensely by the availability of source profiles.
- It is preferred for the area where the receptor is located and synchronised with the existing ambient air measurements.
- The sensitivity of CMB to source profile collinearity obstructs the solution of mass balance equation, so it often becomes essential to group different sources to yield composite profiles.
- Chemical Mass Balance method has gained new stimulus, due to its suitability for application of studies involving molecular markers.

EF:

- EF analysis is simple and elementary assumptions of mass balance equation are fulfilled.
- Where information is easily, accessible EF can be used for initial data screening and supporting assumptions related to receptor and sources.
- But, the presence of exclusive source markers is very unusual; hence, results from EF shall be inferred with restraint.

PCA:

- PCA is sensitive to the relative scaling of original variables.
- For PCA, a data set is normally distributed is the elementary hypothesis. But this may not always be true environmental monitoring data. Location of the variance in the initial components can be partly solved by carrying out orthogonal rotations like varimax.
- A PCA model performs better when applied for an initial assessment or qualitative study.

UNMIX:

- UNMIX does not integrate errors in the analysis but it has few concerns, similar to PCA.
- Limitation of UNMIX is that it is not possible to find a solution for the mass balance equation mathematically.
- UNMIX is capable to determine various intense sources while weak sources demonstrate poor accord between probable and estimated source contributions.

PMF:

- The use of PCA which is unweighted model causes less acceptable factor resolution as compared to PMF which is a recent weighted model.
- PMF needs no previous knowledge about source composition, but any data related to source emissions characteristics helps to differentiate like sources.
- It needs a considerable number of discrete air samples (at least 50) and gives the best result with the big data set.

- PMF has gained impetus due to its ability to handle input and output data uncertainties.

COPREM:

- Hybrid models like COPREM and multilinear engine have been developed to combine the advantages of CMB and factor analysis and reduce their disadvantages.
- An initial profile matrix with the main characteristics of known sources is used in COPREM.
- With a sufficient number of sources, prior information about source characteristics can be used to derive a solution.

MLE:

- Multilinear problems with several constraints can be solved using multilinear engine and involving a script language.
- A multilinear engine program permits to use source composition data to restrain the model.

14.8 Source Signature Selection

Source apportionment of particulate matter involves information about particulate classification and composition. This information is evaluated crucially to ascertain the occurrence of particular species that are assumed to evolve from acknowledged sources, travel due to atmospheric turmoil, and in due course assessed in the receptor site.

Selection of source signature is the most significant parameter in receptor modelling analysis. Source signature denotes a different combination of molecular markers or tracer elements that are used for source categorisation. An extensive array of source profiles has been utilised by the authors, across India for source apportionment. The crucial markers used for source apportionment studies in India are almost similar to those used worldwide. Different researchers have inferred different sources in a different manner. This is one of the significant considerations in synchronisation of source profiles for source recognition and apportionment. Based on literature studies, for various emission sources in India, the key source signatures are discussed below.

14.8.1 Crustal Resuspension or Road Dust

Road dust is a composite mixture of vehicular exhaust emissions, industrial emissions, soil dust, coal combustion, and construction material and from several anthropogenic and biogenic sources. Al, Ca, Fe, Mg, Si, Ti, and Na are the crustal element markers that are classically used as tracers for road dust or crustal resuspension. A complete range of element tracers such as Al, Ca, Co, Cu, Cr, Fe, Ni, Mg, Pb, Si, T, and V have been utilised in India for recognition of crustal resuspension or road dust. The choice of tracer elements for crustal resuspension category is slightly inconsistent in India. Extensive range of elements like Al, Ba, Ca, Na, Ni, Pb, Cu, Fe, K, Mg, Mn, Si, and Zn have been assembled as tracers for crustal markers. Few researchers have counted dust that is resuspended or dust from roads under crustal segments, whereas few others have isolated it. Some of the studies associate road resuspension dust or crustal dust with construction and vehicular activity while some consider them independently. These uncertainties in the selection of source signature limit inter-comparison of results and also regulate policy decisions.

14.8.2 Vehicular Emission Sources

Source profiles related to traffic normally include Cu resulting from brake linings, Pb resulting from gasoline additives, and Zn resulting from tyre wear. Traffic emissions sources can be further augmented with Al, Ba, Ca, Fe, and K from diesel exhausts and wear of brake lining. It may also contain Al from wear and tear of engine pistons; Ba from fuel additives; Fe from the exhaust; Mn from unleaded gasoline; Ni from heavy oil combustion; and Zn from combustion in two-stroke engines. Various basic molecules in different permutations have also been used as signature molecules. Presence of Ba, Mg, Fe, and Zn in major concentration and trace amounts of Al, Ca, Cr, and Mn indicate emissions due to brake lining, while the incidence of Cu, Ba, and Fe reveals brake pad emission. Lead has been used as a common signature marker for vehicular emissions. The utility of leaded petrol is banned from the year 2000; still, its extended residual life renders it pertinent for the airborne particulate matter. Concentrations of Pb are found to be gradually decreasing after the use of unleaded petrol; therefore, Pb is not used as a sole indicator of vehicular emissions. In India, for identification of vehicular emissions, the commonly used tracer element is Pb in addition to elements like V, Mn, Co, and Zn.

14.8.3 Industrial Emissions Sources

Industrial emissions are usually air discharges related to various production processes like pharmaceutical, petrochemical, metallurgy, ceramic, etc. A range of elemental markers such as Co, Cd, As, Cr, and Ni are used for source apportionment of particulate matter in India. Moreover, various trace elements depending on the type of industry, its raw material, and finished products are utilised to recognise explicit industrial emission sources. For instance, Cu, Ni, and Mn are from iron and steel processing industries in Mumbai; Zn and Cr from metal production units in Delhi; Cd, Pb, and V from battery repair units in Delhi; Cr from electroplating units of

Kolkata; Cu, Ni, and Zn from metallurgical industries, galvanising, and electroplating, while Cr from tanneries in Kolkata; Cu, Pb, Ni, and Zn from industrial emissions in Agra; Fe, Mn, Zn, Pb, and SO_4 from smelting units in Ahmedabad; and Ba from oil-fired power plants. Furthermore, source apportionment study taken up by Negi et al. (1987) segregated industrial emission sources as V, Br from the textile industry, Cu, Ni, S, and V from oil refineries, and Cu, Mn, and Zn from non-ferrous industry emissions for Bombay, Bangalore, Nagpur, and Jaipur.

14.8.4 Fuel Incineration

Fuel incineration is a generic term, which applies to incineration of a material that contains potential energy which is released on combustion of that material. Chemical configurations of coal are substantially different in various geographical places that control the choice of explicit marker element. The coal from India has lower sulphur content; however, coal from north-east India normally contains a higher content of sulphur, vitrinite, and volatile matter. Thus, sulphur is the chief contaminant in the coal from north-east India, while As is the main contaminant in Western India coal. As, S, Se, and SO_4 are considered to be the common markers of incineration of coal. However, as per literature survey, among the various marker elements utilised for coal incineration, the most prominent markers are Cr, Cu, Ni, V, K, PAH, Cd, Se, Pb, Zn, As, and Cl^-.

14.8.5 Marine Aerosol Emissions

Marine aerosols control considerable amounts of natural aerosols in the coastal environment. Key markers of marine aerosols contain Na, K, Cl^-, and Mg. Normally, Na^+, K^+, Cl^-, and Mg^{2+} are the tracer markers indicating presence of marine aerosols. However, these markers chemically react with $SO_4{}^{2-}$ and $NO_3{}^-$ and release Cl^- and Br^- to the atmosphere. Presence of $NO_3{}^-$ in fine particulate matter may have marine origin through condensation of HNO_3. Several marker elements are used to identify marine aerosol emissions like K^+, Na^+, and Mg^{2+}. Cl^- in marine aerosols combines with Ca^{2+}, Mg^{2+}, $SO_4{}^{2-}$, and HCO_3. However, the selection of few tracer markers is little bit conflicting and may be influenced by other sources like Cl^- and $SO_4{}^{2-}$ from fuel combustion, K^+ from biomass burning, and Mg^{2+} and Ca^{2+} from crustal emissions. In India, the majority of the studies related to origin distribution have been done for inland cities. Hence, marine aerosols are seldom used as one of the key sources, except in case of coastal cities like Mumbai. Use of K may overlap with wood or biomass combustion and Cl with coal burning.

14.8.6 Biomass and Refuse Burning

In India, biomass incineration is generally used for the combined burning of cow dung and fuelwood, or incineration of agricultural residues after harvesting. Potassium is the generally used inorganic marker for biomass burning in the absence of other organic markers. For crustal dust in coarse range, K is used as a tracer element and finer ranges of particulate matter soluble K from biomass burning is used as a tracer element. Zn, Cr, and Ni are the crucial markers for biomass and refuse burning. Few researchers have grouped refuse burning with hazardous waste disposal, while few researchers have used incineration or refuse burning as individual source. Therefore, establishing the precise influence of various activities is not very easy. Also in Indian cities, waste management is often carried out randomly; hence, predicting the exact nature and contribution of an individual source is bit difficult.

14.9 Source Apportionment of Airborne Particulates for Indian Scenario

In India, various source apportionment studies have been conducted to classify the sources of particulate matter in terms of natural and artificial in addition to quantifying the emissions. Various research papers involving source apportionment studies for different locations in India are studied in detail and logically grouped into four classes, namely North India, South India, Central and East India, and West India, and discussed hereafter to present regional distribution of particulate matter based on the source signatures found.

14.9.1 Source Apportionment Studies in Northern India

Various studies involving CMB, factor analysis using PMF, PCA, and enrichment factor have been carried out for Agra, Chandigarh, Delhi, Kanpur, Moradabad, and Roorkee cities in Northern India (Table 14.1).

14.9.2 Source Apportionment Studies in Southern India

Source apportionment studies have been carried out for Chennai, Hyderabad, Kakinada, Tirupati, and Vishakhapatnam in the southern region by applying CMB and multivariate models (Table 14.2).

Table 14.1 Source apportionment studies in Northern India

Sr. No	North	Receptor model	Year	Author	Particulate matter	Selection of source signature
1	Delhi	PCA	2000	Balachandran and Khillare	RSPM	Vehicular emissions 53.9%, industrial emissions from foundry 19.4%, and soil resuspension 15.7%
2	Delhi	PCA	2002	Balachandran and Khillare	SPM	Vehicular and industrial emissions 60% and crustal source 22%
3	Delhi	PCA	2007	Gupta et al.	SPM	50% from crustal resuspension and 13% from construction material
4	Delhi	PCA	2007	Gupta et al.	RSPM	50% from crustal resuspension, 13% from construction material, and 11% from vehicular emissions
5	Delhi	CMB	2007	Gupta et al.	RSPM	60–89% from traffic emissions
6	Delhi	CMB	2007	Gupta et al.	SPM	51–73% crustal emissions and 24–42% vehicular emissions
7	Delhi	CMB	2007	Chowdhury et al.	RSPM	25–33% from fossil fuel and 7–20% biomass burning
8	Delhi	CMB	2008	Srivastava et al.	RSPM	62% from vehicular emissions and 35% from crustal emissions
9	Delhi	CMB	2008	Srivastava et al.	SPM	64% from crustal dust and 29% from vehicular pollution
10	Delhi	PCA	2008	Srivastava et al.	SPM	68% from crustal resuspension and 23% from vehicular pollution
11	Delhi	PCA	2008	Srivastava et al.	RSPM	86% from vehicular emissions and 10% from crustal resuspension

(continued)

Table 14.1 (continued)

Sr. No	North	Receptor model	Year	Author	Particulate matter	Selection of source signature
12	Delhi	PCA	2010	Chelani et al.	RSPM	42% from auto-exhaust and 23% from industrial emissions
13	Delhi	PCA	2010	Shridhar et al.	Urban	25% from resuspended dust, 20% from industrial emissions, and 10% from biomass and refuse oil burning
14	Delhi	PCA	2010	Shridhar et al.	Rural	35% from resuspended dust, 15% from biomass, and low grade coal burning, 10% from vehicular emission, 7% from wind-blown dust, and 7% from industrial emissions
15	Delhi	PCA	2009	Tiwari et al.	PM2.5	80% from industrial emissions and 6% from crustal resuspension
16	Delhi	Unmix and PMF	2013	Tiwari et al.	RSPM	60% from vehicular emissions and 31% from crustal origin
17	Delhi	–	2014	Tiwari et al.	PM2.5	69% from vehicular emissions and 31% from coal combustion
18	Delhi	PCA-MLR	2012	Khillare and Sarkar	RSPM	49–65% from crustal suspension, 27–31% from vehicular emissions, and 4–21% from industrial emissions
19	Delhi	PMF	2012	Sharma et al.	RSPM	21.7% from secondary aerosols, 207% from soil dust, 17.4% from fossil fuels, 16.8% from vehicular emissions, and 13.4% from biomass burning

(continued)

Table 14.1 (continued)

Sr. No	North	Receptor model	Year	Author	Particulate matter	Selection of source signature
20	Delhi	PMF	2012	Sharma et al.	PM10	22.7% from soil dust, 20.5% by secondary aerosols, 17.0% from vehicle emissions, 15.5% from fossil fuel burning, 12.2% from biomass burning, 7.3% from industrial emissions, and 4.8% from sea salts
21	Delhi	PMF	2012	Sharma et al.	PM2.5	21.3% from secondary aerosols, 20.5% from soil dust, 19.7% from vehicle emissions, 14.3% from biomass burning, 13.7% from fuel combustion, 6.2% from industrial emissions, and 4.3% from sea salt
22	Kanpur	PCA	2010	Shukla	RSPM	23–62% road dust and 4–30% from secondary particles
23	Kanpur	PCA	2010	Chakraborty and Gupta	PM1	Vehicular emissions 50%, industrial emissions 22%, and crustal dust 15%
24	Kanpur	Unmix	2010	Chakraborty and Gupta	PM1	24% from vehicular emissions, 14% from road dust, 39% from inorganic secondary particulates, and 11% from coal combustion
25	Agra	PCA	2009	Kulshrestha et al.	PM10	38% from vehicular emissions, 20% from solid waste dumping, and 18% from industrial emissions

(continued)

Table 14.1 (continued)

Sr. No	North	Receptor model	Year	Author	Particulate matter	Selection of source signature
26	Agra	PCA	2009	Kulshrestha et al.	PM2.5	30% from industrial emissions, 28% from vehicular, and 21% from solid waste dumping
27	Agra	PCA	2012	Singh and Sharma	RSPM	55% from crustal elements, 175 from vehicular emissions, 9% from industrial emissions, and 7% from coal and biomass burning
28	Agra	PCA	2013	Habil et al.	SPM	47% from vehicular emissions, 33% due to combustion, and 18% from garbage burning
29	Chandigarh	PCA	2010	Bandhu et al.	SPM	60% from crustal elements, 20–50% industrial activity, and 14% from refuse burning
30	Chandigarh	CMB	2007	Chowdhury et al.	PM2.5	28% from vehicular emissions and 8% from biomass burning
31	Moradabad	PCA	2013	Pal et al.	RSPM	40% from industrial emissions, 20% vehicular emissions, and 13% from anthropogenic activities

Table 14.2 Source apportionment studies in Southern India

Sr. No.	South	Receptor model	Year	Author	Particulate matter	Summary
1	Hyderabad	CMB	2012	Guttikunda et al.	PM10	30% from direct vehicle exhaust and 30–45% from road dust
2	Hyderabad	CMB	2011	Gummeneni et al.	PM10	40% from resuspended dust, 22% by vehicular emissions, 2% by combustion, 9% by industrial emissions, and 7% by refuse burning
3	Hyderabad	CMB	2012	Gummeneni et al.	PM2.5	31% from vehicular pollution, 26% by resuspended dust, 9% by combustion, 7% by industrial emissions, and 6% by refuse burning
4	Chennai	CMB	2012	Srimuruganandam and Shiva Nagendra	PM10	Diesel emissions contribute 52%, gasoline exhaust contributes 16% followed by cooking emissions 1.5%, and paved road dust 2.3%
5	Chennai	CMB	2012	Srimuruganandam and Shiva Nagendra	PM2.5	Diesel emissions contribute 65%, gasoline exhaust contributes 8% followed by cooking emissions 1.5%, and paved road dust 2.3%
6	Chennai	PCA	2012	Srimuruganandam and Shiva Nagendra	PM10	Marine aerosol 40.4%, secondary PM 22.9%, motor vehicles 16%, biomass burning 0.7%, brake wear 4.1%, soil 3.4%, and other sources 12.7%

(continued)

Table 14.2 (continued)

Sr. No.	South	Receptor model	Year	Author	Particulate matter	Summary
7	Chennai	PCA	2012	Srimuruganandam and Shiva Nagendra	PM2.5	Marine aerosol 21.5%, secondary PM 42.1%, motor vehicles 6%, biomass burning 14%, brake wear 5.4%, soil 4.3%, and other sources 6.8% for urban Chennai
8	Chennai	PCA	2015	Chaudhari and Gajghate	PM10	46% contribution from incinerators and 31% from crustal markers
9	Coimbatore	PCA	2004	Mohanraj et al.	PM2.5	47.4% from vehicular emissions and 27.4% from burning of biomass
10	Tirupati	PCA	2008	Kumar et al.	PM10	38% contribution from traffic emissions and 12.85% from crustal markers
11	Tirupati	PCA	2006	Mouli et al.	PM10	69.41% contribution from crustal material and road dust followed by 11.76% metallurgical and other industrial processes and 6.52% fuel oil combustion
12	Kakinada	PCA	2013	Mohammed et al.	PM10	Crustal road dust to be 52.5% followed by anthropogenic sources 39.5%

(continued)

Table 14.2 (continued)

Sr. No.	South	Receptor model	Year	Author	Particulate matter	Summary
13	Vishakhapatnam	PCA	2015	Rao et al.	PM10	Industrial emissions contributed 53.9%, while the contribution from crustal sources was 17.9% and 13% from industrial sources
14	Vishakhapatnam	PMF	2016	Police et al.	PM10	35% from biomass burning followed by 22.5% contribution from crustal source, coal combustion contributing 14%, sea salt spray 9.7%, metal industry 5.1%, and fuel oil combustion 1.5%
15	Virudhunagar	CMB	2013	Thambavani and Maheswari	PM10	40% from road dust, 32% from vehicular emissions, 20% from biomass burning, and 4–6 from industrial sources
16	Virudhunagar	CMB	2013	Thambavani and Maheswari	PM2.5	32% from road dust, 36% from vehicular emissions, 24% from biomass burning, and 7–9 from industrial sources

(continued)

Table 14.2 (continued)

Sr. No.	South	Receptor model	Year	Author	Particulate matter	Summary
17	Bangalore	CAMx	2013	Guttikunda et al.	PM10	Vehicle exhaust and on-road dust resuspension 70% of total PM10 emissions; followed by 17.8% industries, 11% for open waste burning, and 6.5% domestic cooking
18	Bangalore	CMB	2010	CPCB report by TERI	PM10	50% from road dust, 19% from vehicular emissions, 13% from DG sets, 8.7% from secondary particulates, 4.5% from industrial emissions, and 4.2% from burning of wood
19	Bangalore	CMB	2010	CPCB report by TERI	PM2.5	49.9% from vehicular emissions, 24.7% from DG sets, 12.7% from secondary particulates, 3.5% from road dust, 3.5% from industrial emissions, and 5.8% from burning of wood
20	Bangalore	PCA	2010	CPCB report by TERI	PM10	26% from road dust, 17% from coal combustion, 10% from vehicular emissions, 9% from crustal source, 11% from biomass burning, and 15% from secondary aerosols

14.9.3 Source Apportionment Studies in Eastern and Central India

For Durg, Dhanbad, Jorhat, Jharia, and Kolkata in Eastern and Central India, various receptor model studies have been undertaken in the past years (Table 14.3).

14.9.4 Source Apportionment Studies in Western India

For Western India, maximum studies have been undertaken for Mumbai in addition to Ahmedabad, Anand, Kota, Mithapur, Nagpur, and Pune (Table 14.4).

14.10 Discussions and Conclusion

Various source apportionment studies have been carried out in different cities across India by applying CMB and various multivariate models like PCA, PMF, factor analysis, and enrichment factor. The various inferences and conclusions drawn from the various research papers reviewed have been summarised below:

- Source profile studies involving different geographical areas were found to be unparalleled.
- The inter-comparability of resultant source profile is limited due to inconsistency in selection of marker elements, unavailability of definite source profile, and multi-site studies.
- Non-availability of detailed emissions inventories for a particular area restricts the applicability of advanced receptor models.
- Most studies have focused upon TSP or PM10, therefore not benefiting from the additional insights to be gained from separating coarse from fine particles. There has been insufficient use of size fractionation of particulate matter.
- Data-related fine particulate studies were particularly inadequate in India due to limited emission inventory data.
- Emissions inventory data is very scarce. These need to be spatially and chemically disaggregated. Knowledge of city-specific emissions inventories for specific chemical components would give greater confidence in assigning sources to factors identified through multivariate receptor models.
- There is a lack of multi-site studies. Where these exist, they tend to use multiple sites within a city (e.g. CPCB 2010) rather than using urban/rural contrasts to elucidate the importance of emissions within the city relative to the regional background.
- Restricted use of particulate size data confines using source profile knowledge to recognise contribution from specific sources.

Table 14.3 Source apportionment studies in Eastern and Central India

Sr. No.	East	Receptor model	Year	Author	Particulate matter	Summary
1	Kolkata	PCA	2006	Karar and Gupta	PM10	32% solid waste dumping, 23% vehicular emissions, 15% road dust showing followed by crustal elements with 14%
2	Kolkata	PCA-MLR	2007	Karar and Gupta	PM10	Solid waste dumping contribution 36%, vehicular emissions 26%, coal combustion 13%, cooking 8%, and soil dust 4% at the residential site
3	Kolkata	PCA-MLR	2007	Karar and Gupta	PM10	37% to vehicular emissions, 29% to coal combustion, 18% to electroplating industry, 8% to tyre wear, and 1% to secondary aerosol at the industrial site
4	Kolkata	CMB	2007	Karar and Gupta	PM10	37% contribution from coal combustion, 19% from soil dust, 17% from road dust, and 15% from diesel combustion from residential areas
5	Kolkata	CMB	2007	Karar and Gupta	PM10	36% from soil dust, 17% from coal combustion, 17% from solid waste burning, 16% from road dust, and 7% from tyre wear and tear in industrial areas
6	Kolkata	CMB	2007	Karar and Gupta	PM2.5	Coal combustion 42%, crustal elements 21%, field burning contribution as 7%, and paved road as 1% contribution for residential areas
7	Kolkata	CMB	2007	Karar and Gupta	PM2.5	For industrial areas, vehicular emissions contributed 47%, coal combustion contributed 34%, and metal industry and soil contributed 1%
8	Kolkata	CMB	2007	Chowdhury et al.	RSPM	34–57% from fossil fuel and 13–18% biomass burning

(continued)

Table 14.3 (continued)

Sr. No.	East	Receptor model	Year	Author	Particulate matter	Summary
9	Kolkata	PCA	2015	Das et al.	PM10	Abraded vehicular exhaust gases and municipal waste incineration contributing 35.8%, industrial emissions contributing 31.8%, and coal combustion and non-ferrous metal smelting contributing 16.7%
10	Kolkata	PCA	2015	Das et al.	PM2.5	Road dust and industrial emissions contributing 36.9%, exhaust gases of cars and municipal waste incineration contributing 26.9%, and coal combustion and non-ferrous metal smelting contributing 15.8%
11	Kolkata	PCA	2011	Chatterjee et al.	PM2.5	38% vehicular emission, 27% biomass and fossil fuel burning, 18% road suspension dust, and 11% secondary particulates
12	Kolkata	PCA	2010	Kar et al.	PM2.5	42% anthropogenic sources, 17% electroplating and metallurgy industry, 11% tannery industry, and 8% other industrial emissions
13	Jorhat, Assam	PCA	2010	Khare and Baruah	PM2.5	38% from crustal sources, 26% from fuel accounting 26%, 19% from traffic and industrial emissions, 9% from biomass burning, and 8% from secondary aerosols
14	Jharia	PCA	2014	Pandey et al.	PM10	53.71% coal mining activities, 17.85% vehicular emissions, and 7.54% from wind-blown dust from unpaved roads
15	Durg, Chhattisgarh	CMB	2008	Gadkari et al.	PM10	Ambient emissions 48–73%, soil markers 5–38%, indoor 14–25%, and road 1% for school areas

(continued)

Table 14.3 (continued)

Sr. No.	East	Receptor model	Year	Author	Particulate matter	Summary
16	Durg, Chhattisgarh	CMB	2007	Gadkari et al.	PM10	Ambient emissions 19–26%, soil markers 8–55%, indoor 40–66%, and road 5–10% for residential areas
17	Durg, Chhattisgarh	PCA	2012	Deshmukh et al.	PM10	Anthropogenic sources 58%, crustal sources contribution 22.2%, and biomass burning 12%
18	Dhanbad	PCA	2012	Dubey et al.	PM10	Crustal source 32%, automobile emissions 22%, and road dust 20% for mining areas
19	Dhanbad	PCA	2012	Dubey et al.	PM10	Crustal source 29%, automobile emissions 25%, and road dust 18% from non-mining areas
20	Orissa	PCA	2012	Roy et al.	PM10	18% vehicular emissions and resuspension dust, 17.5% crustal sources, 14.2% from industrial emissions, and 13% from power stations and zinc smelting

Table 14.4 Source apportionment studies in Western India

Sr. No.	West	Receptor model	Year	Author	Particulate matter	Summary
1	Mumbai	FA-MR	2000	Kumar et al.	SPM	Road dust 41%, the marine aerosol 15%, traffic emissions 15%, metal industries 6%, and coal combustion 6%
2	Mumbai	PCA	2004	Tripathi et al.	SPM	63.8% from fine aerosols of soil origin, 12.6% from coarse aerosols of soil origin, and 8.2% from sea salts
3	Mumbai	FA-MLR	2008	Kothai et al.	SPM	Sea salt causing 35%, crustal sources 25%, industrial emissions 14%, vehicular emissions 10%, and fugitive emissions 7%
4	Mumbai	FA-MLR	2008	Kothai et al.	RSPM	Sea salt causing 3%, crustal sources 18%, industrial emissions 23%, vehicular emissions 29%, and fugitive emissions 9%
5	Mumbai	CMB	2007	Chowdhury et al.	RSPM	Fossil fuels 21–36% and biomass burning 7–20%
6	Mumbai	CMB	2008	Chelani et al.	PM10	Traffic emissions 10–23%, industrial emissions 18–50%, and marine aerosols 13–15%
7	Mumbai	PMF	2011	Gupta et al.	PM10	Oil combustion and construction 25%, motor vehicles 23%, marine aerosol and nitrate 19%, and paved road dust 18%
8	Mumbai	PCA	2011	Kothai et al.	SPM	Crustal markers were 52.53%, sea salt 17.43%, combustion sources 6.95%, and industrial sources related to bromine 5.9%

(continued)

Table 14.4 (continued)

Sr. No.	West	Receptor model	Year	Author	Particulate matter	Summary
9	Mumbai	PCA	2011	Kothai et al.	RSPM	45.9% from crustal markers, 22.5% from sea salt, 7.46% from combustion sources, and 7% from industrial sources
10	Mumbai	CMB	2012	Joseph et al.	PM2.5	Organic matter to be 36–52%, secondary inorganic aerosols to be 21–27%, crustal emissions to be 6–12%, non-crustal emissions to be 4–8%, and sea salt to be 6–11%
11	Ahmedabad	PMF	2010	Raman et al.	SPM	Regional dust 57.9% followed by calcium carbonates rich dust infusing 19%, biomass burning or vehicular emissions 8%, secondary nitrate and sulphate 5%, and marine aerosols 4.5%
12	Ahmedabad	PMF	2011	Raman et al.	SPM	Crustal material 44% and sea salt 29.8%, and industry-related nitrate emissions 13.66% and ammonium-rich factor 11.62%
13	Ahmedabad	PMF	2012	Sudheer and Rengarajan	PM2.5	33% from biomass burning sources, 31% from coal-based power stations and vehicular emissions, 11% from industrial or/and incineration emissions, 10% from mineral dust, and about 13% from resuspended dust
14	Pune	FA-CMB		CPCB report	PM10	57% for crustal elements followed by 14.9% construction activities, 10.8% fuel combustions, and 9.8% vehicular emissions

(continued)

Table 14.4 (continued)

Sr. No.	West	Receptor model	Year	Author	Particulate matter	Summary
15	Pune	FA	2013	Yadav and Satsangi	PM10	52% from anthropogenic sources, 15% from traffic emissions, and 14% from crustal markers
16	Pune	FA	2013	Yadav and Satsangi	PM2.5	30.7% from resuspended road dust, 15.45 from traffic emissions, 14.8% from biomass burning, 13.6% from crustal emissions, and 11.9% from anthropogenic sources related to zinc
17	Nagpur	CMB	2014	Pipalatkar et al.	PM2.5	57–65% from vehicular emissions, 12–16% from secondary inorganic aerosol, and 9–15% from biomass burning
18	Alang	PCA	2007	Basha et al.	SPM	80.8% ship-breaking activities, 11.5% for vehicular emissions, and 5% with crustal emissions in coarse particulate matter
19	Anand	PCA	2011	Tanushree et al.	SPM	Traffic emissions contributing 74% and local industrial emissions contributing 24%
20	Mithapur	PCA	2010	Basha et al.	SPM	29.47% impact from industrial emissions, 25.61% impact from resuspended dust, and 11.84% from oil combustion
21	Kota	PCA	2016	Meena et al.	SPM	36–37% from industrial activities and crustal emissions 27–31%
22	Kota	PCA	2016	Meena et al.	RSPM	33–35% from industrial activities and 28–31% from crustal emissions

- Restricted use of organic markers and gas-to-particle conversion reduce the relevance of source profile results for future studies.
- There has, to date, been insufficient use of organic molecular markers. While these alone will not answer all source apportionment questions, they are an important tool in receptor modelling and could help to sharpen up both CMB and multivariate model studies.
- Failure in most cases to distinguish vehicle exhaust from non-exhaust vehicle emissions, particularly resuspension of road dust, and/or inability to differentiate regional crustal sources (e.g. desert dust) from local wind-blown soils and resuspended road dust. Making a distinction between road dust and local soils can be difficult under any circumstances if the soils are polluted by vehicle emissions or the road dust contains a significant soil contribution. However, separating these sources, and in particular, quantifying the vehicle exhaust contribution alone, and differentiating regional crustal sources from local soils and road dust are crucial, as the policy response depends heavily upon these insights.
- Particulate source profile for different geographical areas found largely incomparable and sometimes confusing. Inconsistency in marker selection, unavailability of definite source profile, and multi-site multi-temporal studies critically limit inter-comparability of resultant source profile.
- Extremely limited use of particulate size fraction information limits the applicability of source profile knowledge to identify relative contribution sources to different segments of particulates.
- In most of the cases, the absence of particulate source profile and detailed emissions' inventories limits the applicability of advanced RMs.
- Limited use of organic molecular markers and gas-to-particle conversion reduce the applicability of source profile results for future studies. Little attentions were paid to atmospheric dynamics of sulphate, nitrate, and ammonia which subsequently relates it to secondary regional aerosols and local anthropogenic emissions.
- Source profiles will require periodic review, upgradation, and inclusion of new sources to reflect the increasingly changing urban/natural activities in India.
- Environmental conservation and preservation can only be obtained by systematic planning of emission reduction either through mathematical or by receptor models which are capable of linking the emission source origins to the concentrations at the receptor.

References

Balachandran S, Meena BR, Khillare PS (2000) Particle size distribution and its elemental composition in the ambient air of Delhi. Environ Int 26(1–2):49–54

Bandhu HK, Puri S, Garg ML, Singh B, Shahi JS, Mehta D, Swietlicki E, Dhawan DK, Mangal PC, Singh N (2000) Elemental composition and sources of air pollution in the city of Chandigarh,

India, using EDXRF and PIXE techniques. Nucl Instrum Methods Phys Res Sect B 160(1):126–138

Basha S et al (2007) Heavy metal content of suspended particulate matter at world's largest ship-breaking yard, Alang-Sosiya. India. Water Air Soil Pollut 178(1–4):373–84

Basha S, Jhala J, Thorat R, Goel S, Trivedi R, Shah K, Menon G, Gaur P, Mody KH, Jha B (2010) Assessment of heavy metal content in suspended particulate matter of coastal industrial town, Mithapur, Gujarat. India. Atmos Res 97(1–2):257–265

Chakraborty A, Gupta T (2010) Chemical characterization and source apportionment of submicron (PM1) aerosol in Kanpur Region. India. Aerosol Air Qual Res 10(5):433–445

Chatterjee A, Dutta C, Jana TK, Sen S (2012) Fine mode aerosol chemistry over a tropical urban atmosphere: characterization of ionic and carbonaceous species. J Atmos Chem 69(2):83–100

Chaudhari PR, Gajghate DG, Singh DK (2015) Studies on respirable particulate matter and heavy metal pollution of ambient air in Delhi, India. Am J. Eng Res 4(12):45–57. www.ajer.org

Chelani AB, Gajghate DG, Devotta S (2008) Source apportionment of PM10 in Mumbai, India using CMB model. Bull Environ Contam Toxicol 81(2):190–195

Chelani AB, Gajghate DG, ChalapatiRao CV, Devotta S (2010) Particle size distribution in ambient air of Delhi and its statistical analysis. Bull Environ Contam Toxicol 85(1):22–27

Chowdhury Z, Zheng M, Schauer JJ, Sheesley RJ, Salmon LG, Cass GR, Russell AG (2007) Speciation of ambient fine organic carbon particles and source apportionment of PM in Indian cities. J Geophys Res 112(D15)

Das R, Khezri B, Srivastava B, Datta S, Sikdar PK, Webster RD, Wang X (2015) Trace element composition of PM2.5 and PM10 from Kolkata – a heavily polluted Indian metropolis. Atmos Pollut Res 6(5):742–750

Deshmukh DK, Deb MK, Mkoma SL (2013) Size distribution and seasonal variation of size-segregated particulate matter in the ambient air of Raipur city. India. Air Qual Atmos Health 6(1):259–276

Dubey B, Pal AK, Singh G (2012) Trace metal composition of airborne particulate matter in the coal mining and non-mining areas of Dhanbad Region, Jharkhand, India. Atmos Pollut Res 3(2):238–246

Gadkari NM, Pervez S (2007) Source investigation of personal particulates in relation to identify major routes of exposure among urban residentials. Atmos Environ 41(36):7951–7963

Gummeneni S, Yusup YB, Chavali M, Samadi SZ (2011) Source apportionment of particulate matter in the ambient air of Hyderabad city, India. Atmos Res 101(3):752–764

Gupta AK, Karar K, Srivastava A (2007) Chemical mass balance source apportionment of PM 10 and TSP in residential and industrial sites of an Urban Region of Kolkata, India. J Hazard Mater 142:279–287

Gurugubelli B, Pervez S, Tiwari S (2013) Characterization and spatiotemporal variation of urban ambient dust fallout in central India. Aerosol Air Qual Res 13(1):83–96

Guttikunda SK, Kopakka RV, Dasari P, Gertler AW (2013) Receptor model-based source apportionment of particulate pollution in Hyderabad, India. Environ Monit Assess 185(7):5585–5593

Habil M, Massey DD, Taneja A (2013) Exposure of children studying in schools of India to PM levels and metal contamination: sources and their identification. Air Qual Atmos Health 6(3):575–587

Jain S, Sharma SK, Choudhary N, Masiwal R, Saxena M, Sharma A, Mandal TK, Gupta A, Gupta NC, Sharma C (2017) Chemical characteristics and source apportionment of PM2.5 using PCA/APCS, UNMIX, and PMF at an urban site of Delhi, India. Environ Sci Pollut Res 24(17):14637–14656

Joseph AE, Unnikrishnan S, Kumar R (2012) Chemical characterization and mass closure of fine aerosol for different land use patterns in Mumbai city. Aerosol Air Qual Res 12(1):61–72

Kar S et al (2010) Metallic Components of Traffic-Induced Urban Aerosol, Their Spatial Variation, and Source Apportionment (June 2014)

Karar K, Gupta AK (2006) Seasonal variations and chemical characterization of ambient PM 10 at residential and industrial sites of an urban region of Kolkata (Calcutta), India. Atmos Res 81:36–53

Karar K, Gupta AK (2007) Source apportionment of PM 10 at residential and industrial sites of an urban region of Kolkata, India. Atmos Res 84:30–41

Khare P, Baruah BP (2010) Elemental characterization and source identification of PM2.5 using multivariate analysis at the suburban site of North-East India. Atmos Res 98(1):148–162

Khillare PS, Sarkar S (2012) Airborne inhalable metals in residential areas of Delhi, India: distribution, source apportionment and health risks. Atmos Pollut Res 3(1):46–54

Kothai P, Saradhi IV, Pandit GG, Markwitz A, Puranik VD (2011) Chemical characterization and source identification of particulate matter at an urban site of Navi Mumbai, India. Aerosol Air Qual Res 11(5):560–569

Kulshrestha A, Gursumeeran Satsangi P, Masih J, Taneja A (2009) Metal concentration of PM2.5 and PM10 particles and seasonal variations in urban and rural environment of Agra, India. Sci Total Environ 407(24):6196–6204

Meena M, Meena BS, Chandrawat U, Rani A (2016) Seasonal Variation of Selected Metals in Particulate Matter at an Industrial City Kota, India. Aerosol Air Qual Res 16(4):990–999

Mishra AK, Maiti SK, Pal AK (2013) Status of PM 10 bound heavy metals in ambient air in certain parts of Jharia. Int J Environ Sci 4(2):141–50 http://www.ipublishing.co.in/ijesarticles/thirteen/articles/volfour/EIJES41015.pdf

Mohammed MP, Srinivas N (2013) Trace elemental composition in the atmospheric aerosols of Kakinada City, India. Sustain Environ Res 23(5):315–324

Mohanraj R, Azeez PA, Priscilla T (2004) Heavy metals in airborne particulate matter of urban Coimbatore. Arch Environ Contam Toxicol 47(2)

Mouli PC et al (2006) A study on trace elemental composition of atmospheric aerosols at a semi-arid urban site using ICP-MS technique. Atmos Environ 40:136–146

Negi BS, Sadasivan S, Mishra UC (1987) Aerosol composition and sources in urban areas in India. Atmos Environ 21(6):1259–1266

Pal R, Kumar A, Gupta A, Mahima TA (2014) Source identification and distribution of toxic trace metals in respirable dust (PM10) in Brasscity of India. Glob J Human-Soc Sci 14(5)

Pandey B, Agrawal M, Singh S (2014) Assessment of air pollution around coal mining area: emphasizing on spatial distributions, seasonal variations and heavy metals, using cluster and principal component analysis. Atmos Pollut Res 5(1):79–86

Pipalatkar P, Khaparde VV, Gajghate DG, Bawase MA (2014) Source apportionment of PM2.5 using a CMB model for a centrally located Indian city. Aerosol Air Qual Res 14(3):1089–1099

Police S, Sahu SK, Pandit GG (2016) Chemical characterization of atmospheric particulate matter and their source apportionment at an emerging industrial coastal city, Visakhapatnam, India. Atmos Pollut Res 7(4):725–733

Kumar MP, Mohan VS, Reddy JS (2008) Chemical fractionation of heavy metals in airborne particulate matter (PM) by sequential extraction procedure. Toxicol Environ Chem 90(1):31–41

Raman RS, Ramachandran S (2011) Source apportionment of the ionic components in precipitation over an urban region in Western India. Environ Sci Pollut Res 18(2):212–225

Raman RS, Ramachandran S, Rastogi N (2010) Source identification of ambient aerosols over an urban region in Western India. J Environ Monit 12(6):1330

Rao S, Rajamani NS, Reddi (2015) Assessment of heavy metals in respirable suspended particulate matter at residential colonies of Gajuwaka industrial hub in Visakhapatnam

Roy P, Kumar Sikdar P, Singh G, Kumar Pal A (2012) Source apportionment of ambient PM 10. A case study from a mining belt of Orissa. Atmosfera 25(3):311–324

Sharma SK, Mandal TK, Saxena M, Rashmi, Rohtash, Sharma A, Gautam R (2014) Source apportionment of PM10 by using positive matrix factorization at an urban site of Delhi, India. Urban Climate 10:656–670

Sharma SK, Mandal TK, Jain S, Saraswati, Sharma A, Saxena M (2016) Source apportionment of PM2.5 in Delhi, India using PMF model. Bull Environ Contam Toxicol 97(2):286–293

Shridhar V, Khillare PS, Agarwal T, Ray S (2010) Metallic species in ambient particulate matter at rural and urban location of Delhi. J Hazard Mater 175(1–3):600–607

Shukla SP (2010) Characterization of atmospheric PM 10 of a commercial area in Kanpur city (India). J Environ Res Develop 4(3)

Singh R, Sharma BS (2012) Composition, seasonal variation, and sources of PM 10 from world heritage site Taj Mahal, Agra. Environ Monit Assess 184(10):5945–5956

Srimuruganandam B, Shiva Nagendra SS (2012) Source characterization of PM10 and PM2.5 mass using a chemical mass balance model at urban roadside. Sci Total Environ 433:8–19

Srivastava A, Jain VK (2007) Seasonal trends in coarse and fine particle sources in Delhi by the chemical mass balance receptor model. J Hazard Mater 144(1–2):283–291

Srivastava A, Gupta S, Jain VK (2008) Source apportionment of total suspended particulate matter in coarse and fine size ranges over Delhi. Aerosol Air Qual Res 8(2):188–200

Sudheer AK, Rengarajan R (2012) Atmospheric mineral dust and trace metals over urban environment in Western India during winter. Aerosol Air Qual Res 12(5):923–933

Tanushree B, Chakraborty S, Bhumika F, Piyal B (2011) Heavy metal concentrations in street and leaf deposited dust in Anand city, India. Res J Chem Sci 1(5):61–66

Thambavani DS, Maheswari J (2013) Source apportionment of particulate air pollution and percentage contribution of pM 10 and PM 2.5 using chemical mass balance (CMB) Method. Chem Sc Trans 2(2):614–620

Tiwari S, Srivastava AK, Bisht DS, Bano T, Singh S, Behura S, Srivastava MK, Chate DM, Padmanabhamurty B (2009) Black carbon and chemical characteristics of PM10 and PM2.5 at an urban site of North India. J Atmos Chem 62(3):193–209

Tiwari S, Srivastava AK, Chate DM, Safai PD, Bisht DS, Srivastava MK, Beig G (2014) Impacts of the high loadings of primary and secondary aerosols on light extinction at Delhi during wintertime. Atmos Environ 92:60–68

Tripathi R M et al (2004) Vertical distribution of atmospheric trace metals and their sources at Mumbai, India. Atmos Environ 38:135–146

Winchester JW, Nifong GD (1971) Water pollution in Lake Michigan by trace elements from pollution aerosol fallout. Water Air Soil Pollut 1(1):50–64

Yadav S, Satsangi PG (2013) Characterization of particulate matter and its related metal toxicity in an urban location in South West India. Environ Monit Assess 185 (9):7365–7379

Chapter 15
A Review on Ionic Liquids as Novel Absorbents for SO_2 Removal

Avanish Kumar

Abstract SO_2 emissions are a significant source of atmospheric pollution. The natural sources such as biological decay and sea spray emit about 130 million tons of sulfur per year, and the anthropogenic sources such as coal combustion, petroleum, and smelting operations release an additional 132 million tons of sulfur dioxide annually into the atmosphere. The largest signal contribution to the anthropogenic emission of about 70% is made by coal combustion. The natural sources of sulfur dioxide are probably present in gases emitted all through the volcanic activity. Additionally, SO_2 emissions contribute to the formation of smog, which is a significant human health concern. SO_2 also induces an involuntary coughing reflex. The taste threshold limit is 0.3 ppm while SO_2 produces an unpleasant smell at 0.5 ppm. However, the identification of a material that can selectively and reversibly capture SO_2 has proven to be difficult. This paper critically discusses the recent advances of ionic liquids for SO_2 capture, including the absorption capacity, desorption performance of various other absorbents like wet lime and wet limestone. In addition, some strategies recently developed to enhance the absorption processes have been briefly introduced, such as ionic liquid mixtures, solidified ionic liquids. Moreover, the drawbacks of the industrial application of this technology have been proposed.

Keywords SO_2 · Absorption · Ionic liquids

15.1 Introduction

Increasing concern of the world for environmental protection requires no special mention (De Visscher 2013; Ghosh et al. 2017). The most important reason for this concern on air pollution is certainly related to the increasing energy demand from fossil fuels and also by burning low- to medium-grade high sulfur content lignite coals (Bhoi et al. 2016). These fuels when burnt in fired boilers emit Giga tones of SO_2 to the atmosphere each year (Arif et al. 2015). While in the form of acid

A. Kumar (✉)
Department of Chemical Engineering, Jaipur National University, Jaipur 302017, India
e-mail: avichemical@gmail.com

R. M. Singh et al. (eds.), *Environmental Processes and Management*,
Water Science and Technology Library 91,
https://doi.org/10.1007/978-3-030-38152-3_15

rains, this brings damage to the vegetation, to the buildings, monuments, and aquatic lives; in the gaseous state, it seriously affects human organs (Park and Park 2017). Probably, the first attempt on acid rain reduction was enacted through the clean air act amendment in 1990, although the legislation on SO_2 emission was introduced by the USA and Japan governments in the early seventies of the nineteenth century (Wang et al. 2005). Over the years, many other countries have framed their environmental laws on SO_2 emission control strategies and some countries, e.g., Denmark, even imposed high environmental levy per unit mass of sulfur emitted to the atmosphere (Kinney et al. 2007).

On the basis of the most recent emission estimates, India has surpassed the USA to be the world's second-largest SO_2 emitting country, after China, since 2010 (Lu et al. 2010, 2011). The coal-fired power sector is the single largest contributor, accounting for ~50% of the national SO_2 emissions and ~70% of the emission increment during 1996–2010. In particular, during 2005–2012, SO_2 emissions increased rapidly by 71% from 3354 to 5738 Gg, with an annual average growth rate (AAGR) of 8.0% (Lu et al. 2013).

For controlling the emission of SO_2, various flue-gas desulfurization (FGD) technologies have been evolved through extensive research and development over the last three decades to improve its removal efficiency, the economy of separation, and trouble-free operation (Wang et al. 2005). The existing FGD processes are conventionally classified either as once-through technology if the spent sorbent is disposed of as a waste or utilized as a by-product or as regenerative technology if the absorbent, which is regenerated for reuse and the released SO_2 is either liquefied for sale or converted to other salable products (Pandey et al. 2005). Varieties of solvents, solutions, and slurries in quite different types of contactors have been employed for the removal of sulfur dioxide from gaseous mixtures. It may be observed that the treatment of large tonnages of SO_2 using limestone slurries generates impure gypsum amounting of about three times of the quantity of SO_2 treated annually, and in most of the cases, it causes a serious disposal problem, this being a solid waste. Figure 15.1 represents the schematic representation of a simplified post-combustion SO_2 capture system. Also, the absorption of SO_2 from flue gases in calcium-based slurries is almost exclusively carried out in spray scrubbers in the concerned industries, and there are reports that several critical operational problems arise. In the open literature, experimental performance data of a slurry bubble column contactor for this system is not reported while it is conventionally used for other slurry reactions as mentioned above. In all probability, the use of a three-phase foam-slurry reactor for this system with the addition of a suitable surfactant (Rennie and Evans 1962; Kalekar and Bhagwat 2006) in a very low concentration level would be advantageous, as it may obviate the choking problem. Also, experimental data on the absorption of SO_2 in IL solvents in conventional laboratory-scale process equipment as functions of different operating variables reported in the open literature are meager. Compared to the traditional industrial solvents, ionic liquids present significant advantages, especially in complex systems, as they could be tuned by appropriate selection of the structures of

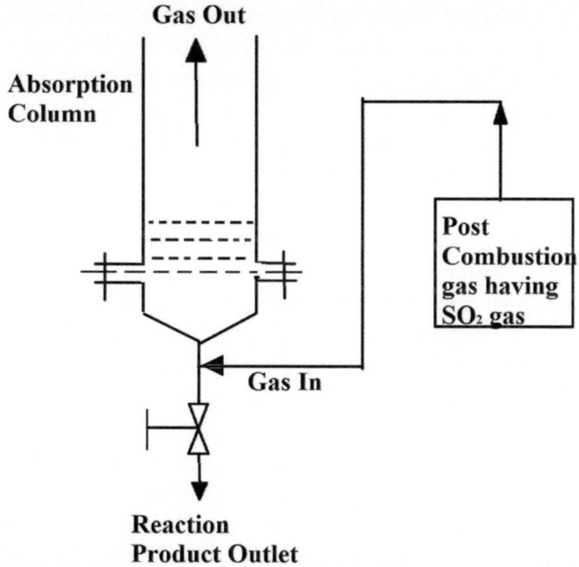

Fig. 15.1 Schematic representation of a simplified post-combustion SO_2 capture system

their cations and anions (Shiflett and Yokozeki 2009). In this work, ionic liquids used for SO_2 capture and separation from flue gas or mixed gas have been briefly summarized. Therefore, this paper reviews the advances in different ionic liquid solvents for SO_2 removal and adsorption mechanism of various ionic liquids.

15.2 SO_2 Absorption from Various Absorbents

15.2.1 Calcium-Based Absorbents

Sada et al. (1977) studied simultaneous absorption of SO_2 and CO_2 into $Ca(OH)_2$ slurry in a stirred-tank reactor with a plane gas–liquid interface both singly and separately. The authors found that in the process of simultaneous absorption of SO_2 and CO_2, the solubility of sulfur dioxide is being more than 25 times larger than carbon dioxide, and then CO_2 could almost be regarded as an inert gas. The reaction of SO_2 with hydroxyl ion was considered to be instantaneous, and the reaction rate was found to be much greater than the corresponding reaction with CO_2. The rate constant for the reaction involving SO_2 was considered to be larger by a factor of 10^4 than that involving CO_2. The authors proposed a two reaction-plane model for absorption of SO_2 in $Ca(OH)_2$ slurry considering that the reaction of various species occurs at two different reaction planes. The authors argued that the model of Uchida et al. (1975) was better in its predictions than the model reported by Ramachandran and Sharma (1969).

Sada et al. (1979) performed the absorption for lean SO_2 and NO_2 by exploiting aqueous slurries of fine $Ca(OH)_2$ and $Mg(OH)_2$ particles in a stirred vessel with a gas–liquid interface at a temperature of 25 °C and 1 atm pressure. That absorption process using $Ca(OH)_2$ slurries was found to be almost gas-film controlled. On the contrary, for $Mg(OH)_2$ slurry system, the absorption process was diffusion controlled across the liquid film, and *the absorption rate increased with solid concentration.*

Sada et al. (1981) performed the absorption of dilute SO_2 into an aqueous slurry of $CaSO_3$ using a stirred vessel with a plane gas–liquid interface. It was assumed that reaction on the solid surface facilitated the SO_2 absorption which helped in deriving the enhancement factor. The reaction between absorbed SO_2 with dissolved solid was assumed to be an irreversible instantaneous reaction, and between absorbed SO_2 with the solid particle, it was irreversible first order.

Sada et al. (1983) investigated CO_2 absorption using $Ca(OH)_2$ and $Mg(OH)_2$ slurries in a bubble column reactor at a temperature of 308 K and 0.1013 MPa pressure. Volumetric mass transfer coefficient and effective gas–liquid interfacial areas were determined. It was found that with an increase in the concentration of slurry liquid-side physical mass transfer coefficient decreased. Effective gas–liquid interfacial area and the rate of chemical absorption reached their maximum values at an optimum concentration of lime reactant particles.

Bravo et al. (2002) studied the absorption of SO_2 from its mixtures with N_2 into limestone slurries. The reactor used was a stirred tank with an almost flat gas–liquid interface of area 67.7×10^{-4} m^2, and the liquid was stirred at 400 rpm. A semi-batch reactor was used for experiments in which the gas phase was passed continuously through a batch of $CaCO_3$ slurry. The experimental data demonstrated a considerable reduction in the rate of sulfur dioxide absorption as the $CaCO_3$ disappeared by the reaction from the slurry and that the flux of SO_2 across the gas–liquid interface did not change significantly with the operating temperature. The SO_2 flux across the gas–liquid interface was also observed to be practically independent of the $CaCO_3$ content by weight in the feed slurry and of the particle size of $CaCO_3$ in it, within the range tested, although the one assayed in this study was limited.

Bjerle et al. (1972) investigated the absorption of SO_2 into $CaCO_3$ slurry using a laminar jet absorber and determined the value of gas-phase mass transfer coefficient from experimental data. The liquid in this type of absorber had very short contact time with the gas, and solid dissolution seemed to not effect on the rate of the absorption. Under the experimental conditions, unsteady-state mass transfer occurred into the liquid jet and the area for mass transfer in a liquid jet being well-defined, penetration theory was used by the authors for the theoretical determination of the rate of gas-phase mass transfer and hence the value of mass transfer coefficient.

Dou et al. (2008) used an electrostatic spraying absorber (ESA) as the reactor, where sulfur dioxide was absorbed into an aqueous slurry of reactive $Ca(OH)_2$. The absorption process was measured by using the two-film theory of mass transfer. The absorption process was reported to be controlled by both liquid and gas side resistances. The ESA characteristics were examined for the applied voltage ranging from -10 to 10 kV at various slurry flow rates. The removal efficiency of sulfur dioxide did not depend on the polarity of the applied voltage. Slightly greater efficiency was achieved with the conduction charging configuration as compared to the induction charging configuration. A model for external mass transfer incorporating an enhancement factor was proposed for the approximation of the absorption efficiency. Predicted values of SO_2 removal efficiency were reported to produce a good agreement with the experimental data.

Dou et al. (2009) performed experiments in a spray tower for the sulfur dioxide removal from flue gases using limestone slurry. Sulfur dioxide concentration in the flue gas was about 2500 ppm. The parameters studied by the investigators included droplet size, pH value of the slurry, gas and liquid phase flow rates, and limestone concentration. Two-film theory was assumed to define the mass transfer mechanism. SO_2 removal efficiency was found to increase as the pH value of the slurry was increased from 5.0 to 6.0. Absorption of SO_2 at lower pH values was reported to increase the utilization efficiency of absorbent and eliminate the sulfite ions responsible for scaling and plugging problems. The SO_2 removal efficiency reduced substantially when the pH was too low and led to very high residual limestone content in the gypsum produced. The authors observed that the SO_2 removal efficiencies greatly increased with increasing the L/G ratios, and it reached 96% at L/G ratios of higher than 13.

Liu and Xiao (2006) performed the absorption of SO_2 into limestone slurry containing suspended reactive particles in a bubbling reactor with continuous feeding of both gas and liquid phases at a constant pH and high temperature (50 °C). An absorption model with a single reaction plane based on the film model was developed. The effects of limestone particle size, concentration of limestone, acetic acid additive, and inlet SO_2 concentration on the concentration distribution of different chemical species in the liquid film and SO_2 absorption rate were investigated. Increasing the concentration of limestone slurry, use of acetic acid additives into the system, decreasing the limestone particle size or inlet SO_2 concentration caused the reaction plane in the liquid film to shift toward the gas–liquid interface. Model predictions were found to fit the experimental data well.

15.2.2 Ammonia-Based Absorbents

Gao et al. (2010) studied the gas–liquid system characteristics for SO_2 absorption in ammonium sulfite solution. A stirred-tank reactor was used for performing the experiments. The authors found that the absorption was controlled by both the gas and liquid films when ammonium sulfite concentration was lower than 0.05 g mol/l and mainly by gas film at a greater temperature. It was observed that the absorption rate was of zero-order in regard to ammonium sulfite and the rate increased as the concentration of sulfur dioxide was increased. The reaction was of 0.6th order with respect to the concentration of SO_2. The absorption rate was also found to increase with an increase in the temperature of operation. Under the experimental conditions, the absorption rate of SO_2 increased with an increase in the $(NH_4)_2SO_3$ concentration from 0.03 to 0.05 g mol/l. From this observation, the authors concluded that the reaction was controlled concurrently by both gas film and liquid film. On the contrary, the absorption rate of SO_2 became nearly constant as the concentration of $(NH_4)_2SO_3$ was raised above 0.05 gmol/l. In the latter case, the reaction might not be influenced by the liquid- film and appeared to be zero-order with regard to the concentration of $(NH_4)_2SO_3$.

Hikita and Konishi (1978) performed the pure SO_2 absorption into aqueous ammonia and aqueous ammonium sulfite solutions in a liquid jet column at a temperature of 25 °C. Penetration theory was used by the authors for the development of a model for determining the rate of absorption. The reaction was considered to be instantaneous, and two reaction-plane model was used to describe the reaction mechanism. The theoretical rate of absorption predicted from the model was found to agree well with the experimental data.

Tang et al. (2004) presented a novel FGD process in which an organic amine was applied to absorb SO_2 from flue gas in a bubbling contactor. The vapor–liquid equilibrium model of absorption of SO_2 from the flue gas by an organic amine was recognized, and the vapor–liquid equilibrium of the sulfur dioxide–ethylenediamine–phosphoric acid–water system was first predicted. For a 0.3 g mol/l ethylenediamine buffer solution system, the predicted values were found to validate with the experimental data well. It was established that their model was significantly good to forecast vapor–liquid equilibrium under the experimental conditions. Using an ethylenediamine–phosphoric acid aqueous solution as the absorbent, some operating conditions including temperature, gas-to-liquid ratio, pH value reaction medium, concentration of the absorbing reagent and liquid flow rate used in the absorption and de-absorption studies were performed in a laboratory packed tower. The optimized conditions calculated for the system studied were stated. The authors claimed that the novel FGD method reported in this paper had levels of high desulfurization efficiency along with low investment.

Li et al. (2013) used a bubble column reactor for studies of the absorption of SO_2 from a gaseous mixture into aqueous NH_4HCO_3 solutions with different concentrations accompanied by an instantaneous irreversible chemical reaction. The investigators also proposed mass transfer models based on Danckwerts surface renewal model as well as the penetration theory for surface stretch proposed by Angelo et al. (1966). They obtained analytical expressions for time-average mass transfer coefficient and enhancement factor. The model predicted values were found to agree well with the experimental results.

15.2.3 Ionic Liquids Used for SO_2 Capture

The technologies for the capture of SO_2 have attracted increasing interests, among which flue-gas desulfurization (FGD) is the most efficient way. The conventional flue-gas desulfurization technologies include wet scrubbing using calcium-based absorbents, dry scrubbing, and absorption by aqueous amines. However, there are large amounts of non-renewable by-products such as calcium sulfate and the consumption of water is huge during the wet scrubbing process. The dry scrubbing process is simple, but limited in industry due to the high ratio of Ca/S, low efficiency in removal of SO_2 and the high cost in recovery of desulfurizer. In addition, secondary pollution is the biggest problem to be solved for the absorption by aqueous amines, as the amines are easy to volatilize to flue gas with water. The reaction product generated in these processes comprises a mixture of calcium sulfite, its sulfate and unreacted lime, not considered to be a useful byproduct (Karatepe 2000). The process suffers from low efficiency and requires disposal of large tonnages of solid waste, and therefore not consistent with the principles of sustainable development. Therefore, the development of recyclable absorbents with low volatility and high capacity is of great significance.

Various regenerative absorbents that are used for SO_2 absorption include magnesium oxide, ammonia-based absorbents, sodium alkali-based absorbents. Aqueous sodium sulfite and sodium bisulfite solutions reported to have been used for the absorption of sulfur dioxide from flue gas. Regeneration of the solvent produced a concentrated SO_2 stream, which was converted into a saleable product such as liquid SO_2, sulfuric acid, or elemental sulfur.

Ionic liquids (ILs) are considered as promising substitutes for traditional organic solvents in many fields due to their excellent properties, such as negligible vapor pressure, tunable structure, and high gas solubility. Ionic liquids are molten organic salts at ambient temperature and consist of big organic cations and inorganic or organic anions. Because of the distinct properties such as negligible vapor pressure, high thermal stability, and excellent solvent power for a varied range of both organic and inorganic materials, ionic liquids have been considered environment-friendly solvents as water and supercritical CO_2. As a result, ILs are considered as a new generation "green solvent." It offers an extensively high potential application in many fields like catalysis (Hallett and Welton 1999), electrolysis (Armand et al. 2009), extraction (Huddleston et al. 1998), membrane separation (Noble and Gin 2011), biocatalysts, nanophase materials, and (Yang and Pan 2005) gas absorption (Bates et al. 2002; Duan et al. 2011; Guo et al. 2011; Shiflett and Yokozeki 2009). As far as the application of ILs in gas absorption is concerned, ILs have been proposed to be more effective and environmentally benign alternative to traditional desulfurization solvents for the separation of SO_2 from flue gas, according to some pioneering works (Huang et al. 2006; Qu et al. 2013; Wu et al. 2004). Table 15.1 represents different types of ionic liquids and their method of preparation along with their absorption mechanism. Table 15.2 represents the comparison of ionic liquids with commercially available solvents in terms of price and SO_2 capture.

15.2.4 Hydroxyl Ammonium Ionic Liquids

Huang et al. (2013) synthesized less viscous IL: hydroxylammoniumdicarboxylate ionic liquids (ILs), dimethyl ethanolammoniumdimalate ([DMEAH][dimalate]),

Table 15.1 Various ionic liquids and their syntheses

S. No.	Ionic liquid	Material required	Method of preparation	Solubility	Regeneration	Absorption mechanism	References
1	Hydroxy ammonium ionic liquid	Base: 2-aminoethanol, tri-ethanol amine Acid: formic acid, lactic acid, acetic acid, Other: ethanol	2-hydroxyethylammonium lactate is described here: 0.5 mol of 2-aminoethanol was dissolved in 100 mL of ethanol to form a liquid mixture and loaded into a 500-mL flask. The flask was placed in a water bath of 298.2 K and equipped with a reflux condenser under vigorous stirring with a magnetic stirrer. A mixture of 0.5 mol of lactic acid dissolved in 100 mL of ethanol was added dropwise to the flask in about **90 min**. The reaction lasted for **2 h**. The solvent was removed by evaporation under vacuum. The resulting crude residue was dissolved in 100 mL of ethanol, treated with active carbon, and filtered. A colorless product was obtained after evaporation and dried under vacuum at 323.2 K for 48 h	tri-(2-hydroxyethyl)ammonium lactate is 0.4957 (mole fraction), corresponding to 0.208 (mass fraction)	Recycled	Chemical	Yuan et al. (2007)
2	Phosphate ionic liquids	1-Methylimidazole, Tributyl phosphate, and Triethyl phosphate	Equimolar alkylphosphonate and 1-ethylimidazole were added into 250-mL flask; the reaction mixture was carried out stirring by a magnetic stirrer for **10 h** at 353.15 K. The resulting liquid was collected and washed with ethyl ether. Light brown liquid was obtained after evaporation and dried under vacuum at 353.15 K for **24 h**	Saturated absorption capacity of SO2 in [Bmim][DBP] = 2.8 [Emim][DEP] = 2.7 mol per mole ILs	Recycled	Physical	Qu et al. (2013)

(continued)

Table 15.1 (continued)

S. No.	Ionic liquid	Material required	Method of preparation	Solubility	Regeneration	Absorption mechanism	References
3	Binary mixture of water and caprolactam tetrabutylammonium bromide ionic liquid (CPL-TBAB IL)	White crystalline caprolactam powder, Tetrabutylammonium bromide	The CPL-TBAB ionic liquid was synthesized as previously reported by Guo et al. (2010). Ionic liquid aqueous solutions were prepared by mixing the ionic liquid and de-ionized water	The solubility (g/100 g solutions) of SO_2 in CPL-TBAB/water (4 mol L^{-1}) is 52.29	Recycled	NA	Duan et al. (2011)
4	Caprolactam tetrabutylammonium bromide ionic liquids	White crystalline caprolactam powder, Tetrabutylammonium bromide	The ILs of different mole ratio of caprolactam and tetrabutylammonium bromide was synthesized following procedures reported elsewhere. The [CPL][TBAB] ILs were dried under vacuum at 323.15 K until the mass remained constant	The mole fraction solubility of SO_2 x in [CPL][TBAB] (1:1) is 0.680	Recycled	NA	Guo et al. (2010)
5	TMG lactate IL	Base: 1,1,3,3-tetramethylguanidine (TMG), Acid: lactic acid	The IL was synthesized by direct neutralization of 1,1,3,3-tetramethylguanidine (TMG) and lactic acid. To synthesize the IL, ethanol (100 mL) and TMG (2.30 g, 20.0 mmol) were loaded into a 250-mL flask in a water bath of 258C. A solution of lactic acid (20.0 mmol) in ethanol (35 mL) was then added into the stirring reaction mixture, and the reaction was allowed to proceed for **2 h**. The solvent was then removed by evaporation under reduced pressure. The resulting crude oily residue was dissolved in ethanol (100 mL), treated with active carbon, filtered; the solvent was removed by evaporation under vacuum	Equilibrium the molar ratio of SO_2 to IL 0.305 g SO_2 g^{-1} IL	NA	Chemical	Wu et al. (2004)

(continued)

Table 15.1 (continued)

S. No.	Ionic liquid	Material required	Method of preparation	Solubility	Regeneration	Absorption mechanism	References
6	$[BMIM][BF_4]$ $[BMIM][PF_6]$ $[TMG][BF_4]$ [TMG]L [MEA]L	([BMIM][BF4]) IL ([BMIM][PF6]) IL	• 1-Butyl-3-methylimidazolium tetrafluoroborate ($[BMIM][BF_4]$) and 1-butyl-3-methylimidazolium hexafluorophosphate ($[BMIM][PF_6]$) were purchased • $[TMG][BF_4]$-neutralization of TMG with fluoroboric acid (HBF4) in ethanolic solution Huang et al. (2006) • [TMG]L refer Wu et al. (2004)	Mole ratio of SO_2 to IL at quilibrium $[BMIM][BF_4]$ 0.713 $[BMIM][PF_6]$ 0.532 $[TMG][BF_4]$ 0.482 [TMG]L 1.621 [MEA]L 0.903	NA	Chemical and Physical	Ren et al. (2010)
7	Caprolactam-based eutectic ionic liquids	CPL-acetamide (1:1) CPL-imidazole (1:1) CPL-furoic acid (1:1) CPL-benzoic acid (1:1) CPL-o-toluic acid (2:1)	CPL-acetamide (1:1) is described here: under the protection of N2, 0.1 mol of acetamide (solid powder) and 0.1 mol of CPL (solid powder) were fully mixed in a 50-mL three-necked flask (they were melted slowly when uniformly mixed), the mixture was stirred at 90 °C for 15–30 min, until the mixture was clear, then dried at 100 °C under vacuum (ca 4.24 kPa) for 12 h, cooled to w.r.t. in a vacuum drier	CPL-acetamide (1:1) = 0.624 CPL-imidazole (1:1) = 0.495 CPL-furoic acid (1:1) = 0.28 CPL-benzoic acid (1:1) = 0.26 CPL-o-toluic acid (2:1) = 0.26 g/g of mass fraction	Completely recycle	CPL-acetamide (1:1) = physical CPL-imidazole (1:1) = Chemical Rest threes of not mentioned in the paper	Liu et al. (2013)

(continued)

Table 15.1 (continued)

S. No.	Ionic liquid	Material required	Method of preparation	Solubility	Regeneration	Absorption mechanism	References
8	Guanidinium-based ionic liquids (GBILs)	Base: 1,1,3,3-Tetramethylguanidine (TMG), 2,2,2-Trifluoroethanol	[TMG][PHE] was synthesized via direct neutralization of TMG and phenol. To synthesize the [TMG][PHE], ethanol (250 mL) and TMG (24.763 g, 0.215 mol) were loaded into a dried 500-mL flask in a water bath of 25 °C. A solution of phenol (20.232 g, 0.215 mol) in ethanol (100 mL) was then added, and the reaction was allowed to proceed for 12 h. The solvent was removed by rotary evaporation under vacuum at 65 °C	[TMG][IM] = 4.1 [TMG][TE] = 3.7 [TMG] = 2.5 molar ratio	Recycle	Both physical and chemical	Shang et al. (2011)
9	(2-hydroxyethylammonium formate	Basic: 2-aminoethanol Acid: formic acid	2-aminoethanol (119.8 g, 0.2 mol) was placed in a two-necked flask equipped with a reflux condenser and a dropping funnel. The flask was mounted in an ice bath. Under vigorous stirring with a magnetic stirring bar, 76 ml (0.2 mol) formic acid was added dropwise to the flask in about 45 min. Stirring was continued for 24 h at room temperature, to obtain a viscous clear liquid	NA	NA	NA	Bicak (2005)
10	Hydroxylammonium dicarboxylate ionic liquids (ILs)	Base: N,N-dimethylethanolamine Acid: malonic acid Other: ethanol	0.1 mol of malonic acid was dissolved in 50 mL of ethanol and loaded into a 250-mL flask. The flask was placed in a water bath of room temperature and equipped with a reflux condenser under vigorous stirring with a magnetic stirrer. Then 0.1 mol of DMEA in 50 mL of ethanol was added dropwise within 2 h. The reaction lasted for 24 h. Ethanol was then removed by evaporation under reduced pressure, and the resulting crude residue was washed several times with diethyl ether to remove unreacted reactants	[DMEAH][glutarate](1:1) = 0.360 [DMEAH][malonate](1.2:1) = 0.181 [DMEAH][malate](1.4:1) = 0.221 Mol/mol IL at 0.004 bar and 313 K	[DMEAH][glutarate](1:1) = 84.21 [DMEAH][malonate](1.2:1) = 95.35 [DMEAH][malate](1.4:1) = 96.25	NA	Huang et al. (2013)

Table 15.2 Comparision of ILs with commercially available solvents

S. No.	Solvent	SO_2 solubility per mole of solvent	Price (solvent cost) per mole of solvent (Rs.)	References
1	tri-(2-hydroxyethyl)ammonium lactate	0.983	329	Yuan et al. (2007)
2	1-Methylimidazole, Tributyl phosphate	2.8	16847.30	Qu et al. (2013)
3	TMG lactate IL	1.026	455	Wu et al. (2004)
4	Caprolactam tetrabutylammonium bromide ionic liquids	2.125	3105	Guo et al. (2010)
5	MgO	1	3868	Mondal (2007)
6	$CaCO_3$	1	52	Pandey et al. (2005)

dimethylethanol ammoniumdimalonate ([DMEAH][dimalonate]), and dimethylethanolammoniumdiglutarate ([DMEAH][diglutarate]). They mixed these with free DMEA and water to form novel hybrid solvents for SO_2 absorption. The mixed absorbents exhibited a low viscosity in the range of 6–13 mPa s at 313.2 K. A certain amount of SO_2 was introduced into the equilibrium cell (with an initial SO_2 partial pressure of about 1.2 bar), and the pressure change in the equilibrium chamber was recorded online until it became constant. The equilibrium solubilities of SO_2 at corresponding SO_2 partial pressures are 0.255 at 0.002 bar for [DMEAH][glutarate] (1.2:1), 0.254 at 0.004 bar for [DMEAH][malonate] (1.2:1), and 0.264 at 0.002 bar for [DMEAH][malate] (1.6:1), respectively.

As far as the regeneration behavior was concerned, 98.1% of SO_2 absorbed in [DMEAH] [malate](1:1) was released within 15 min, whereas the desorption efficiencies of SO_2-[DMEAH][glutarate](1:1) and SO_2-[DMEAH] [malonate](1:1) systems at 15 min approached to 74.1% and 93.0%, respectively.

Cui et al. (2015) synthesized acyl amino-based ILs and used for the absorption of SO_2 at 0.1–1 bar partial pressure. The absorption capacity achieved was 4.5 g mol SO_2 per g mol IL at 1 bar but at 0.1 bar it reduced to 1.7 g mol SO_2 per g mol IL. The effect of water in the ILs on the absorption of SO_2 was investigated at ambient pressure and 20 °C. SO_2 was bubbled through water to obtain SO_2 gas with 100% humidity, which was then bubbled through IL. The SO_2 absorption capacity of [P_{66614}]-[Phth] was 4.32 g mol SO_2 per g mole IL at 20 °C, and 1 bar for SO_2 gas containing 100% humidity, and this value was very close to 4.40 g mol SO_2 per g mole IL for dry SO_2 gas. Spectroscopic investigations and quantum chemical calculations showed that dramatic enhancement in SO_2 absorption capacity was originated from the strong N–S interaction and enhanced C = O–S interactions between acyl amino-based anion and SO_2. The ILs were regenerated by heating or bubbling N_2 through the SO_2-saturated ILs. The desorption of SO_2 by this functionalized ILs [P_{66614}]-[Phth] was also examined at 80 °C, 1.0 bar and 40 mL min^{-1} N_2, and it was found to be 0.62 mol SO_2 per mole IL.

Bicak (2005) prepared 2-hydroxyethylammonium formate ionic liquid by mixing equimolar mixture of formic acid and 2-hydroxyethylamine and determined its physical properties such as viscosity, ionic conductivity, heat stability, and solvation abilities. Its melting temperature was −82 °C. It was *thermally stable* up to 150 °C.

Yuan et al. (2007) synthesized nine different types of hydroxyl ammonium ionic liquids. These authors observed that the solubility of SO_2 in these ILs was quite high at ambient pressure. The solubility of SO_2 in tris-(2-hydroxyethyl) ammonium lactate was 0.4957 g mol fraction, but the solubility of SO_2 sharply decreased as the temperature was increased. They found the absorption and desorption of SO_2 in these ILs were practically reversible. When SO_2-saturated ILs are vacuumed with stirring at 298.2 K, for four times, it was observed that the ILs can be recycled and used repeatedly and that recovery of SO_2 in four times was 97.9%, 95.8%, 93.4%, and 91.3%, respectively.

Ren et al. (2011) used a task-specific IL, monoethanolaminium lactate ([MEAL]), to study the absorption of SO_2 and its oxidation by O_2 present in simulated flue gases (1.65 vol% SO_2 and 11.5 vol% O_2) with and without ash and activated carbon in [MEA]L. It was found that the presence of O_2 in the simulated flue gas did not influence the absorption of SO_2 by [MEAL], but it causes, to a very small extent, the oxidation of SO_2. The authors also observed the effect of sulfuric acid in [MEAL] on the absorption of SO_2 and the regeneration of the exhausted [MEAL]. The main reason was the possible oxidation of SO_2 and its conversion to sulfuric acid when water vapor remained in the flue gas. It was observed that the presence of sulfuric acid could reduce the absorption of SO_2 in [MEAL] greatly and affect the reuse of the IL. However, when NaOH, CaO, or $CaCO_3$ was added into the mixture of [MEAL] and H_2SO_4, sulfuric acid was altered into the corresponding salts, precipitated and separated by filtration, and then [MEAL] could be regenerated (Ren et al. 2012). When SO_2 dissolves in the ILs, it may decrease the interaction between the cation and anion, especially the Coulombic interactions, which are the dominant forces in ILs. Also, the viscosity of ILs arises from the internal friction, which is governed by weak interactions, such as hydrogen bonds, van der Waals forces, π–π interactions, and Coulombic interactions.

Huang et al. (2013) determined the solubility of SO_2 in [N_{2224}][dimalonate], [N_{2224}][disuccinate], and [N_{2224}][dimaleate]. These could absorb 0.2, 0.04, and 0.12 g of SO_2 per gram of IL at pressures as low as 5 kPa, respectively. The IL film was coated on the QCM crystal of 5.0 MHz. During the experiments, the QCM frequency was recorded when SO_2 was injected into the measuring cell of the QCM apparatus and kept at a pressure of 40 kPa. After the frequency was reduced to a steady state, the absorption equilibrium was thought to be realized.

Regeneration of saturated IL was done by heating the sample to 80 °C and evacuating the system to 0.1 kPa for 1 h. The absorption experiments are repeated three times. It can be observed that the regenerated [N_{2224}][dimalonate] and [N_{2224}][dimaleate] can absorb the same amount of SO_2 as the fresh ones (with a maximum difference of 1.7% and 0.7%, respectively), indicating the ASILs can also be completely regenerated by vacuuming and heating, which demonstrates the reversibility of SO_2 absorption at the pressure range from 0.1 to 180 kPa.

15.2.5 Gaunidinium-Based Ionic Liquids

Wu et al. (2004) described a novel method for desulfurization of a simulated flue gas in which an IL: 1,1,3,3-tetramethylguanidinium lactate (TMGL) was used to absorb SO_2. The IL was synthesized by direct neutralization of 1,1,3,3-tetramethylguanidin (TMG) with lactic acid using ethanol as the solvent. To gain some knowledge on the nature of the interaction between the virgin IL and the SO_2 g molecules in the SO_2 absorbed in the IL, the investigators performed ^{H}NMR and FTIR analyses of the absorbent before and after absorption of SO_2. New bands formed at 1230 and 957 cm^{-1} in the product g molecules were assigned to sulfate S = O, and S–O stretches, respectively. The equilibrium SO_2 absorption capacity of this IL for 8% by volume of SO_2 with N_2 at 40 °C and 1 atm total pressure was 0.978 g mol of SO_2 per g mol of IL. For pure SO_2, at 40 °C and 1.2 atm total pressures this ratio was observed to be 1.7. The absorbed SO_2 could be reversibly desorbed, and the regenerated solvent could be reused for absorption of SO_2, and the cycle could be repeated. The practical utility of this IL was limited because of its relatively low thermal stability. Hence, only a fraction of the absorbed gas could be thermally released before degradation of the IL occurred (Duan et al. 2011).

Huang et al. (2008) also performed experiments to absorb SO_2 from a mixture of 10% SO_2 in N_2 in 1,1,3,3-tetramethylguanidine (TMG)-based ILs, [TMGH]BF_4, [TMGHPO]BF_4, [$TMGHPO_2$]BF_4, [TMGH]Tf_2N, and [$TMGHB_2$]Tf_2N and got the absorption capacities of 0.06, 0.131, 0.167, 0.057, and 0.074 g mol fraction, respectively. Desorption of gas from the ILs saturated with SO_2 (1 bar, 20 °C) was carried out by heating the tube with the gas-outlet valve open at fixed temperatures in the interval 20–140 °C (±0.1 °C) in a step-wise manner. SO_2 removal remained consistent after six cycles of desorption.

Huang et al. (2006) examined the absorption of SO_2 gas in TMG (Tetramethyl guanidine), and BMIM (1-butyl-3-methylimidazolium)-based ILs. The SO_2 absorption capacity of the ILs using pure SO_2 gas at 1 bar and 20 °C, g mol of SO_2 per g mol of IL, was found to be 1.33, 1.50, 1.27, 1.18, 1.60 in [BMIM][BTA], [BMIM][BF_4], [TMG][BF_4], [TMG][BTA], and[$TMGB_2$][BTA], respectively. However, when experiments were performed with 10 g mole percent SO_2 in N_2, only 0.007, 0.005, 0.064, 0.061, 0.080 g mol of SO_2 per g mol of IL was absorbed at the same pressure and temperature.

Shang et al. (2011) used new anions for the synthesis of three new guanidine-based ionic liquids (GBILs). These investigators synthesized [1,1,3,3-tetramethylguanidinium][phenol] ([TMG][PHE]),

[1,1,3,3-tetramethylguanidinium][imidazole] ([TMG][IM]), and [1,1,3,3-tetramethylguanidinium][2,2,2-trifluoroethanol] ([TMG][TE]) and determined the SO_2 absorption and desorption characteristics of these ILs. The absorption capacities of ILs were found to be 2.58–4.132 g mol SO_2/g mol IL at 20 °C and 2.24–3.17 g mol SO_2/g mol IL at 40 °C. CO_2 absorption characteristics of these new ILs were also studied. It was observed that only 0.502 g mol CO_2 per g mol IL was absorbed at 20 °C. By performing ^{1}H NMR, ^{13}C NMR, and FTIR analyses, the investigators concluded that the synthesized GBILs could absorb SO_2 by both physical and chemical interactions while physical absorption played an important role at the experimental conditions used by them. Low temperature favored high absorption capacity and could be reused after regeneration. The desorption was carried out at 40 and 100 °C under vacuum, respectively.

Ren et al. (2010) found that task-specific ILs ([TMG]L and [MEA]L) could chemically absorb SO_2 when the g mole ratio of SO_2 to the IL was less than 0.5; when the g mole ratio was greater than 0.5, the IL could physically absorb SO_2. The normal ILs ([BMIM][BF_4], [BMIM][PF_6], and [TMG][BF_4]) could only physically absorb SO_2. For the task-specific ILs to absorb SO_2, before at a g mole ratio of SO_2 to IL less than 0.5, the viscosity and density increase, and the conductivity decrease with an increase of the g mole ratio of SO_2 to IL. After that, the conductivity and density increase, and the viscosity decreases with further increase in the g mole ratio of SO_2 to IL. However, for the normal ILs, the conductivity and density increases and the viscosity decreases with an increase of the g mole ratio of SO_2 to IL.

15.2.6 Caprolactam-Based Ionic Liquids

Guo et al. (2010, 2011) synthesized caprolactam tetrabutylammonium halide ionic liquid by using different g mole ratios of caprolactam (CPL) and tetrabutylammonium bromides (TBAB) and used for the absorption studies of SO_2 and H_2S. The solubility of SO_2 in g mole fraction unit in [CPL][TBAB] (1:1) was reported to be 0.680 at 298.2 K and decreased to 0.351 at 373.2 K. Both absorption and desorption of H_2S gas in the three examined ILs (the mole ratio of CPL and TBAB = 1:1, 3:1, and 5:1) were relatively fast, providing complete absorption in 1 h with pure H_2S gas (10 mL^3 min^{-1}) and complete gas desorption in 50 min at 303.2 K and 10.1 kPa. Because ionic liquid CPL-TBAB itself is of a good thermal stability, it is reasonable that CPL-TBAB has good reusable performance in absorption and desorption of H_2S gas.

Duan et al. (2010) measured the pH of a binary mixture of CPL-TBAB IL and solvents (water, ethanol, and 2-propanol) in the range of ionic liquid concentrations from (5.0×10^{-3} to 0.80) g mol L^{-1} and temperature range from 296.15 to 325.65 K. The results showed that the range of the pH values was from 5.12 to 6.93. The authors also used the pure ionic liquid for the absorption of NO and NO_2 gas (Duan et al. 2011). The binary mixture of CPL-TBAB IL and water was used by the same author for the absorption of SO_2. The CPL-TBAB ionic liquid and water are completely miscible in all proportions under ambient conditions. A second phase appeared when SO_2 was introduced into the solution. The solubility of SO_2 in CPL-TBAB/water (4 g mol l^{-1}) was 0.523 g per g of the solution at 293.2 K. Physical properties of CPL-Halide IL, such as *melting points, heats of fusion, and heat capacities,* were measured by differential scanning calorimeter (DSC) (Jiang et al. 2013).

Liu et al. (2013) prepared five eutectic ionic liquids (EILs) using caprolactam (CPL) and low g molecular weight organic compounds (acetamide, imidazole, furoic acid, benzoic acid, o-toluic acid) which were denoted as CPL-acetamide (1:1), CPL-imidazole (1:1), CPL-furoic acid (1:1), CPL-benzoic acid (1:1), and CPL-o-toluic acid (2:1), respectively. Their properties such as melting point, density, viscosity, conductivity, and decomposition temperature and the solubility behavior of SO_2 in these EILs at 30–70 °C were explored. The viscosity of the EILs was found to decrease in the order: CPL-furoic acid (1:1) > CPL-o-toluic acid (2:1) > CPL-benzoic acid (1:1) > CPL-acetamide (1:1) > CPL-imidazole (1:1) at all the measured temperatures. The solubility of pure SO_2 in CPL-imidazole (1:1) was 0.624 g/g of mixture mass fraction at 30 °C and 1 atm, while it decreased to 0.341 g/g of the mixture at 70 °C. The SO_2 absorption capacities of CPL-organic amines were higher than CPL-organic acids. CPL-acetamide (1:1), as a model EIL, reached saturated absorption in 30 min at 30 °C and 1 atm with a stream of 30 mL/min pure SO_2 and complete desorption in 20 min at 70 °C under vacuum (4.24 kPa).

15.2.7 Pyridinium-Based Ionic Liquids

Anderson et al. (2006) measured the solubility of SO_2 in 1-n-hexyl-3-methylimidazolium bis(trifluoromethylsulfonyl) imide ([hmim][Tf_2N]) and 1-n-hexyl-3-methylpyridinium (bistrifluoromethylsulfonyl) imide ([hmpy]-[Tf_2N]) in the temperature range 25–60 °C and at pressures up to 4 bar. The experiments were carried out in a magnetic suspension balance. It had a working range of a vacuum of 500 bar and –200–350 °C with an accuracy of 2 μg. The absence of any major impurity in the ILs was confirmed by ^1H, and ^{13}C NMR analyses and approximate water content were determined by Karl Fischer titration or by coulometry. The investigators used the virial equation of state for estimation of the density of the SO_2 gas. The authors also showed that although the solubility of SO_2 was much larger than CO_2 in these ILs at a particular temperature and pressure, the numerical value of this quantity is quite close when compared at the same reduced pressure defined as $P_{\text{reduced}} = P_{\text{actual}}/P_{\text{saturation}}$.

Zeng et al. (2014) synthesized a series of thermally stable pyridinium-based ILs, viz. N-butylpyridinium tetrafluoroborate ([C_4Py][BF_4]), N-octylpyridinium tetrafluoroborate ([C_8Py][BF_4]), N-hexylpyridinium tetrafluoroborate ([C_6Py][BF_4]), 1-butyl-3-methylpyridinium tetrafluoroborate ([C_4^3MPy][BF_4]), 1-octyl-3-methylpyridinium tetrafluoroborate ([C_8^3MPy][BF4]), 1-hexyl-3-methylpyridinium tetrafluoroborate ([C_6^3MPy][BF_4]), Nbutylpyridiniumthiocyanate ([C_4Py][SCN]), and N-butylpyridiniumbis (trifluoromethylsulfonyl)imide ([C_4Py][Tf_2N]). It was found that among the investigated ILs [C_2Py][SCN] had the highest absorption capacity of 0.841 g SO_2 per g of IL at 1 atm and 25 °C and claimed that this was much higher than those reported for most of the imidazolium-based ILs. Furthermore, the higher selectivity for SO_2/CO_2, SO_2/N_2, and SO_2/O_2 of the [C_4Py][SCN] indicated that [C_4Py][SCN] had excellent selectivity for SO_2 absorption compared to other imidazolium-based ILs. Especially SO_2/CO_2 selectivity of [C_4Py][SCN] was found to be two times higher than that of [C_4Py][BF]. Meanwhile, the presence of water in the ILs had a slightly negative effect on the SO_2 absorption capacity, but the absorption capacity of [C_4Py][SCN] hardly altered with and without water. The spectroscopic investigations depicted that the SO_2 *was physically absorbed* by the pyridinium-based ILs. The ILs were regenerated by bubbling pure N_2 at 80 °C through the IL that captured SO_2 in the glass container, and the flow rate was 140 mL min^{-1}. The absorption capacity of [C_4Py][BF_4] and [C_4Py][SCN] keeps steady during multiple absorption and desorption cycles.

15.2.8 Imidazolium-Based Ionic Liquids

Shiflett and Yokozeki (2009) developed a ternary equation of state (EOS) model for the CO_2/SO_2/1-butyl-3-methylimidazoliummethyl sulfate ([BMIM][$MeSO_4$]) system. The authors proposed that the enhanced selectivity of [BMIM][$MeSO_4$] for CO_2 over SO_2 was significantly higher than 1-hexyl-3-methylimidazoliumbis(trifluoromethylsulphonyl)imide ([HMIM][Tf_2N]). Through this work, the investigators made a quantitative demonstration for the selectivity and *simultaneous capture* of CO_2/SO_2 mixture.

Mondal and Balasubramanian (2016) examined the anion dependence of SO_2 capture in eight ionic liquids, 1-butyl-3-methylimidazolium ([BMIM]$^+$) as the cation, and chloride ([Cl]$^-$), nitrate ([NO_3]$^-$), tetrafluoroborate ([BF_4]$^-$), hexafluorophosphate ([PF_6]$^-$), triflate ([CF_3SO_3]$^-$), bis(trifluoromethanesulfonyl)imide ([NTf_2]$^-$), acetate ([OAc]$^-$), and thiocyanate ([SCN]$^-$) as anion, using g molecular dynamics simulations and quantum chemical calculations. On the basis of the salvation-free energy and binding energies of SO_2 in ILs, they found the IL containing the *thiocyanate* anion to be the most suited for SO_2 absorption.

Lee et al. (2010a) synthesized BMIm (1-bytul-3-methylimidazolium), HMIm (1-hexyl-3-methyl imidazolium), and EMIm (1-ethyl-3-methyl imidazolium)-based ionic liquid by varying the halide group, viz. [BMIm]Br, [BMIm]Cl, [BMIm]I, [EMIm]Cl, [HMIm]Cl, [OMIm]Cl, [BMPyri]Cl, and [BMPyrr]Cl, and the solubility of SO_2 was found to be 2.06, 2.11, 1.91, 2.03, 2.19, 2.19, 2.22, and 2.22 g mol SO_2 per g mol IL, respectively. The solubility of SO_2 increased in the order of Br > Cl > I. It was proposed that the primary interaction of the halide occurs with the C_2–H of the imidazolium and the S atom of SO_2.

In comparison with conventional ILs, functional ILs, such as [NH_2p-bmim][BF_4] and 2-(2-hydroxyethoxy)ammonium acetate, are highly efficient ILs for CO_2/SO_2 absorption (Zhai et al. 2010). However, the viscosity of the functional ILs is high, which would significantly influence the gas–liquid mass transfer and then limit their eventual use in large-scale gas scrubbing applications.

Li et al. (2017) synthesized amine-functionalized ILs [NH_2emim][OAc] and [NH_2emim][BF_4]. The conventional imidazole-based ILs [bmim][OH] and [bmim][BF_4] with low viscosity were also prepared. Subsequently, the amine-functional ILs were mixed with the conventional ILs to investigate the absorption performance. In this study, the SO_2 absorption capacities of IL mixtures were determined using simulated flue gas with lean SO_2 gas. The investigators observed that the mixture of ionic liquids absorbed SO_2 reversibly, and the absorption capacity *remained unchanged after even 12 cycles*, but the viscosity and density of the mixtures increased by 2–3% in each cycle.

Qu et al. (2013) synthesized two kinds of phosphate ionic liquids and studied the absorption capacity for SO_2. Absorption capacities of [Bmim][DBP] and [Emim][DEP] were found to be 2.8 and 2.7 g mol SO_2 per g mole ILs, respectively at ambient temperature and normal pressure. After four cycles of absorption and desorption, absorption capacity of the ILs was found to remain unchanged. According to the structure of two phosphate-based ionic liquids before and after SO_2 absorption, confirmed by FTIR spectrum and ^{1}H NMR analyses, absorption of SO_2 occurred by purely physical absorption. It was concluded that the anion containing the phosphate-based group might had a great SO_2 philicity, with the free electrons on the oxygen which interact with Lewis acidic sulfur of SO_2. As a result, the absorption capacity of SO_2 in the ionic liquids was improved.

Tian et al. (2014) synthesized two kinds of hydrophobic task-specific ILs, 1-(2-diethyl-aminoethyl)-3-methylimidazolium hexafluorophosphate ([Et_2NEmim][PF_6]) and 1-(2-diethyl-aminoethyl)-1-methylpyrrolidinium hexafluorophosphate ([Et_2NEmpyr][PF_6]). Compared with [Et_2NEmpyr][PF_6], the absorption of SO_2 by [Et_2NEmim][PF_6] could reach the absorption equilibrium rapidly. The investigators observed that [Et_2NEmim][PF_6] had high SO_2 absorption capacity, up to 2.11 g mol SO_2 per g mole IL (for pure SO_2 gas) and 0.94 g mol SO_2 per g mole IL (for 3% SO_2 gas) in the presence of moisture at 30 °C. [Et_2NEmim][PF_6] was found to possess much lower viscosity, substantially higher thermal stability and SO_2 absorption rate than [Et_2NEmpyr][PF_6]. The recycling of [Et_2NEmim][PF_6] with or without water was tested for five cycles and no obvious losses of absorption capacities were determined.

Tian et al. (2014) also studied the effect of water on absorption of pure SO_2 and 3% SO_2 by [Et_2NEmim][PF_6]. It was observed that the effect of water on the absorption of 3% SO_2 was more significant than that on the absorption of pure SO_2. For the absorption of pure SO_2, the absorption equilibrium can be reached under both hydrous and anhydrous conditions. For the absorption of 3% SO_2, the absorption equilibrium can be reached under hydrous conditions while it cannot under anhydrous conditions. The reason is that the chemically absorbed SO_2 results in an extremely high viscosity of IL under anhydrous conditions. Therefore, it was concluded that the SO_2 dissolved in water can only make a minor contribution to the total absorbed SO_2.

Lee et al. (2010a) measured the solubility of SO_2 in [BMIm][OAc] was measured to be 0.6 in mole fraction at 50 °C in a stream of SO_2 gas (Lee et al. 2010b). During the SO_2 absorption, there is a high chance that water is adventitiously introduced into the absorption cell along with SO_2 because most of the commercial SO_2 contains a certain level of water. It has been reported that SO_2 dissolves in water to form sulfurous acid, H_2SO_3, which is known to exist as H^+ and HSO_3^- in aqueous solution. For this reason, acetate ion in [BMIm][OAc] can be replaced by HSO_3^- to produce acetic acid and HSO_3^-. If this postulate is true, acetic acid formed during SO_2 absorption could be easily removed under vacuum at elevated temperature.

Li et al. (2015) studied the effect of the presence of low concentration SO_2 on the capture of CO_2 from flue gases using 1-ethyl-3-methylimidazolium acetate [C_2mim][OAc] ionic liquid. Experiments were performed with a flue gas composition of 15% by volume of CO_2 with 0.2% SO_2. It was found that the absorption capacity of CO_2 in the absence of SO_2 was 0.221 g mol CO_2/g mol IL. But, in the presence of 0.2% SO_2, CO_2 absorption capacity decreased to 0.167 g mol CO_2/g mol IL, which is about 25% of the initial absorption capacity. The investigators explained the microscopic mechanism of the absorption of the two gases through quantum chemical calculation. In comparison with a CO_2 g molecule, a SO_2 g molecule possesses high g molecular polarity and dipole moment that results in strong affinity with the sorbent.

Mohammadi and Foroutan (2014) studied the effects of anion types (PF_6^-, BF_4^-, NO_3^-, Cl^- and Br^-) on SO_2 gas absorption by five ionic liquids (ILs) containing the 1-ethyl-3-methyl imidazole cation using g molecular dynamic simulations. These ILs belong to two different groups: liquid crystals and plastic crystals. It was reported that the diffusion coefficient of the cation in the SO_2 absorbed IL, and in pure, IL was higher than that of the anions and much lesser than SO_2 g molecules. They also observed that the presence of SO_2 in the ILs caused an increase in the conductivity, diffusion coefficients, heat capacity, and density of the ionic species in comparison with those in the pure ILs.

15.2.9 Inverse Supported Ionic Liquids

Supported ionic liquid phase absorbents (SILPs) usually consist of thin layers of ILs, stabilized on the internal (pore) and external surface of porous solids and nanoparticles. The benefits of using SILPs instead of bulk ILs include the solid form of the absorbents which promotes their application in technical, continuous flow processes for flue gas cleaning, and the achievement of the highest possible gas/IL interface which enhances the absorption/desorption rate and makes possible the use of small quantities of a rather expensive type of absorbents such as ILs. A novel type of SILPs, termed as "inverse" SILPs, was proposed by the authors in late 2014 as the inverse analogous of conventional SILPs (Romanos et al. 2014).

Karousos et al. (2016) observed that the "inverse" SILP developed with the physisorbing IL [BMIM][Cl] retains 0.06 mmol/g at 350 mbar having absorbed

0.6 mmol/g at 20 bar (10% of the total CO_2 loading retained during depressurization) while the MEG SILP, developed with the ether functionalized IL, retains the same amount of SO_2 having absorbed 0.87 mmol/g at 20 bar (6.9% of the total CO_2 loading) (Karousos et al. 2016).

15.3 Conclusions

The present study deals with the removal of SO_2 from flue gases using various techniques. The review describes the practical FGD processes, their characteristics, assesses their use, determines which of these are feasible, characterizes the SO_2 removal performance of the processes, and analyses their costs. Processes using limestone slurry as an absorbent are the most widely applied FGD technologies much work has been done on the synthesis and characterization of ionic liquids used for removal of SO_2 as well as the mechanism of the absorption processes. Different kinds of ionic liquids have been prepared, and some of them exhibit a high absorption capacity for SO_2. However, the technology has not been widely used in the industry due to the high cost of ionic liquids, complexity of flue gas, and other factors.

Many ionic liquids so far exhibit high absorption capability for pure SO_2, but low absorption capacity for SO_2 from the simulated flue gases. So the effect of other gases in flue gas on the performance of ionic liquids needs to be investigated so as to design more effective ionic liquids. Thus, the aim of this review paper is to help find the right direction in exploring new solutions.

References

Anderson JL, Dixon JK, Maginn EJ, Brennecke JF (2006) Measurement of SO_2 solubility in ionic liquids. J Phys Chem B 110(31):15059–15062

Angelo JB, Lightfoot EN, Howard DW (1966) Generalization of the penetration theory for surface stretch: Application to forming and oscillating drops. AIChE J 12:751–760

Arif A, Stephen C, Branken D, Everson R, Neomagus H, Piketh S (2015) Modeling wet flue gas desulfurization. In: Conference of the National Association for Clean Air (NACA 2015)

Armand M, Endres F, MacFarlane DR, Ohno H, Scrosati B (2009) Ionic-liquid materials for the electrochemical challenges of the future. Nat Mater Nat Publ Group 8(8):621–629

Bates ED, Mayton RD, Ntai I, Davis JH (2002) CO_2 capture by a task-specific ionic liquid. J Am Chem Soc 124(6):926–927

Bhoi S, Banerjee T, Mohanty K (2016) Beneficiation of Indian coals using Ionic Liquids. Fuel Process Technol 151:1–10

Bicak N (2005) A new ionic liquid: 2-hydroxy ethylammonium formate. J Mol Liq 116(1):15–18

Bjerle I, Bengtsson S, Färnkvist K (1972) Absorption of SO_2 in $CaCO_3$-slurry in a laminar jet absorber. Chem Eng Sci 27(10):1853–1861

Bravo RV, Camacho RF, Moya VM, Garca LAI (2002) Desulphurization of SO_2–N_2 mixtures by limestone slurries. Chem Eng Sci 57:2047–2058

Cui G, Zhang F, Zhou X, Huang Y, Xuan X, Wang J (2015) Acylamido-based anion-functionalized ionic liquids for efficient so_2 capture through multiple-site interactions. ACS Sustain Chem Eng 3(9):2264–2270

De Visscher A (2013) Air dispersion modeling: foundations and applications. Wiley

Dou B, Byun YC, Hwang J (2008) Flue gas desulfurization with an electrostatic spraying absorber. Energy Fuels 22(2):1041–1045

Dou B, Pan W, Jin Q, Wang W, Li Y (2009) Prediction of SO_2 removal efficiency for wet flue gas desulfurization. Energy Convers Manage Elsevier Ltd. 50(10):2547–2553

Duan E, Guo B, Zhang D, Shi L, Sun H, Wang Y (2011) Absorption of NO and NO_2 in caprolactam tetrabutyl ammonium halide ionic liquids. Air Waste Manage Assoc 61(12):1393–1397

Duan E, Guo B, Zhang M, Yang B, Zhang D (2010) pH measurements of caprolactam tetrabutyl ammonium bromide ionic liquids in solvents. J Chem Eng Data 55:3278–3281

Gao X, Ding H, Du Z, Wu Z, Fang M, Luo Z, Cen K (2010) Gas-liquid absorption reaction between $(NH_4)_2SO_3$ solution and SO_2 for ammonia-based wet flue gas desulfurization. Appl Energy 87(8):2647–2651

Ghosh D, Lal S, Sarkar U (2017) Variability of tropospheric columnar NO_2 and SO_2 over eastern Indo-Gangetic Plain and impact of meteorology. Air Qual Atmos Health 10(5):565–574

Guo B, Duan E, Ren A, Wang Y, Liu H (2010) Solubility of SO_2 in caprolactam tetrabutyl ammonium bromide ionic liquids. J Chem Eng Data 55(3):1398–1401

Guo B, Duan E, Zhong Y, Gao L, Zhang X, Zhao D (2011) Absorption and oxidation of H 2 S in caprolactam tetrabutyl ammonium bromide ionic liquid. Energy Fuels 25(1):159–161

Hallett JP, Welton T (1999) Room-temperature ionic liquids. solvents for synthesis and catalysis. Chem Rev 99(8):2071–2083

Hikita H, Asai S, Tsuji T (1978) Absorption of sulfur dioxide into aqueous ammonia and ammonium sulfite solutions. J Chem Eng Jpn 11(3):236–238

Huang J, Riisager A, Berg RW, Fehrmann R (2008) Tuning ionic liquids for high gas solubility and reversible gas sorption. J Mol Catal A Chem 279(2):170–176

Huang J, Riisager A, Wasserscheid P, Fehrmann R (2006) Reversible physical absorption of SO_2 by ionic liquids. Chem Commun 38:4027–4029

Huang K, Lu JF, Wu YT, Hu XB, Zhang ZB (2013) Absorption of SO_2 in aqueous solutions of mixed hydroxylammonium dicarboxylate ionic liquids. Chem Eng J 215–216:36–44

Huddleston JG, Willauer HD, Swatloski RP, Visser AE, Rogers RD (1998) Room temperature ionic liquids as novel media for 'clean' liquid–liquid extraction, 1765–1766

Jiang L, Liguang BAI, Jiqin ZHU, Biaohua CHEN (2013) Thermodynamic Properties of Caprolactam Ionic Liquids. Chin J Chem Eng 21(7):766–769

Kalekar MS, Bhagwat SS (2006) Dynamic behavior of surfactants in solution. J Dispersion Sci Technol 27(7):1027–1034

Karatepe N (2000) A comparison of flue gas desulfurization processes. Energy Sources 22(3):197–206

Karousos DS, Kouvelos E, Sapalidis AA, Pohako K, Bahlmann M, Schulz PS, Wasserscheid P, Siranidi E, Vangeli O, Falaras P, Kanellopoulos NK, Romanos GE (2016) For acidic gas removal from flue gas novel inverse supported ionic liquid absorbents for acidic gas removal from flue gas

Kinney ML, Schoch RM, Yonavjak L (2007) Environmental science: systems and solutions. Fourth edi. Jones and Bartlett Learning, Inc

Lee KY, Kim CS, Kim H, Cheong M, Mukherjee DK, Jung KD (2010a) Effects of halide anions to absorb SO_2 in ionic liquids. Bull Korean Chem Soc 31(7):1937–1940

Lee YK, Kim HS, Kim CS, Jung K (2010b) Behaviors of SO_2 absorption in [BMIm][OAc] as an absorbent to recover SO_2 in thermochemical processes to produce hydrogen. Int J Hydrogen Energy Elsevier Ltd. 35(19):10173–10178

Li W, Liu Y, Wang L, Gao G (2017) Using ionic liquid mixtures to improve the SO_2 absorption performance in flue gas. Energy Fuels. https://doi.org/10.1021/acs.energyfuels.6b02884

Li X, Zhang L, Zheng Y, Zheng C (2015) Effect of SO2 on CO_2 absorption in flue gas by ionic liquid 1-Ethyl-3-methylimidazolium acetate. Ind Eng Chem Res 54(34):8569–8578

Li X, Zhu C, Lu S (2013) Mass transfer of SO_2 absorption with an instantaneous chemical reaction in a bubble column. Braz J Chem Eng 30(3):551–562

Liu B, Zhao J, Wei F (2013) Characterization of caprolactam based eutectic ionic liquids and their application in SO_2 absorption. J Mol Liq 180(3):19–25

Liu SY, Xiao W De (2006) Modeling and simulation of a bubbling SO_2 absorber with granular limestone slurry and an organic acid additive. Chem Eng Technol 29(10):1167–1173

Lu Z, Streets DG, de Foy B, Krotkov NA (2013) Ozone Monitoring Instrument observations of interannual increases in SO_2 emissions from Indian coal-fired power plants during 2005–2012. Environ Sci Technol 47(24):13993–14000

Lu Z, Streets DG, Zhang Q, Wang S, Carmichael GR, Cheng YF, Tan Q (2010) Sulfur dioxide emissions in China and sulfur trends in East Asia since 2000. Atmos Chem Phys 10(13):6311–6331

Lu Z, Zhang Q, Streets DG (2011) Sulfur dioxide and primary carbonaceous aerosol emissions in China and India, 1996–2010. Atmos Chem Phys 11(18):9839–9864

Mohammadi M, Foroutan M (2014) Molecular investigation of SO_2 gas absorption by ionic liquids: effects of anion type. J Mol Liq 193:60–68

Mondal MK (2007) Experimental determination of dissociation constant, Henry's constant, heat of reactions, SO_2 absorbed and gas bubble–liquid interfacial area for dilute sulphur dioxide absorption into water. Fluid Phase Equilib 253:98–107

Mondal A, Balasubramanian S (2016) Understanding SO_2 capture by ionic liquids. J Phys Chem B 120(19):4457–4466

Noble RD, Gin DL (2011) Perspective on ionic liquids and ionic liquid membranes. J Membr Sci Elsevier BV 369(1–2):1–4

Pandey RA, Biswas R, Chakrabarti T, Devotta S (2005) Flue gas desulfurization: physicochemical and biotechnological approaches. Crit Rev Environ Sci Technol 35(6):571–622

Park HW, Park DW (2017) Removal kinetics for gaseous NO and SO_2 by an aqueous $NaClO_2$ solution mist in a wet electrostatic precipitator. Environ Technol 38(7):835–843

Qu G, Zhang J, Li J, Ning P (2013) SO_2 absorption/desorption characteristics of two novel phosphate ionic liquids. Sep Sci Technol 48(18):2876–2879

Ramachandran PA, Sharma MM (1969) Absorption with fast reaction in a slurry containing sparingly soluble fine particles. Chem Eng Sci 24(1):1681–1686

Ren S, Hou Y, Tian S, Wu W, Liu W (2012) Deactivation and regeneration of an ionic liquid during desulfurization of simulated flue gas. Ind Eng Chem Res 51(8):3425–3429

Ren S, Hou Y, Wn W, Liu Q, Xiao Y, Chen X (2010) Properties of ionic liquids absorbing SO_2 and the mechanism of the absorption. J Phys Chem B 114(6):2175–2179

Ren S, Hou Y, Wu W, Jin M (2011) Oxidation of SO_2 absorbed by an ionic liquid during desulfurization of simulated flue gases. Ind Eng Chem Res 50(2):998–1002

Rennie J, Evans F (1962) The formation of froth and foams above sieve plates. Br Chem Eng 7:498

Romanos GE, Schulz PS, Bahlmann M, Wasserscheid P, Sapalidis A, Katsaros FK, Athanasekou CP, Beltsios K, Kanellopoulos NK (2014) CO_2 capture by novel supported ionic liquid phase systems consisting of silica nanoparticles encapsulating amine-functionalized ionic liquids

Sada E, Kumazawa H, Butt MA (1977) Simultaneous absorption of three, reacting gases 13:225–231

Sada E, Kumazawa H, Butt MA (1979) Chemical absorption into a finite slurry. Chem Eng Sci 34(5):715–718

Sada E, Kumazawa H, Hashizume I (1983) Chemical absorption of two gases into a slurry containing fine catalyst particles. Chem Eng J 26(3):239–244

Sada E, Kumazawa H, Hashizume I, Kamishima M (1981) Desulfurization by limestone slurry with added magnesium sulfate. Chem Eng J 22(2):133–141

Shang Y, Li H, Zhang S, Xu H, Wang Z, Zhang L, Zhang J (2011) Guanidinium-based ionic liquids for sulfur dioxide sorption Chem Eng J Elsevier BV 175(1):324–329

Shiflett MB, Yokozeki A (2009) Separation of carbon dioxide and sulfur dioxide using room-temperature ionic liquid [bmim][$MeSO_4$]. Energy Fuels 24(2):1001–1008

Shiflett MB, Yokozeki A (2010) Chemical absorption of sulfur dioxide in room-temperature ionic liquids. Ind Eng Chem Res 49(3):1370–1377

Tang Z, Zhou C, Chen C (2004) Studies on flue gas desulfurization by chemical absorption using an ethylenediamine. Phosphoric Acid Solut: 6714–6722

Tian S, Hou Y, Wu W, Ren S, Qian J (2014) Hydrophobic task-specific ionic liquids: synthesis, properties and application for the capture of SO_2. J Hazard Mater 278:409–416

Uchida S, Koide K, Shindo M (1975) Gas absorption with fast reaction into a slurry containing fine particles. Chem Eng Sci 30(5–6):644–646

Wang LK, Williford C, Chen WY (2005) Desulfurization and emissions control. In: Advanced Air and Noise Pollution Control, Humana Press, pp 35–95

Wu W, Han B, Gao H, Liu Z, Jiang T, Huang J (2004) Desulfurization of flue gas: SO_2 absorption by an ionic liquid. Angewandte Chemie Int Edn 43(18):2415–2417

Yang Z, Pan W (2005) Ionic liquids: green solvents for nonaqueous biocatalysis. Enzyme Microbial Technol 37(1):19–28

Yuan XL, Zhang SJ, Lu XM (2007) Hydroxyl ammonium ionic liquids: synthesis, properties, and solubility of so2. J Chem Eng Data 52(2):596–599

Zeng S, Gao H, Zhang X, Dong H, Zhang X, Zhang S (2014) Efficient and reversible capture of SO2 by pyridinium-based ionic liquids. Chem Eng J 251:248–256

Zhai L, Zhong Q, He C, Wang J (2010) Hydroxyl ammonium ionic liquids synthesized by water-bath microwave: synthesis and desulfurization. J Hazardous Mater Elsevier BV 177(1–3):807–813

Part III
Groundwater Management

Chapter 16
Long-Term Performance Evaluation of Permeable Reactive Barrier for Groundwater Remediation Using Visual MODFLOW

Rahul Singh, Sumedha Chakma and Volker Birke

Abstract Severe groundwater pollution is taking place due to the irregular dumping of liquid and solid waste at the surface. Many conventional groundwater remediation techniques have been developed, over the past few decades, however, permeable reactive barrier (PRB) has emerged as a promising alternative for the in situ treatment of a variety of the groundwater contaminants. This study deals with the long-term performance evaluation of a permeable reactive barrier for a hypothetical site having a study area of 1.04 km^2 by numerically simulating the developed model using Visual MODFLOW. The model has been simulated for a period of 5 years to obtain the real nature of contaminant plume, with chloride as the contaminant, spreading throughout the aquifer. Further, continuous PRB has been installed with optimized design and selected material properties. Activated wood charcoal has been used as a reactive material for the chloride degradation. The simulation of the model provides a performance evaluation criteria for the stability of a barrier configuration over a real contaminated site. Overall, the developed model would be able to provide an optimum and sustainable solution of the reactive barrier for the treatment of certain specific contaminants releasing from the contaminant sites and remediation of the aquifer in the long run.

Keywords Groundwater remediation · Continuous barrier · Visual MODFLOW · Activated wood charcoal

---

R. Singh (✉) · S. Chakma
Department of Civil Engineering, Indian Institute of Technology (IIT) Delhi, Hauz Khas, New Delhi 110016, India
e-mail: Rahul.Singh@civil.iitd.ac.in

S. Chakma
e-mail: chakma@civil.iitd.ac.in

V. Birke
Mechanical Engineering/Process Engineering and Environmental Engineering, Hochschule Wismar—University of Applied Sciences, Technology, Business and Design, Philipp-Müller-Str. 14, 23966 Wismar, Germany
e-mail: volker.birke@hs-wismar.de

R. M. Singh et al. (eds.), *Environmental Processes and Management*,
Water Science and Technology Library 91,
https://doi.org/10.1007/978-3-030-38152-3_16

16.1 Introduction

The last few decades have witnessed an increasing concern over the deterioration of the groundwater quality around the world and especially in developing countries (Naidu et al. 2014). There are various diverse sources of geogenic and anthropogenic activities which lead to contamination of groundwater. Many conventional groundwater remediation techniques have been developed, over the past few decades, among which the permeable reactive barrier (PRB) has emerged as a promising alternative for the in situ treatment of a variety of the groundwater contaminants such as the heavy metals, organics radionuclides, volatile organic carbons (VOCs), chlorinated compounds and many more other complex contaminants (Birke et al. 2003). This technology has proved as one of the passive, low cost, less maintenance and highly efficient method for groundwater remediation in the long run. A schematic diagram is shown in Fig. 16.1 displaying the basic working of the PRB technology. Various studies have investigated the performance of a PRB system for the treatment of different contaminants, on lab-scale experiments or large-scale field assessment, (Blowes et al. 1998; Genç-Fuhrman et al. 2007). However, prior to on-site PRB installation, the modeling of the PRB system aids in analyzing the performance of the PRB system for longer time periods and simulates its behavior under various plausible scenarios by varying factors such as the configuration of PRB, location and orientation of barrier, contaminant properties, etc.

Zeng and Wang (1999) have used the numerical groundwater modeling tools—MODFLOW and MT3DMS for flow and solute transport modeling respectively to analyze the spreading of the contaminant plume in the aquifer system for a longer

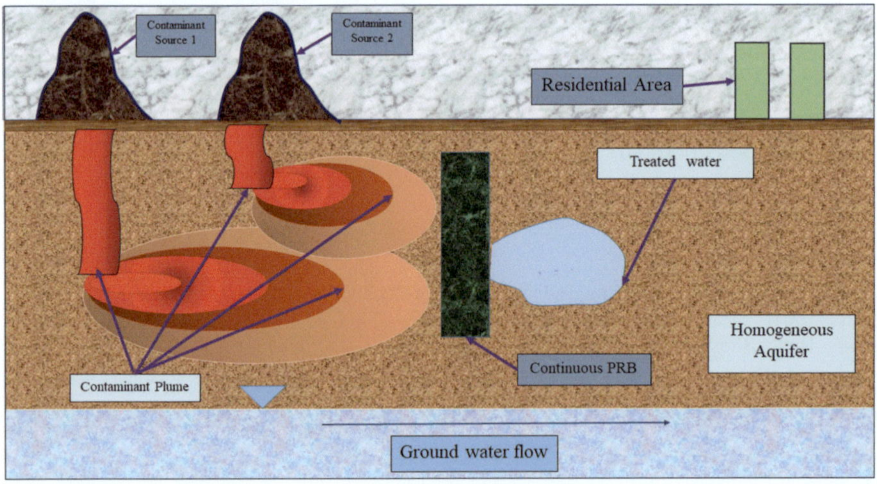

Fig. 16.1 Illustration of various types of PRB for the contaminant plume remediation of the multi-contaminant source site in a homogeneous aquifer (Singh et al. 2020)

duration. The MODFLOW and the MT3DMS tools work on finite difference techniques such as explicit, implicit, and Crank Nicolson finite difference methods. The MODFLOW has been widely used to quantitatively estimate the aquifer response concerning different input parameters (McDonald and Harbaugh 1988; Xu et al. 2012). Scott and Folkes (2000) have introduced reactive barrier in their groundwater models to analyze the behavior of plume migration through the PRB system. Further, few studies have suggested a shift towards multiple permeable reactive barrier (multi-PRB) system instead of a single barrier system for efficient groundwater remediation (Birke et al. 2007).

16.2 Study Area

A hypothetical study area is considered for the development of the flow and solute transport model along with the incorporation of reactive barriers numerical model to analyze the containment of the contaminated plume. Mahar and Data (2000) have introduced this study area for the identification of groundwater pollution sources under transient conditions.

The hypothetical study is a saturated aquifer with an area of 1.3 km × 0.8 km with a depth of 30 m. The aquifer is assumed to be homogeneous and isotropic. A uniform hydraulic conductivity (K) has been assumed in all three dimensions, i.e., $K_{xx} = K_{yy} = K_{zz}$, with a value of 0.000191 m/s. The effective porosity (η) and specific storage (S) of the aquifer are considered to be 0.25 and 0.002, respectively. The dispersivity values in the longitudinal and transverse direction are taken as 40.0 m and 4.0 m, respectively. The pumping rates of the P are varied at an interval of 90 days (one-time step) with the maximum and minimum pumping rates (m^3/day) being 381.024 and 163.296, respectively (Borah and Bhattacharjya 2014). The boundary conditions of the aquifer are considered to be time-invariant. The north and the south boundaries of the study area are considered with no-flow boundaries, while the east and the west boundaries are regarded as the constant head boundaries. The hydraulic head (m) in the east and west boundary is varied from 100.00 to 99.58 and 88.00 to 87.72, respectively as shown in Fig. 16.2. Two potentially active contaminant sources (S1, S2) and one inactive source (S3) are taken into consideration which placed at the upgradient of the study area. The S1 and S2 sources are continuously active for 3600 days, 5 years, with an assumed constant flux of 4,060,800 g/day and 5,080,320 g/day, respectively. It is assumed that the chloride (Cl^-) is the only contaminant produced from all the active contaminant sources. The eight observation wells, C1, C2, C3, C4, C5, C6, C7, and C8, are spread throughout the aquifer.

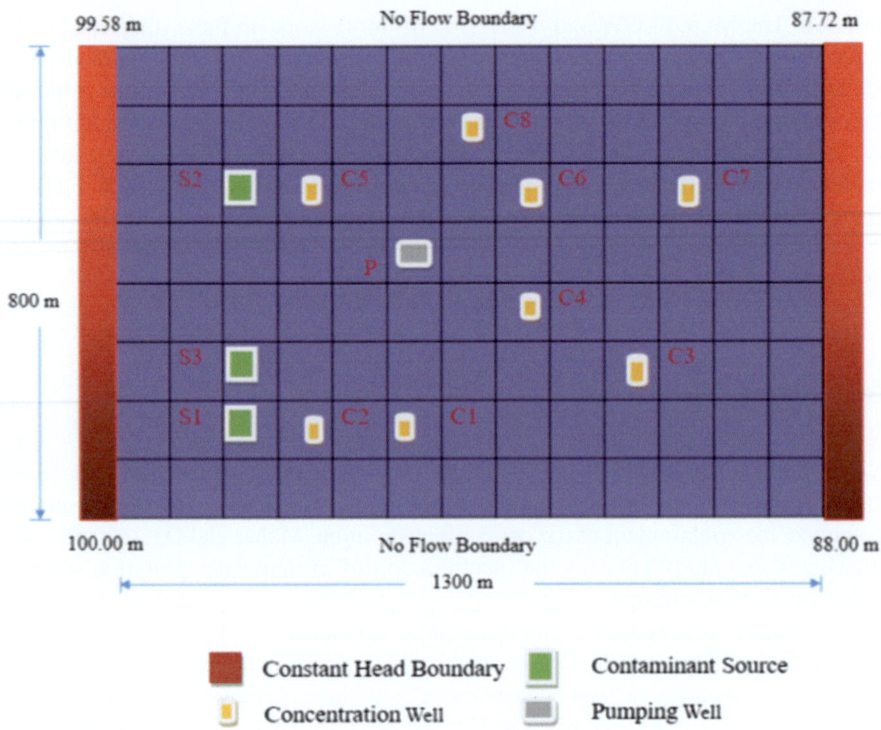

Fig. 16.2 Schematic representation of the hypothetical study area

16.3 Model Development

The general methodology adopted for modeling of PRB(s) is shown in Fig. 16.3 and described below briefly. The first step is the thorough study of the contaminated study area. The subsequent step is the selection and development of the flow and solute transport modeling for the given aquifer system. Based on the identification of highly contaminated zones in the subsequent step, the next step is the installation of single or multiple PRBs as per the aquifer remediation intensity requirements.

16.3.1 Governing Equation for Contaminant Solute Transport

The chemical interaction, specifically adsorption, of the solute with the surfaces of the porous medium, in the PRB, can be defined by the following Eq. (16.1) (Fetter 1992):

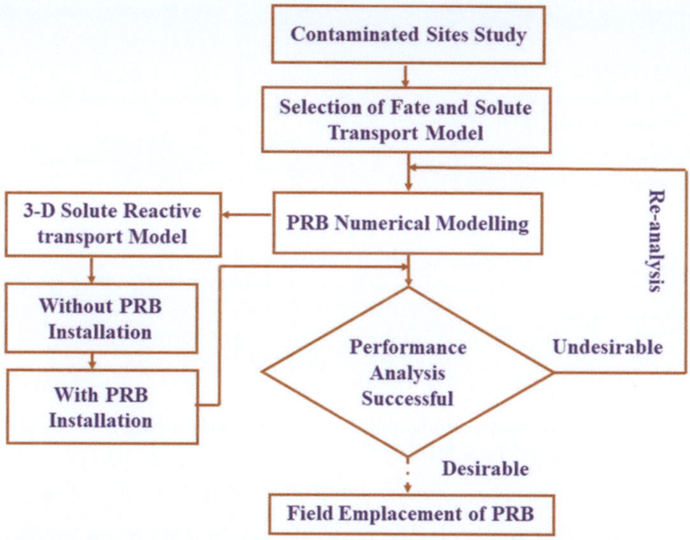

Fig. 16.3 General methodology of the permeable reactive barrier (PRB) modeling for groundwater remediation

$$\frac{\partial C}{\partial t} + v\frac{\partial C}{\partial z} = \frac{\partial}{\partial z}\left(D\frac{\partial C}{\partial z}\right) - \frac{B_d}{\theta}\sum f\left(k_f(C)\right) \quad (16.1)$$

where

B_d bulk density (ML^{-3})
θ porosity of the porous medium (T)
k_f adsorption rate coefficient.

16.3.2 Numerical Modeling

In this study, Visual MODFLOW version 2011.1 was used for the flow and the solute transport simulation in a homogeneous aquifer system. The entire hypothetical study area was disintegrated into smaller rectilinear grid cells of 100 m × 100 m size. A single layer of 30 m depth was considered in all the computational grids of the model. MODFLOW 2000, a three-dimensional finite difference-based groundwater flow modeling tool, was used to simulate the groundwater flow under transient conditions. The contaminant transport simulation was performed using MT3DMS, which followed the flow (MODFLOW) simulation in order to analyze the contaminant spreading throughout the study area.

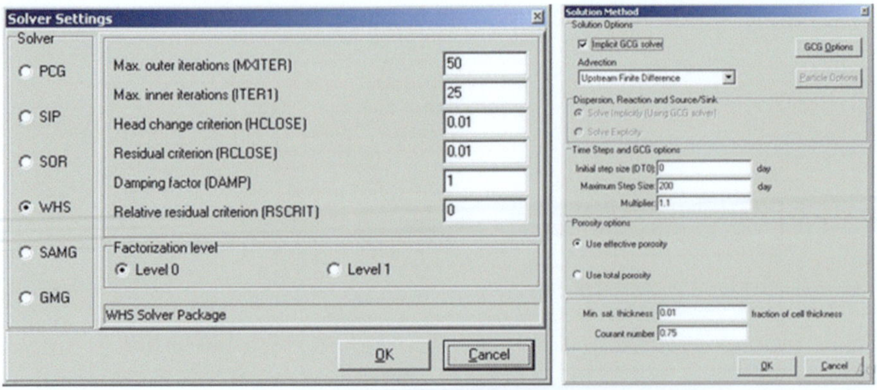

Fig. 16.4 Showcase of "Solver run setting" and "Solution methods" in MODFLOW and MT3DMS respectively for run setting in Visual MODFLOW

The central finite difference method under the implicit state was used to solve the advection-dispersion transport in the model. Further, the flow and transport simulation was run for 20-time steps in the entire five-year period. After a successful model run for the flow and contaminant transport modeling, the installation of PRB systems has been incorporated. The continuous PRB configuration system was adopted for the contaminant remediation which is installed just adjacent to the contaminant source (S1 and S2).

Visual MODFLOW provides a various choices of different solvers to use in solving the advection-dispersion numerical equations for the flow as well as solute transport equations which works on a two-tier approach for the solution at one-time step. In this study, WHS solver has been used for the numerical equation for the flow simulation. Similarly, implicit GCG solver has been used for solute transport modeling. The setting of this solver can be accessed by selecting MODFLOW/solver run setting from the run section of Visual MODFLOW as shown in Fig. 16.4. Similarly, for solute transport modeling MT3DMS/solution method has been selected as shown in Fig. 16.4.

After a successful model run for the flow and contaminant transport modeling, the installation of a continuous PRB system has been incorporated. The continuous PRB configuration system was adopted for the contaminant remediation. The thickness of the PRB was considered as 10 m throughout the width of the study area and installed at an orientation of 90° from the contaminant plume. The activated wood charcoal (AWC) is used as a reactive material in the PRB system for the adsorption of the Cl^- (Mohammed et al. 2012).

16.4 Results and Discussions

The variation in the concentration of chloride with time, in all the observation wells, for all the cases of PRB installation is shown in Fig. 16.5. In case (a) of no PRB installation, the observation wells, C2, C5, and C1, show the steep increase in the contaminant concentration for the initial days of observation period due to continuous emission of the contaminant from the nearby contaminant sources, S1 and S2. The other wells C3, C4, C6, C7, and C8, which are located at far away from source S1 and S2 also show a significant increase in contaminant concentration with respect to time, however, relatively lower than the C2, C5, and C1 concentration wells. Due to the continuity of contaminant sources, the concentration in the observation wells continuously increases with respect to time. However, the slope of the curve is gradually decreasing due to natural attenuation capacity of the aquifer as the plume is passing through the aquifer.

In case (b), i.e., continuous PRB installation, the concentration in all the observation wells drops by more than eight times. Similarly, the rate of increase in contaminant concentration is also decreased in all the wells. This is evident as the time to

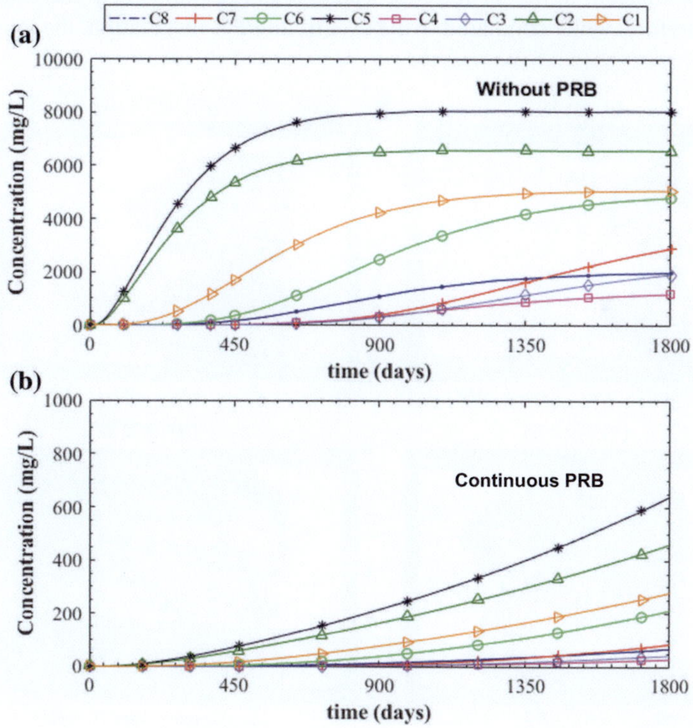

Fig. 16.5 Variation of chloride concentration with time in all observation wells for two cases **a** without PRB; **b** continuous PRB installation

reach the peak concentration is significantly delayed. At the end of the observation period, the wells C2, C5, and C1 show higher concentration, compared to all other observation wells due to higher intensity of the plume containment in the vicinity of the contaminant sources.

The spatial variation of the contaminant plume in the entire aquifer is analyzed at four different time steps—90, 450, 900 and 1800 days for both the cases, i.e., (a) without PRB; (b) continuous PRB installation, as shown in Figs. 16.6 and 16.7. Figure 16.6 shows the spatial distribution of contaminant plume in case of no PRB installation. It is observed that with the passage of time, the plume spreads and moves away from the point source. The concentration intensity of the plume is maximum near the source and it radially decreases while moving away from the source. At the end of 450 days, the plume spread almost reaches the center of the aquifer. At the end of 900 days, the plume spread increases significantly; however, the strength of the plume is reduced. At the end of the simulation period, i.e., 1800 days the overlapped plume has spread entirely in the downstream of the aquifer.

The installation of continuous PRB significantly decreases the contaminant intensity of the plume at all the time steps as shown in Fig. 16.7. However, the spread of plume is reduced up to second time steps only. Due to the placement of the reactive barrier just adjacent to the sources, the containment of plume near the source takes place. Therefore, when compared to the case of no PRB installation, the longitudinal

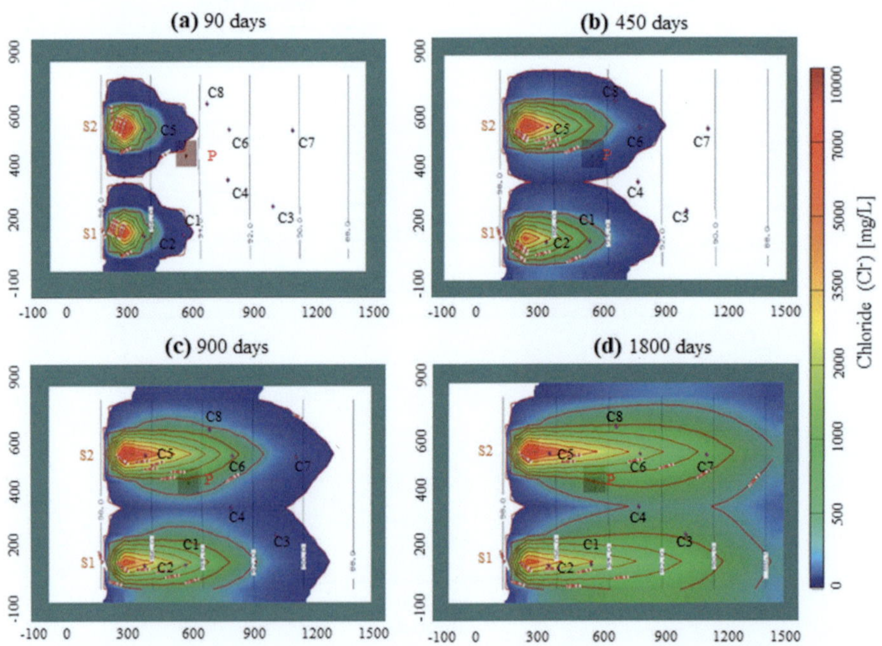

Fig. 16.6 Spatial variation of contaminant plume (mg/L) in case of no PRB installation at time steps of **a** 90 days, **b** 450 days, **c** 900 days and **d** 1800 days

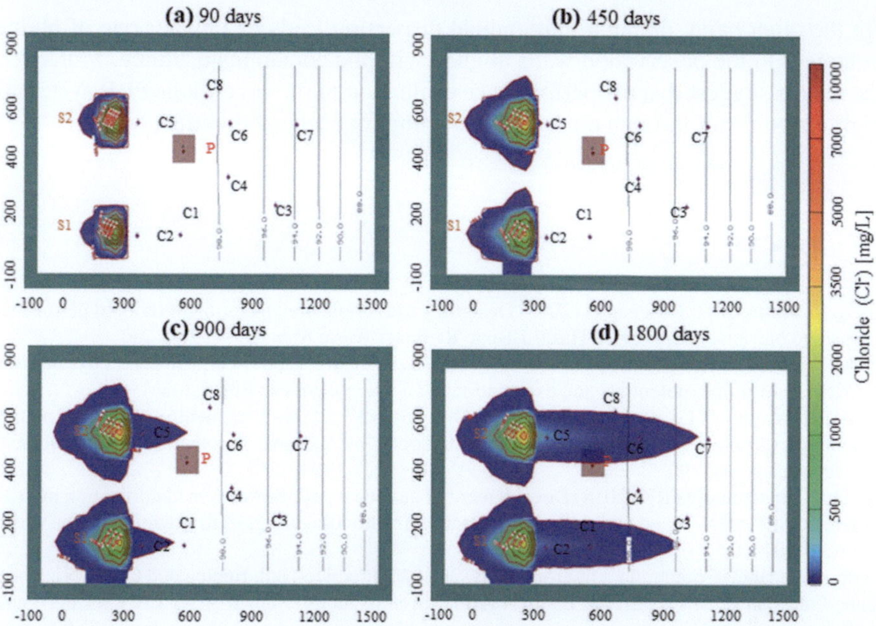

Fig. 16.7 Spatial variation of the contaminant plume in case of continuous PRB installation at time steps of **a** 90 days, **b** 450 days, **c** 900 days and **d** 1800 days

spread of plume is higher within the considered section of the aquifer. Nevertheless, the continuous PRB is highly effective in minimizing the overall spread and the intensity of contamination in the considered aquifer system.

16.5 Conclusion

In this study, It is observed that the performance of continuous PRB system is superior then the natural attenuation process of the aquifer. In the case of no PRB system, the maximum intensity of contaminant plume is significantly higher than the permissible limit. Therefore, with the installation of continuous PRB, adjacent to the contaminant sources, the plume concentration intensity dropped down remarkably. The location of the PRB is a guiding factor in determining the efficiency of PRB in the remediation of the contaminant plume. The PRB, in the proximity of contaminant source, contains the plume more significantly. Further, it is observed that the dispersion plays an important role in governing the treatment rate of the contaminant plume for a longer period of time. The low transverse dispersion coefficient is having lower treatment rate of the plume, reaches the observation wells, not aligned with the point source.

On the other hand, the high longitudinal dispersion leads to a higher rate of plume treatment to the observation wells aligned with the contaminant source. Moreover, the results suggest that the performance evaluation of the continuous PRB system is highly significant through numerical modeling for a longer duration.

References

Birke V, Burmeier H, Rosenau D (2003) Design, construction, and operation of tailored permeable reactive barriers. Pract Period Hazard Toxic Radioact Waste Manag 7(4):264–280

Birke V, Burmeier H, Jefferis S, Gaboriau H, Touze S, Romain C (2007) Permeable reactive barriers (PRBs) in Europe: potentials and expectations. Ital J Eng Geol Environ 1:1–8

Blowes DW, Ptacek CJ, Benner SG, McRae CW, Puls RW (1998) Treatment of dissolved metals using permeable reactive barriers. IAHS Publication (International Association of Hydrological Sciences), pp 483–90

Borah T, Bhattacharjya RK (2014) Development of unknown pollution source identification models using GMS ANN–based simulation optimization methodology. J Hazard Toxic Radioact Waste 19(3):04014034

Fetter CW (1992) Contaminant hydrogeology, 2nd edn. Prentice-Hall, Englewood Cliffs, NJ, p 458

Genç-Fuhrman H, Mikkelsen PS, Ledin A (2007) Simultaneous removal of As, Cd, Cr, Cu, Ni and Zn from stormwater: experimental comparison of 11 different sorbents. Water Res 41(3):591–602

Mahar PS, Datta B (2000) Identification of pollution sources in transient groundwater systems. Water Resour Manag 14(3):209–227

McDonald MG, Harbaugh AW (1988) A modular three-dimensional finite-difference ground-water flow model, vol 6. US Geological Survey, Reston, VA, p A1

Mohammed I, Ariahu CC, Nkpa NN, Igbabul BD (2012) Chlorine adsorption kinetics of activated carbon from selected local raw materials. J Chem Eng Mater Sci 3(2):23–29

Naidu R, Bekele DN, Birke V (2014) Permeable reactive barriers: cost-effective and sustainable remediation of groundwater. In: Permeable reactive barrier. CRC Press, pp 16–39

Scott KC, Folkes DJ (2000) Groundwater modeling of a permeable reactive barrier to enhance system performance. In: Proceedings of the 2000 conference on hazardous waste research. Denver, Colorado, USA

Singh R, Chakma S, Birke V (2020) Numerical modelling and performance evaluation of multi-permeable reactive barrier system for aquifer remediation susceptible to chloride contamination. Groundwater Sustain Dev 10:100317

Xu Z, Wu Y, Yu F (2012) A three-dimensional flow and transport modeling of an aquifer contaminated by perchloroethylene subject to multi-PRB remediation. Transport Porous Media 91(1):319–337

Zheng C, Wang PP (1999) MT3DMS: a modular three-dimensional multispecies transport model for simulation of advection, dispersion, and chemical reactions of contaminants in groundwater systems; documentation and user's guide. Alabama University

Chapter 17
Management of Saltwater Intrusion in Coastal Aquifers: An Overview of Recent Advances

Subhajit Dey and Om Prakash

Abstract The demand for freshwater is very high in the coastal regions due to the high population density in coastal areas. To meet this demand for freshwater, the coastal aquifers are often heavily pumped without any regulation, resulting in saltwater intrusion. Therefore, the biggest challenge in the management of coastal aquifer is to meet the demand for freshwater by pumping the coastal aquifer without causing saltwater intrusion. In this study, a brief overview of various methods for identification, prediction, and management of saltwater intrusion is presented. Detection of saltwater intrusion is largely hindered due to insufficient spatiotemporal monitoring because of budgetary constraints. Application, merits, and demerits of the newer cost-effective techniques as well as conventional techniques for identifying saltwater intrusion are discussed in this chapter. The application of various prediction models and their computational difficulties is also presented in this study. Finally, advanced techniques for identification and sustainable management practice in saltwater intrusion are discussed. Though significant progress has been made in the recent past in the management of coastal aquifers, they still show gaps in addressing real-life scenarios. An attempt has been made to highlight the suitability of a developed methodology and their respective limitations.

Keywords Saltwater intrusion · Coastal aquifers · Density-dependent model · Sharp interface model · Water resources management

S. Dey (✉) · O. Prakash
Department of Civil and Environmental Engineering, Indian Institute of Technology Patna, Bihta, Patna 8001106, India
e-mail: sri.pce15@iitp.ac.in

O. Prakash
e-mail: om.prakash@iitp.ac.in

R. M. Singh et al. (eds.), *Environmental Processes and Management*, Water Science and Technology Library 91,
https://doi.org/10.1007/978-3-030-38152-3_17

17.1 Introduction

From the dawn of civilization, coastal areas continue to be the hub of economic activities and global transport which has led to massive population growth in these coastal areas. It is estimated that about 70% of the world population live near coastal areas (Bear et al. 1999). The high population density in coastal regions results in greater demand for freshwater. Groundwater resources play an essential role in meeting this freshwater demand. These coastal aquifers (aquifers hydraulically connected to the sea) are pumped heavily to meet this demand for freshwater. The natural flow of groundwater is generally toward the sea, thus preventing the flow of saltwater toward the coastal aquifer. Due to heavy pumping, this natural gradient of flow gets disturbed, which allows the saltwater from the sea to flow toward the aquifer, thus causing saltwater intrusion. Once saltwater intrudes into a coastal freshwater aquifer, the reclamation of such aquifers is economically imprudent. Therefore, it is essential to develop a sustainable management policy for preventing saltwater intrusion in coastal aquifers.

The primary pollutant in the case of the coastal aquifer is saline water from the sea, which is already present near the coast. There is a natural balance wherein the lighter freshwater overlays the denser saline water forming a wedge between the two (Fig. 17.1). A mixing zone is present between saltwater and freshwater interface referred to as the transition zone or diffusion zone. In this, zone density of water gradually increases from the density of freshwater to density of saltwater (Shamir et al. 1984). The equilibrium between saltwater and freshwater is maintained due to the natural decreasing gradient of water table toward the sea (Fig. 17.1). This natural gradient is disturbed due to excessive pumping of the groundwater often bringing the water table below the mean sea level (MSL), causing a reversal of the gradient (Fig. 17.2). This allows movement of the denser saltwater from the sea toward the pumping wells. When saltwater reaches the pumping wells, saltwater along with freshwater gets pumped out. This phenomenon is called saltwater intrusion (Bear 2012). If the salt concentration in the pumped water reaches 1% of the total density of the water, then pumped water is deemed unsafe for human consumption without prior treatment (WHO 2004). Such a scenario is highly undesirable, and reversing

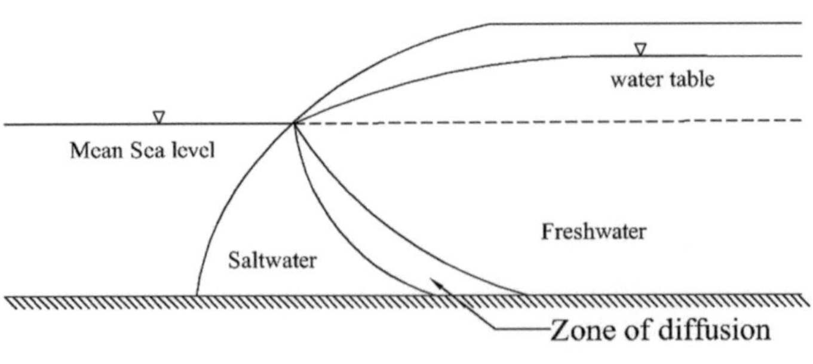

Fig. 17.1 Schematic vertical cross sections of unconfined coastal aquifer in natural condition

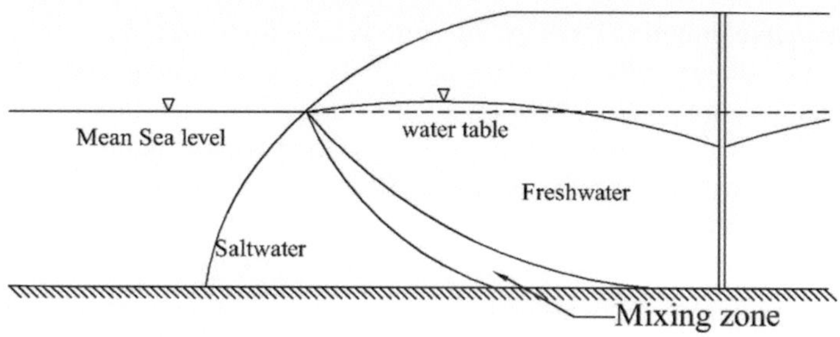

Fig. 17.2 Schematic vertical cross sections of unconfined coastal aquifer while pumping

the impact or reclamation of aquifers polluted by saltwater intrusion is economically unviable.

Thus, the primary challenge in the management of coastal aquifers is to pump large amounts of freshwater from the coastal aquifers to meet the freshwater demands of the growing population, without causing saltwater intrusion. Saltwater intrusion is first detected at an arbitrary water supply well or a group of wells installed for pumping freshwater from the coastal aquifers (Sreekanth and Datta 2015). By the time, saltwater intrusion is detected, and the aquifer is already polluted from the saltwater from the sea. Continuous spatiotemporal monitoring is required for on-time detection of saltwater intrusion. However, long-term spatiotemporal monitoring along with long coastlines is overlooked because of budgetary requirements. Other measures adopted to prevent saltwater intrusion is by hindering the movement of saltwater by the construction of physical barriers along with the coast and increasing water table with the help of artificial recharge of freshwater (Todd 1959).

Management of coastal aquifers requires a proper understanding of saltwater intrusion problem. The movement of the saltwater interface can be understood by developing a density–salinity relationship which is affected by small-scale factors, i.e., beach-scale dynamics (Carey et al. 2009). The density–salinity relationship in saltwater intrusion was first observed by a French teacher named Du Commun (1828), but Ghyben (1888) and Herzberg (1901) provided a mathematical equation relating density and salinity to determine the location of the saltwater–freshwater interface in a coastal aquifer. The relationship given by Ghyben–Herzberg assumed saltwater and freshwater as two immiscible liquids separated by a sharp interface. However, in actual field condition, a mixing zone is present between saltwater and freshwater where the density of water in the mixing zone increases from that of freshwater to saline water (Fig. 17.1). Models using a sharp interface concept are referred to as sharp interface models, whereas the models which consider the variation of density in the mixing zone are classified as density-dependent models (Dagan and Bear 1968).

Various analytical and numerical techniques have been developed for solving both sharp interface and density-dependent model (Barlow and Reichard 2010). In recent times, with increased computational power, complex numerical techniques of solution have gained popularity. These numerical simulation models are used in predicting the saltwater intrusion for different pumping scenarios and thus come up with a sustainable pumping strategy. As there is an infinite number of pumping scenarios possible, sequentially evaluating all possible pumping management scenarios is impractical or unrealistic. Thus, optimization algorithms are linked with simulation models to find an optimal pumping strategy. Optimization algorithm performs an organized search within possible scenarios, iteratively solving the simulation models to find optimal pumping scenario and corresponding location of saltwater interface (Das and Datta 1999). Every iteration of the density-dependent simulation models is computationally taxing (Kopsiaftis et al. 2019), and several iterations of the simulation model within the optimization model may render the linked simulation optimization method inefficient.

This inefficiency of the linked simulation model is overcome by the use of surrogate as simulation models within the optimization framework (Datta et al. 2014). However, surrogate models are computationally efficient but require large datasets for training and validation. In most practical scenarios of saltwater intrusion, such large datasets are seldom available. However, considerable work has been done in the management of coastal aquifers against saltwater intrusion; there are still gaps which need to be addressed. This study attempts to put in perspective the various steps involved in coastal aquifer management, i.e., identification of the saltwater intrusion, prediction of saltwater–freshwater interface, and management of the saltwater intrusion.

17.2 Identification of the Saltwater Intrusion

Chance of saltwater intrusion increases with heavy unregulated pumping of freshwater from coastal aquifers. The first necessary step in any coastal aquifer management is to assess the current state of the aquifer and check if the aquifer is already affected by saltwater intrusion. Identification of saltwater in groundwater aquifers constitutes of two main parts, i.e., determining salt content and finding the origin of the salt in the groundwater samples. Determining the origin of the salt in groundwater is vital as non-costal aquifers can also exhibit high concentrations of salt. Thus, not all salt content in a groundwater aquifer can be attributed to saltwater intrusion. Different geophysical information and geochemical indicators are used in conjunction with Geographical Information System (GIS) to determine saltwater intrusion. Based on the type of data used, the saltwater intrusion identification can be broadly classified as: (1) geochemical techniques (analysis of water quality data); (2) vertical electrical sounding techniques (analysis of resistivity data); and (3) combined approach (analysis of both water quality and resistivity data).

17.2.1 Geochemical Technique

Analysis of water quality data plays a vital role in identifying the origin and concentration of the salt content. Collected water samples from wells are analyzed for total dissolved solids (TDS), electrical conductivity (EC), pH, anions $\left(Ca^{2+}, Mg^{2+}, Na^{+}, K^{+}\right)$, and cations $\left(Cl^{-}, HCO_3^{-}, SO_4^{2-}, NO_3^{-}, PO_4^{3-}\right)$. The concentration of cations and anions, TDS, pH, and EC in the sampled water is analyzed to determine the salinity and origin of the sampled well water. TDS is a good indication for determining the presence of salt in the water. Seawater generally contains about 3.5% salt (Bear 2012), i.e., about 35 g of salt dissolved in one liter of water. Due to this high salt concentration, observed TDS value for seawater ranges between 10,000 and 100,000 mg/l but generally TDS for freshwater is below 1,000 mg/l. TDS value between 1,000 and 10,000 mg/l is considered brackish water (Fetter 2000; Konikow and Reilly 1999; Rao et al. 2017). Therefore, TDS greater than 1,000 mg/l is considered as freshwater affected by salt. Ebraheem et al. (1997) first showed the applicability of TDS in the northern part of the Nile Delta of Egypt by calibrating the geophysical information with the help of TDS. The above-mentioned method of calibration of the geophysical information with the help of TDS was also performed by Khalil (2006), Sherif et al. (2006), Song et al. (2007), Cimino et al. (2008), and Rao et al. (2011) in Abu Zenima area, West Sinai, Egypt, Wadi Ham, UAE, Byunsan, Korea, Northern Sicily, and Godavari Delta Basin, India, respectively. Apart from the calibration of the geophysical information TDS is also used to develop a simulation model based on the sharp interface concept by Ranjbar and Ehteshami (2019). High level of TDS also can be observed in groundwater due to the high concentration of any pollutant (Huang et al. 2019; Jung 2001; Radfard et al. 2019; Shyamala et al. 2008). Therefore, it is challenging to use TDS as a sole identifier for saltwater intrusion, which is the reason TDS should be used only in conjunction with other geochemical methods to identify the presence of the saltwater in the coastal aquifer.

A plot of cations' and anions' concentrations on a piper diagram helps to identify the origin and concentration of salt in groundwater samples. Piper diagram helps to classify the waters sample based on the concentration of $CaHCO_3$, NaCl, $CaNaHCO_3$, CaMgCl, CaCl and $NaHCO_3$ in the water samples (Hoyle 1989; Kelly 2005). As the concentration of the NaCl in saltwater from the sea is very high, therefore a plot of constituents of the water samples on the piper diagram can easily identify the salt concentration and origin of the salt. The above method was applied by Arslan et al. (2012) on Bafra Plain, Turkey, in conjunction with other geochemical indicators for identifying saltwater intrusion. Same approach was used by Wen et al. (2012) in Eastern Laizhou Bay of China, Sola et al. (2013) in Andarax Delta of Spain, Kumar et al. (2014) in the southern portion of Chennai City in India, Askri (2015) in Al Musanaah in Sultanate of Oman, Hamzah et al. (2017) in Terengganu of Malaysia, Sarker et al. (2018) of southwest Bangladesh, and Chidambaram et al. (2018) in Cuddalore District of Tamil Nadu in India. Though piper diagram is a very good identifier of saltwater intrusion, it does not consider the secondary ions in the sample (Pulido-Leboeuf 2004). Also, the use of concentration as a percentage often overlooks the actual level of salinity (Singhal and Gupta 2010). Above-mentioned difficulties were rectified by Tomaszkiewicz et al. (2014) using piper diagram to develop a new method named GQI_{SWI}. The advantages of the GQI_{SWI} will be discussed in the later part of this section.

Another handy indicator for the identification of origin and location of salt is cluster analysis of cations and anions in water samples. In cluster analysis, cations and anions are grouped to find a correlation between them and plotted on a dendrogram. To detect salt in the water, sample cluster analysis was performed by Arslan (2013), Triki et al. (2014), and Askri (2015).

Chlorine (Cl^-) concentration is also an important indicator of saltwater intrusion as saltwater from the sea contains a very high concentration of NaCl. Therefore, the plot of cations and anions especially Br^+ and Na^+ against chlorine concentration helps to categorize the water sample for saltwater intrusion (Sola et al. 2013). Cl^- concentration is also used to calculate the seawater fraction, which is a ratio between chlorine concentration difference in sample water–freshwater and seawater–freshwater. Seawater fraction is extensively used to identify saltwater intrusion by Arslan et al. (2012), Askri (2015), Mondal et al. (2011a), Mondal et al. (2011b), Sola et al. (2013), Tomaszkiewicz et al. (2014), Wen et al. (2012). However, seawater fraction does not identify the main cation exchange that happens due to saltwater intrusion (Appelo and Postma 2004). Generally, saltwater contains a high level of Na^+; therefore, sodium adsorption ratio (SAR) coupled with seawater fraction is used to identify the origin of salt in groundwater samples. High SAR value (>26) is used to identify the presence of a high level of Na^+ for the water sample (Arslan 2013; Kumar et al. 2014; Mondal et al. 2011a; Rao et al. 2017; Triki et al. 2014).

The dominant feature in seawater is the high level of concentration Na^+ and Cl^- which can be easily identified by a multivariate technique called principal component analysis (PCA) (Anderson 1958; Sharma 1995; Singh et al. 2005). PCA helps to identify the dominant components in water samples by establishing a correlation between components of the water samples (Arslan 2013; Askri 2015; Kumar et al. 2014; Mondal et al. 2011b; Triki et al. 2014). Apart from PCA analysis, groundwater quality index (GQI) is also done on various dominant ions present in the water sample. Generally, GQI is calculated from the ratios of various cations and anions in the water sample. Among various GQIs, $GQI_{Piper(mix)}$ is used to classify between freshwater and saltwater and $GQI_{Piper(dom)}$ used to identify high Ca, Cl and $NaHCO_3$ concentrations in water. The concentration of the Na^+, K^+, and HCO_3^- is used to calculate $GQI_{Piper(dom)}$, and the concentration of Ca^{2+}, Mg^{2+}, and HCO_3^- is used to calculate $GQI_{Piper(mix)}$ (Hussien 2015). As discussed, the dominant constituents of saltwater from the sea are of Cl and Na which is not properly described in $GQI_{Piper(mix)}$. Therefore, Tomaszkiewicz et al. (2014) developed a new index named GQI_{SWI}, which is based on $GQI_{Piper(mix)}$ and $GQI_{Piper(dom)}$. GQI_{SWI} reduces difficulties in piper diagram and seawater faction.

The main objective of the geochemical method is to find the relation between two dominant fractions in the water samples. Ionic ratios of cations and anions such as Ca/Mg, Cl/HCO_3, Na/Cl are used for identifying saltwater in the water samples.

The ionic ratio of Ca/Mg, Cl/HCO_3, Na/Cl was used by Hamzah et al. (2017), Cl/Br and Cl/HCO_3 ratio by Sarker et al. (2018), and Cl/HCO_3 ratio by Chidambaram et al. (2018) to identify saltwater intrusion. Stable isotopes of oxygen (O^{18}) and hydrogen (H^2) are also suitable identifiers of origin of saltwater. The isotopic ratios of water sample are compared with international standards for identifying origin and the presence of saltwater in the groundwater (Kanagaraj et al. 2018; Sarker et al. 2018; Sola et al. 2013). Some other notable methods for identifying the origin of saltwater are Gibbs diagram (Rao et al. 2017) and Chadda's hydrographical process plot (Chidambaram et al. 2018). Geochemical methods are applied successfully to understand the presence and origin of saltwater intrusion. Being point source data, geochemical methods are useful in identifying only the horizontal distribution of the saltwater intrusion. It is difficult to quantify the vertical distribution of salt in the water sample using geochemical methods. It is also economically expensive and time-consuming.

17.2.2 Vertical Electrical Sounding

The shortcomings mentioned above are easily tackled by vertical electrical sounding (VES). In the case of VES, a low voltage direct current is passed through the ground by an array of electrodes and resulting change in potential due to the resistivity of the medium is measured via a micro-ohmmeters. Analyzing change in resistivity with different arrays of the electrodes provides information regarding properties of earth crust like density, magnetism, elasticity, and electrical resistivity (Todd 1959). Saltwater is a good conductor of electrical current; therefore, aquifers bearing saltwater pose a very low resistivity (0.2 Ω m) which clearly distinguish saltwater bearing medium from the surrounding area. Swartz (1937) first used VES to identify saltwater intrusion in Hawaiian Islands. Subsequent work by Hallenbach (1953), Flathe (1955), Dam and Meulenkamp (1967), Ginsberg and Levanton (1976), El-Waheidi et al. (1992) showed the applicability of 1-D resistivity method to identify the presence of saltwater in the aquifer.

In the case of 2-D or 3-D models, large array of electrodes are used and computer-driven algorithms automatically measure the change in resistivity by passing direct current to different electrodes. By analyzing different resistivity of different orientations of electrodes, a 2-D or 3-D image is produced. This system is also known as electrical resistivity tomography (ERT) (Werner et al. 2013). ERT was first used in coastal areas by Acworth and Dasey (2003) in a tidal creek of New South Wales, Australia, to identify mixing zone of saltwater with infiltrated rainwater. Day-Lewis et al. (2006) used ERT to identify submarine groundwater discharge. Comte and Banton (2007) used ERT images to validate the variable-density flow model SUTRA. Satriani et al. (2011) used ERT to identify the location of saltwater in coastal areas of Ionian Plain of Southern Italy. Belaval et al. (2003) and Manheim et al. (2004) used ERT method to identify the presence of freshwater under aquifers in offshore waters of South Carolina, Massachusetts, and Delmarva Peninsula (USA).

ERT systems are useful to find the distribution of the saltwater along with the coastal aquifers. The dynamics of saltwater can be observed by time-lapse electrical resistivity tomography (TERT) method. TERT is a modified version of ERT where an array of electrodes is permanently placed in the study area and resistivity information in the specified interval is measured. De Franco et al. (2009) and Poulsen et al. (2010) used TERT system in southern Venice Lagoon and coastal dune in Denmark to identify the behavior of saltwater with rainfall, tidal inundation, and channel stages. An automated TERT system was developed by Ogilvy et al. (2009) to understand saltwater dynamics on lower catchment of river Andarax, Almeria, Spain.

In resistivity method, electrodes are taped on the ground surface. Therefore, a physical ground connection via the electrodes is required to be set up to record the subsurface information, which is time-consuming. In electromagnetic (EM) method, electrodes are laid over the ground surface and do not require any physical ground connections to operate. This results in speedy operation, which increases the popularity of EM methods in recent times. EM methods are also able to detect a better local and near-surface inhomogeneity compared to resistivity method. Besides, EM methods outperform resistivity methods in desert and permafrost conditions (Goldman et al. 1991). EM methods are mainly of two types, i.e., time domain and frequency domain. Time-domain EM methods have been widely used for several decades; but in recent times, frequency-domain EM is gaining popularity (Stewart 1982). Goldman et al. (1991) first used time-domain EM (TDEM) method for detecting saltwater interface in Mediterranean coastal strip of Israel. TDEM was also used by Melloul and Goldenberg (1997) in coastal areas of Israel and by Al-Sayed and El-Qady (2007) in southwestern part of Sinai Egypt. As EM device does not require any physical ground connection, airborne EM measuring system has been developed. Airborne EM measuring system was used for saltwater intrusion by Paine (2003) in Red River Basin in Texas, Fitterman and Deszcz-Pan (1998) in Everglades National Park, Florida, and Viezzoli et al. (2010) in Venice coastal lagoon. The airborne EM system is cost-effective and can be deployed over a large area very quickly.

17.2.3 Combined Method

VES methods are used extensively to determine the location of the saltwater interface. As the observation of salinity is performed through resistivity information, therefore, any other low resistivity geological strata like ore mines, sedimentary rocks may give the same resistance signal as saltwater bearing strata. Therefore, validation of the geophysical data with some physical data is of paramount importance, which is the reason behind the application of VES method in conjunction with a geochemical analysis. Ebraheem et al. (1997) first tried to determine exact resistivity with the help of TDS data in the Nile Delta Region. TDS was also used by Khalil (2006), Sherif et al. (2006), Rao et al. (2011) to calibrate the electrical resistivity data.

Ratios of cations and anions provide comprehensive information about the occurrence and distribution of the salt in groundwater. Song et al. (2007) showed the applicability of ionic ratios to calibrate resistivity survey data in Byunsan, Korea, using ratio of chloride to bicarbonate and calcium to sodium ions. Bromide ion index, ratio of chloride ion to bicarbonate-plus-carbonate ions and ratio of strontium to chloride ion, was used to calibrate resistivity data by Shahsavari et al. (2015). Cimino et al. (2008) determined saltwater through PCA of water samples and used this information to calibrate VES data. As salt in saline water mainly comprises of sodium chloride, chlorine concentration also plays a vital role in determining the presence of saltwater. Samsudin et al. (2008) used chlorine concentration to calibrate VES data to determine the location of the saltwater interface in Kelantan, Malaysia. Stable isotopes of hydrogen and oxygen were used to calibrate VES data and identify the extent of the saltwater intrusion in Chennai, Tamil Nadu, India, by Kanagaraj et al. (2018).

Well log data can provide useful information about the vertical and horizontal distributions of the salinity. Well log data was first used by Edet and Okereke (2001) in Southern Nigeria to calibrate VES data. Hermans et al. (2012) applied this technique in Westhoek, Belgium, Shahsavari et al. (2015) in Kharg Island in Iran, and García-Menéndez et al. (2018) in Spanish Mediterranean coast.

17.3 Prediction Models

The next step of developing a sustainable management model for the coastal aquifer is to use spatiotemporal salt concentration/resistivity data and to assess the extent of saltwater intrusion in the coastal aquifer. Transport of dissolved salts in groundwater is governed by various physical processes such as advection, hydrodynamic dispersion, and density gradient. Numerical models are used to simulate these processes in coastal aquifer. These numerical models are solved forward in time with appropriate initial condition and boundary condition to estimate the extent of saltwater intrusion and to predict the fate and transport of the salt under various pumping stresses.

Most of the salt in groundwater aquifer is transported by advection, which essentially is the carrying of dissolved salt along with the flow of groundwater. The flow of groundwater is caused due to the naturally existing seaward pressure gradient, causing the freshwater to flow from the land toward the sea. The zone where the fresh groundwater meets the saline groundwater, the pressure is constant. This is also referred to as the freshwater–saltwater interface. The movement of this interface is calculated based on the change in pressure throughout the aquifer. Most numerical models can be broadly classified as sharp interface model or density-dependent model. In sharp interface, saltwater and freshwater are considered as two immiscible fluids. Therefore, a sharp interface is present between saltwater and freshwater (Fig. 17.3); but in the actual scenario, a mixing zone is present (Fig. 17.4) which is better described by density-dependent models. Sharp interface models are less accurate compared to density-dependent models as they do not consider mixing between saltwater and freshwater due to hydrodynamic dispersion. Density-dependent model combines flow with transport due to advection and hydrodynamic dispersion making

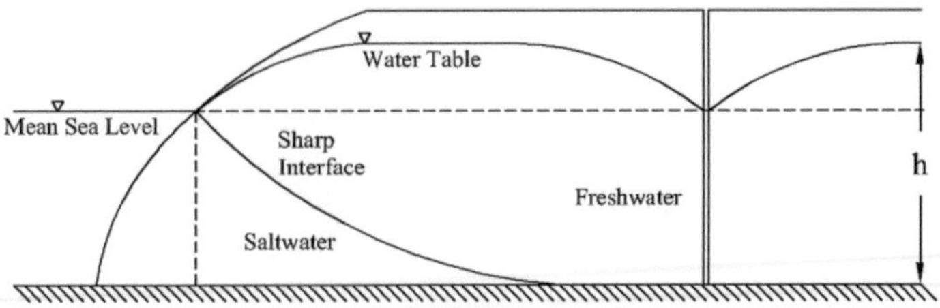

Fig. 17.3 Illustration of sharp interface of an unconfined aquifer

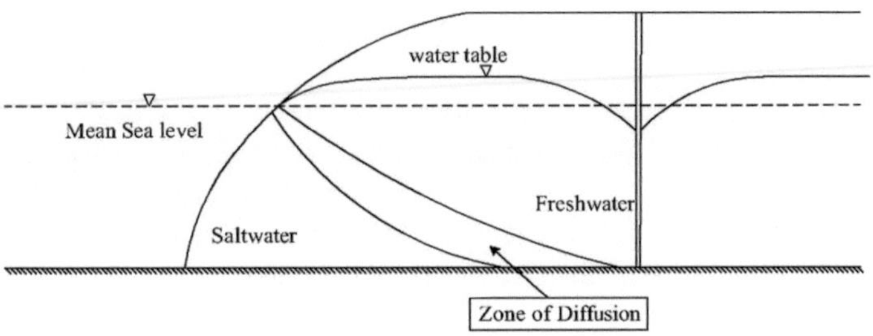

Fig. 17.4 Illustration of a density-dependent model for unconfined aquifer

it more realistic as well as nonlinear and computationally expensive to solve. The solution technique of these simulation models can be broadly classified as graphical methods, analytical methods, and numerical methods.

17.3.1 Graphical Method

Henry (1959) first developed a graphical solution based on sharp interface for single homogeneous aquifer and was first to conceptualize a single velocity potential (φ) for a two-dimensional flow system. This was further extended by Bear and Dagan (1964). Bear and Dagan (1964) introduced the concept of upcoming in saltwater intrusion process. These methods may be applied to very small homogeneous aquifers. However, nonlinearity caused due to change in boundary conditions cannot be handled by these approaches.

17.3.2 Analytical Solution

Dagan and Bear (1968) first developed exact equations for solving saltwater intrusion based on sharp interface model. The developed equations were nonlinear, so Dagan and Bear (1968) used perturbation technique to solve the problem. Haatush

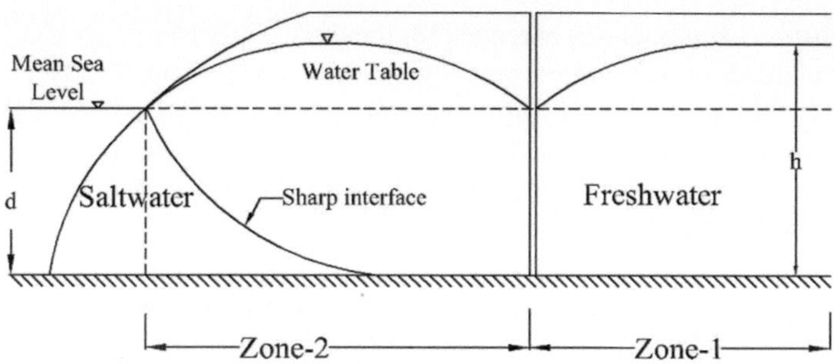

Fig. 17.5 Illustration of an unconfined coastal aquifer according to Strack (1976)

(1968) extended this method for cases where sharp interface between saltwater and freshwater does not reach the bottom of the aquifer. The developed partial differential equation is nonlinear, which only gives an approximate solution of the model. Schmork and Mercado (1969) gave experimental proof supporting this method. However, while incorporating boundary conditions, the above method became nonlinear leading to inaccuracies for large areas.

Instead of deriving the interface equation, Strack (1976) first developed a single potential concept, which reduces nonlinearity. The main advantage of using velocity potential (φ) lies in the fact that the solution can be written as one expression. Besides, Strack (1976) potential can handle three-dimensional problems. Instead of using the whole aquifer as a single system, Strack (1976) divided aquifer into two parts (Fig. 17.5) based on the presence of saltwater and freshwater. In zone 1, the only freshwater is present, and in zone 2, both denser saltwater and freshwater are present. The single potential is continuous throughout the aquifer. Strack (1976) method was extended for layered aquifer by Mualem and Bear (1974). Subsequently, Collins and Gelhar (1971) developed analytical solutions for the layered aquifer.

The inherent problem using sharp interface approach is that, it overestimates the location of the interface. This problem is overcome by the use of density-dependent model-based analytical solutions. Paster et al. (2006) developed a steady-state solution for density-dependent model using boundary layer approach. This was further extended to handle 3-D problem for field applications by Paster and Dagan (2008). The applicability of Strack (1976) approach was enhanced by Pool and Carrera (2011) with the introduction of density in Strack (1976) potential.

17.3.3 Numerical Solution

17.3.3.1 Numerical Solution for Sharp Interface Models

Mercer et al. (1980a) used finite difference technique to solve partial differential equation based on mass and momentum transfer for a two-dimensional single layer aquifer. Further, the model was extended for the three-dimensional problem and integrated with USGS's modular hydrogeological flow model (MODFLOW) by Mercer et al. (1980b). In this case, finite difference scheme was used to solve the partial differential equation. The main drawback of this method was that it could only locate the toe of the interface instead of locating the entire interface. In addition, it lacked stability in convergence. These limitations were overcome by Polo and Ramis (1983) who used a backward difference scheme to solve the problem. However, the developed methods were time-intensive in terms of computation and data preparation. These limitations were overcome by used of boundary integral equation method by Taigbenu et al. (1984) in a Strack potential approach. Mantoglou et al. (2004) solved this method with the help of the finite difference method.

Strack potential-based approach could not handle multilayered aquifer system. Essaid (1990) overcame this problem by solving the partial differential equation given by Mualem and Bear (1974). Authors used central in space and backward in time scheme to solve the partial differential equation. Huyakorn et al. (1996) used the Newton–Raphson method, and Guvanasen et al. (2000) used finite element method to solve the equation.

17.3.3.2 Numerical Solution for Density-Dependent Models

Sharp interface approach generally overestimates the extent of saltwater intrusion as it does not consider hydrodynamic dispersion causing mixing of saltwater and freshwater. This is overcome by the use of flow and transport equations coupled with density coupling term, which increases the complexity in simulation. Pinder and Cooper (1970) first solved Henry's problem for the transient condition using the method of characteristics. Volker and Rushton (1982) used finite difference scheme for estimating saltwater intrusion for a two-dimensional system. Guo and Langevin (2002) developed a three-dimensional finite difference model called SEAWAT to solve the coupled equation. This model combines MODFLOW and MT3DMS into a single program that solves the coupled flow and solute transport equations. This is one of the most widely used and is an efficient density-dependent model.

Lee and Cheng (1974) solved a steady-state two-dimensional problem using finite element-based triangular technique. The above method was solved using Galerkin finite element by Segol (1975) dominated by advective transport. Segol and Pinder (1976) used the above method in an actual field area. Huyakorn and Taylor (1976) presented a model using reference hydraulic head and concentration as dependent variables. Kawatani (1980) presented density-dependent model saltwater intrusion process for an aquifer–aquitard system, which was solved using Galerkin finite element technique by Frind (1982). Voss (1984) developed finite element-based Saturated–Unsaturated TRAnsport (SUTRA) code for the simulation of saltwater intrusion process in two-dimensional coastal aquifers. Galeati et al. (1992) used implicit Eulerian–Lagrangian method to solve the density-dependent model. Lin et al. (1997)

developed three-dimensional finite element-based model, FEMWATER that combines density term with transport term. Sherif et al. (1988) developed finite element model for a leaky confined aquifer.

17.4 Management Model

Sustainable use of water resources requires design and implementation of well-planned management strategies. One of the challenges in management of costal aquifer is to pump a large amount of freshwater without causing saltwater intrusion, which is conflicting in nature. When a coastal aquifer is pumped heavily, water table goes below MSL and saltwater from the sea comes near pumping wells. Therefore, the prime management challenge is to design some barriers or raise water table above MSL such that saltwater does not reach the pumping wells. The barrier can be either a physical barrier or pumping barrier. In addition, the water table can be increased above mean sea level through artificial recharge of freshwater. As a management alternative, pumping of freshwater can be limited such that the saltwater–freshwater wedge never reaches the pumping locations. Most of the above methods for the management of saltwater intrusion can be used in conjunction. Creating physical barriers or artificial recharge along shorelines is economically imprudent, and often the selection of management strategy depends on the availability of funds. Thus, pumping barriers are the most commonly implemented technique used for the management of saltwater intrusion.

Prediction models are used to predict the location of saltwater for different sets of input conditions. The simplest way to find a suitable management alternative is by sequentially evaluating all the possible scenarios, which is unrealistic and impossible. To overcome this problem, numerical simulation models are linked with optimization algorithms. Optimization algorithm performs an organized search for new and improved pumping schemes or other management alternatives. During this search process, the simulation model is run many times to quantify the impact of the pumping on the movement of the saltwater front. Depending upon the type of incorporated simulation model in optimization algorithm, it is classified as: (1) embedding approach, (2) linked simulation optimization approach, (3) surrogate modeling approach, and (4) combined approach using density-dependent model and sharp interface model.

Initially, classical linear and nonlinear optimization techniques were used to solve most of the optimization problems. Aguado and Remson (1974), Alley et al. (1976), Willis and Finney (1988) used a classical optimization algorithm to sequentially evaluate all possible pumping combinations. The main disadvantage of classical methods is that they may converge to a sub-optimal or near-optimal solution. Heuristic optimization techniques such as genetic algorithm (GA), simulated annealing (SA) have gained popularity in recent times due to their capability to find the optimal global solution.

17.4.1 Embedding Techniques

In the embedding technique, the simulation model is incorporated within an optimization model where the flow and transport equations are embedded in the optimization code as binding constraints. An initial attempt at embedding techniques had applied by Aguado and Remson (1974), where a linear response matrix was used to approximate the response of saltwater against pumping. Applicability of embedding technique is shown by Alley et al. (1976), Willis and Finney (1988), Culver and Shoemaker (1993), Das and Datta (1999), Abd-Elhamid (2010), Aguado and Remson (1974), and Alley et al. (1976) using linear programming (LP) in both steady-state and transient conditions. Willis and Finney (1988) first used nonlinear programming techniques and applied the method on an actual field condition. Culver and Shoemaker (1993) used differential dynamic programming with quasi-Newton approximations (QNDDP) and finite element method (FEMWATER) to determine optimal time-varying pumping. Das and Datta (1999) first used nonlinear programming to solve a multi-objective management model. Abd-Elhamid (2010) used the GA to determine the optimum way to recharge freshwater in coastal areas. Embedding techniques are useful for a small-scale problem, but for a large-scale problem, the convergence of the optimization model is the biggest challenge.

The response matrix is based on the principle of superposition and linearity. A linear response matrix is developed by multiple runs of external numerical simulation model prior to the start of optimization. Applicability of this approach on groundwater management was observed by Deininger (1970), Maddock III (1972), Maddock (1974), Rosenwald and Green (1974), Heidari (1982), Willis (1984), Hallaji and Yazicigil (1996), and Kourakos and Mantoglou (2015). Hallaji and Yazicigil (1996) used this approach to the coastal aquifer management problem. Hallaji and Yazicigil (1996) considered seven different groundwater management models to determine the optimal pumping policy for a coastal aquifer in southern Turkey threatened by saltwater intrusion.

17.4.2 Linked Simulation Optimization Approach

In the linked simulation optimization technique, the simulation model is linked with the optimization algorithm externally. The simulation model acts as binding constraint such that any candidate solution from the optimization has to satisfy these constraints. Cheng et al. (2000) show the applicability of the method by linking heuristic optimization, i.e., GA with the analytical model by Strack (1976). Analytical model by Strack (1976) was used by Mantoglou (2003) and Park and Aral (2004) in a linked simulation optimization approach. Mantoglou (2003) used sequential quadratic programming as the optimization algorithm, and Park and Aral (2004) used GA.

Numerical simulation model of sharp interface approach is linked with an optimization algorithm by Willis and Finney (1988), Finney et al. (1992), Emch and Yeh (1998), Mantoglou et al. (2004), Reichard and Johnson (2005), Karterakis et al. (2007). Mantoglou et al. (2004), and Karterakis et al. (2007) optimized pumping and well locations, whereas others used this model to optimize pumping only. Linear programming was used by Reichard and Johnson (2005) and Karterakis et al. (2007). Various heuristic optimization techniques such as quadratic programming (Willis and Finney 1988), projected Lagrangian algorithm MINOS (Emch and Yeh 1998; Finney et al. 1992), sequential quadratic programming (Mantoglou et al. 2004), and evolutionary algorithms (EA) (Karterakis et al. 2007) were used in a linked simulation optimization approach.

17.4.3 Surrogate Models

Density-dependent models can accurately simulate the physical behavior of saltwater intrusion, but due to the nonlinear nature of the density-dependent models, the CPU runtime for a density-dependent simulation model is very large. Due to the computational burden, it is unrealistic to apply density-dependent model linked with an optimization algorithm. To overcome this, problem density-dependent models were replaced entirely with a pattern recognizing algorithm like artificial neural network (ANN). Rao et al. (2003) first showed possibility of ANN as a surrogate in saltwater intrusion problem. ANN as a surrogate is used by Rao et al. (2004), Bhattacharjya and Datta (2005), Kourakos and Mantoglou (2006), Nikolos et al. (2008), and Ataie-Ashtiani et al. (2013). To develop the surrogate, ANN models are trained with pumping as input and saltwater concentration in observation as wells as output. Density-dependent models are used for generating training datasets. Heuristic optimization such as SA was used by Rao et al. (2003), Rao et al. (2004) as optimizer. GA was used by Bhattacharjya and Datta (2005), and Ataie-Ashtiani et al. (2013); sequential quadratic programming was used by Kourakos and Mantoglou (2006), and differential evolution (DE) algorithm was used by Nikolos et al. (2008).

A large number of datasets are required to fit a surrogate model. Generating data using density-depended model is very tiresome (Rao et al. 2004), as the CPU runtime for a density-dependent simulation model is very large. To reduce CPU runtime problem, Dhar and Datta (2009) used an undertrained ANN linked with non-sorting genetic algorithm-II (NSGA-II) as a screening model (metamodel), before linking NSGA-II with a density-dependent model. Another approach to overcome this problem had been addressed by Kourakos and Mantoglou (2009), where the authors used modular neural network (MNN) such that each module are trained for a different decision variable. Kourakos and Mantoglou (2009) used evolutionary annealing-simplex (EAS).

Genetic programming (GP) model was used as a surrogate by Sreekanth and Datta (2010; 2011). As predictions using a single surrogate model may sometimes show eccentric results, thus an ensemble of surrogates based on different realization, i.e., each surrogate is efficient for a segment was developed by Sreekanth and Datta (2010). Sreekanth and Datta (2010) used the nonparametric bootstrap method to generate different realizations. To calculate an optimal number of surrogate models, root-mean-square error (RMSE) of each surrogate was computed, and the coefficient of variation of these RMSE was considered as the measure of uncertainty in the ensemble of the models. Authors compared this result with ANN-based surrogates (Sreekanth and Datta 2011) and reported that this innovative method reduces the computational burden in training the ANN. Authors also applied this method on an actual area to showed the applicability of this method (Sreekanth and Datta 2014).

In the same procedure to reduce the training difficulties, Roy and Datta (2017) used fuzzy logic-based adaptive neuro-fuzzy inference system (ANFIS). The main advantage of fuzzy logic is that instead of using multiple surrogate models for multiple output problems one surrogate is efficient for all regions of solution. Other notable works on surrogate models have been performed by Bhattacharjya and Datta (2009), Papadopoulou et al. (2010), Ataie-Ashtiani et al. (2013), Kourakos and Mantoglou (2013), Hussain et al. (2015), Christelis and Mantoglou (2016b), Christelis et al. (2018), and Lal and Datta (2019).

17.4.4 Combined Approach Using Density-Dependent Model and Sharp Interface Model

Due to its linear nature, sharp interface models are less complicated compared to density-dependent models (Cheng et al. 2000). However, the simplification of the sharp interface model leads to an overestimation of the extent of the saltwater–freshwater wedge (Dausman et al. 2010; Dokou and Karatzas 2012). Results of the numerical model by the sharp interface in the estimation of the saltwater–freshwater wedge can be more realistic if dispersion can be included within the sharp interface model. Sharp interface model is a function of head and density ratio. The head is affected by pumping; therefore, modification of density ratio in sharp interface model based on dispersion such that outcomes of the sharp interface model will be similar as outcome of the density-dependent model (Pool and Carrera 2011). This idea was further extended by Christelis and Mantoglou (2016a), where density ratio is modified with the help of density-dependent model within an optimization framework. Christelis and Mantoglou (2016a) designed density ratio such that the location of sharp interface and iso-salinity contour (calculated by density-dependent model) will match.

17.5 Conclusions

Sustainable exploitation of coastal aquifer requires pumping out a large amount of freshwater keeping the saline water at bay. Managing saltwater intrusion is a complex task due to the conflicting nature of these objectives. A successful management strategy comprises of early detection of saltwater intrusion, predicting the extent and future course of intrusion. Though a majority of the identification technique focuses on the concentration of the salt ions and their origin, coupled approach using geophysical and geochemical approach proves to be better. New detection techniques based on EM method are easy and fast to apply. Airborne EM systems need to be explored more as they can produce reliable results over vast areas in a short span of time.

Predicting the freshwater–saltwater boundary accurately is a major challenge in any saltwater management scenario. Both sharp interface models and density-dependent model predict the boundary but have their own limitations. Sharp interface models are quick to simulate but they overestimate the intrusion. Though density-dependent models are more accurate, it consumes an enormous amount of CPU time in the simulation. Thus, direct use of density-dependent models in a linked simulation optimization is not preferred as a management technique. Surrogate models can be a viable alternative to the use of density-dependent model but require extensive data. MNN and fuzzy logic-based ANN may overcome some of the issues with the density-dependent models but still require training.

To overcome these difficulties, the combined use of density-dependent model and sharp interface model is applied. Capability of fast simulation speeds from sharp interface models, and accuracy of prediction in the density-dependent model is used simultaneously in this methodology. Still considerable work needs to be done for further application of multi-objective problems. Computational burden for seawater intrusion management models is still a challenge, especially when linked to a heuristic optimization algorithm in saltwater intrusion management problems. Use of classical optimization often leads to sub-optimal results. Therefore, further work needs to be done to overcome these limitations.

References

Abd-Elhamid HF (2010) A simulation-optimization model to study the control of seawater intrusion in coastal aquifers

Acworth R, Dasey G (2003) Mapping of the hyporheic zone around a tidal creek using a combination of borehole logging, borehole electrical tomography and cross-creek electrical imaging, New South Wales, Australia. Hydrogeol J 11:368–377

Aguado E, Remson I (1974) Ground-water hydraulics in aquifer management. J Hydraul Div 100:103–118

Al-Sayed E, El-Qady G (2007) Evaluation of sea water intrusion using the electrical resistivity and transient electromagnetic survey: case study at fan of Wadi Feiran, Sinai, Egypt. In: EGM 2007 international workshop

Alley WM, Aguado E, Remson I (1976) Aquifer management under transient and steady-state conditions. J Am Water Resour Assoc 12:963–973
Anderson TW (1958) An introduction to multivariate statistical analysis, vol 2. Wiley, New York
Appelo CAJ, Postma D (2004) Geochemistry, groundwater and pollution. CRC Press
Arslan H (2013) Application of multivariate statistical techniques in the assessment of groundwater quality in seawater intrusion area in Bafra Plain, Turkey. Environ Monit Assess 185:2439–2452
Arslan H, Cemek B, Demir Y (2012) Determination of seawater intrusion via hydrochemicals and isotopes in Bafra Plain, Turkey. Water Resour Manag 26:3907–3922
Askri B (2015) Hydrochemical processes regulating groundwater quality in the coastal plain of Al Musanaah, Sultanate of Oman. J Afr Earth Sci 106:87–98
Ataie-Ashtiani B, Ketabchi H, Rajabi MM (2013) Optimal management of a freshwater lens in a small island using surrogate models and evolutionary algorithms. J Hydrol Eng 19:339–354
Barlow PM, Reichard EG (2010) Saltwater intrusion in coastal regions of North America. Hydrogeol J 18:247–260
Bear J (2012) Hydraulics of groundwater. Courier Corporation
Bear J, Cheng AH-D, Sorek S, Ouazar D, Herrera I (1999) Seawater intrusion in coastal aquifers: concepts, methods and practices, vol 14. Springer Science and Business Media
Bear J, Dagan G (1964) Some exact solutions of interface problems by means of the hodograph method. J Geophys Res 69:1563–1572
Belaval M, Lane J Jr, Lesmes DP, Kineke GC (2003) Continuous-resistivity profiling for coastal groundwater investigations: three case studies SAGEEP proceedings, vol 14. Texas, pp 1–14
Bhattacharjya RK, Datta B (2005) Optimal management of coastal aquifers using linked simulation optimization approach. Water Resour Manag 19:295–320
Bhattacharjya RK, Datta B (2009) ANN-GA-based model for multiple objective management of coastal aquifers. J Water Resour Plan Manag 135:314–322
Carey H, Lenkopane M, Werner A, Li L, Lockington D (2009) Tidal controls on coastal groundwater conditions: field investigation of a macrotidal system. Aust J Earth Sci 56:1165–1179
Cheng AD, Halhal D, Naji A, Ouazar D (2000) Pumping optimization in saltwater-intruded coastal aquifers. Water Resour Res 36:2155–2165
Chidambaram S et al (2018) Assessment of hydrogeochemical status of groundwater in a coastal region of Southeast coast of India. Appl Water Sci 8:27
Christelis V, Mantoglou A (2016a) Coastal aquifer management based on the joint use of density-dependent and sharp interface models. Water Resour Manag 30:861–876
Christelis V, Mantoglou A (2016b) Pumping optimization of coastal aquifers assisted by adaptive metamodelling methods and radial basis functions. Water Resour Manage 30:5845–5859
Christelis V, Regis RG, Mantoglou A (2018) Surrogate-based pumping optimization of coastal aquifers under limited computational budgets. J Hydroinform 20:164–176
Cimino A, Cosentino C, Oieni A, Tranchina L (2008) A geophysical and geochemical approach for seawater intrusion assessment in the Acquedolci coastal aquifer (Northern Sicily). Environ Geol 55:1473
Collins MA, Gelhar LW (1971) Seawater intrusion in layered aquifers. Water Resour Res 7:971–979
Comte JC, Banton O (2007) Cross-validation of geo-electrical and hydrogeological models to evaluate seawater intrusion in coastal aquifers. Geophys Res Lett, 34
Culver TB, Shoemaker CA (1993) Optimal control for groundwater remediation by differential dynamic programming with Quasi-Newton approximations. Water Resour Res 29:823–831
Dagan G, Bear J (1968) Solving the problem of local interface upconing in a coastal aquifer by the method of small perturbations. J Hydraul Res 6:15–44
Dam JV, Meulenkamp J (1967) Some results of the geo-electrical resistivity method in ground water investigations in the Netherlands. Geophys Prospect 15:92–115
Das A, Datta B (1999) Development of management models for sustainable use of coastal aquifers. J Irrig Drain Eng 125:112–121
Datta B, Prakash O, Sreekanth J (2014) Application of genetic programming models incorporated in optimization models for contaminated groundwater systems management. In: EVOLVE-A

bridge between probability, set oriented numerics, and evolutionary computation V. Springer, pp 183–199
Dausman AM, Langevin C, Bakker M, Schaars F (2010) A comparison between SWI and SEAWAT—the importance of dispersion, inversion and vertical anisotropy, vol 21. In: Proceedings of SWIM, pp 271–274
Day-Lewis F, White E, Johnson C, Lane J Jr, Belaval M (2006) Continuous resistivity profiling to delineate submarine groundwater discharge—examples and limitations. Lead Edge 25:724–728
De Franco R et al. (2009) Monitoring the saltwater intrusion by time lapse electrical resistivity tomography: the Chioggia test site (Venice Lagoon, Italy). J Appl Geophys 69:117–130
Deininger RA (1970) Systems analysis of water supply systems. J Am Water Resour Assoc 6:573–579
Dhar A, Datta B (2009) Saltwater intrusion management of coastal aquifers. I: linked simulation-optimization. J Hydrol Eng 14:1263–1272
Dokou Z, Karatzas GP (2012) Saltwater intrusion estimation in a karstified coastal system using density-dependent modelling and comparison with the sharp-interface approach. Hydrol Sci J 57:985–999
Du Commun J (1828) ART. XXV—On the cause of fresh water springs. Fountains, &c. Am J Sci Arts (1820–1879) 14:174
Ebraheem AAM, Senosy MM, Dahab KA (1997) Geoelectrical and hydrogeochemical studies for delineating ground-water contamination due to salt-water intrusion in the northern part of the Nile Delta, Egypt. Groundwater 35:216–222
Edet A, Okereke C (2001) A regional study of saltwater intrusion in southeastern Nigeria based on the analysis of geoelectrical and hydrochemical data. Environ Geol 40:1278–1289
El-Waheidi M, Merlanti F, Pavan M (1992) Geoelectrical resistivity survey of the central part of Azraq basin (Jordan) for identifying saltwater/freshwater interface. J Appl Geophys 29:125–133
Emch PG, Yeh WW (1998) Management model for conjunctive use of coastal surface water and ground water. J Water Resour Plan Manag 124:129–139
Essaid HI (1990) A multilayered sharp interface model of coupled freshwater and saltwater flow in coastal systems: model development and application. Water Resour Res 26:1431–1454
Fetter CW (2000) Applied hydrogeology. Prentice Hall
Finney BA, Samsuhadi, Willis R (1992) Quasi-three-dimensional optimization model of Jakarta basin. J Water Resour Plan Manag 118:18–31
Fitterman DV, Deszcz-Pan M (1998) Helicopter EM mapping of saltwater intrusion in Everglades National Park, Florida. Explor Geophys 29:240–243
Flathe H (1955) Possibilities and limitations in applying geoelectrical methods to hydrogeological problems in the coastal areas of north west Germany. Geophys Prospect 3:95–109
Frind EO (1982) Seawater intrusion in continuous coastal aquifer-aquitard systems. Adv Water Resour 5:89–97
Galeati G, Gambolati G, Neuman S (1992) Coupled and partially coupled Eulerian-Lagrangian model of freshwater-seawater mixing. Water Resour Res 28:149–165
García-Menéndez O, Ballesteros BJ, Renau-Pruñonosa A, Morell I, Mochales T, Ibarra PI, Rubio FM (2018) Using electrical resistivity tomography to assess the effectiveness of managed aquifer recharge in a salinized coastal aquifer. Environ Monit Assess 190:100
Ghyben WB (1888) Notes on the probable results of well drilling near Amsterdam Tijdschrift van het Koninklijk Inst van Ingenieur. Hague 9:8–22
Ginsberg A, Levanton A (1976) Determination of saltwater interface by electrical resistivity sounding. Hydrol Sci Bull 21:561–568
Goldman M, Gilad D, Ronen A, Melloul A (1991) Mapping of seawater intrusion into the coastal aquifer of Israel by the time domain electromagnetic method. Geoexploration 28:153–174
Guo W, Langevin CD (2002) User's guide to SEAWAT: a computer program for simulation of three-dimensional variable-density ground-water flow
Guvanasen V, Wade SC, Barcelo MD (2000) Simulation of regional ground water flow and salt water intrusion in Hernando County Florida. Groundwater 38:772–783

Haatush MS (1968) Unsteady movement of fresh water in thick unconfined saline aquifers. Hydrol Sci J 13:40–60
Hallaji K, Yazicigil H (1996) Optimal management of a coastal aquifer in southern Turkey. J Water Resour Plan Manag 122:233–244
Hallenbach F (1953) Geo-electrical problems of the hydrology of west german areas. Geophys Prospect 1:241–249
Hamzah Z, Aris AZ, Ramli MF, Juahir H, Narany TS (2017) Groundwater quality assessment using integrated geochemical methods, multivariate statistical analysis, and geostatistical technique in shallow coastal aquifer of Terengganu, Malaysia. Arab J Geosci 10:49
Heidari M (1982) Application of linear system's theory and linear programming to ground water management in Kansas JAWRA. J Am Water Resour Assoc 18:1003–1012
Henry HR (1959) Salt intrusion into fresh-water aquifers. J Geophys Res 64:1911–1919
Hermans T, Vandenbohede A, Lebbe L, Martin R, Kemna A, Beaujean J, Nguyen F (2012) Imaging artificial salt water infiltration using electrical resistivity tomography constrained by geostatistical data. J Hydrol 438:168–180
Herzberg A (1901) DieWasserversorgung einiger Nordseeber (The water supply of parts of the North Sea coast in Germany): Z Gasbeleucht Wasserversorg
Hoyle BL (1989) Ground-water quality variations in a silty alluvial soil aquifer. Oklahoma Groundwater 27:540–549
Huang KZ, Xie YF, Tang HL (2019) Formation of disinfection by-products under influence of shale gas produced water. Sci Total Environ 647:744–751
Hussain MS, Javadi AA, Ahangar-Asr A, Farmani R (2015) A surrogate model for simulation—optimization of aquifer systems subjected to seawater intrusion. J Hydrol 523:542–554
Hussien R (2015) Groundwater quality index studies for seawater Intrusion in coastal aquifer Ras Sudr Egypt using geographic information system. Sciences 5:209–222
Huyakorn P, Taylor C (1976) Finite element models for coupled groundwater flow and convective dispersion. In: Gray WG et al (eds) Proceedings of 1st international conference on finite elements in water resource. Pen-tech Press, Princeton University, London, pp 1131–1151
Huyakorn P, Wu Y, Park N (1996) Multiphase approach to the numerical solution of a sharp interface saltwater intrusion problem. Water Resour Res 32:93–102
Jung MC (2001) Heavy metal contamination of soils and waters in and around the Imcheon Au–Ag mine, Korea. Appl Geochem 16(11–12):1369–1375
Kanagaraj G, Elango L, Sridhar S, Gowrisankar G (2018) Hydrogeochemical processes and influence of seawater intrusion in coastal aquifers south of Chennai, Tamil Nadu, India. Environ Sci Pollut Res, 1–23
Karterakis SM, Karatzas GP, Nikolos IK, Papadopoulou MP (2007) Application of linear programming and differential evolutionary optimization methodologies for the solution of coastal subsurface water management problems subject to environmental criteria. J Hydrol 342:270–282
Kawatani T (1980) Behavior of seawater intrusion in layered coastal aquifers finite elements in water resources, 1
Kelly F (2005) Seawater intrusion topic paper: Island country health department
Khalil MH (2006) Geoelectric resistivity sounding for delineating salt water intrusion in the Abu Zenima area, west Sinai, Egypt. J Geophys Eng 3:243
Konikow L, Reilly T (1999) Seawater intrusion in the United States. Seawater intrusion in coastal aquifers-concepts, methods and practices. Springer, pp. 463–506
Kopsiaftis G, Christelis V, Mantoglou A (2019) Comparison of sharp interface to variable density models in pumping optimisation of coastal aquifers water resources management, pp 1–13
Kourakos G, Mantoglou A (2006) Pumping optimization of coastal aquifers using 3-d density models and approximations with neural networks. In: XVI international conference on computational methods in water resources, Copenhagen
Kourakos G, Mantoglou A (2009) Pumping optimization of coastal aquifers based on evolutionary algorithms and surrogate modular neural network models. Adv Water Resour 32:507–521

Kourakos G, Mantoglou A (2013) Development of a multi-objective optimization algorithm using surrogate models for coastal aquifer management. J Hydrol 479:13–23

Kourakos G, Mantoglou A (2015) An efficient simulation-optimization coupling for management of coastal aquifers. Hydrogeol J 23:1167–1179

Kumar PS, Elango L, James E (2014) Assessment of hydrochemistry and groundwater quality in the coastal area of South Chennai, India. Arab J Geosci 7:2641–2653

Lal A, Datta B (2019) Multi-objective groundwater management strategy under uncertainties for sustainable control of saltwater intrusion: Solution for an island country in the South Pacific. J Environ Manag 234:115–130

Lee CH, Cheng RTS (1974) On seawater encroachment in coastal aquifers. Water Resour Res 10:1039–1043

Lin H-CJ, Richards DR, Yeh G-T, Cheng J-R, Cheng H-P (1997) FEMWATER: a three-dimensional finite element computer model for simulating density-dependent flow and transport in variably saturated media. Army engineer waterways experiment station vicksburg MS coastal hydraulics Lab

Maddock T III (1972) A ground-water planning model: a basis for a data collection network, paper presented at the international symposium on uncertainties in hydrologic and water resource systems. Int Assoc Sci Hydrol, Univ of Ariz, Tuscon

Maddock T (1974) The operation of a stream-aquifer system under stochastic demands. Water Resour Res 10:1–10

Manheim FT, Krantz DE, Bratton JF (2004) Studying ground water under Delmarva coastal bays using electrical resistivity. Groundwater 42:1052–1068

Mantoglou A (2003) Pumping management of coastal aquifers using analytical models of saltwater intrusion. Water Resour Res, 39

Mantoglou A, Papantoniou M, Giannoulopoulos P (2004) Management of coastal aquifers based on nonlinear optimization and evolutionary algorithms. J Hydrol 297:209–228

Melloul AJ, Goldenberg L (1997) Monitoring of seawater intrusion in coastal aquifers: basics and local concerns. J Environ Manag 51:73–86

Mercer JW, Larson S, Faust CR (1980a) Finite-difference model to simulate the areal flow of saltwater and fresh water separated by an interface. US Geological Survey

Mercer JW, Larson SP, Faust CR (1980b) Simulation of salt-water interface motion. Groundwater 18:374–385

Mondal N, Singh V, Saxena V, Singh V (2011a) Assessment of seawater impact using major hydrochemical ions: a case study from Sadras, Tamilnadu, India. Environ Monit Assess 177:315–335

Mondal NC, Singh VP, Singh S, Singh VS (2011b) Hydrochemical characteristic of coastal aquifer from Tuticorin, Tamil Nadu, India. Environ Monit Assess 175:531–550

Mualem Y, Bear J (1974) The shape of the interface in steady flow in a stratified aquifer. Water Resour Res 10:1207–1215

Nikolos IK, Stergiadi M, Papadopoulou MP, Karatzas GP (2008) Artificial neural networks as an alternative approach to groundwater numerical modelling and environmental design. Hydrol Process 22:3337–3348

Ogilvy R et al (2009) Automated monitoring of coastal aquifers with electrical resistivity tomography near surface. Geophysics 7:367–375

Paine JG (2003) Determining salinization extent, identifying salinity sources, and estimating chloride mass using surface, borehole, and airborne electromagnetic induction methods. Water Resour Res, 39

Papadopoulou MP, Nikolos IK, Karatzas GP (2010) Computational benefits using artificial intelligent methodologies for the solution of an environmental design problem: saltwater intrusion. Water Sci Technol 62:1479–1490

Park C-H, Aral MM (2004) Multi-objective optimization of pumping rates and well placement in coastal aquifers. J Hydrol 290:80–99

Paster A, Dagan G (2008) Mixing at the interface between two fluids in aquifer well upconing steady flow. Water Resour Res, 44
Paster A, Dagan G, Guttman J (2006) The salt-water body in the northern part of Yarkon-Taninim aquifer: field data analysis, conceptual model and prediction. J Hydrol 323:154–167
Pinder GF, Cooper HH (1970) A numerical technique for calculating the transient position of the saltwater front. Water Resour Res 6:875–882
Polo JF, Ramis FJR (1983) Simulation of salt water–fresh water interface motion. Water Resour Res 19:61–68
Pool M, Carrera J (2011) A correction factor to account for mixing in Ghyben-Herzberg and critical pumping rate approximations of seawater intrusion in coastal aquifers. Water Resour Res, 47
Poulsen SE, Rasmussen KR, Christensen NB, Christensen S (2010) Evaluating the salinity distribution of a shallow coastal aquifer by vertical multielectrode profiling (Denmark). Hydrogeol J 18:161–171
Pulido-Leboeuf P (2004) Seawater intrusion and associated processes in a small coastal complex aquifer (Castell de Ferro, Spain). Appl Geochem 19:1517–1527
Radfard M, Yunesian M, Nabizadeh R, Biglari H, Nazmara S, Hadi M, Yousefi N, Yousefi M, Abbasnia A, Mahvi AH (2019) Drinking water quality and arsenic health risk assessment in Sistan and Baluchestan, Southeastern Province, Iran. Hum Ecol Risk Assess Int J 25(4):949–965
Ranjbar A, Ehteshami M (2019) Spatio-temporal mapping of salinity in the heterogeneous coastal aquifer. Appl Water Sci 9:32
Rao NS, Vidyasagar G, Rao PS, Bhanumurthy P (2017) Chemistry and quality of groundwater in a coastal region of Andhra Pradesh, India. Appl Water Sci 7:285–294
Rao S, Sreenivasulu V, Bhallamudi SM, Thandaveswara B, Sudheer K (2004) Planning groundwater development in coastal aquifers/Planification du développement de la ressource en eau souterraine des aquifères côtiers. Hydrol Sci J 49:155–170
Rao S, Thandaveswara B, Bhallamudi SM, Srinivasulu V (2003) Optimal groundwater management in deltaic regions using simulated annealing and neural networks. Water Resour Manage 17:409–428
Rao VG, Rao GT, Surinaidu L, Rajesh R, Mahesh J (2011) Geophysical and geochemical approach for seawater intrusion assessment in the Godavari Delta Basin, AP, India. Water Air Soil Pollut 217:503–514
Reichard EG, Johnson TA (2005) Assessment of regional management strategies for controlling seawater intrusion. J Water Resour Plan Manag 131:280–291
Rosenwald GW, Green DW (1974) A method for determining the optimum location of wells in a reservoir using mixed-integer programming. Soc Petrol Eng J 14:44–54
Roy DK, Datta B (2017) Fuzzy c-mean clustering based inference system for saltwater intrusion processes prediction in coastal aquifers. Water Resour Manag 31:355–376
Samsudin AR, Haryono A, Hamzah U, Rafek A (2008) Salinity mapping of coastal groundwater aquifers using hydrogeochemical and geophysical methods: a case study from north Kelantan, Malaysia. Environ Geol 55:1737–1743
Sarker MMR, Van Camp M, Islam M, Ahmed N, Walraevens K (2018) Hydrochemistry in coastal aquifer of southwest Bangladesh: origin of salinity. Environ Earth Sci 77:39
Satriani A, Loperte A, Proto M (2011) Electrical resistivity tomography for coastal salt water intrusion characterization along the Ionian coast of Basilicata region (Southern Italy). In: Fifteenth international water technology conference
Schmork S, Mercado A (1969) Upconing of fresh water—sea water interface below pumping wells, field study. Water Resour Res 5:1290–1311
Segol G (1975) A Galerkin-finite element technique for calculating the transition position of the salt-water front. Water Resour Res 11:347–373
Segol G, Pinder GF (1976) Transient simulation of saltwater intrusion in southeastern Florida. Water Resour Res 12:65–70
Shahsavari AA et al (2015) Determination of origin and distribution of saline water in the aquifer of Kharg Island, Iran. Arab J Geosci 8:3129–3137

Shamir U, Bear J, Gamliel A (1984) Optimal annual operation of a coastal aquifer. Water Resour Res 20:435–444
Sharma S (1995) Applied multivariate techniques. Wiley
Sherif M, El Mahmoudi A, Garamoon H, Kacimov A, Akram S, Ebraheem A, Shetty A (2006) Geoelectrical and hydrogeochemical studies for delineating seawater intrusion in the outlet of Wadi Ham UAE. Environ Geol 49:536–551
Sherif MM, Singh VP, Amer AM (1988) A two-dimensional finite element model for dispersion (2D-FED) in coastal aquifers. J Hydrol 103:11–36
Shyamala R, Shanthi M, Lalitha P (2008) Physicochemical analysis of borewell water samples of Telungupalayam area in Coimbatore district Tamilnadu, India. J Chem 5:924–929
Singh KP, Malik A, Singh VK, Mohan D, Sinha S (2005) Chemometric analysis of groundwater quality data of alluvial aquifer of Gangetic plain, North India. Anal Chim Acta 550:82–91
Singhal BBS, Gupta RP (2010) Applied hydrogeology of fractured rocks. Springer Science and Business Media
Sola F, Vallejos A, Moreno L, Geta JL, Bosch AP (2013) Identification of hydrogeochemical process linked to marine intrusion induced by pumping of a semiconfined mediterranean coastal aquifer. Int J Environ Sci Technol 10:63–76
Song S-H, Lee J-Y, Park N (2007) Use of vertical electrical soundings to delineate seawater intrusion in a coastal area of Byunsan Korea. Environ Geol 52:1207–1219
Sreekanth J, Datta B (2010) Multi-objective management of saltwater intrusion in coastal aquifers using genetic programming and modular neural network based surrogate models. J Hydrol 393:245–256
Sreekanth J, Datta B (2011) Comparative evaluation of genetic programming and neural network as potential surrogate models for coastal aquifer management. Water Resour Manag 25:3201–3218
Sreekanth J, Datta B (2014) Stochastic and robust multi-objective optimal management of pumping from coastal aquifers under parameter uncertainty. Water Resour Manag 28:2005–2019
Sreekanth J, Datta B (2015) Review: simulation-optimization models for the management and monitoring of coastal aquifers. Hydrogeol J 23:1155–1166
Stewart MT (1982) Evaluation of electromagnetic methods for rapid mapping of salt-water interfaces in coastal aquifers. Groundwater 20:538–545
Strack O (1976) A single-potential solution for regional interface problems in coastal aquifers. Water Resour Res 12:1165–1174
Swartz J (1937) Resistivity-studies of some salt-water boundaries in the Hawaiian Islands Eos. Trans Am Geophys Union 18:387–393
Taigbenu AE, Liggett JA, Cheng AHD (1984) Boundary integral solution to seawater intrusion into coastal aquifers. Water Resour Res 20:1150–1158
Todd DK (1959) Ground water hydrology. Wiley, New York
Tomaszkiewicz M, Najm MA, El-Fadel M (2014) Development of a groundwater quality index for seawater intrusion in coastal aquifers. Environ Model Softw 57:13–26
Triki I, Trabelsi N, Zairi M, Dhia HB (2014) Multivariate statistical and geostatistical techniques for assessing groundwater salinization in Sfax, a coastal region of eastern Tunisia. Desalin Water Treat 52:1980–1989
Viezzoli A, Tosi L, Teatini P, Silvestri S (2010) Surface water—groundwater exchange in transitional coastal environments by airborne electromagnetics: the Venice Lagoon example. Geophys Res Lett 37
Volker R, Rushton K (1982) An assessment of the importance of some parameters for seawater intrusion in aquifers and a comparison of dispersive and sharp-interface modelling approaches. J Hydrol 56:239–250
Voss CI (1984) A finite-element simulation model for saturated-unsaturated, fluid-density-dependent ground-water flow with energy transport or chemically-reactive single-species solute transport. US Geological Survey, Reston, VA
Wen X, Diao M, Wang D, Gao M (2012) Hydrochemical characteristics and salinization processes of groundwater in the shallow aquifer of Eastern Laizhou Bay. China Hydrol Process 26:2322–2332

Werner AD et al (2013) Seawater intrusion processes, investigation and management: recent advances and future challenges. Adv Water Resour 51:3–26
WHO (2004) Guidelines for drinking-water quality: recommendations, vol 1. World Health Organization
Willis R (1984) A unified approach to regional groundwater management. Groundwater Hydraul, 392–407
Willis R, Finney BA (1988) Planning model for optimal control of saltwater intrusion. J Water Resour Plan Manag 114:163–178

Chapter 18
Numerical Errors Associated with Groundwater Models and Improving the Reliability of Models in Environmental Management Issues

K. V. Sruthi, Kim Hyun Su, Anupma Sharma and N. C. Ghosh

Abstract Improving numerical accuracy of the finite difference (FD) models of groundwater transport is achieved here by removing the truncation error associated with advection–dispersion equation with first-order reaction and sink/source (ADERS). This chapter presents theoretical and numerical truncation error associated with ADERS for the first time. The truncation errors associated with the FD models of the ADERS are formulated from Taylor series analysis. The error expressions are based on a general form of the corresponding FD equation. A temporally and spatially weighted parametric approach is applied to differentiate among the various FD models. The study revealed that all the FD models (explicit, Crank–Nicolson, implicit) suffer from truncation errors and formulated an expression for error from sink/source term for the first time. The effects of these truncation errors on the solution of ADERS are demonstrated by comparison of numerical solution from different FD models with the analytical solution. The results revealed that these errors are not negligible and correcting the FD schemes for truncation error can result in a more accurate solution in groundwater transport models which are applied for environmental management as well as hydrological investigations.

Keywords Truncation error · Finite difference method · ADERS subsurface transport models · Numerical accuracy

---

K. V. Sruthi (✉) · A. Sharma · N. C. Ghosh
Groundwater Hydrology Division, National Institute of Hydrology, Roorkee, Uttarakhand, India
e-mail: kvsruthi@gmail.com

K. V. Sruthi
KSCSTE—Centre for Water Resources Development and Management, Kozhikode, Kerala, India

K. H. Su
Department of Earth and Environmental Sciences, Chonbuk National University, Jeonju, Republic of Korea

The Earth and Environmental Science System Research Center, Chonbuk National University, Jeonju, Republic of Korea

R. M. Singh et al. (eds.), *Environmental Processes and Management*,
Water Science and Technology Library 91,
https://doi.org/10.1007/978-3-030-38152-3_18

18.1 Introduction

Mathematical models play a major role in describing the contamination of groundwater and soil water which are widely recognized as most critical environmental problems of recent times. The emergence of this notion witnessed the development of an increasing number of mathematical models (MODFLOW, MT3DMS, FEFLOW, PHT3D) describing the flow and transport processes of groundwater. These models are used as tools for decision making in the management of a water resource system. They may also be used to predict future groundwater scenarios. Therefore, groundwater models are now an important part of many hydrogeological investigations.

An equation describing a groundwater transport model is a partial differential equation (advection–dispersion equation, ADE). It can be solved mathematically by analytical or numerical solutions. Analytical solutions are very difficult to apply because they require specific parameters and boundaries that should be highly idealized. Therefore, numerical models are used in groundwater modeling as it yields approximate solutions to the governing equations through discretization of space and time. One of the main types of numerical models that are accepted for solving the groundwater transport equation is the finite difference method (FDM), an approach to computational fluid dynamics (CFD) and very effective in groundwater modeling (Anderson and Woessner 1992; Igboekwe et al. 2008). In this method, continuous variable is replaced by discrete variables that are defined at grid blocks. Also, the continuous differential equations which define the variable in the system are replaced by a finite number of variables at different grids. Ultimately, FDM seems to be more popular to solve ADE mainly due to the ease of implementation and its relative simplicity (Ataie-Ashtiani et al. 1999a, b; Sheu et al. 2000).

In all of these applications, an understanding of model accuracy is essential. Several approaches have been developed previously in order to improve the numerical accuracy of the models. One factor affecting the accuracy of the FDM is the numerical error, which occurs in all computational simulations. Numerical error can lead to quantitative and even qualitative changes in simulation results, potentially affecting the management of field sites (Simmons et al. 1999; Woods et al. 1998, 1999). There are many types of numerical errors. For instance, if the chosen grid spacing and time step length are too large, small errors may grow to dominate part of that simulation, resulting in numerical instability (Ferziger and Perić 1999; Noye 1978). This often leads to physically unreasonable results and problems with convergence. Another kind of numerical error is the truncation error (Gresho and Sani 1998; Noye and Hayman 1985). Approximating differential equations in the FDM by discretization introduces truncation error. Truncation error limits the use of numerical finite difference approximations in order to solve the partial differential equations. In case of ADE, numerical dispersion is the well-known consequence of truncation error. It results in an artificial dispersion, velocity, reaction term often denoted as numerical diffusion, numerical velocity, and numerical reaction coefficient. Numerical dispersion is insidious because it mimics the hydrodynamic dispersion (heuristic description of various physical processes) (Bear 1972) producing smooth results that may seem plausible.

In previous studies, truncation errors for the FDM were first evaluated theoretically and then spot-checked with numerical calculations (Lantz 1971). Chaudhari (1971) applied an explicit-backward space FDM and showed that the addition of a term to the dispersion coefficient could reduce the smearing of a front by generating non-oscillatory numerical solutions. This study quantified the numerical dispersion as a second-order error through the examination of the truncated Taylor series approximation of an explicit FD solution of one-dimensional ADE. Pinder and Gray (1977) adopted Fourier analysis to examine the behavior of the numerical error in time for FD schemes. Their work provided a valuable insight into the nature of numerical dispersion. Unfortunately, it did not yield easily applicable criteria for the control of numerical dispersion in real-world situations. Campbell et al. (1981) presented criteria for the control of numerical dispersion in a solution using the FD formulation for the time derivative. Several other schemes have been proposed in order to minimize the effects of numerical dispersion through the application of dispersion coefficient corrections in the transport equation (Bresler 1973; Chaudhari 1971; Lantz 1971; Van Genuchten and Wierenga 1974). Several studies have also considered the effect of numerical dispersion associated with the ADE during their simulations (May and Noye 1984; Noye 1990; Van Genuchten and Gray 1978).

Typically, the above-mentioned studies have considered the effect of numerical dispersion because it is the only truncation error in case of ADE (De Smedt and Wierenga 1977; Dudley et al. 1991; Moldrup et al. 1992, 1994a, b; Notodarmojo et al. 1991; Van Genuchten and Gray 1978). However, general transport equation should include truncation errors from all physical process terms such as advection, dispersion, reaction, and sink/source term. Ataie-Ashtiani et al. (1996) estimated the truncation error from dispersion, advection, and reaction which were termed as numerical dispersion, numerical velocity, and numerical reaction coefficient, respectively. A correction method for the numerical truncation errors of an explicit centered in space scheme was proposed (Ataie-Ashtiani et al. 1996). Also, zero- and first-order truncation errors in the ADE with reaction (ADER) were quantified for all widely applied numerical models (Ataie-Ashtiani et al. 1999a, b). Further, these studies were carried out in order to assess the effects of these truncation errors on the numerical solution of a two-dimensional advection–dispersion equation with a first-order reaction (Ataie-Ashtiani and Hosseini 2005a, b). For instance, the well-known groundwater modeling software 'MT3DMS' which is widely being used by the groundwater community (Zheng 1999), applied the standard FDM in order to solve the ADE, in which the FDM suffers truncation error. Therefore, it is very important to estimate the truncation error resulting from various physical process terms in the ADE in order to avoid the numerical inaccuracy in groundwater transport models based on FDM. However, the truncation errors arising due to the sink/source term (zero-order production) in the ADER have not been considered in the previous studies. In other words, no previous studies estimated the truncation error due to the sink/source term in the ADERS.

Therefore, we estimate the analytical expressions for all truncation errors resulting especially from the sink/source term for the general FD form of the ADERS. The numerical model section includes the development of different FD models by the elimination of truncation error followed by results and discussion. The study covers explicit, Crank–Nicolson, and implicit schemes to reveal that none of the widely used FD scheme models have complete accuracy. This chapter also aims to provide the user more than just a qualitative effect for the importance of truncation error for all terms such as dispersion, advection, reaction, and sink/source. In addition, we also present the numerical results after eliminating the truncation errors specifically resulting from the sink/source term, which improves the results of the FD models based on ADERS and can lead to a more accurate numerical solution.

18.2 Numerical Approach

In the field of geosciences, the partial differential equation describing one-dimensional transport of a solute with sink/source term through a homogeneous subsurface medium is

$$\frac{\partial C}{\partial t} = D\frac{\partial^2 C}{\partial x^2} - U\frac{\partial C}{\partial x} - kC + Q \tag{18.1}$$

where C is the solute concentration [ML^{-3}]; t is time [T]; x is the horizontal coordinate [L]; U is the Darcy flux [LT^{-1}]; D is the physical dispersion coefficient [L^{2}T^{-1}], Q is the sink/source term.

A general form of the FD model using ω and α as the temporal and spatial weighting parameters, respectively, can be expressed as

$$\begin{aligned}\frac{C_i^{n+1} - C_i^n}{\Delta t} &= D\left[\omega\frac{C_{i+1}^{n+1} - 2C_i^{n+1} + C_{i-1}^{n+1}}{\Delta x^2} + (1-\omega)\frac{C_{i+1}^n - 2C_i^n + C_{i-1}^n}{\Delta x^2}\right] \\ &\quad - U\left[\begin{array}{l}\omega\dfrac{(1-\alpha)C_i^{n+1} + \alpha C_{i+1}^{n+1} - (1-\alpha)C_{i-1}^{n+1} - \alpha C_i^{n+1}}{\Delta x} \\ +(1-\omega)\dfrac{(1-\alpha)C_i^n + \alpha C_{i+1}^n - (1-\alpha)C_{i-1}^n - \alpha C_i^n}{\Delta x}\end{array}\right] \\ &\quad - k\left[\omega C_i^{n+1} + (1-\omega)C_i^n\right] + Q\end{aligned} \tag{18.2}$$

where the superscript n refers to the time level; the subscript i refers to the node point, Δx is the spatial increment of grid [L] and Δt is the temporal increment [T]. Here, uniform time and space increment is applied.

A Taylor series expansion of C about any grid point is used to determine the form of the truncation errors (Chaudhari 1971; Lantz 1971). By neglecting the third- and higher-order spatial derivatives, the following formulations are obtained

$$C_i^{n+1} \approx C_i^n + \sum_{m=1}^{\infty} \frac{\Delta t^m}{m!} \frac{\partial^m C}{\partial t^m} \tag{18.3}$$

$$C_{i+1}^{n+1} \approx C_i^{n+1} + \Delta x \frac{\partial C^{n+1}}{\partial x} + \frac{\Delta x^2}{2} \frac{\partial^2 C^{n+1}}{\partial x^2} \tag{18.4}$$

$$C_{i-1}^{n+1} \approx C_i^{n+1} - \Delta x \frac{\partial C^{n+1}}{\partial x} + \frac{\Delta x^2}{2} \frac{\partial^2 C^{n+1}}{\partial x^2} \tag{18.5}$$

$$C_{i+1}^{n} = C_i^n + \Delta x \frac{\partial C}{\partial x} + \frac{\Delta x^2}{2} \frac{\partial^2 C}{\partial x^2} \tag{18.6}$$

$$C_{i-1}^{n} = C_i^n - \Delta x \frac{\partial C}{\partial x} + \frac{\Delta x^2}{2} \frac{\partial^2 C}{\partial x^2} \tag{18.7}$$

18.2.1 Determine the Expression for C_i^{n+1} in Terms of Spatial Derivative

In order to change Eq. (18.3) in terms of a spatial derivative, the following formulations are applied. The second- and the higher-order temporal derivatives of C are written in terms of spatial derivatives using the differentiated form of Eq. (18.1) as the following

$$\frac{\partial^2 C}{\partial t^2} = \frac{\partial}{\partial t}\left(\frac{\partial C}{\partial t}\right) = \frac{\partial}{\partial t}\left(D\frac{\partial^2 C}{\partial x^2} - U\frac{\partial C}{\partial x} - kC + Q\right) \tag{18.8}$$

$$\frac{\partial^2 C}{\partial t^2} = D\frac{\partial^2}{\partial x^2}\left(\frac{\partial C}{\partial t}\right) - U\frac{\partial}{\partial x}\left(\frac{\partial C}{\partial t}\right) - k\left(\frac{\partial C}{\partial t}\right) + \frac{\partial Q}{\partial t} \tag{18.9}$$

In order to express Eq. (18.9) only in spatial terms, the temporal terms are eliminated. For the elimination of temporal terms, Eq. (18.1) is substituted into Eq. (18.9) as the following

$$\begin{aligned} &= D\frac{\partial^2}{\partial x^2}\left(D\frac{\partial^2 C}{\partial x^2} - U\frac{\partial C}{\partial x} - kC + Q\right) \\ &\quad - U\frac{\partial}{\partial x}\left(D\frac{\partial^2 C}{\partial x^2} - U\frac{\partial C}{\partial x} - kC + Q\right) \\ &\quad - k\left(D\frac{\partial^2 C}{\partial x^2} - U\frac{\partial C}{\partial x} - kC + Q\right) + \frac{\partial Q}{\partial t} \end{aligned} \tag{18.10}$$

The higher-order derivatives are neglected. Here, Q is considered as a constant value. Therefore, the spatial and temporal derivative terms of Q become zero.

$$\frac{\partial^2 C}{\partial t^2} \approx \left(U^2 - 2kD\right)\frac{\partial^2 C}{\partial x^2} + 2kU\frac{\partial C}{\partial x} + k^2 C - kQ \tag{18.11}$$

Similarly, the higher-order temporal derivative can be formulated as follows

$$\frac{\partial^3 C}{\partial t^3} = \left(-3kU^2 + 3k^2 D\right)\frac{\partial^2 C}{\partial x^2} - 3k^2 U\frac{\partial C}{\partial x} - k^3 C + k^2 Q \tag{18.12}$$

$$\frac{\partial^4 C}{\partial t^4} = \left(6k^2U^2 - 4k^3 D\right)\frac{\partial^2 C}{\partial x^2} + 4k^3 U\frac{\partial C}{\partial x} + k^4 C - k^3 Q \tag{18.13}$$

$$\frac{\partial^5 C}{\partial t^5} = \left(-10k^3U^2 + 5k^4 D\right)\left(\frac{\partial^2 C}{\partial x^2}\right) - 5k^4 U\frac{\partial C}{\partial x} - k^5 C + k^4 Q \tag{18.14}$$

From Eqs. (18.11), (18.12), (18.13), and (18.14), the following general formula could be generated, i.e. for $m \geq 2$

$$\begin{aligned}\frac{\partial^m C}{\partial t^m} \approx\ & (-1)^m\left(\frac{m(m-1)}{2}k^{m-2}U^2 - mk^{m-1}D\right)\frac{\partial^2 C}{\partial x^2} \\ & + (-1)^m mk^{m-1}U\frac{\partial C}{\partial x} + (-1)^m k^m C \\ & + (-1)^{m-1}k^{m-1}Q\end{aligned} \tag{18.15}$$

Therefore, Eq. (18.3) could be written as the following

$$\begin{aligned}C_i^{n+1} \approx\ & C_i^n + \Delta t\frac{\partial C}{\partial t} \\ & + \sum_{m=2}^{\infty}\frac{\Delta t^m}{m!}\left[\begin{array}{r}(-1)^m\left(\frac{m(m-1)}{2}k^{m-2}U^2 - mk^{m-1}D\right)\frac{\partial^2 C}{\partial x^2} \\ +(-1)^m mk^{m-1}U\frac{\partial C}{\partial x} + (-1)^m k^m C + (-1)^{m-1}k^{m-1}Q\end{array}\right]\end{aligned} \tag{18.16}$$

Similarly, the formula for C_{i+1}^{n+1} and C_{i-1}^{n+1} in terms of the spatial derivative could be written as the following

$$\begin{aligned}C_{i+1}^{n+1} =\ & C_i^n + \Delta t\frac{\partial C}{\partial t} \\ & + \sum_{m=2}^{\infty}\frac{\Delta t^m}{m!}\left[\begin{array}{r}(-1)^m\left(\frac{m(m-1)}{2}k^{m-2}U^2 - mk^{m-1}D\right)\frac{\partial^2 C}{\partial x^2} \\ +(-1)^m mk^{m-1}U\frac{\partial C}{\partial x} + (-1)^m k^m C + (-1)^{m-1}k^{m-1}Q\end{array}\right] \\ & + \Delta x\frac{\partial C}{\partial x} + \Delta x\Delta t\left(-U\frac{\partial^2 C}{\partial x^2} - k\frac{\partial C}{\partial x}\right)\end{aligned}$$

$$+ \Delta x \sum_{m=2}^{\infty} \frac{\Delta t^m}{m!} \begin{bmatrix} (-1)^m m k^{m-1} U \frac{\partial^2 C}{\partial x^2} \\ +(-1)^m k^m \frac{\partial C}{\partial x} \end{bmatrix}$$

$$+ \frac{\Delta x^2}{2} \frac{\partial^2 C}{\partial x^2} - \Delta t \frac{\Delta x^2}{2} k \frac{\partial^2 C}{\partial x^2} + \frac{\Delta x^2}{2} \sum_{m=2}^{\infty} \frac{\Delta t^m}{m!} \left[(-1)^m k^m \frac{\partial^2 C}{\partial x^2} \right] \quad (18.17)$$

$$C_{i-1}^{n+1} = C_i^n + \Delta t \frac{\partial C}{\partial t}$$

$$+ \sum_{m=2}^{\infty} \frac{\Delta t^m}{m!} \begin{bmatrix} (-1)^m \left(\frac{m(m-1)}{2} k^{m-2} U^2 - m k^{m-1} D \right) \frac{\partial^2 C}{\partial x^2} \\ +(-1)^m m k^{m-1} U \frac{\partial C}{\partial x} + (-1)^m k^m C + (-1)^{m-1} k^{m-1} Q \end{bmatrix}$$

$$- \Delta x \frac{\partial C}{\partial x} - \Delta x \Delta t \left(-U \frac{\partial^2 C}{\partial x^2} - k \frac{\partial C}{\partial x} \right)$$

$$- \Delta x \sum_{m=2}^{\infty} \frac{\Delta t^m}{m!} \begin{bmatrix} (-1)^m m k^{m-1} U \frac{\partial^2 C}{\partial x^2} \\ +(-1)^m k^m \frac{\partial C}{\partial x} \end{bmatrix}$$

$$+ \frac{\Delta x^2}{2} \frac{\partial^2 C}{\partial x^2} - \Delta t \frac{\Delta x^2}{2} k \frac{\partial^2 C}{\partial x^2} + \frac{\Delta x^2}{2} \sum_{m=2}^{\infty} \frac{\Delta t^m}{m!} \left[(-1)^m k^m \frac{\partial^2 C}{\partial x^2} \right] \quad (18.18)$$

Substitution of all the Taylor series expansion for C about any grid point in the finite discretized approximation of solute transport equation in order to estimate the truncation error will result in the following equation. By inserting Eqs. (18.6), (18.7), (18.16), (18.17), and (18.18) into Eq. (18.2), the final expression for $\frac{\partial C}{\partial t}$ would be obtained in the following form

$$\frac{\partial C}{\partial t} = \frac{\partial^2 C}{\partial x^2} \left\{ \begin{array}{l} D - 2D\omega \Delta t k + \left(\alpha - \frac{1}{2} \right) u\omega \Delta x \Delta t k + \left(\frac{1}{2} - \alpha \right) U \Delta x + U^2 \omega \Delta t \\ + (-1 - k\omega \Delta t) \left[\sum_{m=2}^{\infty} \frac{\Delta t^{m-1}}{(m-1)!} (-1)^m \left(\frac{(m-1)}{2} k^{m-2} U^2 - k^{m-1} D \right) \right] \\ + (-U\omega) \left[\sum_{m=2}^{\infty} \frac{\Delta t^m}{m!} (-1)^m m k^{m-1} U \right] \\ + \sum_{m=2}^{\infty} \frac{\Delta t^m}{m!} [(-1)^m k^m] * \left(\omega D - \alpha u \omega \Delta x + u\omega \frac{\Delta x}{2} \right) \end{array} \right\}$$

$$- \frac{\partial C}{\partial x} \begin{bmatrix} U - 2U\omega \Delta t k \\ +U\omega \sum_{m=2}^{\infty} \frac{\Delta t^m}{m!} (-1)^m k^m + (1 + k\omega \Delta t) \left(\sum_{m=2}^{\infty} \frac{\Delta t^{m-1}}{(m-1)!} (-1)^m k^{m-1} U \right) \end{bmatrix}$$

$$-C\left[k-\omega\Delta tk^2+(1+k\omega\Delta t)\sum_{m=2}^{\infty}\frac{\Delta t^{m-1}}{m!}(-1)^m k^m\right]$$
$$\left[Q-k\omega\Delta tQ+(-1-k\omega\Delta t)\sum_{m=2}^{\infty}\frac{\Delta t^{m-1}}{m!}(-1)^{m-1}k^{m-1}Q\right] \quad (18.19)$$

18.2.2 Derivation of the Truncation Error Formula

By comparing the above Eq. (18.19) with the original governing Eq. (18.1), four forms of truncation errors due to discretization are observed. It can be formulated as follows

Second-order truncation error or numerical dispersion

$$\begin{aligned}D_{\text{num}} &= -2D\omega\Delta tk+\left(\alpha-\frac{1}{2}\right)U\omega\Delta x\Delta tk+\left(\frac{1}{2}-\alpha\right)U\Delta x+U^2\omega\Delta t\\&+(-1-k\omega\Delta t)\left[\sum_{m=2}^{\infty}\frac{\Delta t^{m-1}}{(m-1)!}(-1)^m\left(\frac{(m-1)}{2}k^{m-2}U^2-k^{m-1}D\right)\right]\\&+(-U\omega)\left[\sum_{m=2}^{\infty}\frac{\Delta t^m}{m!}(-1)^m mk^{m-1}U\right]\\&+\sum_{m=2}^{\infty}\frac{\Delta t^m}{m!}[(-1)^m k^m]*(\omega D-\alpha U\omega\Delta x+U\omega\frac{\Delta x}{2})\end{aligned} \quad (18.20)$$

First-order truncation error or numerical water velocity

$$\begin{aligned}U_{\text{num}} &= -2U\omega\Delta tk+\sum_{m=2}^{\infty}\left[\frac{\Delta t^{m-1}}{(m-1)!}(-1)^m k^{m-1}U\right](1+k\omega\Delta t)\\&+\sum_{m=2}^{\infty}\left[\frac{\Delta t^m}{m!}(-1)^m k^m\right]U\omega\end{aligned} \quad (18.21)$$

Zero-order truncation error or numerical reaction coefficient

$$k_{\text{num}} = -\omega\Delta tk^2+\left[\sum_{m=2}^{\infty}\frac{\Delta t^{m-1}}{m!}(-1)^m k^m\right](1+k\omega\Delta t) \quad (18.22)$$

Sink/Source term truncation error

$$Q_{\text{num}} = -k\omega\Delta t Q + (-1 - k\omega\Delta t)\sum_{m=2}^{\infty}\frac{\Delta t^{m-1}}{m!}(-1)^{m-1}k^{m-1}Q \qquad (18.23)$$

The truncation error due to the sink/source term (Q_{num}) is quantified in the above Eq. (18.23). It is to be noted that no previous studies have attempted to quantify the error due to the sink/source term.

18.2.3 Reformulation of the Truncation Error Corrected Subsurface Transport Equation

In order to eliminate the truncation error due to the numerical dispersion, numerical velocity, numerical reaction coefficient, and the source/sink truncation, the derived formula for these terms will be subtracted from the physical dispersion, velocity, reaction coefficient, and the sink/source term. The resulting terms are inserted in Eq. (18.1)

$$\frac{\partial C}{\partial t} = D*\frac{\partial^2 C}{\partial x^2} - U*\frac{\partial C}{\partial x} - k*C + Q* \qquad (18.24)$$

where D^*, U^*, k^*, Q^* denotes the truncation error corrected forms.

$$D^* = D - D_{\text{num}} \qquad (18.25)$$

$$U^* = U - U_{\text{num}} \qquad (18.26)$$

$$k^* = k - k_{\text{num}} \qquad (18.27)$$

$$Q^* = Q - Q_{\text{num}} \qquad (18.28)$$

18.3 Results and Discussion

In order to study the effect of numerical error due to numerical truncation of Taylor series expansion (numerical dispersion, numerical velocity, numerical reaction coefficient, numerical sink/source term) on FD model such as explicit scheme, Crank–Nicolson scheme, and implicit scheme, the study compared the numerical simulation results of truncation error corrected and non- corrected scheme with analytical solution. The analytical solution is adopted from van Genuchten and Alves (1982). The analytical solution for solute transport equation for the following initial and boundary condition

$$C(x,0) = C_i \quad t = 0 \; x > 0$$

$$C(0,t) = C_o \quad t > 0 \; x = 0$$

$$\frac{\partial C}{\partial x}(\infty, t) = 0$$

is given as

$$C(x,t) = \frac{Q}{k} + \left(C_i - \frac{Q}{k}\right)A(x,t) + \left(C_o - \frac{Q}{k}\right)B(x,t)$$

where C_i represents initial concentration of solute in transport medium $[ML^{-3}]$, C_o is incoming concentration $[ML^{-3}]$.

$$A(x,t) = \exp(-kt)\left\{1 - \frac{1}{2}\text{erfc}\left[\frac{x - Ut}{2(Dt)^{1/2}}\right] - \frac{1}{2}\exp\left(\frac{Ux}{D}\right)\text{erfc}\left[\frac{x + Ut}{2(Dt)^{1/2}}\right]\right.$$

$$B(x,t) = \frac{1}{2}\exp\left[\frac{(U - v)x}{2D}\right]\text{erfc}\left[\frac{x - vt}{2(Dt)^{1/2}}\right] + \frac{1}{2}\exp\left[\frac{(U + v)x}{2D}\right]\text{erfc}\left[\frac{x + vt}{2(Dt)^{1/2}}\right]$$

And

$$v = U\left(1 + \frac{4kD}{U^2}\right)^{1/2}.$$

18.3.1 Comparison of Numerical Solution of Corrected and Non-corrected Truncation Error for Different FD Schemes with Analytical Solution

In order to compare the accuracy of truncation error corrected FD models, a numerical problem is formulated. The numerical problem is composed of a semi-infinite column, where $U = 5$ cm/h; $D = 100$ cm^2/h; $k = 0.1$ h^{-1}; source concentration = 100 mg/L; incoming concentration = 1000.0 mg/L. Here, a space increment of 20 cm and temporal increment of 1.0 h are applied. And the study compared the numerical solution with analytical solution at time of 20 h. Also, arbitrary units can be used for the numerical parameters. Figure 18.1 displays the comparison between the numerical simulation results (truncation error corrected and non-corrected) and analytical solution of numerical problem for different FD models. The results include the correction of all truncation error terms associated with advection—dispersion equation with first-order reaction and zero-order production term. The comparison between the results shows that none of the explicit schemes have the numerical accuracy without truncation error correction (Fig. 18.1). The numerical results from Crank–Nicolson method show that the centered scheme in space has negligible truncation error compared to other methods. The analytical solution and simulation results (truncation error corrected and non-corrected) for Crank–Nicolson centered scheme well matched each other. In Fig. 18.2, numerical results of implicit scheme reveal that implicit upstream and centered scheme without truncation error correction deviates

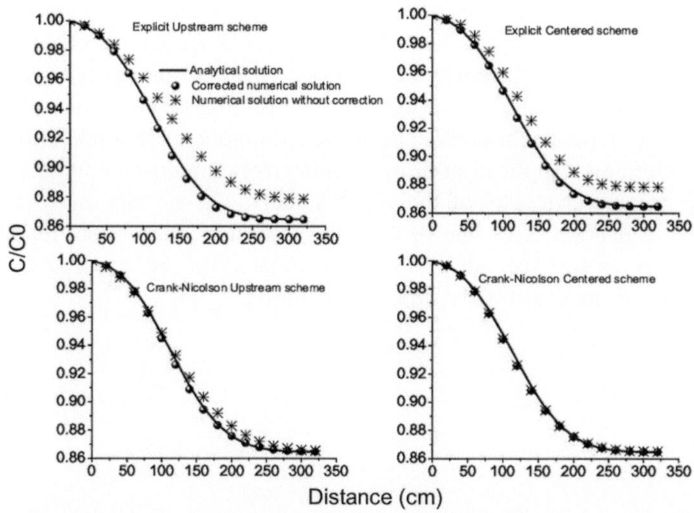

Fig. 18.1 Comparison of corrected and non-corrected numerical solution with analytical solution for explicit and Crank–Nicolson schemes with upstream and centered in space scheme

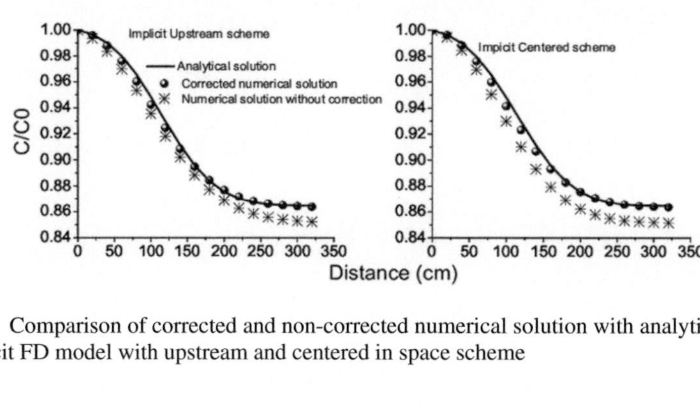

Fig. 18.2 Comparison of corrected and non-corrected numerical solution with analytical solution for implicit FD model with upstream and centered in space scheme

significantly from the analytical solution. The numerical results from truncation error corrected implicit scheme well corresponded with analytical solution (Fig. 18.2). From Figs. 18.1 and 18.2, it is absolutely clear that none of the FD models are free of truncation error; but the Crank–Nicolson centered scheme yields numerical solution which has negligible truncation error compared to other FD models.

18.3.2 Estimation of Relative Error in the Numerical Results of Truncation Error Corrected and Non- Corrected FD Models

The study estimated the relative error for truncation error corrected and non-corrected numerical solution of different FD models. Our results revealed that numerical error is decreased drastically by the removal of truncation error from FD models (Fig. 18.3). The maximum error limit is reduced from 3 to 0.5% after truncation error correction (Fig. 18.3). It is very significant to study the truncation error correction of FD model because the groundwater model such as MT3D applies FD models to solve the numerical problems. Application of truncation error correction term can reduce the error from numerical results of these FD models. The present study could shed light on the truncation error due to the advection term, dispersion, reaction term, and sink/source term. Especially, the study estimated for the first time a numerical formula for truncation error from sink/source term (Q_{num}).

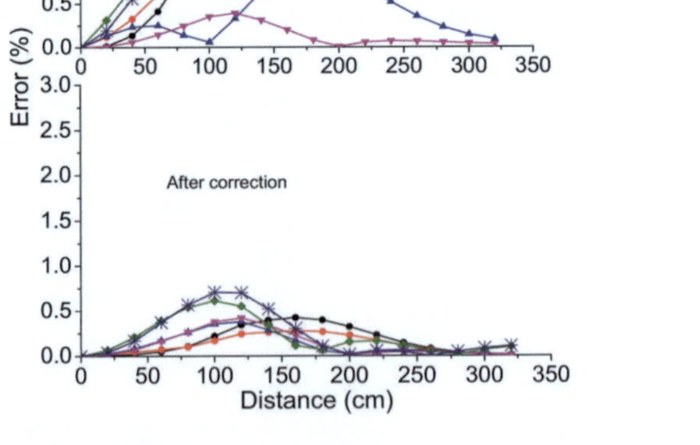

Fig. 18.3 Comparison of estimated error in numerical solution for different FD model before and after removal of truncation error

The present study reveals the importance of the removal of truncation error due to the different terms in ADERS. The study estimates the truncation error resulting from sink/source term and its effect on numerical accuracy of different FD models. It is revealed that numerical truncation error correction has a significant impact on improving solution accuracy of the finite difference models. Also, the numerical solution without error correction has shown a significant deviation from analytical solution in the case of all finite difference models except Crank–Nicolson-centered scheme. It showed a less deviation, even in the absence of truncation error correction. Here, the study estimates for the first time the truncation error term due to sink/source part in advection–dispersion equation. None of the studies have quantified the truncation error due to the sink/source term. Ultimately, the study reveals that none of the finite difference models are free of truncation error and the numerical accuracy is affected by several truncation errors which result from advection, dispersion, reaction, and sink/source term.

18.4 Conclusions

This chapter explains the estimation of the truncation error associated with the FDM models based on ADERS by applying the Taylor series expansion. The study analyzes the truncation error both theoretically and numerically. It reveals that modification or subtraction of numerical truncation error term significantly increases the solution accuracy of the FD model which is being applied widely in popular subsurface transport models such as MODFLOW, MT3DMS. Moreover, the study estimates the truncation error due to the sink/source term in the ADERS for the first time. The study compared the solution accuracy of the FD models with and without truncation error correction. The results showed that none of the FD models were free of truncation errors which ultimately lead to misinterpretation during hydrological investigations by affecting the accuracy. Here, the least truncation error was exhibited by Crank–Nicolson-centered scheme. Also, the study estimated the error in the numerical solution (both with correction and without correction), and it was observed that the maximum level of relative error reduced from 3 to 0.5% by eliminating the truncation error. The comparative study of numerical solution of FD models revealed that the truncation error correction can improve the solution accuracy of the FD models significantly. We suggest from this study that the application of truncation error removal method is most significant to increase the solution accuracy in different FD models, which is widely applied in the groundwater community.

Acknowledgements We would like to thank the Director, National Institute of Hydrology, Roorkee, India for constant encouragement and support.

References

Anderson MP, Woessner WW (1992) Applied groundwater modeling. Academic Press, San Diego, C. A, Simulation of Flow and Advective Transport

Ataie-Ashtiani B, Lockington D, Volker R (1996) Numerical correction for finite-difference solution of the advection–dispersion equation with reaction. J Contam Hydrol 23:149–156

Ataie-Ashtiani B, Volker R, Lockington D (1999a) Numerical and experimental study of seepage in unconfined aquifers with a periodic boundary condition. J Hydrol 222:165–184

Ataie-Ashtiani B, Lockington D, Volker R (1999b) Truncation errors in finite difference models for solute transport equation with first-order reaction. J Contam Hydrol 35:409–428

Ataie-Ashtiani B, Hosseini S (2005a) Error analysis of finite difference methods for two-dimensional advection–dispersion–reaction equation. Adv Water Resour 28:793–806

Ataie-Ashtiani B, Hosseini S (2005b) Numerical errors of explicit finite difference approximation for two-dimensional solute transport equation with linear sorption. Environ Modell Softw 20:817–826

Bear J (1972) Dynamics of fluids in porous media. Elsevier Science, New York

Bresler E (1973) Simultaneous transport of solutes and water under transient unsaturated flow conditions. Water Resour Res 9:975–986

Campbell JE, Longsine DE, Reeves M (1981) Distributed velocity method of solving the convective-dispersion equation: 1. introduction, mathematical theory, and numerical implementation. Adv Water Resour 4:102–108

Chaudhari NM (1971) An improved numerical technique for solving multidimensional miscible displacement equations. Soc Petrol Eng J 11:277–284

De Smedt F, Wierenga P (1977) Simulation of water and solute transport in unsaturated soils. In: Proceedings of 3rd international symposium in hydrology. Fort Collins

Dudley L, McLean J, Furst T, Jurinak J (1991) Sorption of cadmium and copper from an acid mine waste extract by two calcareous soils: Column studies. Soil Sci 151:2

Ferziger JH, Perić M (1999) Computational methods for fluid dynamics. Springer, New York

Gresho P, Sani R (1998) Incompressible flow and the finite element method: advection-diffusion and isothermal laminar flow, vol 1. Wiley, New York

Igboekwe MU, Gurunadha Rao VVS, Ok-wueze EE (2008) Groundwater flow modelling of Kwa Ibo River watershed Southeastern Nigeria. Hydrol Process 22:1523–1531

Lantz R (1971) Quantitative evaluation of numerical diffusion (truncation error). Soc Petrol Eng J 11:315–320

May R, Noye BJ (1984) The numerical solution of ordinary differential equations: initial value problems. In: Noye B (ed) Computational techniques for differential equations. North-Holland, NewYork

Moldrup P, Yamaguchi T, Hansen J, Rolston D (1992) An accurate and numerically stable model for one-dimensional solute transport in soils. Soil Sci 153:261–273

Moldrup P, Yamaguchi T, Ralston EVK, Hansen J (1994a) Removing numerically induced dispersion from finite difference models for solute and water transport in unsaturated soils. Soil Sci 157:153–161

Moldrup P, Paulsen T, Rolston D, Yamaguchi T, Hansen J (1994b) Integrated flux model for unsteady transport of trace organic chemicals in soils. Soil Sci 157:137–152

Notodarmojo S, Ho G, Scott W, Davis G (1991) Modelling phosphorus transport in soils and groundwater with two-consecutive reactions. Water Res 25:1205–1216

Noye BJ, Hayman K (1985) Accurate finite difference methods for solving the advection-diffusion equation. In: Noye B, May R (eds) Computational techniques and applications: CTAC-85. North-Holland, NewYork, pp 137–157

Noye BJ (1978) Finite difference techniques. In: Noye B (ed) Numerical simulation of fluid motion. North-Holland, NewYork, pp 1–112

Noye BJ (1990) A new third-order finite-difference method for transient one-dimensional advection–diffusion. Commun Appl Numer M 6:279–288

Pinder GF, Gray WG (1977) Finite element simulation in surface and subsurface hydrology. Academic Press, New York

Sheu TW, Wang S, Lin R (2000) An implicit scheme for solving the convection–diffusion–reaction equation in two dimensions. J Comput Phys 164:123–142

Simmons CT, Narayan KA, Wooding RA (1999) On a test case for density-dependent groundwater flow and solute transport models: the salt lake problem. Water Resour Res 35:3607–3620

Van Genuchten MTh, Wierenga PJ (1974) Simulation of one-dimensional solute transfer in porous media. AES Bull, New Mexico State University, Las Cruces, New Mexico, p 628

Van Genuchten MTh, Gray WG (1978) Analysis of some dispersion corrected numerical schemes for solution of the transport equation. Int J Numer Meth Eng 12:387–404

Van Genuchten MT, Alves W (1982) Analytical solutions of the one dimensional convective-dispersive solute transport equation. Technical Bulletin, No. 1661, US. Department of Agriculture, p 155

Woods J, Simmons CT, Narayan KA, Tuck E, Stott J (eds) (1998) Verification of black box groundwater models. In: Third biennial engineering mathematics and applications conference: EMAC98, Australia, pp 523–526

Woods JM, Teubner D, Simmons CT, Narayan KA (1999) Numerical inaccuracy in groundwater modelling: diagnosis and pathology. In: Joint congress of the 25th hydrology and water resources symposium and the 2nd international conference on water resources and environmental research "Water 99", Australia

Zheng C (1999) MT3D: a modular three-dimensional transport model for simulation of advection, dispersion and chemical reactions of contaminants in groundwater systems. Report to U.S. Environmental Protection Agency, Ada, OK, p 170

Chapter 19
Groundwater Salinity in Northwestern Region of India: A Critical Appraisal

Gopal Krishan, Mamta Bisht, N. C. Ghosh and Gokul Prasad

Abstract Groundwater plays a pivotal role in India, particularly in hard rock and semi-hard rock regions of the country to support domestic, agricultural, and industrial requirements of water in addition to environmental needs. Rising demands of groundwater for rapidly increasing population, developmental activities and urbanization have resulted in unsystematic over extractions of groundwater that led to the decline of groundwater levels in many parts of the world including India. Many areas have even no proper groundwater development program that has given rise to problems of waterlogging and salinity. Approximately, 25% area of the Indo-Gangetic basin has saline water with TDS more than 1000 mg/L (as per WHO 2004) or conductivity more than 1500 μS/cm. In Indo-Gangetic basin, the problems of salinity in Indus and upper Gangetic parts covering northwestern states of India are different than the salinity of coastal areas. The major problem of groundwater salinity in the northwestern states, namely Delhi, Haryana, Punjab, and Rajasthan, is of terrestrial origin. The order of salinity affected state, in terms of magnitude, is Rajasthan > Haryana > Punjab > Delhi. Over-abstraction of groundwater can also spread saline water into the freshwater zones due to the presence of evaporative sequences from deeper to shallower depths. Modern irrigation practices with dense canal distribution network may lead to very shallow water tables, hence may increase waterlogging and salinization due to the leakage. The impacts of groundwater salinity largely hamper the crop productivity and thereby can affect the food security. Management of groundwater salinity is thus essential for the sustainability of food security and to remediate health hazards and ecosystem services in the northwestern region of India. To obviate the problem and for better management of groundwater resources in the area, one has to know the interaction between the aquifers and surface water sources.

G. Krishan (✉) · M. Bisht · G. Prasad
Groundwater Hydrology Division, National Institute of Hydrology, Roorkee, Uttarakhand 247667, India
e-mail: drgopal.krishan@gmail.com

N. C. Ghosh
Bengal Institute of Technology, Kolkata 700150, West Bengal, India

R. M. Singh et al. (eds.), *Environmental Processes and Management*, Water Science and Technology Library 91,
https://doi.org/10.1007/978-3-030-38152-3_19

Keywords Groundwater resources · Salinity · Northwestern region · India · Assessment

19.1 Introduction

Groundwater is the largest freshwater resource on the earth and is usually reserved beneath the water table in soils and in geologic formations that are fully saturated (Central Ground Water Board—CGWB 2011), called aquifers. The groundwater is a key to sustainable development and supports industry, agriculture, ecosystems, and human lives and their livelihoods. The critical role of groundwater is to maintain important surface water systems, riparian, and other types of vegetation at a particular area. Central Ground Water Board (CGWB 2012) reported that groundwater has emerged as the prime source in India as drinking water and irrigation and estimated that 92% withdrawal of groundwater is being used for agricultural purposes that contribute largely to the food security of the country. Chakraborti et al. (2009) reported that groundwater has a vital role in sustaining aquatic ecosystems and provides a reliable discharge to streams, rivers, and wetlands sustaining flows during dry seasons and droughts. In humid areas, groundwater generally discharges to rivers, sustaining their aquatic ecosystems. In arid areas, the relationship becomes more complex and rivers often discharge their water to the groundwater system through seepage, infiltration, and flood events.

In recent years, the use of groundwater has become unsustainable and their supplies are diminishing in some regions due to its excessive use, and as a result of which 20% of the world's aquifers being categorized as over-exploited (Gleeson et al. 2012). Groundwater overexploitation is defined as a situation in which, over a period of time, average extraction rate from aquifers is greater than the average recharge rate. In India alone, farmers have installed number of groundwater abstraction pumps. Higher groundwater abstraction leads to adverse impacts on groundwater resources. Because of increasing food demand, the agricultural technology is being updated with indiscriminate exploitation of groundwater resources. In India, 89% of groundwater extracted is used in the irrigation sector, making it the highest category user in the country and followed by 9% of the extracted for domestic, 2% for Industrial (MoWR, Annual Report 2013). Shah et al. (2000) reported that 70% of the irrigation depends on groundwater. Irrigation through groundwater is a major component of the extraction of groundwater and led to problems such as declining water table, deterioration of groundwater quality, waterlogging, and soil salinity.

Groundwater salinity is prominent in many parts of the country including northwestern region (NWR) of India which is one of the agriculture-dominant areas. In this part, occurrence, origin, distribution, and effect of groundwater salinity have not been studied in detail. The objective of this chapter is mainly focused on northwestern region (NWR) of India where the salinity range is beyond the permissible limits. The impacts of groundwater salinity largely hamper the crop productivity. Therefore, the management of groundwater salinity is essential for sustaining life in the northwestern region of India. There is also a need to find out the interaction between the aquifers and surface water sources for better management of groundwater resources.

19.2 Groundwater Salinity

Human and ecological uses of groundwater depend on ambient water quality. Ewusi et al. (2013) reported that the suitability of groundwater for agriculture and domestic purposes largely depends on the climatic variations as well as the residence time of water within the aquifer and anthropogenic activities. The problems associated with the use of poor quality water include reduction in the infiltration rate and toxicity of soil due to certain ions and excessive nutrients (Ayers and Westcot 1994). The physicochemical composition of groundwater is a measure of its suitability for agriculture (irrigation), industrial, and domestic purposes. Thus, it is necessary to evaluate the hydro-chemical characteristics of groundwater before using it in any field. One of the important chemical constituents that affect the suitability of water for agricultural purposes is salinity. Groundwater salinity is mainly caused by higher range of electrical conductivity as a result of increased concentrations of dissolved salts in soil and water due to dissolution of salt-bearing strata during mineral weathering—evaporative concentrations—may be imported into a basin by deposition of marine aerosols and membrane filtration of clays, siltstones, mudstones, and shales in the basins (Jolly et al. 2008; Salama et al. 1999; Xun et al. 1997; Hem 1985). Shallow groundwater is salinized due to evaporation of residual brines, groundwater coming in contact with evaporate sediment, and leaching of salts after groundwater table rise (Datta et al. 1996; Gilfedder et al. 2000; Cooper 2002; Salama et al. 1999).

Groundwater contains a high amount of salts which is able to conduct electricity. The more electrical conductivity of the water shows more saline water. It is measured by electric conductivity (EC) probe and measure of total mineral contents of dissolved solids in water. The standard unit for conductivity is mho cm^{-1} but is also represented by S m^{-1}. Electrical conductivity is usually used for indicating the total concentration of the ionized constituents of natural water. If EC of groundwater water is high, it will affect soil, plant, and human beings. In plants, higher the EC, the lesser water available to plants because plants can only transpire pure water. It is considered as a good indicator of water suitability for various uses and arises from natural processes and over-irrigation in Punjab and Haryana. The EC classification is indicated in Table 19.1. The use of saline groundwater for agricultural purposes can alter accessibility of water to the crops. It can increase the osmotic pressure of the soil water by the excess salts and reduces the absorption of water by plant roots. Even though the soil appears to have enough moisture, the plants may wilt because the roots do not absorb sufficient water to compensate for the water loss by transpiration. Electrical conductivity varies with concentration, degree of ionization of the constituents, and temperature.

Table 19.1 Norms of groundwater water quality based on EC (S/m) (Wilcox 1955)

Classification of water	EC (S/m)
Excellent (C1)	<0.025
Good (C2)	0.025–0.075
Permissible (C3)	0.075–0.2
Doubtful (C4)	0.2–0.3
Unsuitable (C5)	>0.3

19.2.1 Groundwater Salinity in India

The availability of freshwater supplies to agriculture sector in the future is likely to reduce in the Asian countries due to the increasing growth of population, improved living standards and inter-sector competition for water. The estimate for India shows this reduction to be 10–12% by 2025 (Sharma and Minhas 2005). The groundwater surveys in India indicate that poor quality of water being utilized by most of the states where the total groundwater development is between 32 and 84% (CGWB 2015a, b). Many more Indian states have a good quality of aquifers but due to over-extraction; these aquifers are being endangered now. The groundwater of arid and semiarid region is largely saline or sodic in nature. In India so far, no systematic attempts have been made to reach at the estimate of saline groundwater. However, some prediction about groundwater salinity in various states is given in Fig. 19.1. The CGWB (1977) approximated that total area underlain with saline groundwater (EC > 0.4 S/m^{-1}) is 193,438 km^2 with the annual replenishable recharge of 11,765 MCM $year^{-1}$ leaving aside minor patches.

In central alluvial plains of India, the inland salinity is continuously increasing due to the expansion of canal network and arid climatic conditions. These factors led to excessive evapotranspiration rate in that area and concentrate the salt over the soil and escalating the groundwater salinity. Misra and Mishra (2007) reported that salinity in deep and shallow aquifers is continuously escalating. The deep aquifers (borewells) are found more saline as compared to the shallow aquifers (dugwells). One of the major effects of inland salinity in this region is from saline groundwater, which is reaching the land surface and causing soil salinizations and waterlogging.

It has been estimated that approximately 25% area of the Indo-Gangetic basin have saline water over 1000 mg/L total dissolved solids (Bonsor et al. 2017; MacDonald et al. 2016) which should be less than 1000 mg/L or conductivity less than 1500 μS/cm for drinking (WHO 2004) and 2000 mg/L or conductivity less than 3000 μS/cm for drinking (IS: 10500: 2012); however, there are no strict limits for irrigation. In Indo-Gangetic basin, the problems of salinity in Indus and upper Gangetic part covering northwestern states of India are different than the salinity of coastal regions. The major problem of groundwater salinity occurs in northwestern region of India in the states of Delhi, Haryana, Punjab, and Rajasthan where the salinity is reported to be of terrestrial origin (Bonsor et al. 2017). Order of salinity affected states is: Rajasthan > Haryana > Punjab > Delhi. The groundwater is declining at an

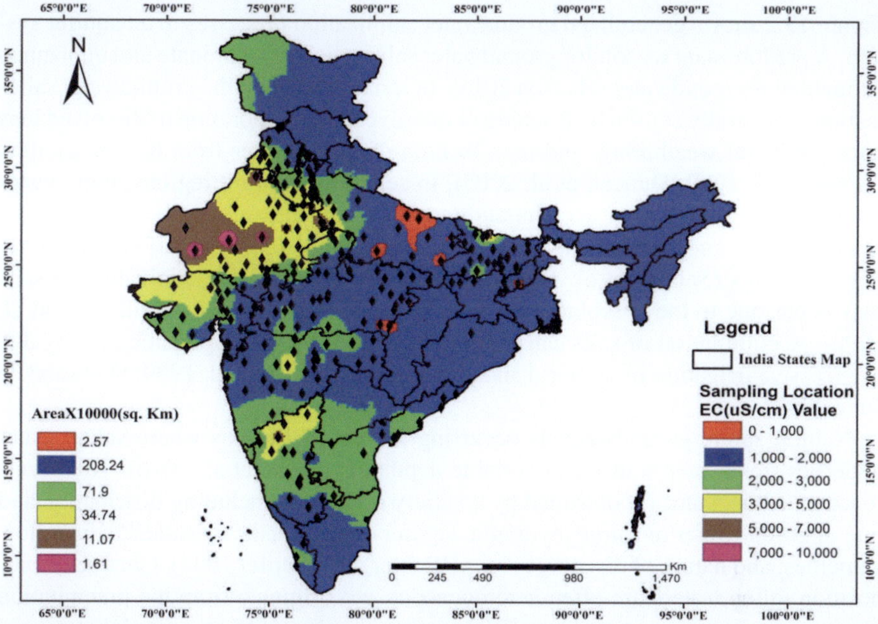

Fig. 19.1 Groundwater salinity in India (CGWB 1997)

alarming rate in freshwater regions, and the southwestern parts of Punjab are facing problems of severe waterlogging and salinization (Misra and Mishra 2008; Chopra and Krishan 2014; Krishan et al. 2017).

The yellow to brown and pink areas occupying more than 40 km^2 of states of Rajasthan, Haryana, Gujarat in Fig. 19.1 indicate high saline zones in southwestern states which are facing severe problems of waterlogging and resultant soil salinity. These states are irrigated by the various canals and twin-pronged strategy to manage its groundwater resources resulted in waterlogging and soil salinization. Waterlogging causes depletion of oxygen and increase of carbon dioxide in the root zone of crops which causes loss of plant nutrients and the loss of useful microorganisms at the expense of the growth of harmful ones. It also causes chemical degradation due to the accumulation of salts at the soil surface leading to an ecological imbalance (CGWB 1997).

19.2.2 Origin of Groundwater Salinity

Groundwater salinization is resulted from geogenic sources such as seawater intrusion in coastal aquifers, contact with salt deposits, and upcoming of deep natural saline water and anthropogenic activities. It may occur locally or regionally.

Figure 19.2 shows generalized groundwater salinization processes in an aquifer system. A predominant reason for groundwater salinity is indiscriminate and unplanned extraction of groundwater (Hanson 2011). In semiarid region, the groundwater salinization is generally caused by leaching of dissolved ion, evaporation of dissolved ions during mineral weathering, and mobilization of saline water from deeper aquifers (Grube et al. 1999; Duncan et al. 2005). In semiarid and arid regions, more evaporation can lead to higher accumulation of salts over the soil and become a major reason for soil and groundwater salinity. Salts accumulated on top of soil in poorly drained basins (Salama et al. 1999; Misra and Mishra 2007). Salinization of salts may occur due to the percolation of salts in shallow groundwater (Gilfedder et al. 2000). Accumulation of salts can also be introduced by anthropogenic activity due to excessive irrigation in semiarid and arid regions (Salama et al. 1999; Datta and de Jong 2002).

Natural saline groundwater is occurring on regional scales where saline water underlies freshwater aquifers at variable depths (Priyanka et al. 2016). The occurrence of saline water is controlled by a variety of factors, including distribution and rate of groundwater recharge, hydraulic aquifer characteristics, residence time, flow velocities, and nature of discharge areas (Richter and Kreitler 1993). Chemical composition saline waters are often heterogeneous and originate from the precipitation of evaporates solution of rock salt. Richter and Kreitler (1986) reported that natural

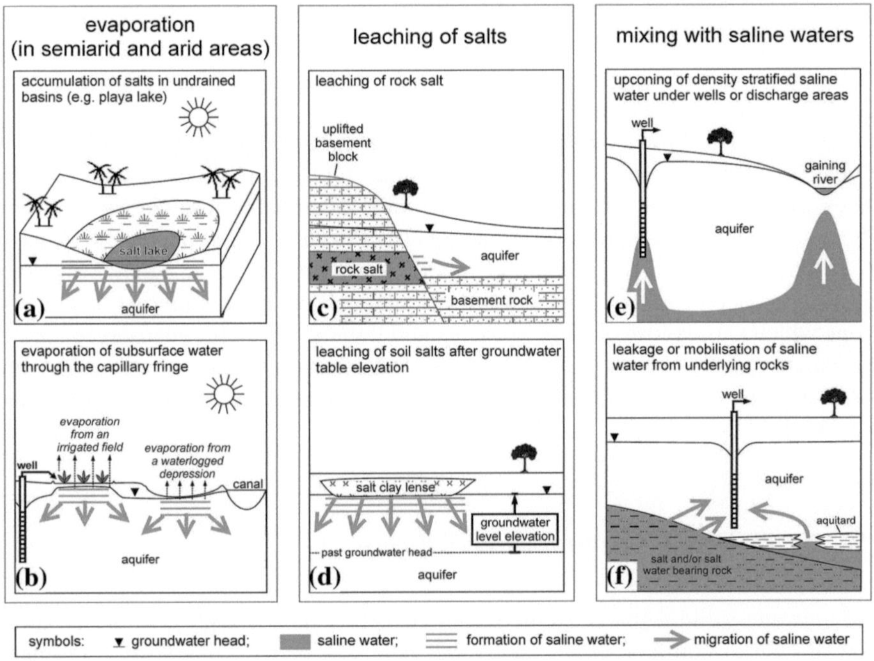

Fig. 19.2 Schematic sketches of common sources of groundwater salinity (Gunnar et al. 2012)

salinization of freshwater occurred where saltwater from saline aquifers discharges at the land surface or mixes with freshwater in the subsurface. Freshwater and saltwater are usually separated by a transition zone of variables thickness. Many sedimentary basins contain large salt deposits occurring either at great depths or close to the land surface.

Shallow salt deposits pose a higher potential for freshwater salinization than deeper ones. Besides geogenic salinization of groundwater, anthropogenic also has to be considered as potential groundwater salinity sources (Grube et al. 1999).

- abandoned industrial sites—waste deposits
- infiltration of wastewater—damaged sewers
- bank filtration from rivers with a high mineral content
- agriculture and forestry road-salting
- oxidation of artificially drained peat and carbon deposits.

The human activities have different effects on the magnitude and spatial distribution of salt concentration. Depending on the kind and duration of the source, the concentration may vary considerably.

19.3 Groundwater in NWR

Groundwater is an important source of water for irrigation in the northwestern part of India, where the insufficient and limited surface water have occurred during dry months of the year. The northwestern region forms a large part of the Indo-Gangetic Plain (Misra and Mishra 2007). The NWR is characterized by the subtropical monsoonal type of climatic condition which is differentiated into three different seasons: winter (cool and dry), pre-monsoon (hot and dry), and monsoon (hot and wet). The average annual rainfall and humidity are approximately 1600 mm and 78%. This region shows higher evapotranspiration rates causing high water losses. The northwestern states have a diversified geological, climatological, and topographic set up giving rise to divergent groundwater situations in different parts of the state. Therefore, the climatic and topographic variability of northwestern states is causing groundwater salinity. The prevalent rock formation ranging in age from Archaeans to the recent which controls occurrence and movement of groundwater is widely varied in composition and structure (Villholth and Sharma 2006). The geographical conditions of northwestern region are favored of the concentration of salts on the top of soil. High potential evaporation is resulted in the warm and dry climate of northwestern region (Datta et al. 1996; Misra and Mishra 2007). It is generally accepted that the construction of the canal network and subsequent water table elevation, waterlogging, and intensive irrigation are a major source of groundwater salinity in this region (Singh and Kothari 2004). The geo-physiographical features of NWR are described in Table 19.2.

Most affected northwestern states are Rajasthan, Punjab, Haryana, and Delhi, where groundwater consumption has been increasing over the past decades, along

Table 19.2 Geo-physiography of NWR of India

	Punjab	Haryana	Delhi	Rajasthan
Latitude	29° 32′ and 32° 28′ N	27° 30′ and 30° 35′ N	28° 24′ and 28° 53′ N	23° 03′ and 30° 12′
Longitude	73° 50′ and 77° 00′ E	74° 28′ and 77° 36′ E	76° 50′ and 77° 20′ E	69° 30′ and 78° 17′
Area (km^2)	50,362 km^2	44,212 km^2	1483 km^2	3,42,239 km^2
Climate	Tropical, semiarid, and subtropical monsoon type	Semiarid	Dry with the intensely hot summer and cold winter	Dry hot to cold
Annual rainfall	780 mm	313 mm in southwestern parts of the state over 862 mm in the northeastern region	612 mm	580 mm
Rivers	Ravi, Beas, Satluj, and Ghaggar	Ghaggar and Yamuna	Najafgarh drain, Yamuna River, Barapullah drain, Shahdara drain, Bawana drain basin	Chambal, Yamuna–Ganga, Narmada, Mahi
Neighboring states	Himachal Pradesh, Jammu and Kashmir, Haryana, Rajasthan	Delhi, Punjab, Rajasthan, Uttar Pradesh	Punjab, Rajasthan, Haryana, Uttar Pradesh	Haryana, Punjab, Delhi, Gujarat, Madhya Pradesh
Hydrogeology	Three subdivisions: Kirana (bedrock hills): Central Rechna Doab Pabbli hills—northernmost part of Chaj Doab formed by an anticline in the Siwalik group composed by sandstone, siltstone, and conglomerate. Piedmont found in alluvial plains and the mountainous areas of the Himalayan foothills. Alluvial plain's dominant parts consist of sand, silt, and clay	Quaternary-Holocene alluvial sediments which are deposited by rivers and about 5% area is covered by Proterozoic hard rocks Tertiary group-outer Siwalik system composed mainly of sandstones, clay and boulders. The rocks of pre-Cambrian age—part of the Aravalli hill ranges	Quartzite with Mica observed by unconsolidated Quaternary to recent sediments	Aravalli hill ranges—main water divide in Rajasthan elevation varies from about 600 m to over 900 m above mean sea level (m amsl) Supergroup of rocks ranging in age from Archaean to Proterozoic

Source Datta and Tyagi (2004), Datta et al. (1991)

Table 19.3 Groundwater resources availability, utilization, and stages of development in NWR (CGWB 2011)

NWR	Annual replenishable groundwater resource BCM/year	Natural discharge during non monsoon season BCM/year	Irrigation BCM/year	Domestic and industrial uses BCM/year	Total BCM/year	Stage of groundwater development (%)
Haryana	10.48	0.68	11.71	0.72	12.43	127
Punjab	22.56	2.21	33.97	0.69	34.66	170
Rajasthan	11.86	1.07	12.86	1.65	14.52	135
Delhi	0.31	0.02	0.14	0.26	0.40	138

Source http://www.cgwb.gov.in

with population growth, agricultural revolution, and industrialization (Ambast et al. 2006; Villholth and Sharma 2006). In the NWR states including Haryana, Delhi, Punjab, Rajasthan, the stage of groundwater development is more than 85%. The groundwater resources, availability, utilization, and stages of development in NWR, are described in Table 19.3.

19.3.1 Groundwater Salinity in NWR

Groundwater salinity is a widespread problem and a challenge to water resources management. It has become a national concern in the alluvial plains of Delhi and neighboring Haryana State. Groundwater salinity can be defined as high concentration of dissolved salts in water more than permissible limits for drinking and irrigation water use. Once groundwater gets saline, it is very difficult to remediate as it needs very scientific program with enough time and finance. The groundwater salinity endangers ecosystems that degrade the productivity of agricultural land, health, and livelihood of individuals (Villholth and Sharma 2006; Tyagi 1988). This region shows high population density, where large regions are dealing with saline groundwater. Most affected are the semiarid to arid northwestern states of Rajasthan, Punjab, Haryana, and Delhi, where groundwater consumption has been increasing over the past decades, along with population growth, agricultural revolution, and industrialization (Daga 2003; Ambast et al. 2006). In NWR, the annual consumption of groundwater is more than annual groundwater recharge. The demand of groundwater for irrigation started increasing in Punjab and Haryana due to the advent of the Green Revolution in the 1960s. The climate of northwestern regions is best suited for wheat but farmers give preference to grow rice crop in their fields. This changed cropping system led to over-exploitation of groundwater resources (CGWB 2011).

In terms of extraction of groundwater Punjab comes on top (about 94%), followed by Haryana (84%), Rajasthan (51%), and Delhi (34%).

Besides, that there are several causes of groundwater salinity in NWR such as the concentration of dissolved salts through evaporation, the leaching of salts or the mobilization of saline groundwater from deeper aquifer sections (Salama et al. 1999; Tyagi 1988). The surface soil salts dissolve and percolate to groundwater during the infiltration of rain or irrigation water (Duncan et al. 2005). Salama et al. (1999) reported that excessive irrigation in semiarid and arid regions also lose the soil structure and fertility and become more susceptible to waterlogging. Grube et al. (1999) have observed the mixing of freshwater to deep saline groundwater through leakage is the other region. It has been reported that the accumulation of salts in soil profiles or groundwater can be controlled by climate, morphology, surface, and subsurface drainage of NWR. Thus, the management of groundwater has become essential that support to achieve sustainable use of groundwater sources (Singh and Tewari 1998). Figure 19.3 described the affected areas by groundwater salinity.

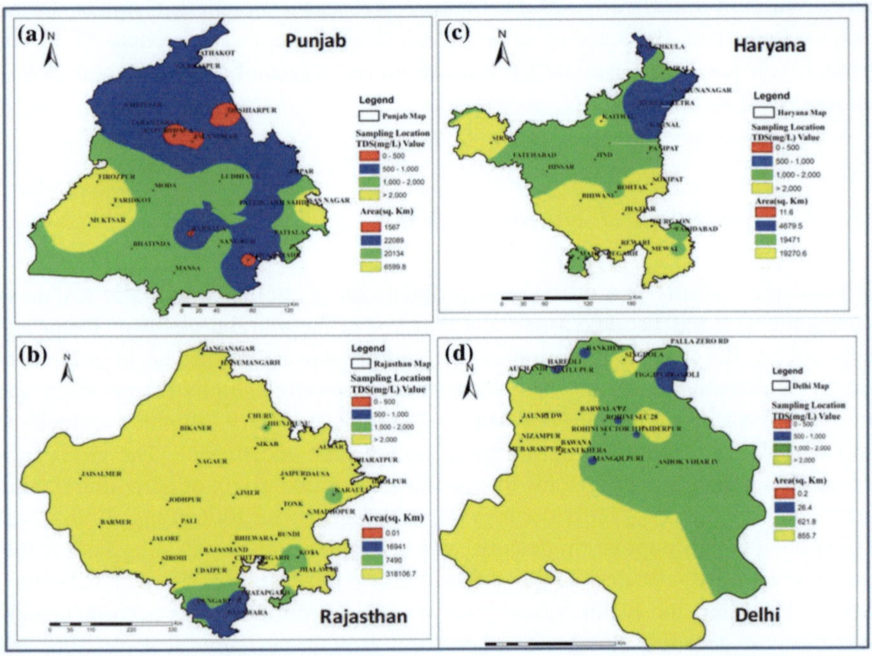

Fig. 19.3 Districts affected by salinity in groundwater in NWR, **a** Punjab, **b** Rajasthan, **c** Haryana, **d** NCT Delhi (CGWB 1997, 2006, 2011, 2014)

19.3.2 Reclamation of Groundwater Salinity in Northwestern Region

There are number of ways to reclaim the groundwater salinity in more efficient manner in northwestern region of India. Punjab State is an agricultural state of India formed of a large part of Indo-Gangetic alluvial plain. This land is highly fertile due to depositions carried by Indo-Gangetic Rivers (Hira and Khera 2000). Sustainable use of groundwater involves the adoption of norms related to enhancing recharge through the protection of the recharge area, controlling the depth of wells, efficiency of pumps used, water-saving irrigation methods, and overall regulation of cropping pattern to rationalize water use in agriculture (Beltrán 1999). For sustainable use of groundwater resources, we need to build a disaggregated picture by mapping aquifers along with describing variation in aquifer typologies in hydrogeological and socioeconomic contexts.

According to the planning commission report, Punjab is not in favor of any water legislation and policies because such steps will create hardship for farmers. Therefore, this state has banned the formation of new tube wells and restricted to 10 horsepower (HP) pumps so that the deeper aquifers are not tapped (Planning commission 2007a, b). Instead, to tackle over-exploitation of groundwater resources, the Punjab Government have been taking various initiatives such as diversification of crops, constructing a large scale of the artificial recharge well, and promoting micro-irrigation practices to conserve excess use of water. State government is providing the subsidy to individual farmer to lay down an underground pipeline, drip, and sprinkler systems (CGWB 2011; Planning commission 2007a, b).

Haryana, which is also a part of the semiarid region, is facing higher groundwater salinity due to over-exploitation of groundwater and more evapotranspiration of natural concentration of salts (Pradhan and Chandrasekharan 2009). It is further degraded by overuse of submersible pumps, extensive canal system, and application of fertilizers in agriculture. This state has been adopting many groundwater management plans including drip irrigation (Keller and Bliesner 1990; Goldberg et al. 1976), growing salt-tolerant crops like cotton, wheat, guar, chickpea, soybeans, sugarcane, and sunflower in agricultural belts (Kang 1998), use subsurface planting, and furrow irrigation method (SPFIM) for brackish-water-based agroforestry systems on degraded soils (Haryana Kisan Ayog 2012; Goel et al. 1977).

Rajasthan is the largest state and arid region of the country where Aravalli hill ranges form the main water divide. This state is also facing high salinity like Haryana State. The cause of higher salinity is more evapotranspiration due to the higher temperature. The arid condition has resulted from natural concentration of salts and lack of drainage (CGWB 2015a, b). To overcome these challenges, state has established Intensive Awareness Programmes (IAP) for educating the local people for the judicious use of groundwater assets. Some ancient water harvesting technologies such as Khadins and Nadis are being used at village level. In village level, some local level regulatory bodies such as panchayat have created (Mathur 2007). This state is also practicing intensive afforestation in the catchment area and cultivate salt-tolerant plants (sunflower, millets, mustard, cotton) in the salt-affected land (Kang 1998; Malash et al. 2005).

Like others state, Delhi is involved in groundwater management program to tackle the groundwater salinity. The major reason of salinity is due to the drain of finer sediments in the aquifer, improper flushing of groundwater, and longer residence time of water in the aquifer (CGWB 2016). For the management of groundwater resources, the entire NCT has been categorized into four zones, viz. Zone I, Zone II, Zone III, and Zone IV. Zone-wise groundwater management plans have been presented in Table 19.4.

Table 19.4 Zone-wise groundwater management in Delhi (CGWB 2015)

Zone	Aquifer management plan
Zone I (Narela, Saraswati Vihar, Civil Lines, Punjabi Bagh, Najafgarh, eastern parts of Model Town, Kotwali, Daryaganj, Preet Vihar, and Seelampur)	• Groundwater withdrawal recommended • No additional artificial recharge interventions required • Poor quality water can be used for growing salt-tolerant crops and can also be used after blending for uses other than drinking
Zone II (area of Najafgarh Tehsil falling in the west of Najafgarh drain and western part of Saraswati Vihar	• Recommended groundwater and artificial recharge • Poor quality water can be used for growing salt-tolerant crops and can also be used after blending for uses other than drinking
Zone III (Delhi Cantt., Hauz Khas, Kalkaji, Chanakyapuri, Connaught Place, Karol Bagh, Paharganj, Rajouri Garden, Defense Colony, Punjabi Bagh, Preet Vihar, Vivek Vihar)	• No further groundwater development • Regulation of existing groundwater withdrawal • Artificial recharge may be taken up • Possibility of using tertiary treated wastewater for recharge may be explored
Zone IV (buffer zone up to 1 km on each side of the drain falling in Najagfgarh, area around landfill sites and industrial belts)	• Groundwater unsuitable for drinking and domestic water use • May be used only after proper treatment • Landfill sites should be selected based on hydrogeological surveys and scientifically designed so as to avoid groundwater pollution • Ensuring proper disposal of industrial solid waste and effluents after removal of various pollutants

19.4 Groundwater Management Plans

Groundwater resources' management deals with balancing the increasing demands of water in terms of quantity, quality, and surface water interactions. The groundwater resources do not usually arise until a quality of water cannot affect the large group of people. The exploitation and restoration of fresh groundwater in arid and semiarid region should form a part of integrated water management, comprising surface water and groundwater—both in terms of water quantity and water quality—and taking into account the demands (The World Bank 2010). This requires cooperation, information, study, planning, and legislation (Kumar 2006). Groundwater management plans involve the following three general principles: (i) development of technologies that will enhance the storage capacity of groundwater reservoirs, (ii) protection of groundwater quality, and (iii) utilization of groundwater resources, for higher or most valuable use to society (Singh and Tewari 1998). Salinity problems in agriculture occur due to naturally saline soils, evaporative concentration of salts in irrigation water, or capillary rise of a shallow groundwater table and subsequent evaporative concentration (Provin and Pitt 2017). The organizational structure for groundwater management varies from regional to regional. Integrated water management can be attained only by cooperation between all authorities at different levels. The measures required for proper groundwater management which are described followed.

19.4.1 Artificial Recharge

Artificial recharge through rainwater harvesting is being practiced in northwestern region of the country. It was observed that the selection of sites and type of recharge structures are not always compatible with hydrological and hydrogeological conditions. As a result, the desired benefits have not been realized. CGWB initiated 165 artificial recharge schemes under central sector during the Ninth Plan with the active involvement of state government/union territories (CGWB 2016). The artificial recharge structure projects have been taking up in water-scarce areas along with best suited to different agro-climatic and hydrogeological conditions. The major concern of artificial recharge is non-uniformity and largely dependent on the source of water availability and total rainfall on the particular area. The construction of percolation tanks, check dams, recharge shafts, and subsurface barriers is more suitable structures in alluvial areas having efficient geophysical properties. It is observed that the cost of recharging structure is dependent on-site locations, the range of rainfall, and agro-climatic conditions (Planning commission 2007a, b).

CGWB has prepared a 'Master Plan for Artificial Recharge to Groundwater in India' bringing the areas for artificial recharge to groundwater reservoir, wherein schemes need to be implemented as s top-most priority to ameliorate the water scarcity problems. A total area of 0.45 million km^2 were identified in the country which needed artificial recharge of groundwater. The master plan envisages construction of 225,000 artificial recharge structures in rural areas. The break-up includes 37,000 percolation tanks 110,00 check dams, nala bunds, cement plugs, weirs, anicuts 48000 recharge shafts, dugwell recharge, around 1000 revival of ponds, and 26,000 gully plugs and gabion structures (CGWB 2002). Total cost of the implementation of master plan is Rs. 245 billion. The state-wise feasibility and cost estimates are given in Table 19.5.

Table 19.5 State-wise feasibility and cost estimates of artificial recharge structures in India (CGWB 1997, 2002)

Name of states	Area identified for artificial recharge (sq.km)	Quantity of surface water to be recharged (MCM)	Type and number of artificial recharge of structures	Cost (Rs. billion)
Delhi	693	444	23 percolation tanks, 23 existing dugwells, 10 nala bunds, 2496 roof top rainwater harvesting	2.57
Haryana	16,120	1408	15,928 recharge shafts and recharge trenches	3.32
Punjab	22,750	1200	40,030 recharge shafts and recharge trenches 12,800 roof top harvesting structures in urban areas	5.28
Rajasthan	39,120	861	3228 percolation tanks 1291 anicuts 2871 recharge shafts Rooftop rainwater harvesting structure (0.4 M houses)	11.40

19.4.2 Cropping Sequence

For successful utilization of saline water, crops which are semi-tolerant to tolerant such as mustards, wheat, and cotton as well as those with low water requirement are recommended. Malash et al. (2008) reported the impact of higher salinity irrigation has decreased the leaf area index, plant dry weight, fruit total yield, and individual fruit weight of tomato (*Lycopersicon esculentum*). In addition, a few days after saline irrigation, the soil moisture content was decreased. Crops, such as rice, sugarcane which require less water, should be restricted. In low rainfall area, i.e., western region of India, mono-cropping is recommended for maintains salt balances. Cropping sequence is another critical step in mitigating saline conditions of soil. The recommended cropping sequence for saline conditions is pearl millet–barley, pearl millet–wheat, pearl millet–mustard, sorghum–wheat, barley–sorghum–mustard, etc. These cropping systems are more remunerative in saline soil. In semiarid region, mustard can be replaced with wheat in the cropping sequence since its water requirement is low compared to wheat. Manchanda et al. (1985) reported that the tolerance limit of crop is highly restricted by high concentration of chloride ion in groundwater. More salts tend to accumulate in soils when irrigated with water of high sodium adsorption ration (SAR) and thus tend to reduce the tolerance limit of crop. All crops do not tolerate salinity equally at different stages of their growth.

Singh (2018) had observed that growing fruit crops in salt-affected soils can broadly be grouped into four categories, viz. ‘salt-tolerant crops and cultivars,’ ‘improving the root zone conditions,’ ‘reducing crop evapotranspiration losses,’ and ‘irrigation management for reducing salt hazard.’ The strengths and weaknesses of such salinity management techniques and practices are discussed. Depending on factors like crop species, climate and salinity level, a combination of techniques is likely to give better results than a single intervention in the long run.

19.4.3 Micro-Irrigation Method

The distribution of water and salts in soil varies with the method of irrigation. A shift toward micro-irrigation practices such as drip and sprinklers can be used to control salinity in soil (Sharma and Minhas 2001). The increasing demand for irrigation water to secure food for growing populations with limited water supply suggests rethinking the use of non-conventional water resources. Malash et al. (2008) reported that saline drainage water (EC > 0.4 S m^{-1}) was used for irrigation of tomato (*Lycopersicon esculentum*) using drip and furrow irrigation. It was observed that drip irrigation played an efficient role to manage the higher fruit yield. Drip irrigation has the potential to increase crop yields with less irrigation water, and under saline conditions, it has additional advantages over furrow and sprinkle irrigation systems. Hanson

(2011) studied that drip irrigation has potential to increase crop yield in saline condition, and it has an additional advantage over furrow irrigation and sprinkler irrigation systems.

19.4.4 Community Participation

The community participation is also emphasized recently by launching Jal Shakti Abhiyan (JSA) under the Ministry of Jal Shakti. The central scheme has got the approval for the grant from the World Bank. The other proposed scheme Atal Bhujal Yojana aims to improve groundwater management in priority areas in the country through community participation. The priority areas are identified under the scheme fall in the states of Haryana, Rajasthan, and Uttar Pradesh. These states represent about 25% of the total number of over-exploited, critical, and semi-critical blocks in terms of groundwater in India. The scheme emphasizes the active participation of the communities in groundwater management such as the formation of water user associations, monitoring and disseminating groundwater data, water budgeting, preparation and implementation of gram panchayat-wise water security plans and IEC activities related to sustainable groundwater management.

19.5 Conclusion

This article presents an overview of the status of groundwater salinity and its confronted issue for the northwestern region (NWR) of India. In northwestern states, groundwater being important for socioeconomic and sociocultural harmony, its sustainable development, and proper utilization can be achieved by understating hydrogeology and regular investigation of the aquifer's system. The largest component of groundwater use in the states is for irrigation. Because of large-scale extraction along with arid and semiarid climates accumulation of salt on soil surface in NWR is prevalent. In NWR region, salinity in groundwater has been observed more (>0.3 S m^{-1}). The effective measures can help make the groundwater management plan more accurate, holistic, and sustainable. Instead of putting ban on over-exploitation, the government should give incentives for the reconstruction of new wells, recharge structures, and for the use of micro-irrigation technology along with energy-saving devices. The main reason of groundwater salinity in the northwestern region is the predominant use of groundwater in agricultural purposes, nearly 96%. Indus Basin that covers the major part of Punjab has witnessed huge development in agricultural production that resulted in enormous exploitation of groundwater resources, leading to the problems of water table decline, waterlogging, and salinization. In other states, namely Haryana, Rajasthan, and Delhi, which have the groundwater extraction less than Punjab which also faces higher groundwater salinity due to over-exploitation and higher rate of evapotranspiration of dissolved salts. There are no effective water management policies, which address the equity and management of groundwater water resources. The integrated water resources management (IWRM) can help in developing water sustainable program. The ongoing 'Jal Shakti Abhiyan (JSA)' and proposed Atal Bhujal Yojana of Government of India that emphasize efficient use of groundwater resources along with public participation can be a promising way forward toward achieving an inclusive groundwater development program.

References

Ambast SK, Tyagi NK, Raul SK (2006) Management of declining groundwater in the Trans Indo-Gangetic Plain (India): Some options. Agric Water Manag 82:279–296

Annual Report (2013–14) Ministry of water resources. River Development and Ganga Rejuvenation

Ayers RS, Westcot DW (1994) Water quality for agriculture. FAO Irrigation and Drainage

Beltrán JM (1999) Irrigation with saline water: benefits and environmental impact. Agric Water Manag 40:183–194

Bonsor HC, MacDonald AM, Ahmed KM, Burgess WG, Basharat M, Calow RC, Dixit A, Foster SSD, Gopal K, Lapworth D, Lark RM, Moench M, Mukherjee A, Rao MS, Shamsudduha M, Smith L, Taylor R, Tucker J, van Steenbergen F, Yadav SK (2017) Hydrogeological typologies of the Indo-Gangetic basin alluvial aquifer, South Asia. Hydrogeol J 25(5):1377–1406

CGWB (1997) Inland groundwater salinity in India. GOI, Ministry of Water resource, central groundwater board (CGWB), Faridabad, 62 p

CGWB (2002) Master plan for artificial recharge to groundwater in India. Central Ground Water Board, Faridabad, India

CGWB (2006) Hydrogeological framework and groundwater management plan of NCT Delhi

CGWB (2011) Dynamic ground water resources of India. Central Ground Water Board, Faridabad, India

CGWB (2012) Dynamic ground water resources of India (as on 31 March 2009). Central Ground Water Board Ministry of Water Resources Government of India Faridabad November 2011 (pages 243)

CGWB (2014) A concept note on geogenic contamination of ground water in India with special reference to nitrate

CGWB (2015a) Draft report on dynamic ground water resources of NCT Delhi (as on 2013)—Central Ground Water Board, Delhi

CGWB (2015b) Report on central ground water board ministry of water resources. River Development and Ganga Rejuvenation in North West Chandigarh

CGWB (2016) Report on central ground water board ministry of water resources. River Development and Ganga Rejuvenation in Haryana state

Chakraborti D, Ghorai SK, Das B, Pal A, Nayak B, Shah BA (2009) Arsenic exposure through groundwater to the rural and urban population in the Allahabad-Kanpur track in the upper Ganga plain. J Environ Monit 11:1455–1459

Chopra RPS, Krishan G (2014) Analysis of aquifer characteristics and groundwater quality in Southwest Punjab, India. J Earth Sci Eng 4:597–604

Cooper AH (2002) Halite karst geohazards (natural and man-made) in the United Kingdom. Environ Geol 42(5):505–512

Daga S (2003) Private supply of water in Delhi in research internship papers 2003. Centre for Civil Society, New Delhi, pp 172–182

Datta KK, de Jong C (2002) Adverse effect of water logging and soil salinity on crop and land production in northwest region of Haryana, India. Agric Water Manag 57:223–238

Datta PS, Tyagi SK (2004) Evidences of climatic uncertainties linked groundwater variability in peri-urban agricultural areas of Delhi. In: Proceedings of the 1st Asia Oceania geosciences society (AOGS) conference, Singapore, 5–8 Jul 2004

Datta PS, Tyagi SK, Chandrasekharan H (1991) Factors controlling stable isotope composition of rainfall in New Delhi, India. J Hydrol 128:223–236

Datta PS, Bhattacharya SK, Tyagi SK (1996) ^{18}O studies on recharge of phreatic aquifers and groundwater flow-paths of mixing in Delhi area. J Hydrol 176:25–36

Duncan R, Bethune M, Christen E, Hornbuckle J (2005) A review of salt mobilisation and management in irrigated areas of the Murray-Darling Basin. Cooperative Research Centre for Catchment Hydrology, Monash University, Victoria, Australia

Ewusi A, Obiri-Yeboah S, Hans-Jurgen V, Asabere BS, Bempah CK (2013) Groundwater quality assessment for drinking and irrigation purposes in Obuasi municipality of Ghana, a preliminary study. Res J Environ Earth Sci 5(1):6–17

Gilfedder M, Mein RG, Connell LD (2000) Border irrigation field experiment. II: salt transport. J Irrig Drain 126:92–97

Gleeson T, Wada Y, Bierkens MF, van Beek LP (2012) Water balance of global aquifers revealed by groundwater footprint. Nature 488(7410):197–200

Goel PD, Datta PS, Tanwar BS (1977) Measurement of vertical recharge to groundwater in Haryana State (India) using tritium tracer. Nord Hydrol 8:211–224

Goldberg D, Gornat B, Rimon D (1976) Drip irrigation-principles. Drip Irrigation Scientific Publications, Israel, Design and Agricultural Practices

Grube A, Schmalz B, Nachtigall K, Wichmann K (1999) Provisional results of investigations on groundwater salinization with emphasis on the region of Lübeck,Northern Germany. Naturweed Tiddschr 79:164–171

Gunnar L, Sprenger C, Baudron P, Gupta D, Pekdeger A (2012) Origin and dynamics of groundwater salinity in the alluvial plains of western Delhi and adjacent territories of Haryana state, India. Hydrol Process 26:2333–2345

Hanson B (2011) Drip irrigation salinity management for row crops, agricultural and natural sciences

Haryana Kisan Ayog (2012) Report on natural resource management in Haryana, CCS HAU campus, Government of Haryana

Hem JD (1985) Study and interpretation of the chemical characteristics of natural water. US Geol Surv Water Supply, p 2254

Hira GS, Khera KL (2000) Water resource management in Punjab under rice-wheat production system. Department of soils, Punjab Agricultural University, Ludhiana, Resarch Bulletin

Jolly ID, McEwan KL, Holland KL (2008) A review of groundwater—surface water interactions in arid/semi-arid wetlands and the consequences of salinity for wetland ecology. Ecohydrology 1:43–58

Kang YH (1998) Microirrigation for the development of sustainable agriculture. 251–255

Keller J, Bliesner RD (1990) Sprinkle and trickle irrigation. Van Nostrand Reinhold, New York, pp 463–465

Krishan G, Rao MS, Kumar CP, Kumar S, Loyal RS, Gill GS, Semwal P (2017) Assessment of salinity and fluoride in groundwater of semi-arid region of Punjab, India. Curr World Environ 12(1):34–41

Kumar CP (2006) Management of groundwater in salt water ingress coastal aquifers

MacDonald Alan, Bonsor Helen, Ahmed Kazi, Burgess William, Basharat Muhammad, Calow Roger, Dixit Ajaya, Foster Stephen, Krishan Gopal, Lapworth Daniel, Lark Murray, Moench Marcus, Mukherjee Abhijit, Rao MS, Shamsudduha Mohammad, Smith Linda, Taylor Richard, Tucker Josephine, van Steenbergen F, Yadav S (2016) Groundwater depletion and quality in the Indo-Gangetic Basin mapped from in situ observations. Nature Geosciences. 9:762–766

Malash N, Flowers TJ, Ragab R (2005) Effect of irrigation system and water management practices using saline and non-saline water on tomato production. Agric Water Manag 78:25–38

Malash NM, Flowers TJM, Ragab R (2008) Effect of irrigation methods, management and salinity of irrigation water on tomato yield, soil moisture and salinity distribution. Irrig Sci 26(4):313–323

Manchanda HR, Sharma SK, Singh JP (1985) Effect of increasing levels of residual sodium carbonate in irrigation water on exchangeable sodium percentage of a sandy loam soil and crop yields. J Indian Soc Soil Sci 33:366–371

Mathur LN (2007) Geoscientific studies and managing lakes of arid and semi-arid regions of Rajasthan. In: World Lake Conference, pp 1928–1932

Misra AK, Mishra A (2007) Escalation of salinity levels in the quaternary aquifers of the Ganga alluvial plain, India. Environ Geol 53:47–56

Misra AK, Mishra A (2008) Study of quaternary aquifers in Ganga Plain, India: focus on groundwater salinity, fluoride and fluorosis. J Hazardous Mater 144(1–2)

Planning Commission (2007a) Report of expert group on groundwater management and ownership. Government of India

Planning Commission (2007b) Report of the export group on ground water management and ownership. Planning Commission of India

Pradhan S, Chandrasekharan H (2009) Effect of monsoon rain on quality of groundwater for irrigation in Gohana block of Haryana. Journal of Agricultural Physics 9:38–43

Priyanka Krishan G, Sharma LM, Yadav BK, Ghosh NC (2016) Analysis of water level fluctuations and TDS variations in the groundwater at Mewat (Nuh) district, Haryana (India). Curr World Environ 11(2):388–398

Provin T, Pitt JL (2017) Managing soil salinity. The Texas A&M University System, Internet Resource

Richter BC, Kreitler CW (1986) Geochemistry of salt water beneath the rolling plains, North-Central Texas. Ground Water 24:735–742

Richter BC, Kreitler CW (1993) Geochemical techniques for identifying sources of ground-water salinization

Salama RB, Otto CJ, Fitzpatrick RW (1999) Contributions of groundwater conditions to soil and water salinization. Hydrogeol J 7:46–64

Shah T, Scott C, Narayana P, Kishore A, Sharma A (2000) The water—energy nexus in India: Approaches to Agraian prosperity with a viable power industry. IWMI, Research report 70, Anand: International water management institute

Sharma GS, Minhas PS (2001) Response of rice-wheat to alkali water irrigation and gypsum application. J Indian Soc Soil Sci 49:324–327

Sharma RB, Minhas PS (2005) Strategies for managing saline/alkali waters for sustainable agricultural production in South Asia. Agric Water Manag 78:36–151

Singh A (2018) Soil and water management practices for fruit cultivation in salt affected soils. Advances in salinity and sodicity management under different agro-climatic regions for enhancing farmers' income. Publisher, ICAR-CSSRI, Karnal, pp 172–181

Singh J, Kothari M (2004) Waterlogging and salinity studies for the irrigated tracts of South West Punjab using satellite remote sensing and geographical information system. M.Sc. Thesis. Maharana Pratap University of Agriculture and Technology, Udaipur

Singh SK, Tewari L (1998) An Assessment of institutional managements impact on the use of groundwater resources for sustainable agriculture development in Uttar Pradesh. Indian journal of Agricultural Economics 53(3)

The World Bank (2010) Deep wells and prudence: towards pragmatic action for addressing ground water overexploitation in India

Tyagi NK (1988) Managing salinity through conjunctive use of water resources. Ecol Model 40:11–24

Villholth KG, Sharma BR (2006) Creating synergy between groundwater research and management in South and South East Asia. In: Proceedings of IWMI-ITP-NIH international workshop on "creating synergy between groundwater research and management in South and Southeast Asia. Roorkee, India. 8–9 Feb 2005

Wilcox L (1955) Classification and uses of irrigation waters. USDA Circular No. 969, Washington DC

World Health Organization (WHO) (2004) Guidelines for drinking water quality, 2nd edn, vol 1. Geneva, WHO, pp 130

Xun Z, Cijun L, Xiumin J, Qiang D, Lihong T (1997) Origin of subsurface brines in the Sichuan basin. Ground Water 35:53–58

Chapter 20
Interaction Between Groundwater and Surface Water and Its Effect on Groundwater Quality

S. K. Pramada and Sowmya Venugopal

Abstract The unscientific disposal of wastes into rivers and canals can cause surface water pollution. The surface water and groundwater are fundamentally interconnected and thus one can contaminate the other. In some cases, surface water systems gain water and solutes from groundwater and in others the surface water body is a source of groundwater recharge and causes changes in groundwater quality. For managing the water resources, it is essential to study, how the surface affects the groundwater systems. Mathematical models have been widely used in planning and management of water resources. This paper presents a study where a surface water and groundwater interaction model is developed and applied to a case study. A finite-difference code was developed for the modeling of river water quality. A program is written in MATLAB using the explicit finite difference method. In this study, MODFLOW is used to model the groundwater flow and MT3DMS is used to model the contaminant transport flow in groundwater. Finally, the interaction between the surface water and groundwater was studied from the surface and groundwater quality models.

Keywords MODFLOW · MT3DMS · Groundwater · Contaminant transport · River water quality

20.1 Introduction

Water pollution is a major problem in the global context. When waste materials enter lakes, rivers, oceans, and other water bodies, they get dissolved or suspended in water or get deposited on the bed. This causes the pollution of surface water bodies. Naturally, surface water contains a wide variety of substances, and human activities inevitably add to this mixture. There are many sources of surface water pollution.

S. K. Pramada (✉) · S. Venugopal
Department of Civil Engineering, NIT Calicut, Kozhikode 673601, India
e-mail: pramada@nitc.ac.in

S. Venugopal
e-mail: sowmya031@gmail.com

R. M. Singh et al. (eds.), *Environmental Processes and Management*,
Water Science and Technology Library 91,
https://doi.org/10.1007/978-3-030-38152-3_20

The main sources of pollution are the sewage and industrial waste. The facilities to treat wastewater are not adequate in many cities in India. Presently, only about 10% of the wastewater generated is treated; the rest is discharged into our water bodies (Radha 2008). Due to this, pollutants enter groundwater, rivers, and other water bodies. Groundwater contains mineral ions naturally. Human activities can affect the natural mineral composition of groundwater through the disposal waste matter at the land surface or through surface water. Since one of the sources of groundwater is surface water, high concentrations of chemical parameters in surface water can cause pollution to groundwater. Since both surface and groundwater are the sources of drinking water, this polluted water can seriously affect the health of the people consuming it. Thus, it is essential to study both surface water and groundwater pollution.

Since surface water and groundwater are fundamentally interconnected, one can contaminate the other. The ability to link groundwater to surface water bodies makes it possible to predict the migration of contaminants from surface water to groundwater or vice versa. Inadequate and incompetent management of water resources causes water-borne diseases and related health problems. Thus, water quality management is an important issue of relevance in the context of present times. Hence the interaction of surface water and groundwater has to be deeply studied for better management of water resources. Many researchers carried out work on stream aquifer interactions (Yan and Smith 1994; Hussein and Schwartz 2003; Sophocleous 2002; Serdar Korkmaz et al. 2009; Rodríguez et al. 2006; Kim et al. 2008; Zhou et al. 2019). In this paper, the stream water quality model is developed using finite difference approach and groundwater quality is modeled using MT3DMS. The objectives of the present study are set as follows (1) To develop a surface water quality model (2) To develop a groundwater quality model, and (3) To analyze the interaction between surface water and groundwater through the above models.

20.2 Study Area and Data Collection

The study area is located along the Canoly Canal in Kozhikode, Kerala. The Kozhikode district falls within latitudes 11° 08′ and 11° 42′ and longitudes 75° 31′48″ and 75° 49′30″ and is situated along the southwest coast of India. The coast of the district is about 71 km and it covers an area of 91 km^2. This district has a humid tropical climate and an annual rainfall is estimated to be about 3000 mm.

The Canoly Canal is a manmade canal constructed in 1848. This canal connects the Korapuzha River in the north and the Kallai River in the south. The canal is 11.4 km long. The width of the canal ranges from 6 to 20 m and the water depth in the peak rain period varies from 0.5 to 2 m. The canal is a part of the West Coast Canal System. There are a lot of industrial activities such as coir retting, log setting and other kinds of timber industries around the southern end of the canal. Most residential areas and several hospitals along the canal are letting out their wastewater into the canal or the sea via ditches without any treatment. In addition to the liquid waste, there are also

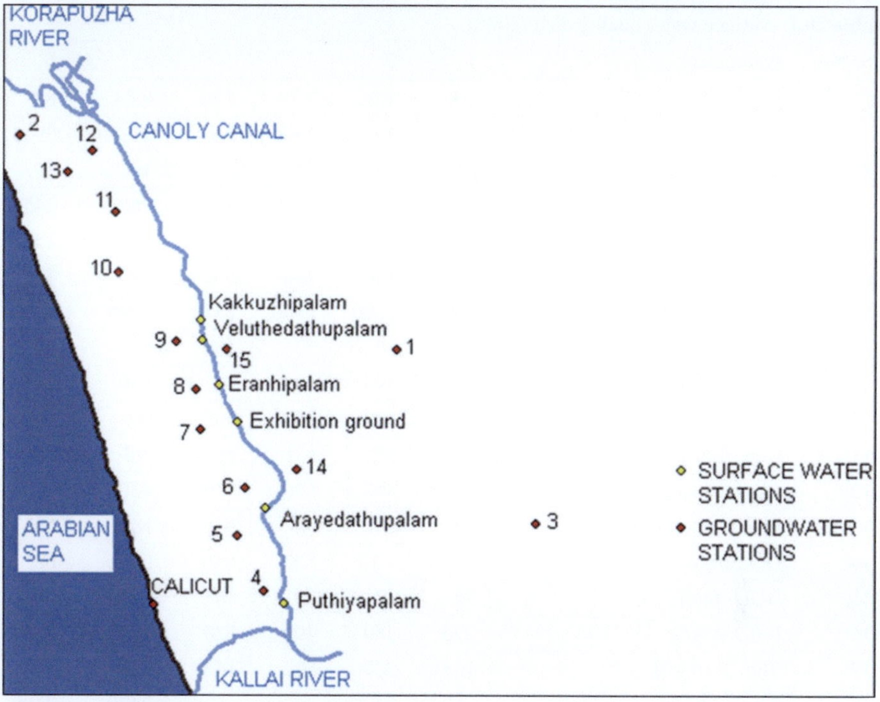

Fig. 20.1 Map showing wells in the study area

considerable amounts of solid waste dumped into the Canoly Canal, both domestic as well as industrial waste. There is lining made of stones along the sides of the canal but incomplete and collapsed at some locations. Toward the junction of Canoly Canal with Kallai River in the south, there are some sections where there has not been any lining constructed. In many places, trees and bushes are also present in the canal together with a lot of water living plants such as water hyacinths on the surface. Thus, the water flow is low in the middle stretch of the canal. The data required for the study were collected from Centre for Water Resources Development and Management, Kozhikode (CWRDM, Kozhikode) and Groundwater department, Kozhikode during the period 2002–2008. The surface water sampling site and groundwater observation wells in the study area are shown in Fig. 20.1.

20.2.1 Surface Water Quality

The Canoly Canal is heavily polluted from all the surrounding activities that let out their sewage into the canal water. Among the present activities are hospitals, hotels, garages, timber industries, coir retting, slaughterhouses as well as big residential

Table 20.1 Surface water quality data (2005)

S. No.	Parameters	1	2	3	4	5	6
1	pH	7.8	7.96	7.77	7.03	8.04	8.1
2	EC (micro siemens/cm)	8521	4039	4466	8361	10,744	3953
3	Turbidity (NTU)	26	18	35	13	15	19
4	TDS (mg/l)	5283	2504	2769	5184	6661	2451
5	Total alkalinity (mg/l)	132	109	104	92	120	136
6	Total hardness (mg/l)	400	120	240	670	340	140
7	Sodium (mg/l)	1590	1000	1800	1370	2090	1200
8	Potassium (mg/l)	125	68	70	165	45	47
9	Magnesium (mg/l)	240	390	109	267.3	230	104
10	Calcium (mg/l)	128	156	260	192	96	200
11	Iron (mg/l)	Nil	Nil	Nil	Nil	Nil	Nil
12	Chloride (mg/l)	3200	890	530	3190	4200	900
13	Phosphate (mg/l)	Nil	Nil	Nil	Nil	Nil	Nil
14	DO (mg/l)	7.05	7.13	6.87	6.9	5.9	6.2
15	BOD (mg/l)	30	22	23	18	12	20
16	COD (mg/l)	390	389	400	320	423	230
17	Sulfate (mg/l)	155	209	512	145	366	201
18	Total coliform (MPN/100 ml)	2400	2400	2400	2400	2400	2400

areas. Many drainage outlets are connected to the canal and all together they drain almost the whole city from stormwater, household gray water and also sewage. This wastewater has not been treated so far. These activities contribute to the poor condition of the surface water. The city has problems with epidemics of typhoid, hepatitis, cholera, and other water-borne diseases regularly due to the lack of potable water. Table 20.1 shows the surface water quality parameters for 6 sampling sites during 2005 collected from CWRDM, Kozhikode.

20.2.2 Groundwater Level and Quality

The topography in Kerala is generally sloping from the Western Ghats in the east toward the Arabic Sea in the west making the groundwater flow being orientated in a westerly direction. The available water level data and quality data over the period 2002–2008 were collected from Groundwater Department, Kozhikode. The water level plots for representative three wells are given in Figs. 20.2, 20.3 and 20.4. The groundwater quality data obtained from Groundwater Department, Kozhikode is depicted in Table 20.2. The water in a large number of wells is not potable due to high bacteriological content.

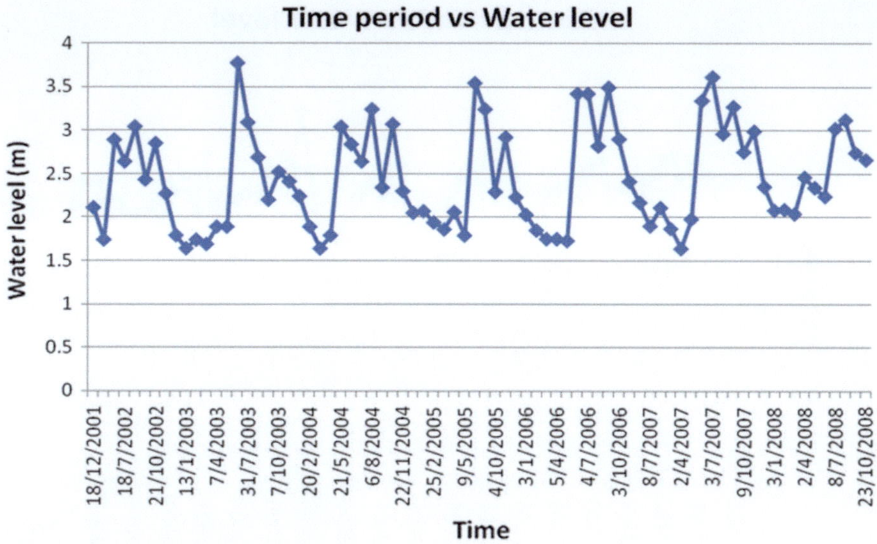

Fig. 20.2 Water level data—well no. 1

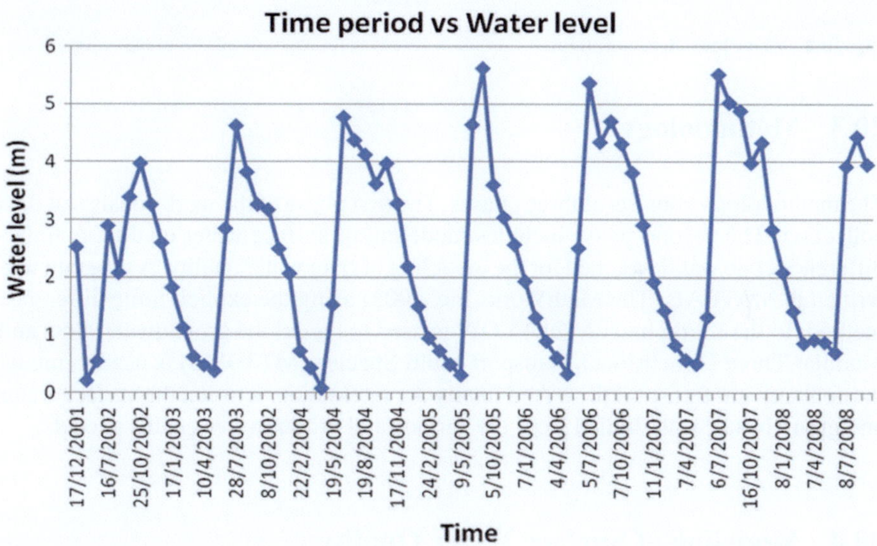

Fig. 20.3 Water level data—well no. 2

20.2.3 Rainfall Data and Lithology

The lithology data was available for the four wells and is given in Table 20.3. The rainfall data during the period 2002–2008 is given in Table 20.4.

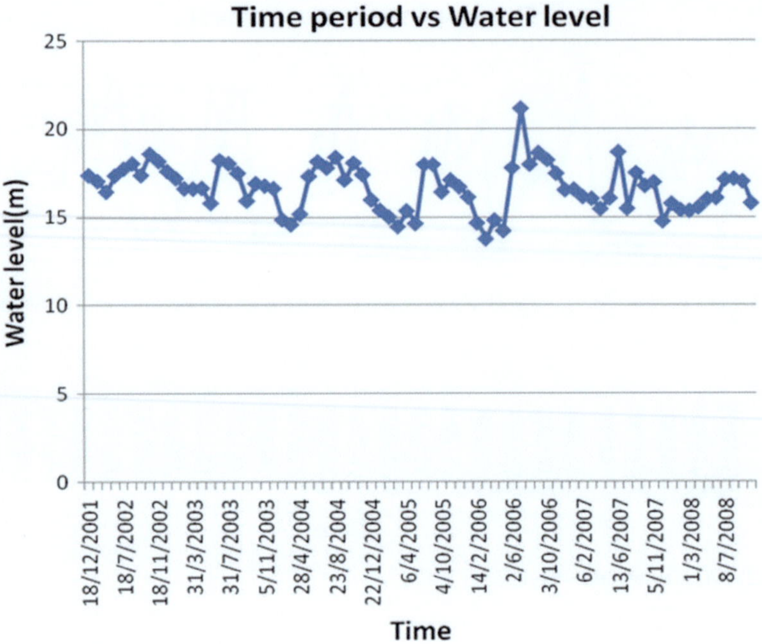

Fig. 20.4 Water level data—well no. 3

20.3 Methodology

The methodology consists of three phases. The first phase of the work consists of data collection. The second phase includes modeling of surface water quality. A finite-difference code was developed for the modeling of river water quality. A program was written in MATLAB (The MathWorks Inc. 2003) using the explicit finite difference method. In the third phase, MODFLOW is used to model the groundwater flow and Modular Three Dimensional Transport Multi Species (MT3DMS) is used to model the contaminant transport flow. And finally the interaction between the surface water and groundwater was studied from the surface and groundwater quality models.

20.4 Modeling of Surface Water Quality

The surface water quality of Canoly Canal is modeled using the explicit finite difference method. To apply finite difference method, the problem domain is divided into a finite difference grid.

The governing equation for one-dimensional solute transport can be expressed as

Table 20.2 Groundwater quality (2002)

Well No	pH	EC (μS/cm)	TDS (mg/l)	Calcium (mg/l)	Magnesium (mg/l)	Sodium (mg/l)	Potassium (mg/l)	Chloride (mg/l)	Sulfate (mg/l)	Phosphate (mg/l)
Well 1	8.1	694	416	121	2.4	22.5	1.8	37	16	0.002
Well 2	7.8	953	572	137	7.6	52.4	2.3	63	23	0.016
Well 3	8	171	103	17	6	7.6	0.5	7.1	6	0.055
Well 4	8.1	2700	1900	113.6	37.9	380	30	800	86.6	0.069
Well 5	8	186	130	20.8	2.9	19.6	8.4	50	22.4	0.018
Well 6	7.9	370	250	34.4	7.8	40	5.8	90	24.68	0
Well 7	8	520	380	98.2	18.7	35.2	16.2	45	67.52	0.011
Well 8	8.1	600	360	121	4.3	22.5	3.9	70	51.68	0.238
Well 9	8.1	320	190	124	7.6	52.4	3.6	4	12.36	0.202
Well 10	7.2	750	450	117	6.3	7.6	4.8	100	88.2	0.025
Well 11	7.4	720	430	113.6	37.9	37.4	7.7	110	97.2	0.026
Well 12	6.9	285	170	29.4	2.9	19.6	6.7	60	52.8	0.024
Well 13	7.8	953	572	137	7.6	52.4	2.3	63	23	0.016
Well 14	8	1925	1155	17	6	7.6	0.5	7.1	6	0.055
Well 15	8.1	320	190	124	7.6	52.4	3.6	4	12.36	0.202

Table 20.3 Lithology

Well	Depth up to (m)	Soil type
Well no. 1	34.75	Laterite
	35.10	Granite weathered
Well no. 3	3.25	Top soil
	6.20	Laterite
Well no. 6	12.19	Laterite with clay and sand
	15.24	Gneissic rock weathered
Well no. 14	18.29	Laterite with clay and sand
	22.25	Gneiss weathered

$$\frac{\partial C}{\partial t} = D_x \frac{\partial^2 C}{\partial x^2} - v_x \frac{\partial C}{\partial x} + q_s C_s \tag{20.1}$$

where

C aqueous concentration of the solute
D dispersion coefficient
v average linear velocity of flow
q_s velocity
C_s concentration of solute.

The first term on the right-hand side represents the dispersion of the solute, the second term is the advection term, the third term is the source/sink term and the term on the left-hand side denotes rate of change solute mass within the control volume.

On applying Taylor's series expansion to Eq. 20.1

$u(x_i, t_j + \Delta t)$ gives $u(x_i, t_j + \Delta t) = u(x_i, t_j) + u_t(x_i, t_j)\Delta t + \mathrm{O}((\Delta t)^2)$.

The forward approximation is given by

$$(u_t)_{i,j} \approx \frac{u_{i,j+1} - u_{i,j}}{\Delta x} \tag{20.2}$$

The backward approximation is given

$$(u_t)_{i,j} \approx \frac{u_{i,j} - u_{i,j-1}}{\Delta x} \tag{20.3}$$

The central approximation is given by

$$(u_x)_{i,j} \approx \frac{u_{i+1,j} - u_{i-1,j}}{2\Delta x}$$

The spatial and temporal derivatives in the governing equation are written in the finite difference form as

Table 20.4 Rainfall data in mm (2002–2008)

	Jan	Feb	Mar	Apr	May	Jun	Jul	Aug	Sep	Oct	Nov	Dec
2002	Nil	0.4	6.1	27.5	4500.7	666.9	666.9	343.6	418.4	128.4	Nil	Nil
2003	Nil	5.1	1.9	198.6	90.3	919.3	515.8	210.1	165	243.8	114.2	Nil
2004	1.8	0.4	2.8	97.6	863.6	850.8	438.9	429.8	125.8	33.1	35.6	0.4
2005	18.3	Nil	Nil	56.3	56.1	697	610.9	207.6	334.9	190.7	155.2	17.7
2006	Nil	Nil	14.5	20.8	636.6	983.6	599.9	531.9	707.1	331.6	996	Nil
2007	Nil	0.4	12.4	81.2	379.4	1030.1	1078	596.3	671.1	353.3	141.3	Nil
2008	Nil	Nil	266.5	151.1	80.4	698.3	485.5	247	414.3	Nil	Nil	Nil

$$\frac{\partial C}{\partial t} = \frac{C_{i,j+1-}C_{i,j}}{\Delta t}$$
$$\frac{\partial C}{\partial x} = \frac{C_{i,j-}C_{i-1,j}}{\Delta x}$$
$$\frac{\partial^2 C}{\partial x^2} = \frac{C_{i-1,j-}2C_{i,j+}C_{i+1,j}}{\Delta x^2}$$

Substituting in the advection—dispersion equation, for a fully explicit temporal discretization, leads to the following form:

$$C_i^{j+1} = a(C_{i-1}^j - 2C_i^j + C_{i+1}^j) - b(C_i^j - C_{i-1}^j) + C_i^j + q_s C_s \Delta t \qquad (20.4)$$

where

$$a = \frac{D_x \Delta t}{\Delta x^2}$$
$$b = \frac{v_x \Delta t}{\Delta x}$$

Equation 20.4 is written for each active node, with initial and boundary conditions leading to an equation with one unknown. A program is written in MATLAB for explicit finite difference solution. The unknown concentration at any node i at present time level depends on the concentration at the adjacent nodes at the previous time level. Initial and boundary conditions of the model were assigned based on the field data. The finite difference model, when applied to the case study by considering total dissolved solids (TDS) and indicator of contamination, the concentration of TDS was found to be 6661 mg/l for the surface water at Puthiyapalam.

20.5 GroundWater Quality Modeling

In order to study the effect of surface water contamination on groundwater, a groundwater flow and transport model were developed. The groundwater flow model (MODFLOW) (Harbaugh et al. 2000) is used for the development of groundwater flow model and MT3DMS (Zheng et al. 1999) was used for contaminant transport model. The dimension of the study area was selected as 11,000 m × 12,000 m. The grid size is 50 m × 50 m. Based on the lithological data obtained, the geology of the study area is divided into two layers. The pumping rate from fifteen wells in the area is obtained from Groundwater Department, Kozhikode.

Thirteen observation wells were identified for the calibration of model parameters. The available water levels of the wells were imported to the model. The water level data of January 2002 were interpolated in the study area and were assigned as the initial head. For contaminant transport model, total dissolved solids (TDS) was taken as an indicator. The TDS concentration data observed in January 2002 around the

Table 20.5 Aquifer parameters

Parameter	Layer 1	Layer 2
Conductivity-Kx	20 (m/day)	10 (m/day)
Conductivity-Kz	2 (m/day)	1 (m/day)
Specific storage	1×10^{-5} (1/m)	5×10^{-5} (1/m)
Specific yield	0.15	0.1
Porosity	0.2	0.2

study area was interpolated and assigned as initial concentration. The seaside is represented with constant head of 0 m and constant TDS concentration of 35,000 mg/l. The Canoly Canal is represented as the river boundary condition. The data during 2002–2008 were considered for model calibration. During calibration, the aquifer parameters were slightly modified to match the observed head and concentration to that of simulated values. Table 20.5 shows the finally adopted aquifer parameters for two layers of the aquifer system. For contaminant transport model, dispersivity was varied to match the simulated and observed TDS values. The calibrated longitudinal dispersivity was found to be 50 m. Figure 20.5 shows the comparison between observed and computed heads for an observation well 2. Figure 20.6 shows the calibration plot for TDS concentration for all wells for a time period. Figure 20.7 shows the head and velocity contour during 2009. Figure 20.8 shows the concentration contour during 2009. It can be seen that there is reasonable agreement between the

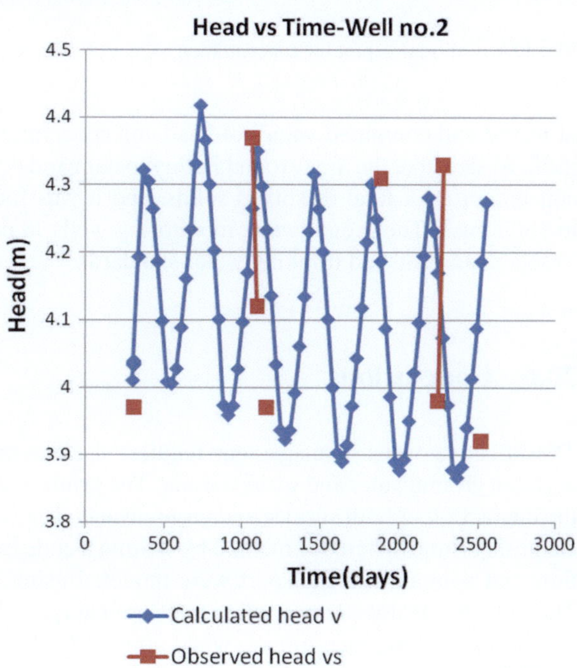

Fig. 20.5 Simulated and Observed water level in well no. 2

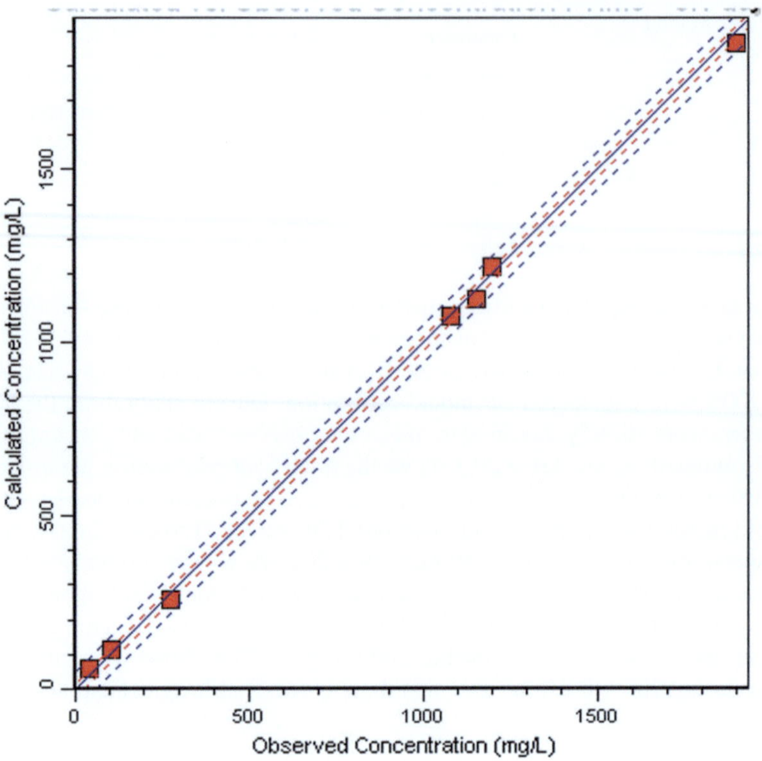

Fig. 20.6 Calibration plot for concentration

observed and computed values of head and concentration. From the obtained plots, it can be seen that the area around Puthiyapalam and Arayedathupalam shows higher concentration of total dissolved solids, and it was found to be varying from 1000 to 1800 mg/l. The groundwater monitoring wells in between Korapuzha River and Eranhipalam satisfied drinking water standards.

20.6 Conclusions

The effective water management requires a clear understanding of the linkages between groundwater and surface water. The primary goal of this study was to assess the interaction of both surface and groundwater. The surface water quality was modeled using finite difference method by writing a code in MATLAB. The groundwater flow and contaminant transport were modeled using MODFLOW and MT3DMS. The models developed are applied to a case study.

The finite difference model, when applied to the case study, the concentration of TDS was found to be 6661 mg/l for the surface water at Puthiyapalam. The

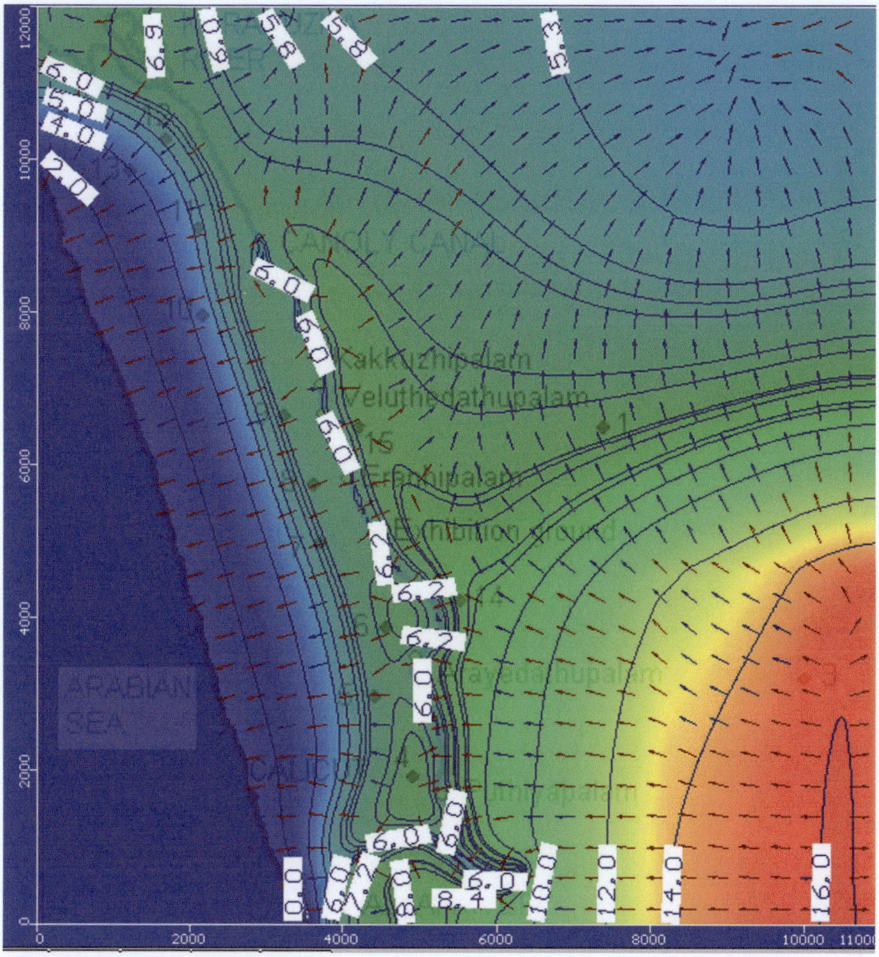

Fig. 20.7 Head and velocity contour during 2009

concentration of TDS for groundwater in the area around Puthiyapalam was obtained as 1500 mg/l, which revealed that the groundwater was also polluted. From the results obtained, it is evident that Canoly Canal is highly polluted. It can be concluded that the surface water—groundwater interaction is significant in this area. It is high time that this pathetic picture of the canal to be noticed. Strict vigil is required to prevent people from dumping the wastes into the canal. The developed models can be used for decision making, specifically to decide how much treatment is required before discharging the waste into surface water system and also the treatment required for groundwater for specific uses.

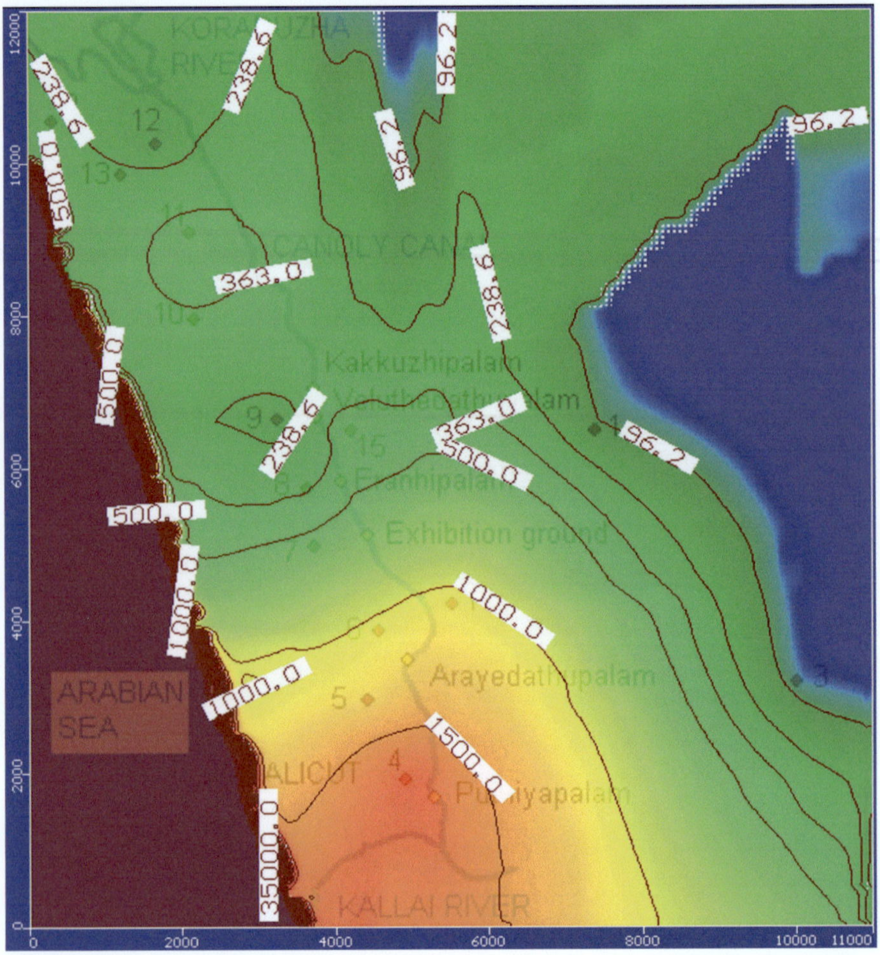

Fig. 20.8 Concentration contour during 2009

References

Harbaugh AW, Banta ER, Hill MC, McDonald MG (2000) MODFLOW-2000, the US Geological Survey modular ground-water model—user guide to modularization concepts and the ground water flow process. In: US geological survey open-file report 00-92, 121

Hussein M, Schwartz FW (2003) Modeling of flow and contaminant transport in coupled stream-aquifer systems. J Contam Hydrol 65(1–2):41–64

Kim NW, Chung IM, Won YS, Arnold JG (2008) Development and application of the integrated SWAT-MODFLOW model. J Hydrol 356(1–2):1–16

Korkmaz S, Ledoux E, Onder H (2009) Application of the coupled model to the Somme river basin. J Hydrol 366:1–4

Radha R (2008) Indian business environment, Mumbai, published by Nirali Prakashan

Rodríguez LP, Cello PA, Vionnet CA (2006) Modeling stream-aquifer interactions in a shallow aquifer, Choele Choel Island, Patagonia, Argentina. Hydrogeol J 14(4):591–602
Sophocleous M (2002) Interactions between groundwater and surface water: the state of the science. Hydrogeol J 10(1):52–67
The MathWorks Inc. (2003) MATLAB users manual, Natick, MA, US
Yan J, Smith KR (1994) Simulations of integrated surface water and groundwater systems—model formulation. Water Res Bullet Am Water Resour Assoc 30(5):879–890
Zheng C, Wang PP (1999) MT3DMS—a modular three-dimensional multispecies transport model for simulation of advection, dispersion and chemical reactions of contaminants in ground-water systems. Documentation and User's Guide, Jacksonville, Fla, US Army Corps of Engineers Contract Report SERDP-99-1
Zhou D, Zhang Y, Gianni G, Lichtner P, Engelhardt I (2019) Numerical modelling of stream–aquifer interaction: Quantifying the impact of transient streambed permeability and aquifer heterogeneity. Hydrol Process 32(14):2279–2292

Chapter 21
Assessment of Groundwater Vulnerable Zones of Nagpur City Using DRASTIC and Susceptibility Index Method

Sahajpreet Kaur Garewal, Avinash D. Vasudeo and Aniruddha D. Ghare

Abstract Increased exploitation, environmental degradation and usages of chemicals have threatened the surface and groundwater quality and presently the aforesaid are the major issues faced by water resources engineers over the world. Assessment of groundwater vulnerability can act as a significant tool for making effective policies for planning, management and taking necessary remedial action for protecting the groundwater from further contamination. In the present study, a GIS-based overlay method DRASTIC and Susceptibility Index (SI) method have been used to identify the groundwater vulnerability of the study area. The overall methodology has been performed using overlay analysis in ArcGIS 10.0 software. The resultant groundwater vulnerability maps are validated using actual field groundwater quality data of the study area. In the present study, it was found that the SI method correlates well with the actual field condition and manages to produce a more reliable result.

Keywords Groundwater vulnerability · DRASTIC · Susceptibility Index · GIS

21.1 Introduction

Urbanization, agriculture, industrial activities and climatic change threatens to cause major alteration to the hydrological cycle. Continuous regarding quality of surface water increases the reliability on groundwater which stresses the aquifer system (WaterAid 2016). Understanding the role of groundwater in the hydrological cycle is important for making effective policies for its protection measure. While making efforts to fulfill the required quantity of water, quality of water is an important aspect. The declining quality of groundwater is a major concern due to its adverse impact on human health, other living creatures and to the overall environment (Rahman 2008).

S. K. Garewal (✉)
National Institute of Technology, Raipur, India
e-mail: sahaj012@gmail.com

A. D. Vasudeo · A. D. Ghare
Visvesvaraya National Institute of Technology, Nagpur, India

R. M. Singh et al. (eds.), *Environmental Processes and Management*, Water Science and Technology Library 91,
https://doi.org/10.1007/978-3-030-38152-3_21

Consecutively, we are endeavoring to maintain stability between the management of water resources and the constraint imposed by the environment. Due to the invisibility of groundwater, its remediation is costly and a difficult task or sometimes impossible considering the restriction imposed by the layers of media above it. There is no direct measure to groundwater contamination, it can be somewhat handled in regional scale by monitoring the local groundwater.

Groundwater vulnerability assessment has proven to be an essential tool by many researchers for the management of groundwater (Wang et al. 2012). It is basically a sensitivity of groundwater quality to highly heterogeneous hydro-geological settings and effects of anthropogenic activities. Groundwater vulnerability assessment identifies the area setting less and more vulnerable to contamination and helps for proper planning and management of regional groundwater.

A variety of techniques and methodologies have been developed in past decades which can be used to provide information about groundwater pollution and are generally divided into three major categories, i.e., overlay and index method, process-based simulation model and statistical method. The selection of the method depends upon the availability of regional data. Overlay and index method is one of the most extensively used approaches for groundwater vulnerability assessment using models such as DRASTIC (Aller et al. 1987), GOD (Foster 1987), SINTACS (Civita 1999), Susceptibility Index (Ribeiro 2000), etc. The advantage of these methods is they provide a simple algorithm to handle a large amount of spatial data which can be integrated into a map that classifies the different vulnerability level.

Nagpur (Maharashtra state, India) being a highly urbanized city, faces the problem of over-exploitation and quality degradation of groundwater in various regions. Nagpur city is selected as a study area, based on the previous studies (Pujari and Deshpande 2005; Pujari et al. 2007; Puri et al. 2011) and groundwater quality report (CGWB 2009, 2011, 2013; MPCB 2016), which states that the city is affected by a higher concentration of contaminants. The main reason behind the increasing groundwater contamination in the city is disposal of untreated domestic waste, industrial waste, etc., directly into rivers (Nag river) which is turning into being a sewage disposal drain (Jain and Sharma 2011) and leaching from the waste disposal site (Bhandewadi) of the city (Pujari et al. 2007).

Before implementing any groundwater vulnerability assessment tool in the city, the groundwater quality of the region is analyzed. The groundwater quality data of Nagpur city is collected from Central Ground Water Board (CGWB), Nagpur and the quality parameters, such as Electrical Conductivity (EC), Total Dissolved Solids (TDS), Nitrate (NO_3), Magnesium (Mg), Sodium (Na), were checked against the permissible limit allowed by Indian standard code for drinking water (IS 10500:1991, Revised 2003). Various zones of the city are found to be affected by a higher concentration of contaminants, especially nitrate. The higher concentrations of other quality parameters are well within the city limit except few isolated pockets. Groundwater vulnerability assessment in the city can act as an essential tool for identifying high-risk potential area and can help in groundwater management of the city.

In the present study, a GIS-based overlay and index method have been used to identify the groundwater vulnerability of the city. DRASTIC and Susceptibility Index

(SI) methods are implemented in the study area. The overall methodology has been performed using overlay analysis in ArcGIS 10.0 software. The resultant groundwater vulnerability maps are validated using actual field groundwater quality data of the study area.

21.2 Materials and Methodology

21.2.1 Study Area

Nagpur city is located at the geographical center of India marked by zero milestone (Fig. 21.1). The city is named after the Nag River flowing within the urban setting of the city. It lies between 21° 00′ and 21° 15′ North latitudes and 79° 00′–79° 15′ East longitude, at 310 m above mean sea level in the eastern part of the Maharashtra state in an area known as Vidarbha (CGWB 2011). The city covers an area of 217.56 km^2

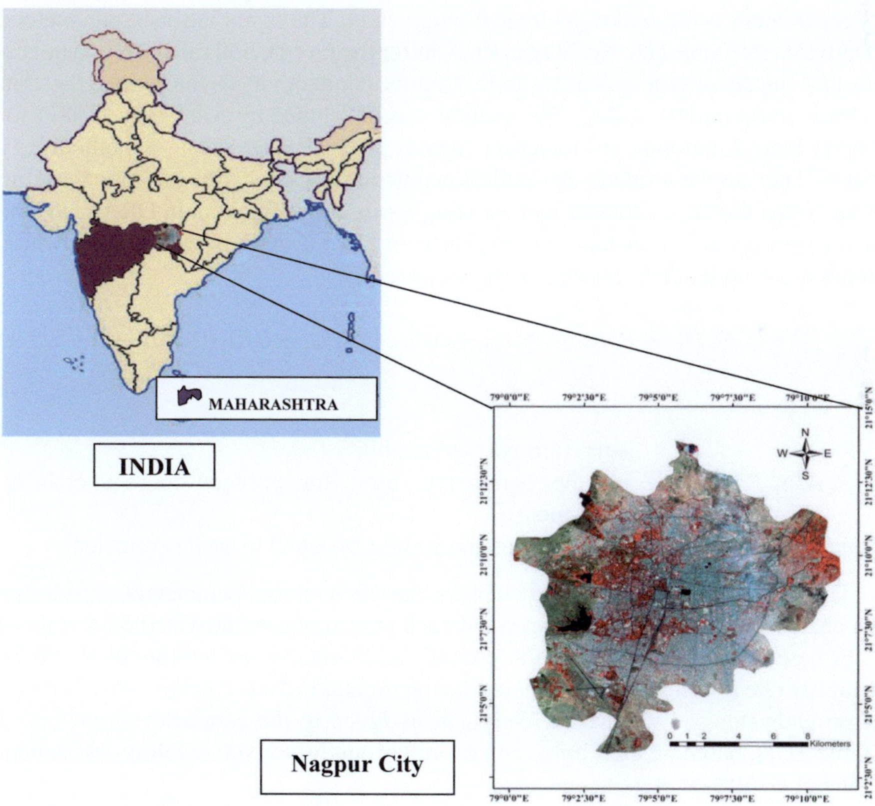

Fig. 21.1 Map showing the location of study area

governed by Nagpur Municipal Corporation (NMC). The two main rivers of the city are Nag River and Pilli River. Nag River originated from Ambazari Lake at western end which runs through middle of the city toward east, and Pilli River starts from Gorewada Lake at northwest side and runs toward east. Pilli River meets Nag River before pouring into Kanhan River in Wainganga, which happens to be the sub-basin of Godavari. Deccan trap basalt and granite gneiss (crystallines) are the main consolidate rock setting occurring in the city. The city follows the natural gradient topography, i.e., from west to east. The city experiences summer from March to June where the temperature rises to 45–48 °C, especially during May. The winter season lasts from November to February, during which temperatures may drop below 10 °C. The average annual rainfall in the city is 1100 mm mostly during monsoon, i.e., June–September.

21.2.2 DRASTIC

DRASTIC is an overlay and index method used for intrinsic groundwater vulnerability assessment using hydro-geological parameters. The seven intrinsic parameters; Depth to water table (D), Recharge (R), Aquifer media (A), Soil media (S), Topography (T), Impact of vadose zone (I) and Hydraulic conductivity (C) are coined together to form acronym DRASTIC. The method was developed by Aller et al. (1987) for United State Environment Protection Agency (US EPA), to create a methodology that will permit the groundwater pollution potential of any hydro-geological setting to be systematically evaluated with existing intrinsic information. In DRASTIC, the hydro-geological parameters are integrated to form a map using Eq. 21.1, which shows areas under different degree of vulnerability.

$$\mathrm{IVI} = D_r D_w + R_r R_w + A_r A_w + S_r S_w + T_r T_w + I_r I_w + C_r C_w \tag{21.1}$$

where

IVI	is the Intrinsic Vulnerability Index
D, R, A, S, T, I and C	are the parameters used for groundwater vulnerability assessment
r and w	are the rating and weight assigned to each parameter

The involved controlling parameters are classified into subparameters considering the characteristics of the study area and each subparameters are rated on a scale of 1–10 based on their contribution in groundwater contamination (Aller et al. 1987). A higher rate is assigned to parameter having greater influence in the contamination of groundwater and vice versa. Weight is assigned to the parameters from 1 to 5 (Table 21.1), based on their impact on overall groundwater vulnerability assessment (Aller et al. 1987).

Table 21.1 Weight of the DRASTIC and SI parameters

S. no.	Parameters	DRASTIC weight	SI weight
1	Depth to water table (*D*)	5	0.186
2	Recharge (*R*)	4	0.212
3	Aquifer media (*A*)	3	0.259
4	Soil media (*S*)	2	–
5	Topography (*T*)	1	0.121
6	Impact of vadose zone (*I*)	5	–
7	Hydraulic conductivity (*C*)	3	–
8	Land use (Lu)	–	0.222

21.2.3 Susceptibility Index (SI)

The Susceptibility Index (SI), an adaptation of the DRASTIC method developed by Ribeiro (2000) by addition of land use (Lu) parameter abandoning the concept of a purely intrinsic vulnerability assessment method. The index name is in harmony with the definition of susceptibility, i.e., the lack of ability to resist the impact of contaminants on the quality of groundwater, provided by Vrba and Zoporozec (1994). In SI, five parameters are involved to define different classes of vulnerability; four of them are identical to DRASTIC, i.e., Depth to water table (*D*), Recharge (*R*), Aquifer media (*A*) and Topography (*T*), and one additional L and use (Lu) parameter is incorporated in the analysis. The five parameters were overlayed together using Eq. 21.2, by applying the weight described in Table 21.1.

$$\mathrm{SI} = D_r D_w + R_r R_w + A_r A_w + T_r T_w + \mathrm{Lu}_r \mathrm{Lu}_w \tag{21.2}$$

where

SI is the Susceptibility Index.
D, *R*, *A*, *T* and Lu are the parameters used for susceptibility assessment.
r and *w* are the rating and weight assigned to each parameter.

The four parameters as used in DRASTIC were assigned ratings ten times the rating of DRASTIC. The principal types of land use and their assigned ratings provided by a team of Portuguese scientists (Ribeiro 2000) are used in the analysis.

All the required data to define different classes of groundwater vulnerability are collected from a different government organization (Central Ground Water Board (CGWB), India Meteorological Department (IMD), National Bureau of Soil Survey (NBSS) and official Web site (Bhuvan). The parameter maps and vulnerability classification maps were prepared using GIS technique in ArcGIS 10.0 software and the overall framework of the methodology adopted for the study is described in Fig. 21.2. The weights of the parameters were assigned by Aller et al. (1987)

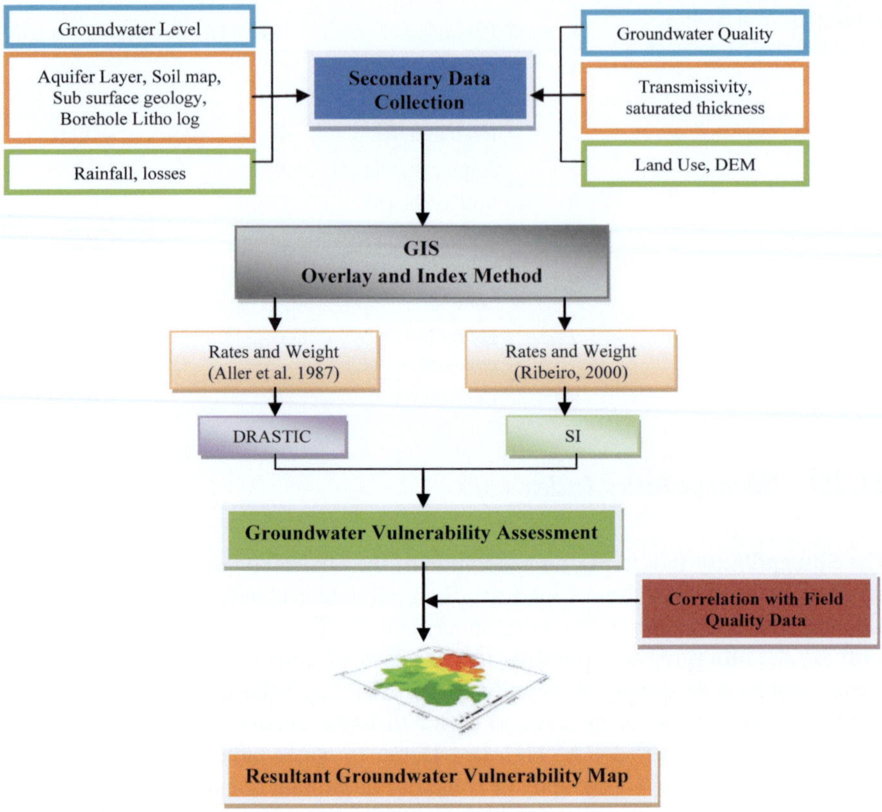

Fig. 21.2 Methodology framework adopted in this study

(DRASTIC) and Ribeiro (2000) (SI) (Table 21.1) and rates using Delphi technique proposed by Aller et al. (1987) (Table 21.2). To study the application of applied approaches, vulnerability maps were validated using field quality parameters of the city.

21.2.4 Importance and Generation of Parameters Map Involved in Groundwater Vulnerability Assessment

The data collected from various goverment organization and official website are in different format which needs to be proceeded before using as DRASTIC Layer. The digital data in raster maps are obtained from satellite, aerial images and scanning map. The scanning map does not contain information like where this image fit in the earth surface. The location information delivered by the satellites or aerial images is often inadequate to perform analysis or display it in proper alignment with other data.

Table 21.2 Rates of the controlling parameters (Aller et al. 1987)

1. Depth to groundwater (m)		2. Recharge (mm)		3. Aquifer media	
Subparameter	*r*	Subparameter	*r*	Subparameter	*r*
0.70–2.62	10	396–404	3	Intertrapean	1
2.62–3.87	9	405–410	5	Massive basalt	4
3.87–4.69	8	411–416	7	Amgaon-gneiss complex	7
4.69–5.95	6	417–422	8	Unclassified-gneiss-tirods	8
5.95–7.86	4	423–433	9		
7.86–10.77	2				
10.77–15.2	1				

4. Soil media		5. Topography (%)		6. Impact of vadose zone (m)		7. Hydraulic conductivity (m/s)	
Subparameter	*r*	Subparameter	*r*	Subparameter	*r*	Subparameter	*r*
Clay loam	3	<2.7	10	0.6–3.2	8	$<10^{-6}$	5
Clayey	7	2.7–5	9	3.2–3.9	7	10^{-5}–10^{-6}	6
Alluvial	8	5–7.9	7	3.9–4.5	6	10^{-4}–10^{-5}	8
		7.9–11	5	4.5–5	5	10^{-3}–10^{-4}	9
		11–16	4	5–5.9	4		
		16–23	3	5.9–7	3		
		23>	1	7–10.8	2		

r rating assigned to the classification of parameters

Before performing overlay and index analysis of parameters, all the collected data are georeferenced which defines where the data lies in map coordinate and reprojected to UTM projection system (WGS_1984_UTM_Zone_44).

Depth to water table (D)

It is the depth that a contaminant must travel before reaching the groundwater (Aller et al. 1987). It is an important parameter, determine the amount of time during which contaminant is in contact with surrounding media. Greater the depth, the chance of attenuation will be higher as travel time of contaminant increases which reduces the groundwater contamination. Lower the depth, nearer to the land surface and more will be the contamination.

The data of 45 monitoring wells showing depth to groundwater table (bgl) is collected from CGWB, Nagpur. The collected data are point data showing depth at a particular location which is capable to be interpolated. Kriging interpolation tool is used to form a surface map of depth to water level showing a different range of depth to water table within the city limit. The depth to water table map is classified into seven classifications varies from 0.7 to 15.2 m. Each classification is rated using rates defined by Aller et al. (1987) (Table 21.2) to form a thematic map which can be used for DRASTIC proceeding.

Recharge (R)

It is the amount of water that infiltrates and percolate to meet groundwater table. The primary source of recharge to groundwater is rainfall. Recharge can increase or decreases the contamination in groundwater depending on the property of water recharging. It helps in dilution and dispersion of contamination. Recharge is considered as the principal medium for transporting and leaching solid or liquid contaminant to the water table (Amadi et al. 2014). Therefore, greater the recharge higher will be the chances of groundwater contamination (Aller et al. 1987).

As rainfall is the fundamental source of recharge in the study area, 30–40% of mean rainfall is considered as recharge (Gupta 2014). The rainfall data of 14 rain gauge stations are collected from IMD, Nagpur. Some rain gauge station is located within the NMC boundary and few at the immediate boundary. The data are point data which are interpolated to develop a surface map of recharge. The recharge map for the study area varies from 396 to 433 mm is classified into five classifications, which are then rated to form a thematic map (Table 21.2).

Aquifer Media (A)

Aquifer media refers to the composition of unconsolidated and consolidated media like sand, clay, loam, sandstone, etc., which serve as an aquifer. Water is stored and move slowly through these geological formations. The migration and quality of water depend upon the characteristic of aquifer media. In general, larger the grain size, more the fracture opening, lower attenuation and higher permeability; consequently greater the pollution potential of the aquifer (Aller et al. 1987).

To understand the aquifer media of the study area, it is necessary to recognize the sub-surface lithological condition. The aquifer media for the study area was obtained using sub-surface geology map and lithology data at a various section within the study area and nearby boundary collected from CGWB, Nagpur. The study area is mostly covered with hard rock formation such as Amgaon-gneiss complex (metamorphic rock), unclassified-gneiss-tirodi gneissic complex (metamorphic rock), massive basalt (igneous rock) and Intertrapean. The sub-classifications are rated to form a thematic map for DRASTIC (Table 21.2).

Soil Media (S)

It is the uppermost layer of the earth surface characterized by significant biological activities. It is an important parameter, as it acts as a protection layer and restricts the infiltration of contaminant through recharge in groundwater. The properties of protection layer such as grain size, clay percentage, swell/shrink potential of clay, etc., principally control the contamination potential. In general, larger the grain size and more the shrink/swell property of clay more will be the contamination potential of the area (Aller et al. 1987).

Soil map for the study area is collected from the NBSS, Nagpur as a hard copy of 1:50,000 scale. The map is processed in GIS (scanned, digitized, rasterized) to convert in an appropriate format to processed in the DRASTIC analysis. The map is classified and rated considering its contribution to groundwater contamination (Table 21.2).

Topography (T)

It is considered as the slope and slope variability of the land surface. The degree of the slope will decide whether the contaminant will flow as runoff or remain on the land surface and infiltrate into the media. Steeper the slope higher will be the runoff, lesser will be the infiltration and contamination potential. Flatter the slope more will be the infiltration, more will be the time of retention and groundwater contamination.

Slope (%) for the study area is extracted from the Digital Elevation Model (DEM) downloaded from Bhuvan. The slope (%) of the study area varies from 2.7 to 23%, which are rated to form a thematic map which can be integrated with other controlling parameters (Table 21.2).

Impact of Vadose Zone (I)

It the unsaturated zone extended between the soil media and aquifer media. It is an important parameter as various chemical, physical and microbiological reactions occurs within this zone. The thickness of vadose zone influences the time available for the contaminant to undergo attenuation, which affects the path length and direction of contaminant. In general, greater the thickness of vadose zone lesser will be the chance of groundwater contamination.

The thickness of the vadose zone map was generated using unconfined depth to water level and DEM of the study area. The methodology explained by Li and Zhao (2011) was applied to obtain the thickness map of vadose zone. The data was collected from CGWB department; normally kriging interpolation tool is used to interpolate the depth to water level data which does not consider the topographical variation of the area. To improve the accuracy, the depth to water table is subtracted from the DEM. The thickness of vadose zone varies from 0.6 to 10.8 m in the study area, which is rated on the basis of their contribution to groundwater contamination (Table 21.2).

Hydraulic Conductivity (C)

Hydraulic conductivity is the potential of the aquifer material to transmit water. It is influenced by the property of media through which it is flowing, such as grain size, inter-granular porosity and fractures. Higher the hydraulic conductivity, greater will be the contamination migration.

The hydraulic conductivity map is obtained using the well log data, transmissivity and saturated thickness collected from CGWB, Nagpur. The surface map of transmissivity and saturated thickness are prepared. The resultant hydraulic conductivity map is obtained by dividing saturated thickness from transmissivity. It varies from 10^{-6} to 10^{-4} m/s and is rated in the order to their contribution to groundwater contamination (Table 21.2).

Land Use (Lu)

Land use map shows the area covered under different land surface activity. Lu is considered in the SI approach for groundwater vulnerability assessment. It is an important parameter which includes the effect of anthropogenic activity in vulnerability assessment along with the intrinsic property of the study area.

In the current study, the land use map is prepared using LISS-III data of Indian Remote Sensing (IRS) satellite downloaded from Bhuvan (Fig. 21.3). The satellite data is registered and rasterized in the ArcGIS environment. The study area is classified into five classifications built-up, agriculture, wasteland, water bodies and forest. The classifications are rated on the basis of Ribeiro (2000) as agriculture (9), built-up area (7), water bodies (5), wasteland (3) and forest (1).

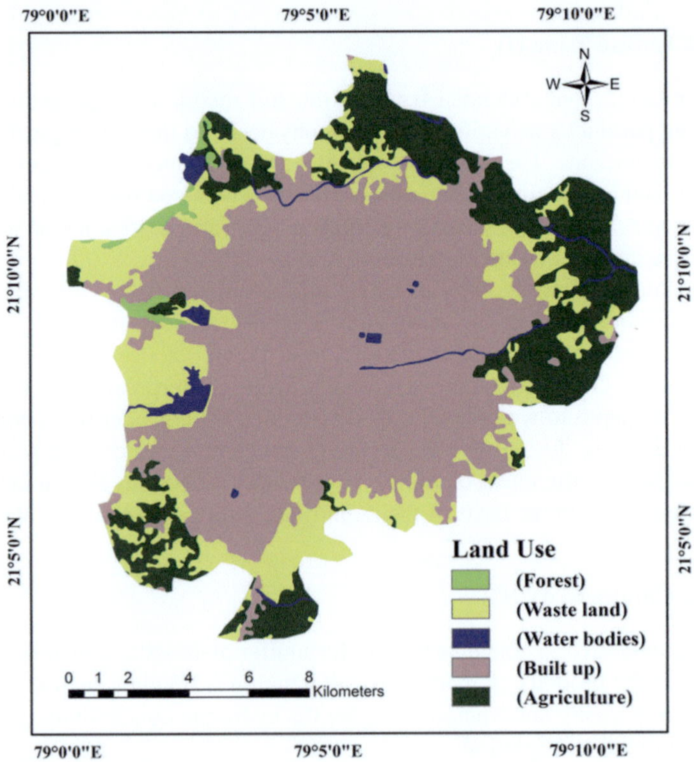

Fig. 21.3 Land use map of Nagpur city

21.3 Results and Discussion

21.3.1 DRASTIC

Intrinsic groundwater vulnerability map is obtained by the linear combination of thematic maps of seven hydro-geological parameters using Eq. 21.1. The weight (Table 21.1) and rate (Table 21.2) developed by standard DRASTIC (Aller et al. 1987) are assigned to parameter. The obtained intrinsic groundwater vulnerability map (Fig. 21.4a) is classified into five classifications varying from very high to very low vulnerability index. From the resultant intrinsic vulnerability map (Fig. 21.4a), it was observed that the southwest region of the city shows moderate vulnerability, while south part of the city is safe from the contamination as it is having the least vulnerability index. The north, northeast and central region of the city is under high risk of contamination, having very high to high vulnerability index, except few small patches in this zones falls under moderate vulnerability index. This shows a collective influence of higher recharge, lower depth to groundwater and aquifer media, for higher vulnerability index.

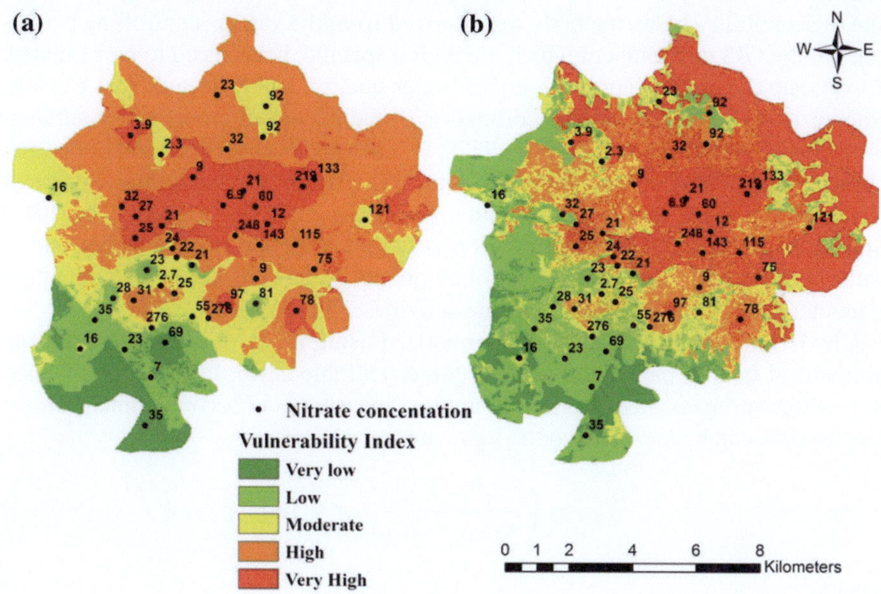

Fig. 21.4 Groundwater vulnerability map; **a** DRASTIC index, **b** Susceptibility index

21.3.2 Susceptibility Index (SI)

The Susceptibility Index (SI) of the city is generated by combining the thematic map of four hydro-geological parameters and Lu parameter using Eq. 21.2. Weight is assigned to parameters by Ribeiro (2000) (Table 21.1) and rates of four hydro-geological parameters are assigned ten times the rating of DRASTIC (Table 21.2) and Lu as Ribeiro (2000). The obtained Susceptibility Index (SI) map (Fig. 21.4b) is classified into five zones showing very high to very low vulnerability index. From the resultant Susceptibility Index (SI) map (Fig. 21.4b) it was observed that west to south zone of the city is less vulnerable of contamination having very low to low Vulnerability index. The center to east and north zone of the city is highly vulnerable to groundwater contamination having very high to high vulnerability index, except few isolated pocket showing moderate vulnerability index.

21.3.3 Validation

The groundwater vulnerability map developed in the present study using DRASTIC and Susceptibility Index methods were derived from the various controlling parameters in the GIS environment which are region specific, hence need to be validated. To validate the applied approach groundwater quality data collected from CGWB, Nagpur is used (Fig. 21.5). The correlation of the generated groundwater vulnerability maps is obtained with the No_3, TDS, EC, Mg and Na concentration of the city. Nitrate is considered in the analysis, as in natural groundwater the primary source of nitrate is minor and in low concentration (Williams et al. 1998; Javadi et al. 2011; Shirazi et al. 2012). If its concentration is obtained higher, it is generally due to anthropogenic activities. Presence of other quality parameters can be due to natural characteristics as well as anthropogenic activities.

The DRASTIC and SI maps are normalized using Eq. 21.3 and validated using normalized quality parameter maps of the city (Table 21.3). In the normalization step, different layers having different units of measurement become dimensionless and the data can be measured at the same numerical scale.

$$X_{\text{nor}} = \left(\frac{X - X_{\min}}{X_{\max} - X_{\min}} \right) \times 100 \tag{21.3}$$

where

X_{nor} is the normalized data
X is the data layer
$X_{\min}$ and $X_{\max}$ are minimum and maximum index value of the respective data layer.

The Intrinsic Vulnerability Index (IVI) map obtained using the technique of Aller et al. (1987), involving seven hydro-geological parameters defining the characteristic of the study area shows less correlation with the actual field quality data of the

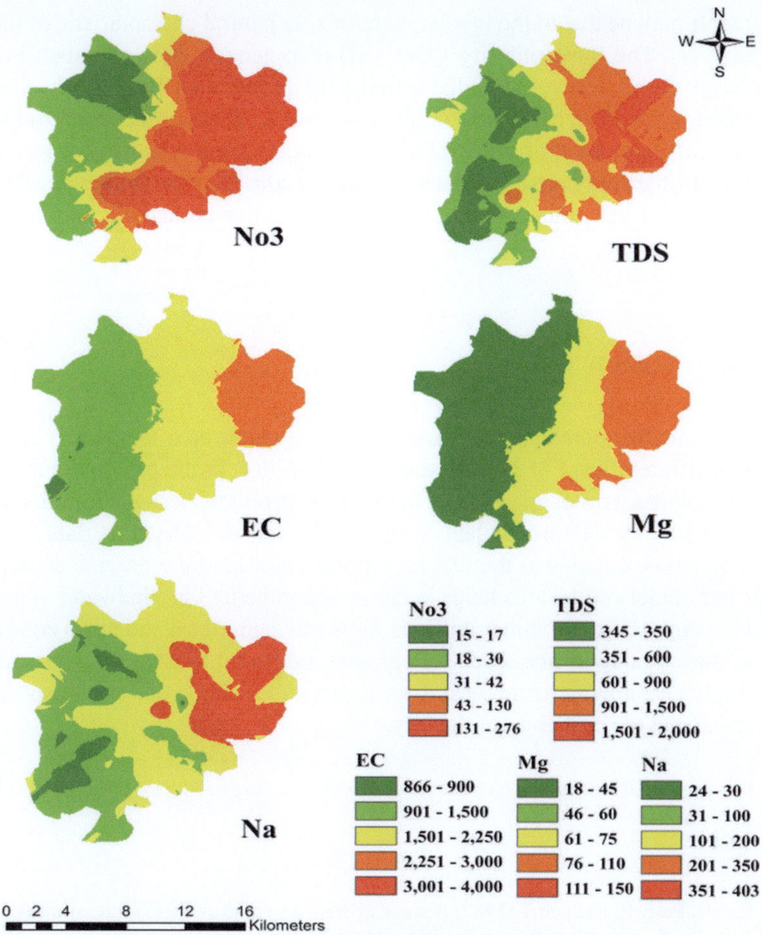

Fig. 21.5 Spatial variation of field quality parameters in Nagpur city

Table 21.3 Correlation of DRASTIC and SI approach with field quality parameters

S. no	Field quality parameters	DRASTIC	SI
1	No_3	0.206	0.535
2	TDS	0.327	0.627
3	EC	0.306	0.590
4	Mg	0.338	0.612
5	Na	0.377	0.609

study area. It may be due to the involvement of sole natural characteristic of the area in the analysis. The Susceptibility Index (SI) map generated using Ribeiro (2000) methodology correlates well with the actual field quality data of the study area. The involvement of land use parameter in the assessment enhances the result to identify area under different level of vulnerability and shows good result in comparison to DRASTIC for the study area. Land use parameter considers the influence of anthropogenic activities on groundwater, which helps is better groundwater vulnerability assessment.

21.4 Conclusion

DRASTIC and SI approaches are applied in the study area to identify area setting under different level of groundwater vulnerability. In the present study, the SI approach provides a more appropriate map for the groundwater vulnerability assessment of Nagpur city, which correlates well with the actual field quality data. From the analysis, it can be concluded that the incorporation of land use parameters with the intrinsic parameters of the area helps in better assessment of groundwater vulnerable zones. In Nagpur city, the vulnerability is higher at center-east and north zone of the city, it is may be due to one-sided topography, i.e., west to east, the contaminated river (Nag River) flowing from west toward east and location of waste disposal city (Bhandewadi) at the east zone of the city.

References

Aller L, Lehr J, Petty R, Bennett T (1987) A standardized system to evaluate groundwater pollution using hydrogeologic setting. J Geol Soc India 29(1):23–37

Amadi AN, Olasehinde PI, Nwankwoala HO, Dan Hassan MA, Okoye NO (2014) Aquifer vulnerability studies using drastica model. Int J Eng Sci Invention 3(3):1–10

CGWB (2009) Report no. 1608/DBR/2009 on ground water information Nagpur district. Central Ground Water Board, Nagpur

CGWB (2011) Report on Contamination of ground water by Sewage. Central Ground Water Board, Faridabad

CGWB (2013) Report on ground water information Nagpur district. Central Ground Water Board, Nagpur

Civita M (1999) Le carte della vulnerabilita degli acquiferi all'inquinamiento: teoria e pratica [Contamination Vulnerability Mapping of the Aquifer: Theory and Practice]. Quaderni di Tecniche di Protezione Ambientale, Pitagora, Italy, p 344

Foster S (1987) fundamental concepts in aquifer vulnerability, pollution risk and protection strategy. In: Netherlands Organization for Applied Scientific Research, The Hague, pp 69–86

Gupta N (2014) Groundwater vulnerability assessment using DRASTIC method in Jabalpur District of Madhya Pradesh. Int J Recent Technol Eng 3(3):36–43. https://doi.org/10.4236/ijg.2016.74043

Jain SK, Sharma V (2011) Contamination of ground water by Sewage. Central Ground Water Board Ministry of Water Resources Government of India, Faridabad

Javadi S, Kavehkar N, Mousavizadeh MH, Mohammadi K (2011) Modification of DRASTIC model to map groundwater vulnerability to pollution using nitrate measurements in agricultural areas. J Agr Sci Tech 13:239–249
Li R, Zhao L (2011) Vadose zone mapping using geographic information systems and geostatistics a case study in the Elkhorn River Basin, Nebraska, USA. In: Proceedings of international symposium on water resource and environmental protection, IEEE. Xi'an, China, pp 3177–3179
MPCB (2016) Water quality status of Maharashtra 2015–16. Maharashtra Pollution Control Board, Mumbai, India
Pujari PR, Deshpande V (2005) Source apportionment of groundwater pollution around landfills site in Nagpur, India. Environ Monit Assess 111:43–54. https://doi.org/10.1007/s10661-005-8037-4
Pujari PR, Pardhi P, Muduli P, Harkare P, Nanoti MV (2007) Assessment of pollution near landfill site in Nagpur, India by resistivity imaging and GPR. Environ Monit Assess 131:489–500. https://doi.org/10.1007/s10661-006-9494-0
Puri PJ, Yenkie M, Sangal S, Gandhare N, Sarote GB, Dhanorkar DB (2011) Surface water (Lakes) quality assessment in Nagour city (India) based on water quality index (WQI). RASAYAN J. Chem 4(1):43–48
Rahman A (2008) A GIS based DRASTIC model for assessing groundwater vulnerability in shallow aquifer in Aligarh, India. Appl Geogr 28(1):32–53. https://doi.org/10.1016/j.apgeog.2007.07.008
Ribeiro L (2000) SI: a new index of aquifer susceptibility to agricultural pollution. ERSHA/CVRM, Instituto Superior Técnico, Lisboa, Portugal
Shirazi S, Imran H, Akib S (2012) GIS-based DRASTIC method for groundwater vulnerability assessment: a review. J Risk Res 15(8):991–1011. https://doi.org/10.1080/13669877.2012.686053
Vrba J, Zaporozec A (eds) (1994) Guidebook on mapping groundwater vulnerability, vol 16. In: International contributions to hydrogeology. Heise, Hannover, Germany, pp 131
Wang J, He J, Chen H (2012) Assessment of groundwater contamination risk using hazard quantification, a modified DRASTIC model and groundwater value, Beijing Plain, China. Sci Total Environ 432: 216–226. https://doi.org/10.1016/j.scitotenv.2012.06.005
WaterAid (2016) FSM—Urban Wash: an assessment of faecal sludge management policies and programmes at the national and select states level. WaterAid, India
Williams AE, Lund LJ, Johnson JA, Kabala ZJ (1998) Natural and anthropogenic nitrate contamination of groundwater in a rural community, California. Environ Sci Technol 32:32–39. https://doi.org/10.1021/es970393a

Chapter 22
Integrated Planning and Management of Water Resources in Sheonath River Basin of Chhattisgarh State India

Ishtiyaq Ahmad and Mukesh Kumar Verma

Abstract Chhattisgarh State of India is bestowed with rich water resources, and about 80% of the rainfall occurs during four monsoon months from June to September. Because of time and spatial variability of rainfall, a number of water storages have been constructed to tap the available water so that it can be utilized in accordance with requirement. Sheonath River Basin, a sub-basin of Mahanadi river, comprises of 26 number of different command areas. Out of the total agriculture zone of 22,702 km^2, only 11,903 km^2 is irrigated through networks of canal, which is only 52% of the total agriculture area. The basic objective of this study is to plan a suitable number of water storage sites for achieving maximum possible irrigation for the remaining 48% of the agriculture area which is not falling under the command area of the existing schemes. In order to assess the suitable number sites, GIS-based multi-criteria evaluation method analytic hierarchy process (AHP) is used to determine the percentage importance of the parameters used in the identification of suitable sites for water storage in accordance with the guidelines laid by integrated mission for sustainable development (IMSD) (1995). Six numbers of suitable sites are assessed using an integrated approach of GIS and AHP, situated outside the existing command area. These sites are assessed for water storage capacity at possible heights as per the topographical conditions. Digital elevation model (DEM) is used to determine the maximum water surface area as well as the potential capacity at maximum elevation. All these six reservoirs have potential of about 524.20 million cubic meters (MCM). This amount of storage is sufficient for fulfilling the needs of irrigation for the agriculture land falling outside the existing command area of the basin. In the present study, Hydrologic Engineering Centre-Hydrologic Modelling System (HEC-HMS) is used for estimating the availability of water in the basin. The model uses Soil Conservation Services Curve Number (SCS-CN) method to estimate the loss from the basin.

Keywords Runoff · SCS curve number · Hydrologic soil group · Remote sensing · GIS · Analytic hierarchy process

I. Ahmad (✉) · M. K. Verma
Department of Civil Engineering, National Institute of Technology Raipur, Raipur, India
e-mail: iahmad.ce@nitrr.ac.in

R. M. Singh et al. (eds.), *Environmental Processes and Management*,
Water Science and Technology Library 91,
https://doi.org/10.1007/978-3-030-38152-3_22

22.1 Introduction

Water is one of the basic needs of mankind for all socio-economic development. Water is called as 'Indrajal' in mythology—the nature's gift through rainfall; it not only satisfies the thirst of human beings but also gives food and sustains the life of human beings, plants and animals. The constantly increasing scarcity of water due to population growth has created immense pressure on the available water resources. This problem can only be solved by harvesting the additional water potential through the development of new water resource projects and/or by formulating strategies for optimal utilization of available resource in existing system.

Most of the world's land surface, apart from the most arid and cold areas, is divisible into river basins (Barrow 1998; Saha et al. 2008). River basins have been used for planning and management since the 1930s (Barrow 1998). A river basin can be defined as the geographical area demarcated by the topographic limits of the water system. There is a strong interaction between land and water resources. Hence, river basins are important elements in water resources planning and management. The upstream basin characteristics influence the availability of water for various purposes.

The planning and management of water resources is also imperative due to regional imbalances in the available water. For proper planning and management, it is necessary to accurately measure the quantity of available water in the basin, through the study of runoff. The need for accurate measurement of parameters involved in the hydrological process has grown rapidly due to acceleration in water resources planning and management. Numerous models have been developed to predict runoff under various management regimes. Among these physically based models, HEC-HMS is designed to simulate the precipitation-runoff processes. It is designed to be applicable in a wide range of geographic areas for solving a broad range of problems. This includes large river basin water supply and flood hydrology to natural watershed runoff. Several applications of HEC-HMS have been shown promising results (Chu and Steinman 2009; Sherif et al. 2011; Ahmad and Verma 2015). In these studies, the model was tested mainly on a monthly and annual basis for predicting runoff. In this model, application of Soil Conservation Services Curve Number (SCS-CN)is used for transformation of rainfall to runoff. It is basically a coefficient that reduces the rainfall to runoff. The SCS-CN is a quantitative description of land use/land cover/soil complex characteristics of a basin. This model is a widely used hydrological model for estimating runoff using rainfall and curve number (CN).

With the advent of remote sensing (RS) and geographic information system (GIS) technologies, it becomes easier to study the most recent land use pattern and manage a huge set of spatial data. There are a number of strengths that GIS technologies bring to water resources research. The objectives of many hydrological studies include watershed segmentation, identification of drainage divides and the networks of channel, characterization of terrain slope and aspect, catchment configuration and routing of the flows of water. Obtaining these variables has been difficult to do from paper maps and aerial photographs. These traditional methods are subjected to errors related to manual operations (Lyon 2003). The tediousness and time-consuming nature of extraction of these basin parameters can be eliminated by means of RS and GIS in addition to obtain high accuracy. The digital elevation model (DEM) can be used to successfully extract several watershed parameters using GIS. These techniques can provide more precise and reproducible measurements than the traditional manual techniques applied to topographic maps.

Water storage is one of the important tasks for managing the available water in an efficient manner. Runoff can be stored by constructing suitable structures (Ramakrishnan et al. 2009). For this purpose, a detailed understanding and analysis of rainfall-runoff-related parameters are required (Prasad et al. 2014). In recent years, it has been observed that numbers of storage structures are constructed without considering the effect of existing structures. Occasionally, isolated planning for water resources project has been resulted in effecting other existing reservoirs in the absence of integrated planning. For example, if a reservoir is already constructed on a river and some other reservoir is planned on upstream of a catchment, then it will be reducing the capacity of the previous reservoir and the existing reservoir may not fill up to its maximum capacity. This type of scenario necessitates that river basin should be planned as integrated, considering the effects of upstream also. Integrated river basin planning of water resources projects for a hydrological unit is a systematic approach for the development of new schemes and allocation of water in an integrated manner. Integrated river basin management is the coordinated use and management of land, water and other natural resources and activities within a basin.

River Mahanadi is known as the lifeline of Chhattisgarh State. Sheonath River is one of the major river tributaries of Mahanadi River. Almost all the major cities, agriculture land and industries are coming under the catchment of Sheonath River. As per GIS analysis, the total agriculture area covered by the Sheonath River Sub-Basin is about 22,702 km^2 which is 73% of the total area. The study area comprises of 26 major and medium (excluding the Mahanadi Reservoir Project and Minimata Bango Project) schemes, supplying water through networks of canal having command area of about 10,606 km^2 which is only 47% of the total agriculture area irrigated. About 48% of the agriculture area is not coming under the command area of existing projects. This area gets water by rainfall, sub-surface resources or by small tanks. Thus, there is need for new schemes to be established for sustainable development of the agriculture area. As the Sheonath River Sub-Basin comprises of existing projects, it becomes necessary to plan and manage the new projects in an integrated manner.

The major objective of this study is to provide irrigation water to the agriculture area which is not falling under the existing command area of the major and medium projects. The source of water for these agriculture lands is rainfall and subsurface water. This objective can be achieved by planning new schemes for harvesting more water. For this purpose, study of water availability in the basin is necessary. The remote sensing and GIS technology will help in quantifying the suitable optimal number of water storage structures using the analytic hierarchy process (AHP) (Ahmad and Verma 2016, 2018) providing irrigation to these agriculture lands so as to achieve maximum possible irrigation in an integrated manner.

22.2 Materials and Methodology

22.2.1 Study Area

Chhattisgarh State of India is located between 17° 50′00″ N and 24° 08′00″ N latitude and 80° 15′00″ E and 84° 13′00″ E longitude covering an area of about 136 lakh hectares of land, out of which 58.81 lakh hectares is cultivable land area and 60.76 lakh hectares is forest land area. Chhattisgarh State being situated in central part of India is surrounded by Uttar Pradesh in the north, Jharkhand in the north-east, Odhisa in the south-east, Madhya Pradesh in the north-west, Maharashtra in the south-west and Andhra Pradesh in the south. Chhattisgarh State has geographical area of about 1,35,097 km^2 having portions of five major river basins, viz. Brahmani, Ganga, Godavari, Mahanadi and Narmada. In the present research work, Sheonath River Basin in Chhattisgarh State, a sub-basin of Mahanadi River, is selected for fulfilling the objectives of the study. As per GIS, total catchment area of the basin is 31066.60 km^2 (30716.46 km^2 in Chhattisgarh, 350.14 km^2 in Maharashtra and a very small area in Madhya Pradesh State). Sheonath Basin covers parts of Bilaspur, Dhamtari, Durg, Raipur, Rajnandgaon, Kanker, Kawardha, Janjgir-Champa and Korba districts of the Chhattisgarh State. The location map of the study area is presented in Fig. 22.1. The study area extends between latitude 20° 00′00″ N and 23° 00′00″ N, and longitudes 80° 00′00″ E and 83° 00′00″ E. The river is centrally located in the basin and has almost flat gradient throughout the plains. The basin elevation ranges between 133 and 1146 m of the average mean sea level. Sheonath River traverses a length of about 380 km. The Sheonath River Basin is divided into 16 sub-basins, i.e. Kharun, Jamunia, Khorsi, Lilagar, Arpa, Maniyari, Sakri, Karua, Dotu, Surhi, Amner, Sukha-Ghumariya, Dalekasa, Kharkhara, Tandula and Sheonath main.

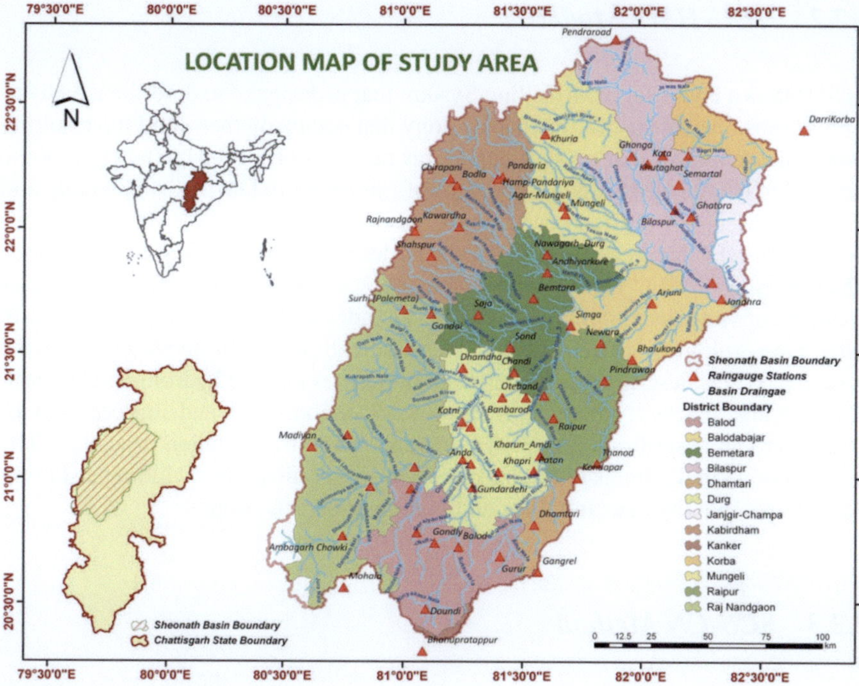

Fig. 22.1 Location map of study area

There are about 65 rain gauge stations, located in and around the study area, considered for the study. The data pertaining to these stations were collected from the office of the Deputy Director, Hydrometeorology Division No. 4, Chhattisgarh Water Resources Department (WRD), Raipur (C.G.) and office of the Chief Engineer, Mahanadi & Eastern Rivers Organizations, Central Water Commission (CWC), Bhubaneswar (Odissa). Rainfall data collected for a period of 33 years, i.e. from 1980 to 2013.

Topographic sheets in the scale 1:50,000 were procured from Geospatial Data Centre, Survey of India (SOI), Raipur, for preparing the base data, i.e. location of villages, town, cities, preparing the drainage network, etc. Drainage network prepared using SOI topographic sheet is presented in Fig. 22.1. The basin is covering about 65 numbers of topographic sheets. Land use data were collected from National Remote Sensing Centre (NRSC), Hyderabad. Digital elevation model (DEM) of 30 m resolution was collected from the Japan Space Agency Portal known as ASTER GDEM (USGS 2016). It is available in 30 m × 30 m tiles (size of one pixel).

Soil data of the Sheonath River Basin was prepared from the literature of National Bureau of Soil Survey & Land Use Planning (NBSS & LUP), Nagpur. The shape files of the soil data were collected from the Chhattisgarh Council of Science & Technology (CCOST), Raipur. On the basis of soil properties, the basin is divided into hydrological soil group (HSG).

22.2.2 HEC-HMS Model

HEC-HMS is a Hydrologic Modelling System that is designed to describe the physical properties of river basins, the meteorology that occurs on them, and the resulting runoff and streamflow that are produced. This model is produced by the Hydrologic Engineering Center of the US Army Corps of Engineers in Davis, CA (Fleming and Doan 2009; Merwade 2012). A GIS pre-processor called HEC-GeoHMS is also available from the HEC for creating the basin model describing the physical characteristics of the watershed in HEC-HMS. Hydrograph produced by the program is used directly for studies of water availability. HEC-HMS includes three main components: basin model; meteorological model; and control specification. The basin model stores the physical data sets describing the catchment properties and the hydrologic elements, i.e. sub-basins, reaches, junctions, reservoirs, diversions, sink and sources. Meteorological model describes the rainfall event and evapo-transpiration processes. The time span of a simulation is controlled by control specifications including a starting date and time, ending date and time, and computation time step.

22.2.3 SCS-CN Method

In this study, loss from the basin was computed by the Soil Conservation Services Curve Number (SCS-CN) model also known as runoff curve number model. This model simulates surface runoff volumes for the given rainfall amounts. This model was developed by United States Department of Agriculture (USDA) Soil Conservation Services (SCS) in the year 1972 (Nagarajan and Poongothai 2012; Subramanya 2013). Curve number is essentially a coefficient that reduces total precipitation to runoff potential after losses, i.e. evaporation, absorption, transpiration and surface storage (Chow et al. 1988; Mockus 2007). Basically, it is a quantitative descriptor of land use, soil characteristic and antecedent moisture condition. This method is based on the following two concepts:

1. The ratio of the actual amount of runoff to maximum potential runoff is equal to the ratio of actual infiltration to the potential maximum retention and is expressed as:

$$(P - I_a - R)/S = R/(P - I_a) \tag{22.1}$$

where

P precipitation in millimetres ($P \geq R$);
R runoff in mm;
S potential maximum retention in mm;
I_a Initial abstraction in mm.

2. The second concept is that the amount of initial abstraction is some fraction of the potential maximum retention and thus expressed as:

$$I_a = \lambda S \quad (22.2)$$

for Indian conditions, λ = 0.3S (Subramanya 2013).
Where

$$S = 25400/\text{CN} - 254 \quad (22.3)$$

Solving Eqs. 22.1 and 22.2, we get

$$R = (P - I_a)^2/(P - I_a + S) \text{ for } P \geq 0.3S. \quad (22.4)$$

Equation (22.4) is the required equation for computing runoff potential of a basin. The runoff curve numbers vary with the specific combination of hydrologic soil cover complexes, hydrologic soil group (HSG), land use and treatment practices, hydrologic surface condition and antecedent moisture condition (AMC).

22.2.4 Terrain Pre-processing

GIS approach towards hydrologic analysis requires a terrain model which should be hydrologically corrected. For most of the hydrological analysis, the first and foremost task is to get the boundary of the hydrological unit. The boundary of the Sheonath Basin is to be demarcated by the process known as delineation. This involves the application of 30 m resolution ASTER Global Digital Elevation Model (GDEM) and the drainage lines, digitized with the help of topographic sheets in the scale 1:50,000 procured from Survey of India (SOI). For demarcation of sub-basin boundary, HEC-GeoHMS tool is utilized. The process involves DEM reconditioning, fill sink, flow direction, flow accumulation, stream definition, stream segmentation, catchment grid delineation, catchment polygon processing, drainage line processing and watershed aggregation, adopted to demarcate the sub-basin boundary (Merwade 2012).

22.2.5 Analytic Hierarchy Process

Analytic hierarchy process (AHP) is one of the multi-criteria decision-making methods that was originally developed by Saaty (1980, 1987, 1995), Ahmad and Verma (2017). This method is used to determine the percentage importance of the parameters used in the identification of suitable sites for water storage in accordance with the guidelines laid by IMSD (1995). The AHP procedure involves performing comparison of pairs of parameters within a set of reciprocal matrices. In comparing pairs

Table 22.1 AHP scale for pair-wise comparison

Intensity of importance	Description
1	Equal importance—two parameters contribute equally to the objective
3	Moderate importance—slightly favours one parameter over another
5	Strong importance—strongly favours one parameter over another
7	Very strong importance—very strongly favours one parameter over another
9	Extremely strong importance—highest possible order of importance of one parameter over another
2, 4, 6, 8	Intermediate values between the intensity of importance
Reciprocal of above numbers	If an activity has one of the above numbers assigned to it, when compared with a second activity, then the second activity has the reciprocal value when compared to the first

of factors, the AHP scale of relative importance is used in the scale 1–9. The AHP scale of paired comparison with its description is listed in Table 22.1 (Saaty 1987; Mishra et al. 2007; Singh et al. 2009).

The number of comparison can be determined using

$$\text{no. of comparison } = n(n-1)/2 \tag{22.5}$$

where n = number of parameter.

After the formation of pair-wise comparison, matrix priority vector is computed, which is the normalized eigenvector of the matrix. The pair-wise comparison matrix is normalized by dividing the values by the sum of each column. A new matrix is formed and the normalized principal eigenvector (or the priority vector) can be obtained by averaging across the rows. The relative importance given to the parameters one over another is acceptable if the consistency ratio (CR) is less than 10%. If it increases 10%, a new value is assigned in the pair-wise comparison matrix. CR is computed as:

$$\text{CR} = \text{CI}/\text{RI} \tag{22.6}$$

where

$$\begin{aligned} \text{CR} &= \text{Consistency ratio} \\ \text{CI} &= \text{Consistency index} \\ &= (\lambda\text{max} - n)/(n-1) \\ \lambda\text{max} &= \text{Principal Eigen value} \\ &= \text{value obtained from the summation of products between each elements} \end{aligned}$$

of eigen

vector and sum of columns of reciprocal matrix.

RI = Randomness Index.

RI is derived from a sample of size 500 of a randomly generated reciprocal matrix using the scale 1/9, 1/8 … 1 … 8, 9, is given by the size of the matrix (or the number of parameters, n, in the comparison matrix). When n is 1, 2, 3, 4, 5, 6, 7, 8, 9, 10 corresponding RI will be 0, 0, 0.58, 0.9, 1.12, 1.24, 1.24, 1.32, 1.41, 1.45 and 1.49, respectively.

22.3 Results and Discussion

22.3.1 Potential Runoff

In HEC-GeoHMS, HMS Project set-up is responsible for extracting data for the number of sub-basins created in terrain pre-processing and that will be used to develop necessary information to create HEC-HMS Project. A new project is set up by selecting a project point using the cell in stream segmentation layer, acting as the outlet of the basin. Several topographic characteristics of the basin like river length, river slope, basin slope, basin area, longest flow path, etc. were derived with the help of basin processing tool to assist in estimating the hydrological parameters for each sub-basin.

Average slope is necessary for computing the basin lag time and time of concentration. For determining the slope, the digital elevation model is used as a base data for creating the slope file. As DEM is in raster format, the slope is computed for each cell comparing the eight cells around that particular cell. The land use data collected from NRSC Hyderabad is available in the raster format. The land use data is clipped with the help of demarcated boundary of the study area. Further, these are converted to shape file so as to compute its attribute data of the study area. Table 22.2 presents the areal extent of land use classified as build-up, plantation, forest, wasteland, agriculture (rabi, kharif, zaid, double/triple, current fallow) and water bodies. The same is presented in Fig. 22.2.

The soil data is collected from NBSS & LUP, Nagpur. The data collected is digitized for the study area, and shape file is created. Based on the infiltration capacity of the soil described by USDA National Engineering Handbook, NEH (2004), the soil is classified into four groups A, B, C and D known as hydrologic soil group (HSG). The main characteristic of HSG is shown in Table 22.3. For the study area, soil group is demarcated (NRCS 2004a, b; Tamgadge et al. 2002) and presented in Fig. 22.3.

Land use map, soil map and sub-basins maps of the study area are used for the determination of curve number. These maps were merged using GIS and then statistical information is extracted. The information is then entered to get the area-weighted

Table 22.2 Land use of the study area

Land use	Land use code	Area extent (km^2)	% Area
Build-up	1	290.65	0.94
Kharif only	2	8506.00	27.38
Rabi only	3	1021.98	3.29
Zaid only	4	24.08	0.08
Double/triple	5	9696.42	31.21
Current fallow	6	3452.73	11.11
Plantation/orchard	7	2.28	0.01
Deciduous forest	9	5048.12	16.25
Scrub/deg. forest	10	1884.18	6.07
Other wasteland	13	180.28	0.58
Scrubland	15	290.54	0.94
Water bodies	16	667.50	2.15
Total		31064.76	100.00

(or composite) curve number of the study area based on the standard table of curve numbers for the Indian conditions (Subramanya 2013). One of the important parameters in the estimation of curve number is antecedent moisture condition (AMC). It represents the hydrologic condition of the surface prior to the rainfall event (i.e. total rainfall in the 5-day period preceding a storm). There are three levels of AMC conditions, i.e. AMC I, AMC II and AMC III. AMC I has the lowest runoff potential with the soil being dry, AMC II has an average runoff potential and AMC III has the highest runoff potential with the surface practically saturated from antecedent rainfall (Seth et al. 1997). The limits of these three AMC classes are based on the rainfall magnitude of the previous five days and season (dormant season and growing season) (Subramanya 2013). The curve number used is based on the AMC II condition. The curve number basically ranges from 1 to 100. Value of '1' implies that there will be no runoff and value of '100' implies that the runoff is equal to the rainfall. Area-weighted composite curve number for various conditions of land use and hydrologic soil conditions are computed as follows:

$$\mathrm{CN} = (\mathrm{CN}_1 \times A_1) + (\mathrm{CN}_2 \times A_2) + \ldots + (\mathrm{CN}_n \times A_n)/A \tag{22.7}$$

where $A_1, A_2, A_3, \ldots$, An represent areas of each sub-basin having CN values CN_1, $\mathrm{CN}_2, \mathrm{CN}_3, \ldots, \mathrm{CN}_n$, respectively, and A is the total area of the basin. The curve number for all the grids were computed and presented in Fig. 22.4. The curve number ranges from 26 to 100.

SCS curve number loss method uses three parameter, namely curve number, initial abstraction and percent imperviousness. Percentage imperviousness grid is prepared using the land use and sub-basin. The percentage imperviousness means the amount of impervious area in each sub-basin, i.e. total built-up area in each sub-basin. It is computed as:

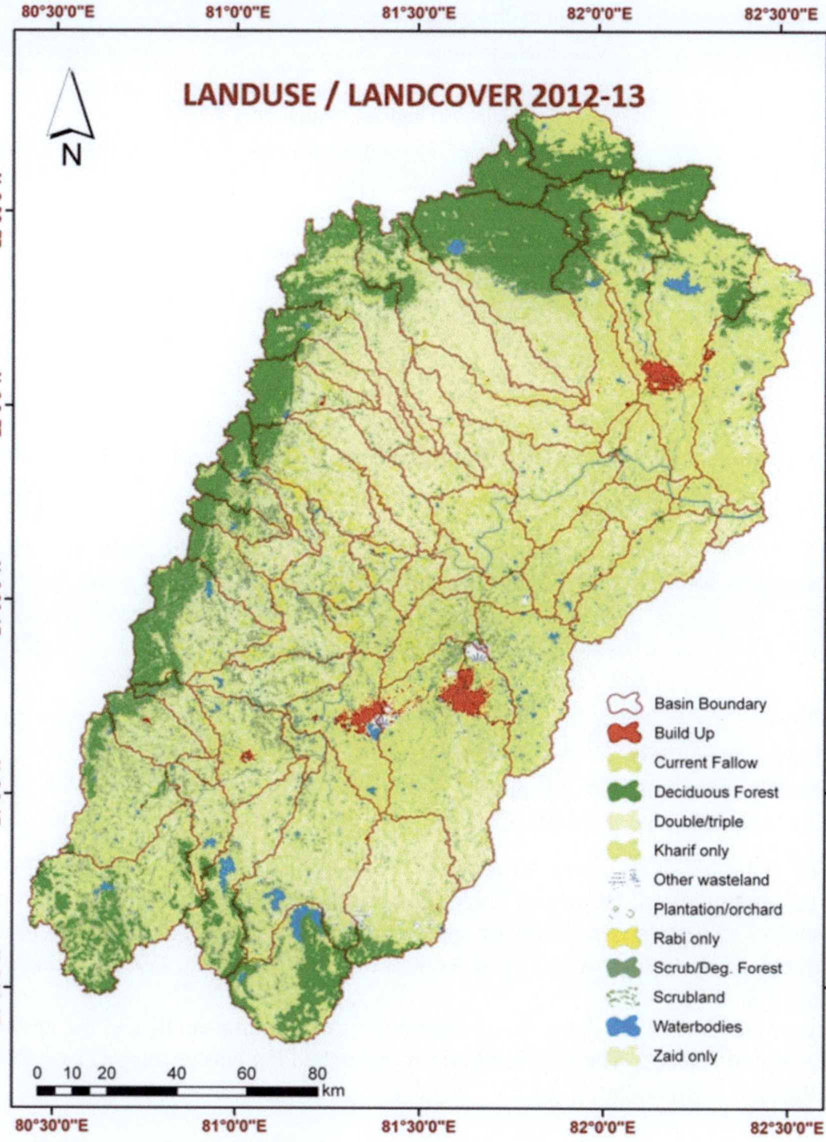

Fig. 22.2 Land use in the study area

Table 22.3 Main characteristics of hydrological soil groups

Hydrological soil group	Main characteristics	Areal extent (km^2)
A	Sand, loamy sand or sandy loam soils with low runoff potential and high infiltration rates	1448.42
B	Silt loam or loam soils with a moderate infiltration rates	9177.00
C	Sand clay loam soils with low infiltration rates	2689.06
D	Clay loam, silty clay loam, sandy clay, silty clay or clay soils with very high runoff potential and low infiltration rates	17475.80

$$\%\ \text{Imperviousness} = (\text{Total built up area})/(\text{Total area of the sub}-\text{basin}) \times 100 \tag{22.8}$$

As per the National Engineering Handbook (NEH) (NRCS 2004, b), initial abstraction is some fraction of surface retention (or surface storage) given by:

$$I_a = \lambda S \tag{22.9}$$

where

I_a Initial abstraction
λ coefficient of initial abstraction
S Surface retention = (25400/CN) − 254 in mm.

The value of λ varying in the range $0.1 \le \lambda \le 0.4$ has been documented in a number of studies (Subramanya 2013). For use in Indian conditions, $\lambda = 0.1$ and 0.3 subject to certain constraints of soil type and AMC type. As suggested in the Handbook of Hydrology, Ministry of Agriculture, Govt. of India, 1972, λ is taken as 0.3 and 0.2.

Basin lag time is the delay time between the maximum rainfall and the amount of the peak discharge. The SCS method for watershed lag was developed by Mokus in 1961 and computed as:

$$L = \{l^{0.8}(S+1)\}/1900y^{0.5} \tag{22.10}$$

where

L Basin lag time in hours
l longest flow length in ft
S Max. potential retention in inches = (1000/CN) − 100
y Average watershed slope (%).

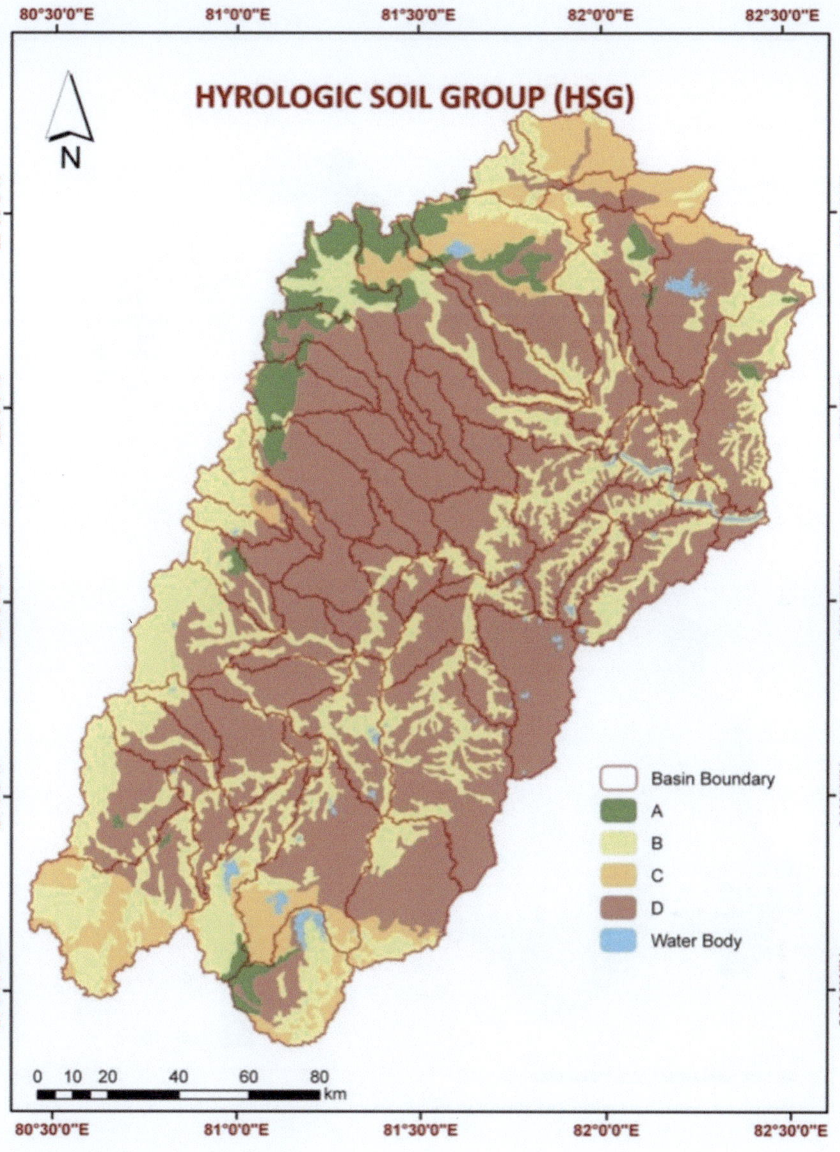

Fig. 22.3 HSG map of the study area

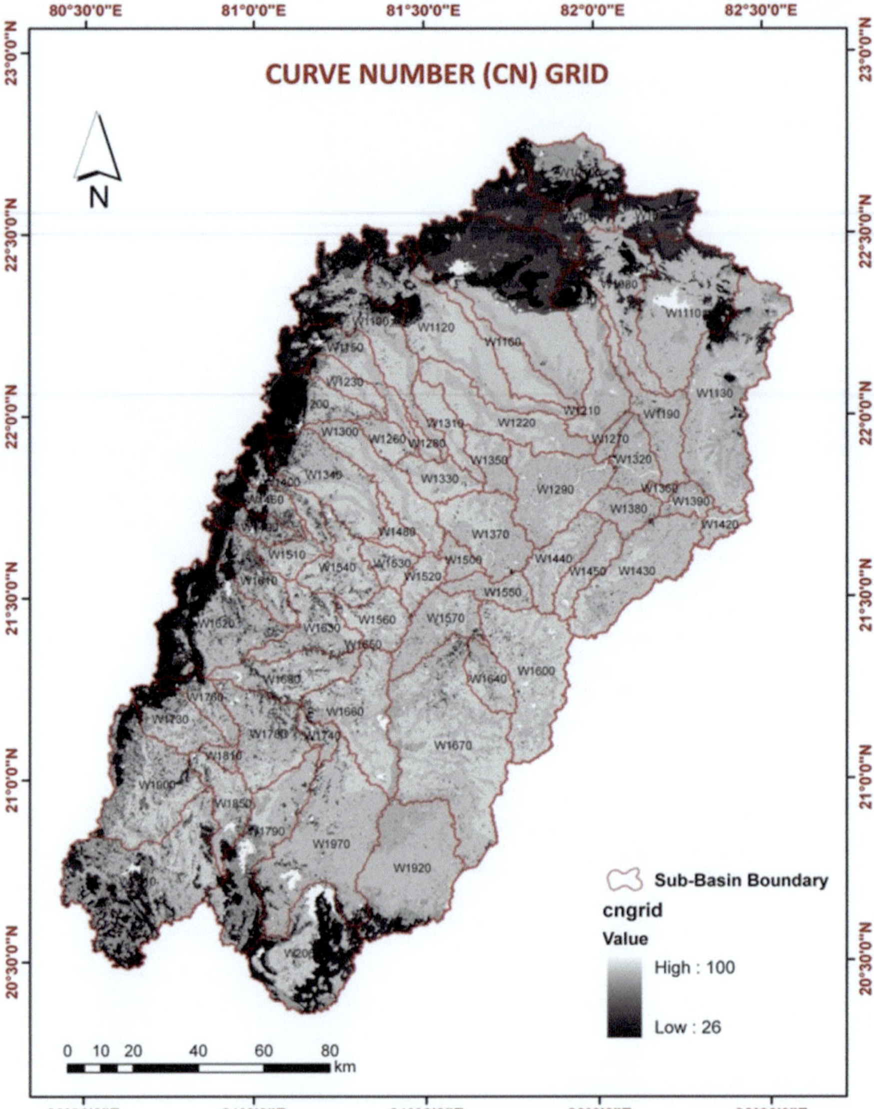

Fig. 22.4 Curve number (CN) grid of the study area

Time of concentration is defined as the time required for a drop of water to travel from most hydrologically remote point in the sub-basin to the collection point, i.e. outlet (NRCS 2010). It is computed as:

$$T_C = L/0.6 \tag{22.11}$$

where

T_C Time of concentration in hours
L Basin lag time in hours.

Routing is a technique used to predict the changes in the shape of a hydrograph as water moves through a river channel (Chow et al. 1988). Muskingum routing method is a commonly used routing method. It enhances the accuracy of the presentation of the basin response to the rainfall event. The key parameters in Muskingum routing are K (travel time of wave) and x (weighting coefficient). Muskingum routing equation is written as:

$$S = K\left[xI^m + (1-x)O^m\right] \tag{22.12}$$

where

S storage
I inflow
O outflow
K and x Muskingum parameter (or storage parameters)
m coefficient; 0.6 for rectangular channel and 1.0 for natural channels.

In our study, the Muskingum routing method is applied. It required three parameters 'K', 'x' and 'n', they are computed as:

$$K = L_s/3600V_s \tag{22.13}$$

'n' is the number of sub-reaches and is computed as:

$$n = \ln[2 \times ((L_s/60V_s)/\Delta t)] + 1 \tag{22.14}$$

where

L_s Length of the channel in the sub-basin in meters
V_s Average velocity of flow in the channel; 0.3–1.5 m/s for natural channels
Δt Analysis time step (or time of simulation).

In the natural stream, x ranges from 0.0 to 0.5 with a mean value nearer to 0.2 (Chow et al. 1988).

Theissen polygon method is applied to estimate the average rainfall occurring in the influencing polygon of the rain gauge station. Since the study, the area consists of 65 numbers of rain gauges and the study area is divided into 69 sub-basins, it is necessary to determine the weightage of each rain gauge station corresponding to the sub-basin. Weightage is based on the area of the rain gauge polygon in a particular sub-basin. The influence area, i.e. Theissen polygon of each rain gauge station in each sub-basin is shown in Fig. 22.5. The evapo-transpiration measurement monthly average method is used. Potential evapo-transpiration (PET) is measured

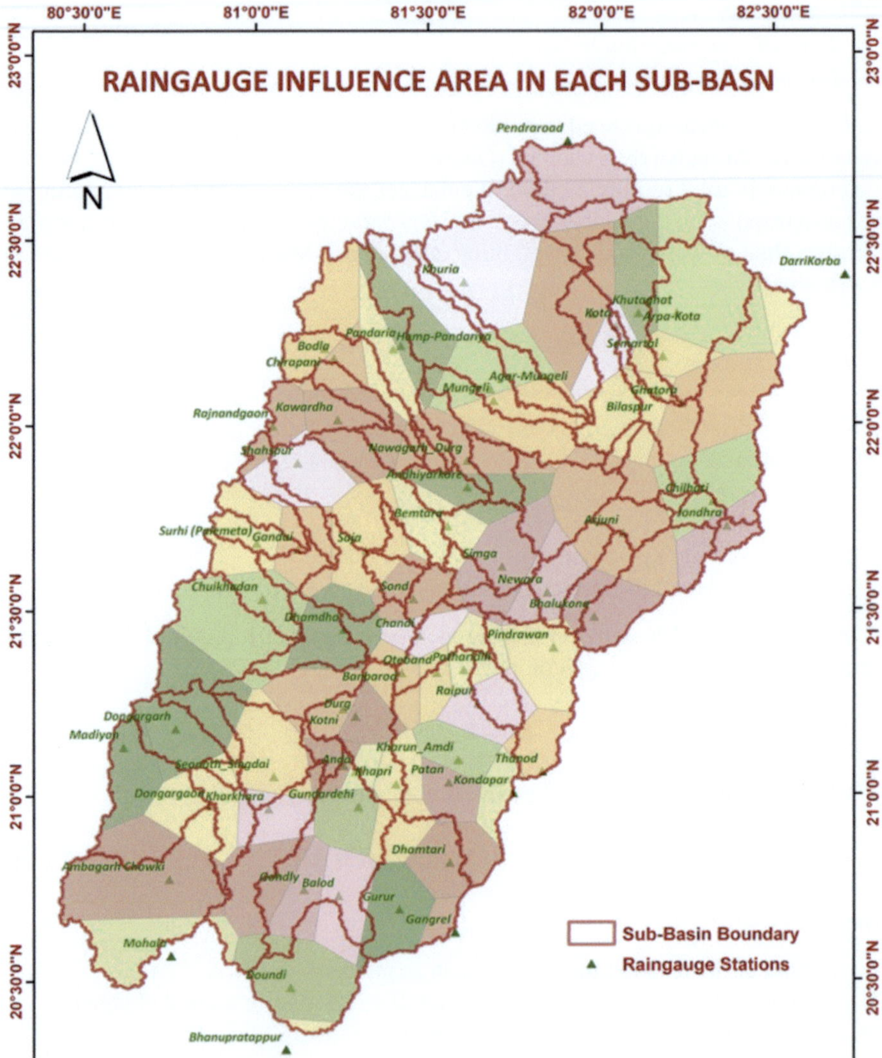

Fig. 22.5 Rain gauge influence area

at Labhandi station. The schematic of the basin model is prepared for visualizing different elements of the model and the same is presented in Fig. 22.6.

Runoff for each cell of the study area is computed with the help of runoff curve number loss method (Ahmad and Verma 2015a, b; Ahmad et al. 2015). In this method, parameters used are precipitation, land use and soil type along with its antecedent moisture condition (AMC). Average yearly rainfall data from the year 1980 to 2013

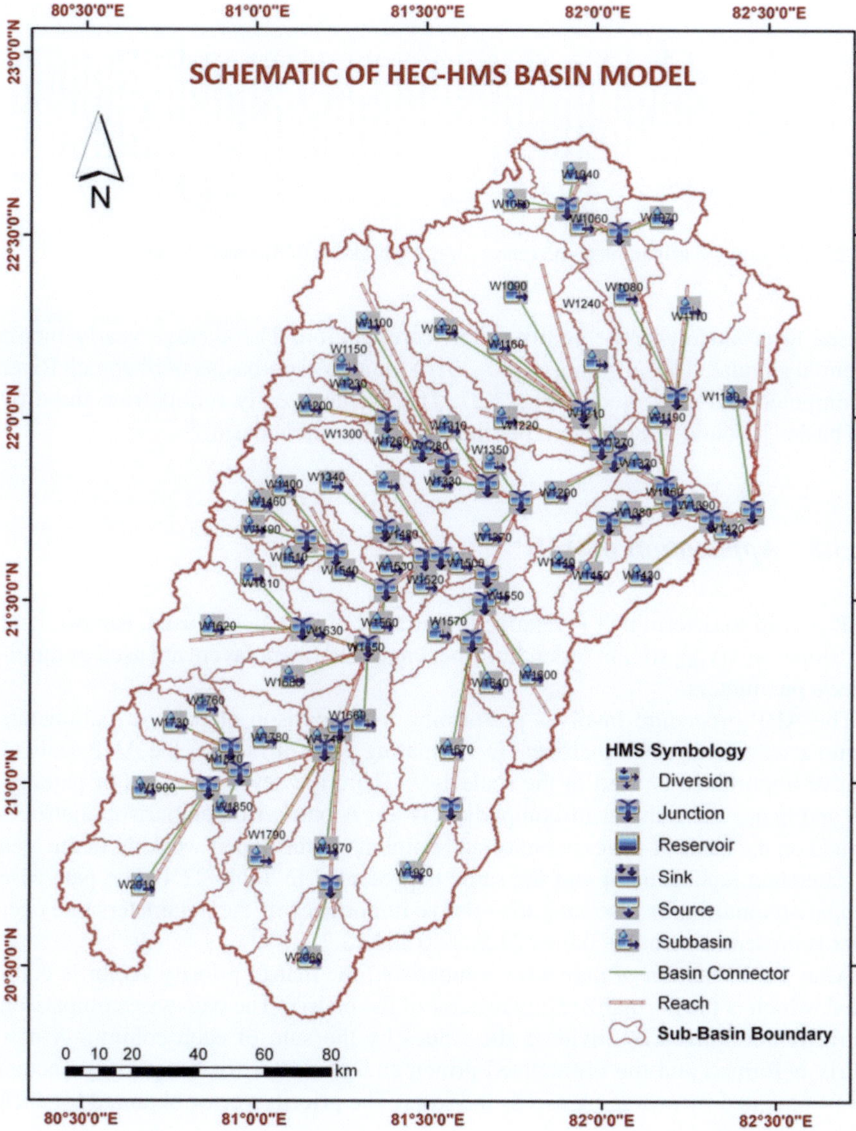

Fig. 22.6 Schematic of HEC-HMS model

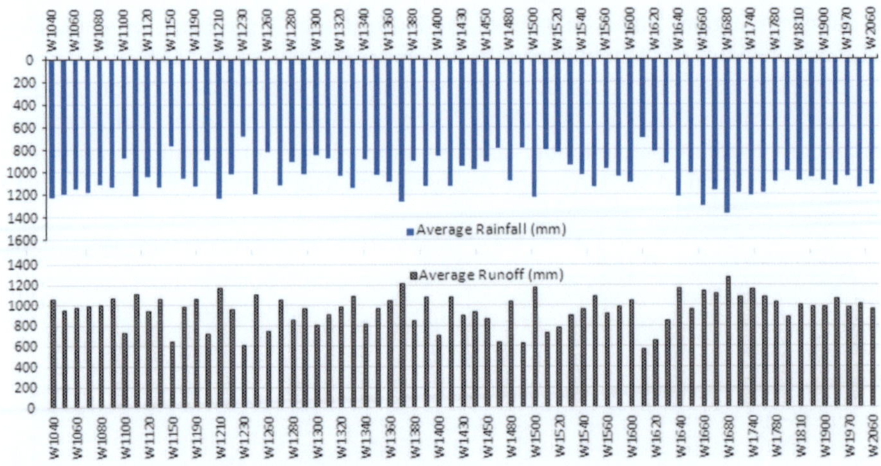

Fig. 22.7 Average yearly rainfall and runoff for the sub-basins of Sheonath River

is used to produce average yearly runoff computation. The average yearly runoff potential simulated for 33 years (1980–2013) for all 69 sub-basins of Sheonath River is computed and presented in Fig. 22.7. The average yearly runoff from these 69 sub-basins is found to 86.07 mm per month for the whole basin.

22.3.2 Application of AHP

AHP is used to determine the relative importance of parameters used. Runoff, land use, slope, soil type, stream order, lineament and settlement layers are used as multi-criteria parameters.

The AHP procedure involves performing a comparison of pairs of parameters within a set of reciprocal matrices. In comparing pairs of factors, the AHP scale of relative importance is used in the scale 1–9. There are seven numbers of parameters and hence the number of comparison is 21. A pair-wise comparison matrix is formed on the basis of the experience and opinions of the expert working in the area of watershed management and the same is presented in Table 22.4. The pair-wise comparison matrix for assessing the relative importance of the parameters one over other is presented through Tables 22.5, 22.6 and 22.7.

After the formation of pair-wise comparison, the matrix priority vector is computed, which is the normalized eigenvector of the matrix. The pair-wise comparison matrix is normalized by dividing the values by the sum of each column. A new matrix is formed and the normalized principal eigenvector (or the priority vector) can be obtained by averaging across the rows. The priority vector obtained for each parameter is listed in Table 22.7.

Table 22.4 Pair-wise comparison matrix

S. No	Parameter	Intensity of importance									Parameter
		LHS				Equal importance	RHS				
		9	7	5	3	1	3	5	7	9	
1	Runoff	9	–	–	–	–	–	–	–	–	Land use
2	Runoff	9	–	–	–	–	–	–	–	–	Slope
3	Runoff	9	–	–	–	–	–	–	–	–	Soil type
4	Runoff	9	–	–	–	–	–	–	–	–	Stream order
5	Runoff	9	–	–	–	–	–	–	–	–	Lineament
6	Runoff	9	–	–	–	–	–	–	–	–	Settlement
7	Land use	–	–	–	–	–	3	–	–	–	Slope
8	Land use	–	–	–	–	–	3	–	–	–	Soil type
9	Land use	–	–	–	–	–	3	–	–	–	Stream order
10	Land use	–	–	–	3	–	–	–	–	–	Lineament
11	Land use	–	–	–	–	1	–	–	–	–	Settlement
12	Slope	–	–	–	–	1	–	–	–	–	Soil type
13	Slope	–	–	–	–	1	–	–	–	–	Stream order
14	Slope	–	–	5		–	–	–	–	–	Lineament
15	Slope	–	–	–	3	–	–	–	–	–	Settlement
16	Soil type	–	–	–	–	–	–	–	–	–	Stream order
17	Soil type	–	–	–	3	–	–	–	–	–	Lineament
18	Soil type	–	–	–	3	–	–	–	–	–	Settlement
19	Stream order	–	–	–	3	–	–	–	–	–	Lineament

(continued)

Table 22.4 (continued)

S. No	Parameter	Intensity of importance									Parameter
		LHS				Equal importance	RHS				
		9	7	5	3	1	3	5	7	9	
20	Stream order	–	–	–	3	–	–	–	–	–	Settlement
21	Lineament	–	–	–	–	–	–	–	–	–	Settlement

LHS—left-hand side; *RHS*—right-hand side

Table 22.5 Pair-wise comparison for computation of priority vector

Parameter	Runoff	Land use	Slope	HSG	Stream order	Lineament	Settlement
Runoff	1	9	9	9	9	9	9
Land use	1/9	1	1/3	1/3	1/3	3	1
Slope	1/9	3	1	1	1	5	3
Soil type	1/9	3	1	1	1	3	3
Stream order	1/9	3	1	1	1	3	3
Lineament	1/9	1/3	1/5	1/3	1/3	1	1
Settlement	1/9	1	1/3	1/3	1/3	1	1
Sum	1.67	20.33	12.87	13.00	13.00	25.00	21.00

Table 22.6 Dividing each element of matrix with the sum of its column

Parameter	Runoff	Land use	Slope	HSG	Stream Order	Lineament	Settlement
Runoff	0.60	0.44	0.70	0.69	0.69	0.36	0.43
Land use	0.07	0.05	0.03	0.03	0.03	0.12	0.05
Slope	0.07	0.15	0.08	0.08	0.08	0.20	0.14
Soil type	0.07	0.15	0.08	0.08	0.08	0.12	0.14
Stream order	0.07	0.15	0.08	0.08	0.08	0.12	0.14
Lineament	0.07	0.02	0.02	0.03	0.03	0.04	0.05
Settlement	0.07	0.05	0.03	0.03	0.03	0.04	0.05
Sum	1.00	1.00	1.00	1.00	1.00	1.00	1.00

Table 22.7 Take the sum of each row dividing by number of parameter

Parameter	Runoff	Land use	Slope	Soil type	Stream order	Lineament	Settlement	Sum	Priority vector	% Priority
Runoff	0.60	0.44	0.70	0.69	0.69	0.36	0.43	3.92	0.56	56
Land use	0.07	0.05	0.03	0.03	0.03	0.12	0.05	0.36	0.06	6
Slope	0.07	0.15	0.08	0.08	0.08	0.20	0.14	0.79	0.11	11
Soil type	0.07	0.15	0.08	0.08	0.08	0.12	0.14	0.71	0.10	10
Stream order	0.07	0.15	0.08	0.08	0.08	0.12	0.14	0.71	0.10	10
Lineament	0.07	0.02	0.02	0.03	0.03	0.04	0.05	0.24	0.03	3
Settlement	0.07	0.05	0.03	0.03	0.03	0.04	0.05	0.28	0.04	4
Sum								7	1	100

The relative importance given to the parameters one over another is acceptable if the consistency ratio (CR) is less than 10%. If it increases 10%, a new value is assigned in the pair-wise comparison matrix. CR is computed as:

$$\begin{aligned}
\text{CR} &= \text{Consistency ratio} \\
\text{CI} &= \text{Consistency index} \\
&= (\lambda_{\max} - n)/(n - 1) \\
\lambda_{\max} &= \text{Principal Eigen value} \\
&= 7.75 \\
n &= \text{number of parameters} = 7 \\
\text{CI} &= 0.13 \\
\text{RI} &= \text{Randomness Index} = 1.32 \text{ (for 7 numbers of parameter)}
\end{aligned}$$

Checking the consistency of the importance given to each parameter:

$$\begin{aligned}
\text{Principal Eigen vector} &= 7.75 \\
\text{Consistency Index(CI)} &= (\lambda_{\max} - n)/(n - 1) \\
&= 0.13 \\
\text{RI} &= 1.32 \text{ (for number of parameters = 7)} \\
\text{Consistency Ratio(CI/RI)} &= 9.8\%
\end{aligned}$$

Consistency ratio (CR) is computed and found to be within 10%, i.e. 9.8%. Hence, the % priority listed in Table 22.7 is used as a weightage for the given parameter. There are 18 number of locations found to be highly suitable for creating water storage sites. Since Sheonath River Sub-Basin is already existing number of storage structures, it is required to have optimum number of storage structures. For this purpose, suitable sites were checked with the constraints suggested for selecting the optimal number of sites following criteria.

1. The sites were outside the existing command area.
2. The site is outside a buffer of 10 km from the existing reservoirs and outside the existing command area.
3. The site is outside a buffer of 5 km from the gauge & discharge sites and outside the existing command area.
4. The site is outside a buffer of 2 km from the town and 1 km from the village.

Based on the above criteria, buffers are created around the mentioned objects and the sites coming within these buffers are removed. The remaining numbers of sites are considered as optimal number of suitable sites. Out of eighteen sites, six numbers of the sites are found to be the most suitable sites for creating water storage for the agriculture area not falling under existing command area. The location of sites is shown in Fig. 22.8.

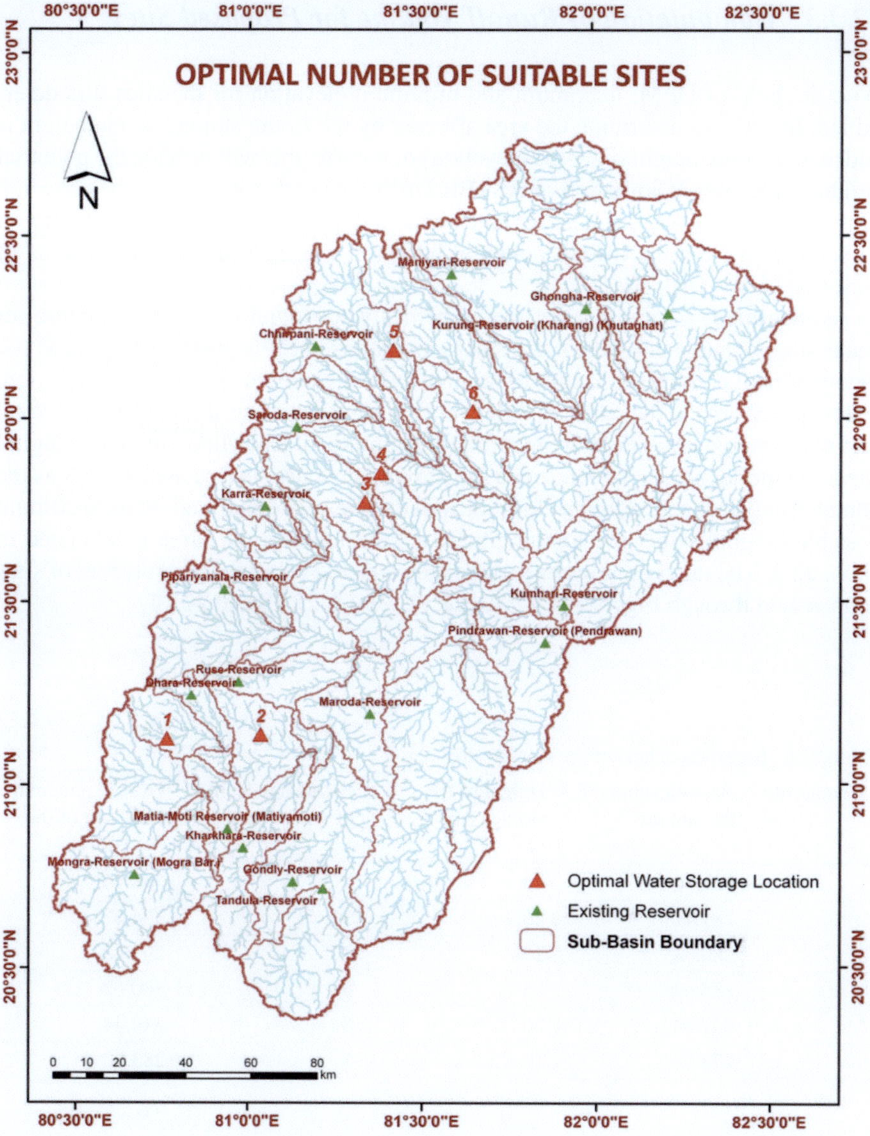

Fig. 22.8 Optimal number of suitable sites for water storage

22.3.3 Computation of Runoff Volume for Proposed Sites

With the help of DEM, maximum and minimum elevation for the sites was determined. In order to determine the area affected by the future storage, some height is added to the base height (i.e. base elevation) of the site, this will provide the potential height of the site. A limitation of ASTER DEM data is that it supplies the elevation of the top of the river, considering an estimated depth of 3 m of the streams; it will be subtracted from the overall potential height. Table 22.8 provides the details of elevation set for each site.

As identified, six numbers of new schemes were found to be most suitable for water storage and analysed for different parameters such as possible heights as per topographical conditions, length of storage axis at a particular elevation, elevation, water spread area at the particular elevation, storage capacity at that elevation, total area of submergence including forest area and number of land village affected. Digital elevation model is used to determine the maximum water surface area as well as the potential capacity at maximum elevation. All the sites were analysed for its maximum possible heights as per the topographical conditions, and the same is tabulated in Table 22.9. The maximum water surface area covered by all the six numbers of sites is presented through Figs. 22.9, 22.10, 22.11, 22.12, 22.13 and 22.14.

Table 22.8 Elevation details of the storage sites

Storage site	Base elevation of the site (m)	Height of the storage (m)	Estimated depth of stream (m)	Overall potential elevation of the site (m)
1	321.56	20	3	338.56
2	287.47	20	3	304.47
3	280.68	20	3	297.68
4	288.62	15	3	300.62
5	327.44	20	3	344.44
6	273.93	10	3	283.93

Table 22.9 Comparison of six water storage sites at maximum limits

Water storage site	Location		Maximum limits of elevation (m)	Catchment area on μ/s of the site (km²)	Maximum water surface area (km²)	Potential capacity (MCM)
	Longitude	Latitude				
1	80.76	21.13	338.56	135.26	16.53	87.90
2	81.34	21.77	304.47	188.01	27.46	59.90
3	81.38	21.85	297.68	385.71	23.45	165.50
4	81.42	22.19	300.62	190.94	16.53	70.30
5	81.65	22.02	344.44	588.64	11.73	60.50
6	81.03	21.14	283.93	175.53	19.90	80.10
Total				1664.09	115.60	524.20

22.3.4 *Irrigation Water Requirement*

Paddy is the major crop growing in the agriculture area of Sheonath River Sub-Basin. The total water requirement for paddy crops in kharif and rabi seasons is 1325.70 mm per hectare. As per the GIS analysis, the agriculture area which is outside the existing command area in the basin is about 12,096.00 km^2. The net agriculture area is taken as 20% of the total area under consideration, i.e. 9676.00 km^2, out of this area 34.82% is Kharif, 5.88% is Rabi and 59.30% for others crops. As per the report on Mahanadi Component, total irrigation water requirement for paddy crop in different seasons is 1425.70 mm/ha. The total water requirement is found to be 284.20 MCM presented in Table 22.10. Potential capacity of water stored in six storage sites is 524.20 MCM which is sufficient to provide irrigation to the paddy crops of the area under consideration, i.e. outside the existing command area.

22.4 Conclusion

For developing the additional potential of irrigation water, sufficient numbers of storage sites are required to collect the surface runoff. On the basis of water availability study, suitable numbers of sites are identified for creating water storage. Multi-criteria evaluation process is applied for this purpose. Various parameters, viz. runoff, land use, slope, soil (HSG), stream order, lineament and settlement are used for site suitability analysis. Pair-wise comparison has been done using analytic hierarchy process (AHP). The optimum number of suitable sites was identified based on various criteria. Six numbers of optimum sites are identified for water storage. The potential capacity of these reservoirs is worked out as 524.20 MCM is sufficient to meet the requirements of water for irrigation.

SCS-CN model is successfully applied to compute the water availability in the basin with initial abstraction taken as 0.3 times the surface retention. This curve

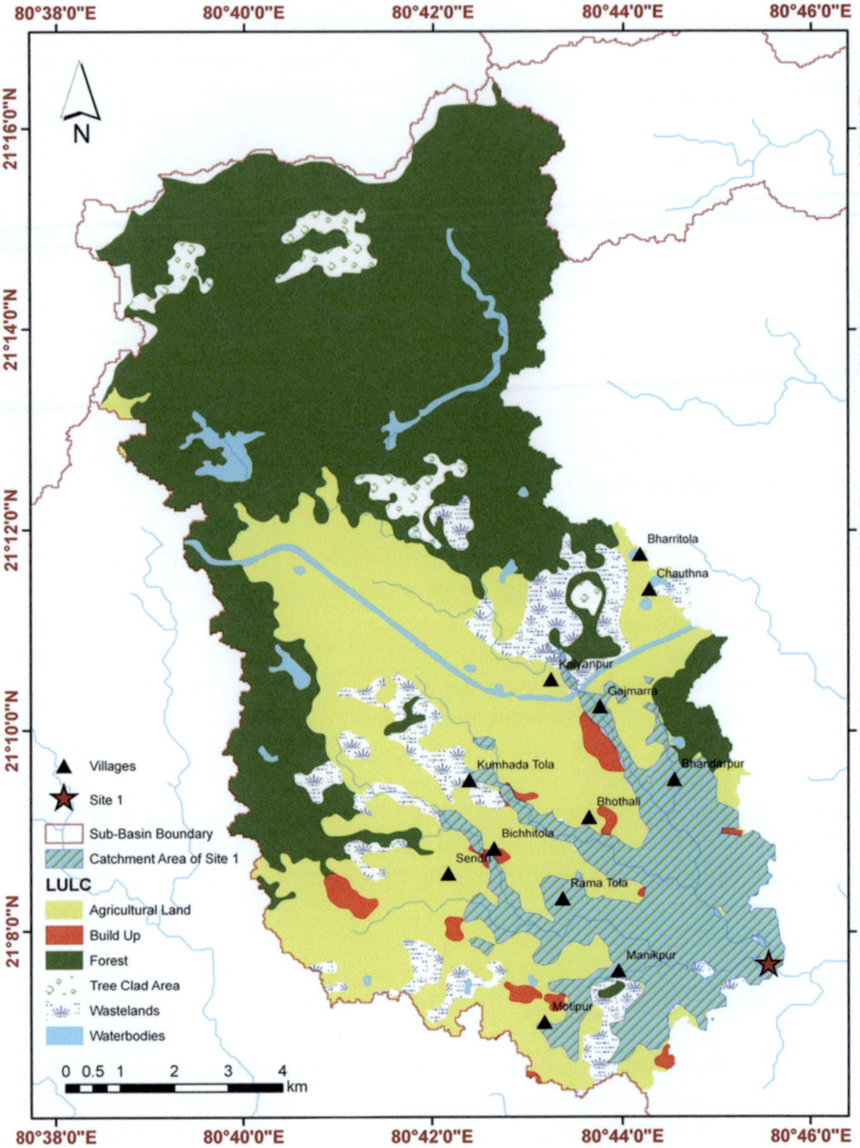

Fig. 22.9 Catchment area with maximum water surface area of storage site 1

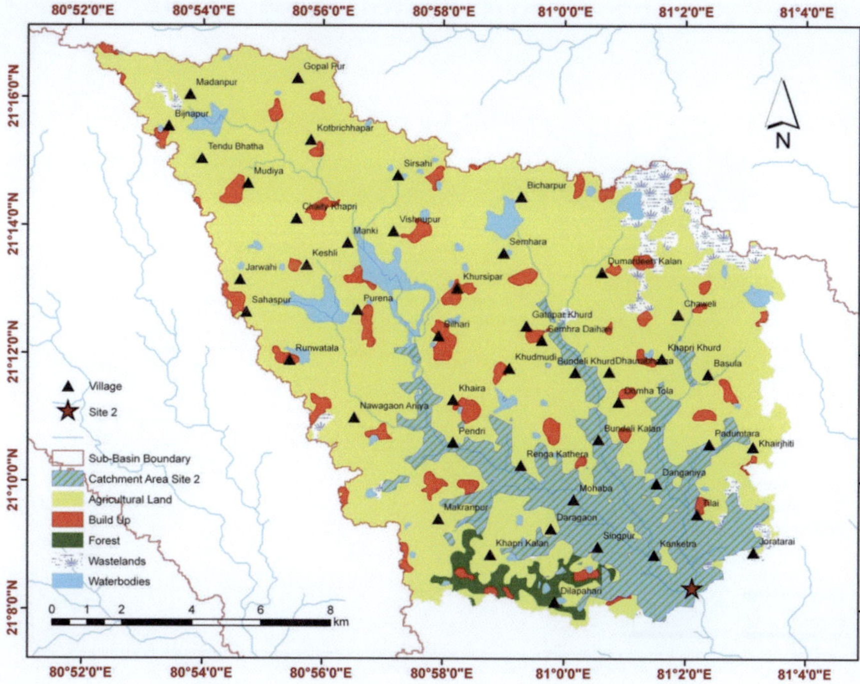

Fig. 22.10 Catchment area with maximum water surface area of storage site 2

number loss model is simple and reliable as it is a quantitative description of land use, hydrologic soil group and ground conditions. The study area comprises of all the four types of hydrologic soil group *A*, *B*, *C* and *D* in which Group *B* and Group *D* are dominant. Using these three parameters, the curve number grid is prepared. The curve numbers are found to be in the range of 26–100.

Through the experience of the present research work, it has been observed that SCS-CN loss model can be applied efficiently to basin level for determination of water availability accurately, as it depends upon three parameters, viz. land use, soil and ground conditions. Looking into the availability of these parameters, this method is recommended for water availability study.

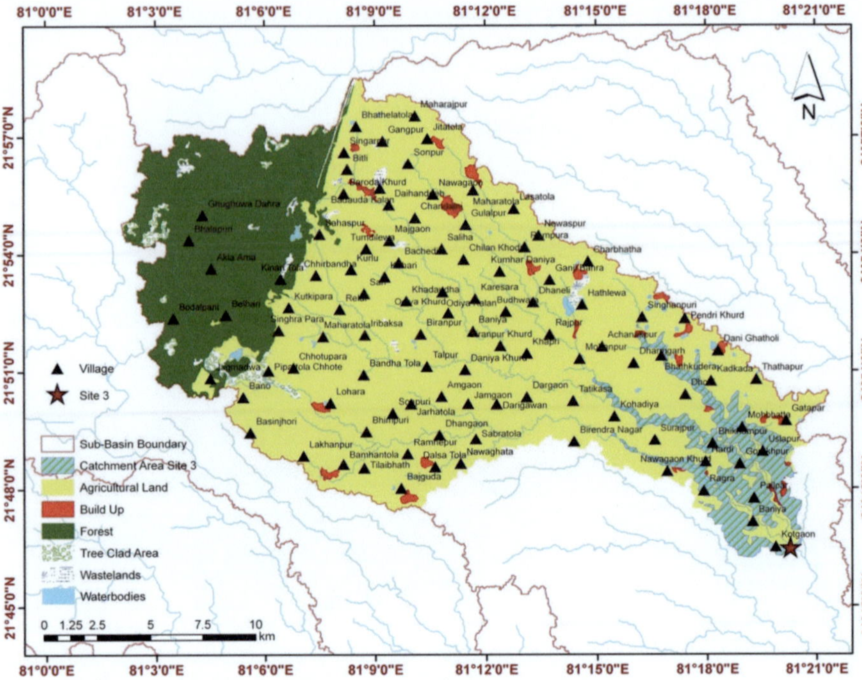

Fig. 22.11 Catchment area with maximum water surface area of storage site 3

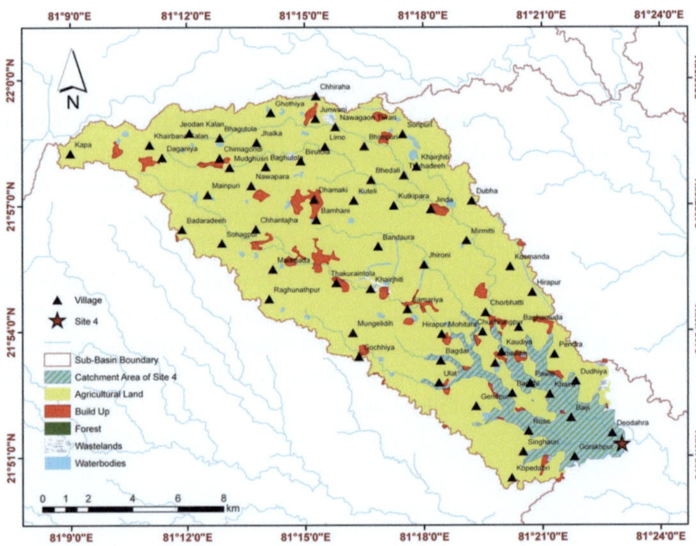

Fig. 22.12 Catchment area with maximum water surface area of storage site 4

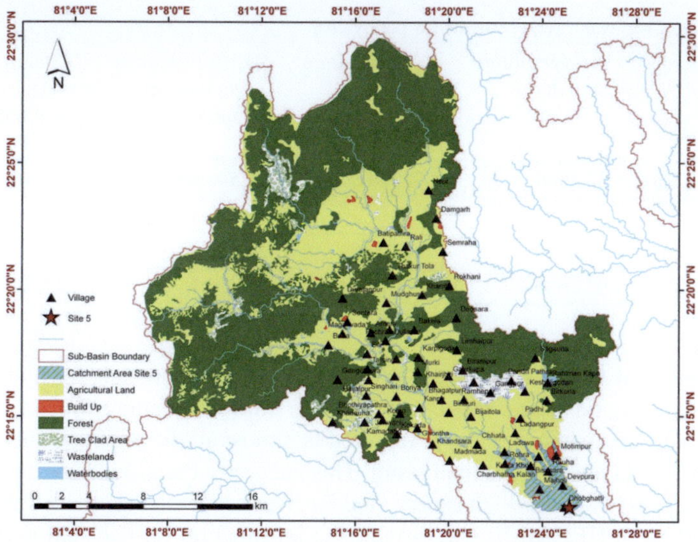

Fig. 22.13 Catchment area with maximum water surface area of storage site 5

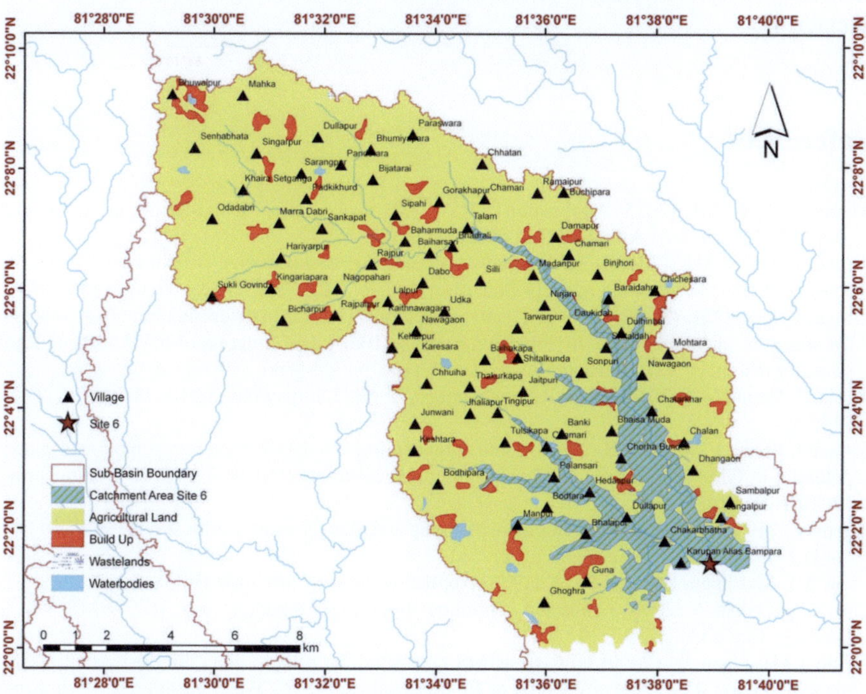

Fig. 22.14 Catchment area with maximum water surface area of storage site 6

Table 22.10 Crop water requirements for paddy crop

S. No.	Crops	Net agriculture area (ha)	% Agriculture area	Irrigation water requirement (mm/ha)	Total irrigation water requirement (MCM)
1	Kharif	33,6960	34.82%	540.60	182.16
2	Rabi	56,880	5.88%	785.10	44.66
3	Others	573,840	59.30%	100.00	57.38
Total		967,680	100	1425.70	284.20

The power of GIS is to integrate various mathematical models for hydrological analysis to generate inputs and outputs for the entire modelling processes. In the present study, application of GIS is required in almost all the steps starting from the preparation of database to the final result. This integration is utilized in establishing a decision support system which is very vital for river basin management. The integration of GIS with HEC-HMS will provide the information for decision-making and finally establishing a decision support system. The information and findings from the simulation results are used for river basin planning and management in an integrated manner.

References

Ahmad I, Verma MK (2015) Application of RS & GIS in estimation of sub-basin runoff potential using HEC HMS. Int J Appl Eng Res 10(20):41243–41248

Ahmad I, Verma MK (2016) Site suitability mapping for water storage structures using remote sensing & GIS for Sheonath basin in Chhattisgarh state. Int J ApplEng Res 11(6):4155–4160

Ahmad I, Verma MK (2017) GIS based analytic hierarchy process in determination of suitable site for water storage application. In: Proceedings of the 10th world congress of EWRA, Panta Rhei on water resources and environment, Athens, Greece, pp 753–760, 5–9 July 2017

Ahmad I, Verma MK (2018) Water Resour Manage 32: 5093. https://doi.org/10.1007/s11269-018-2135-x

Ahmad I, Verma V, Verma MK (2015) Application of curve number method for estimation of runoff potential in GIS environment. In: 2nd international conference on geological civil engineering IPCBEE 80, pp 16–20

Barrow CJ (1998) River basin development planning and management: a critical review. World Dev 26(1):171–186

Chow VT, Maidment DR, Mays LW (1988) Applied hydrology. McGraw-Hill, Singapore

Chu X, Steinman A (2009) Event and continuous hydrologic modeling with HEC-HMS. J Irrig Drainage Eng 135(1):119–124

Fleming MJ, Doan JH (2009) HEC-GeoHMS geospatial hydrologic modelling extension: user's manual version 4.2. US Army Corps of Engineers, Institute for Water Resources, Hydrologic Engineering Centre, Davis, CA

IMSD (1995) Integrated mission for sustainable development: technical guidelines. In: NRSA. Hyderabad, India

Lyon JG (2003) GIS for water resource and watershed management. CRC Press

Merwade V (2012) Terrain processing and HMS-model development using GeoHMS. Lafayette, Indiana
Mishra SK, Babu PS, Singh VP (2007) SCS-CN method revisited. In: Singh VP (ed) Advances in hydraulics and hydrology. Water Resources Publication (WRP), Colorado, 36 p
Mockus V (2007). Estimation of direct runoff from storm rainfall, chapter 10. US Department of Agriculture, SCS National Engineering Handbook, Section, 4
Nagarajan N, Poongothai S (2012) Spatial mapping of runoff from a watershed using SCS-CN method with remote sensing and GIS. J Hydrol Eng 17:1268–1277
NRCS (2004a) Estimation of direct runoff from storm rainfall, national engineering handbook, part 630 hydrology, chapter 10. United States Department of Agriculture, Natural Resources Conservation Service, Washington, DC
NRCS (2004b) Hydrologic soil-cover complexes, national engineering handbook, part 630 hydrology, chapter 9. United States Department of Agriculture, Natural Resources Conservation Service, Washington, DC
NRCS (2010) Time of concentration, national engineering handbook, part 630 hydrology, chapter 15. United States Department of Agriculture, Natural Resources Conservation Service, Washington, DC
Prasad HC, Bhalla P, Palria S (2014) Site suitability analysis of water harvesting structures using remote sensing and GIS—a case study of Pisangan watershed, Ajmer District, Rajasthan. Int Arch Photogrammetry Remote Sens Spat Inf Sci 40(8):1471
Ramakrishnan D, Bandyopadhyay A, Kusuma KN (2009) SCS-CN and GIS-based approach for identifying potential water harvesting sites in the Kali watershed, Mahi river basin. India J Earth SystSci 118(4):355–368
Saaty TL (1980) The analytic hierarchy process. McGraw-Hill, New York
Saaty RW (1987) The analytic hierarchy process—what it s and how it is used. Math Model 9(3):161–176
Saaty TL (1995) Decision making for leaders: the analytic hierarchy process for decisions in a complex world, 3d edn. RWS Publications, Pittsburgh, PA
Saha AK, Arora MK, Csaplovics E, Gupta RP (2008) Land cover classification using IRS LISS III image and DEM in a rugged terrain: a case study in Himalayas. Geocarto Inte 20(2):33–40
Seth SM, Kumar Bhism, Thomas T, Jaiswal RK (1997–98). Rainfall-Runoff modelling for water availability study in ken River basin using SCS-CN model and remote sensing approach. Technical Reports, National Institute of Hydrology, Roorkee, No. CS/AR-12/97-98
Sherif MM, Mohamed MM, Shetty A, Almulla M (2011) Rainfall-runoff modeling of three wadis in the northern area of UAE. J Hydrol Eng 16(1):10–20
Singh JP, Singh D, Litoria PK (2009) Selection of suitable sites for water harvesting structures in Soankhad watershed, Punjab using remote sensing and geographical information system (RS & GIS) approach—a case study. J Indian Soc Remote Sens 37:21–35
Subramanya K (2013) Engineering hydrology. McGraw-Hill, Delhi, India
Tamgadge DB, Gajbhiye KS, Pandey GP (2002) Soil series of chhattisgarh state, vol 85. National Bureau of Soil Survey and Land Use Planning in co-operation with Dept. of Agriculture, Govt. of Chhattisgarh, Raipur
USGS (2016) ASTER data. http://earthexplorer.usgs.gov/

MIX
Papier aus verantwortungsvollen Quellen
Paper from responsible sources
FSC® C105338

If you have any concerns about our products,
you can contact us on
ProductSafety@springernature.com

In case Publisher is established outside the EU,
the EU authorized representative is:
Springer Nature Customer Service Center GmbH
Europaplatz 3, 69115 Heidelberg, Germany

Printed by Libri Plureos GmbH
in Hamburg, Germany